Build Your Own Metaverse with Unity

A practical guide to developing your own cross-platform Metaverse with Unity3D and Firebase

David Cantón Nadales

BIRMINGHAM—MUMBAI

Build Your Own Metaverse with Unity

Group Product Manager: Rohit Rajkumar
Publishing Product Manager: Nitin Nainani
Content Development Editor: Abhishek Jadhav
Technical Editor: Simran Ali
Copy Editor: Safis Editing
Project Coordinator: Aishwarya Mohan
Proofreader: Safis Editing
Indexer: Subalakshmi Govindhan
Production Designer: Jyoti Chauhan
Marketing Coordinator: Nivedita Pandey

First published: September 2023

Production reference: 1110823

Packt Publishing Ltd
Grosvenor House
11 St Paul's Square
Birmingham
B3 1RB, UK.

ISBN 978-1-83763-173-5

www.packtpub.com

To my sons, Leo and Luca, the best of my life. To my wife, Nerea, for always believing in me and supporting me at all times. To my parents, José and Reyes, who have always worked so hard to bring us forward. To my in-laws, Chary and Manuel, who are like my second parents, always there, in the best and worst moments. To all my family, my blood brothers and sisters at Bluekiri, the best workmates I have ever had. But above all, I would like to dedicate this book to all the people who never believed in me.

– David Cantón Nadales

Contributors

About the author

David Cantón Nadales is a software engineer from Seville, Spain, with more than 20 years of experience. He is currently a technical leader at Grupo Viajes El Corte Inglés, a leading travel company in Europe. He has worked on a multitude of projects and games with Unity, VR with Oculus/Meta Quest 2, HoloLens, HTC VIVE, Daydream, and Leap Motion.

He was an ambassador of the Samsung community *Samsung Dev Spain* and organizer of *Google Developer Group Sevilla*. He has led more than 100 projects throughout his career. As a social entrepreneur, he created the app *Grita*, a social network that emerged during the confinement of COVID-19 that allowed people to talk to others and help each other psychologically. In 2022, he won the Top Developer Award by Samsung.

I would like to dedicate this book to my colleagues, the Bluekiri BSE team, Nacho, Jesús, Jordi, Asier, Ferran, Alberto (Zhuo Han), and Sebas.

About the reviewer

Mohit Chandrashekhar Attarde is an XR Metaverse developer with a passion for 3D graphics and a focus on gamification principles in XR Metaverse applications. His dedication to advancing learning and training within XR has resulted in significant contributions to various metaverse projects.

With expertise in Unity, C#, and Visual Studio, Mohit enjoys building cross-platform metaverse experiences that push the boundaries of immersive technology. Additionally, his proficiency in WebGL and other web frameworks further enhances his ability to create captivating virtual experiences on the web. Through mentorship and community support, he inspires others to delve into the vast potential of XR technology.

His commitment to innovation continues to shape the landscape of virtual reality and its applications.

Table of Contents

3

Preparing Our Home Sweet Home: Part 1 77

4

Preparing Our Home Sweet Home: Part 2 137

5

Preparing a New World for Travel 175

Part 2: And Now, Make It Metaverse!

6

Adding a Registration and Login Form for Our Users 241

7

Building an NPC That Allows Us to Travel 287

8

Acquiring a House 333

9

Turning Our World into a Multiplayer Room 371

10

Adding Text and a Voice Chat to the Room 399

Part 3: Adding Fun Features Before Compiling

11

Creating an NPC That Allows Us to Change Our Appearance 425

12

Streaming Video Like a Cinema 465

13

Adding Compatibility for the Meta Quest 2 477

14

Preface

Welcome to a world of infinite possibilities, where reality and imagination converge in a digital ether. This book is an invitation to explore the fascinating universe of metaverses and virtual worlds, a window into a new era of interaction and creativity.

At the heart of this technological odyssey is Unity, the platform that has revolutionized the creation of interactive experiences. With its powerful and versatile tools, Unity has enabled developers to bring to life unimaginable worlds and dreams that transcend the boundaries of the tangible.

The purpose of this book is to guide you through the exciting journey of building metaverses and virtual worlds, equipping you with the skills and knowledge you need to shape your own digital realities. From the fundamentals to advanced techniques, we will take you by the hand on this journey of discovery and learning.

But a world is not complete if it is not populated by beings, by presences that inhabit it and give life to its corners. This is where Firebase comes in. By integrating Firebase services into your creations, you will be giving your metaverses the intelligence and connectivity needed to create dynamic and personalized experiences. Firebase is the thread that will allow you to weave the web of interactions between users, the bridge that links their actions and gives them a unique and unforgettable experience.

However, the magic of a truly captivating metaverse is at its best when it is shared with others. This is where Photon, with its powerful network SDK, takes over. Through Photon, your creations come to life in the multiplayer realm, connecting people from all over the world and giving you the ability to share experiences, emotions, and adventures as if you were face to face.

With this guide in your hands, you will challenge the boundaries of conventional reality and venture into a territory without borders. This book is a map that will lead you to explore the vast universe of metaverses and virtual worlds, but it is you who will determine the course and magnitude of your creations.

So, take a seat on the ship of creativity and get ready for a journey that will challenge your imagination, sharpen your skills, and, above all, open the door to a new dimension, where the possibilities are endless.

What will you learn on this fascinating journey?

- The fundamental concepts of metaverses and virtual worlds
- How to plan and design an engaging and captivating metaverse
- Using Unity to build realistic and immersive virtual environments and scenarios

- Implement impressive and dynamic visual effects in the metaverse
- Leverage Firebase's capabilities to create authentication and data storage systems for a personalized user experience
- Integrate and synchronize data between devices using Firebase Cloud Firestore
- Develop game mechanics for meaningful interactions in the metaverse
- Use Photon to implement multiplayer, chat, and voice chat functionality, allowing users to share their experience with friends and other online players
- Optimize the performance of the metaverse to provide a smooth and efficient experience
- Resolve common problems and debug bugs in the creation of virtual worlds
- Ethical and security considerations in the development and deployment of metaverses and virtual worlds
- Integrate the Oculus SDK to export your project and run it on Meta Quest 2 goggles

This book is designed to take you from the basics to advanced levels of creation and development, allowing you to explore new frontiers in the age of metaverses and open up endless possibilities for creativity and innovation.

Get ready to embark on an exciting journey of knowledge and skill in building digital universes that transcend the imagination!

Who this book is for

This book is designed especially for you, a game developer or programmer who wants to enter the exciting world of metaverses and virtual worlds using Unity, Firebase, and Photon. If you are a student or apprentice interested in learning about advanced technologies for creating immersive, multiplayer experiences, you will find this book a valuable guide.

If you are a designer or creative artist, you will be able to expand your skills in creating stunning and immersive virtual environments. If you work in the field of **Virtual Reality** (**VR**) or **Augmented Reality** (**AR**), you will discover how to build virtual worlds for these technologies and how to implement multiplayer interactions.

If you are a tech entrepreneur with a business vision, you will understand the emerging trends in metaverses to develop new projects or start-ups. Even if you are a technology and video game enthusiast interested in creating video games or virtual worlds, this book will give you technical knowledge and a deeper insight into the creative process.

What this book covers

Chapter 1, Getting Started with Unity and Firebase, lays the foundations of the knowledge necessary to work with Unity 3D. You will learn in depth about all that the Firebase suite offers and you will create your first scene.

Chapter 2, Preparing Our Player, teaches you how to create a 3D character fully controllable by you, for both desktop and mobile platforms. We will also see how to download new appearances and movements in Mixamo.

Chapter 3, Preparing Our Home Sweet Home: Part 1, is where we will design a beautiful virtual village with houses, pavements, trees, and buildings, a perfect setting to welcome the players of our metaverse.

Chapter 4, Preparing our Home Sweet Home: Part 2, teaches you how to optimize the scene built in the previous chapter to guarantee good performance for your players. We will also make our first foray into the Firebase Firestore database.

Chapter 5, Preparing a New World for Travel, helps you build a world with houses created and connected to the Firestore database. This scene will have the capacity to load, in real time and in a totally dynamic way, existing buildings in the database.

Chapter 6, Adding a Registration and Login Form for Our Users, provides an identification service for users, who will be able to log in or create a new account, through screens designed with the Unity GUI. Firebase Authentication will be the protagonist in this chapter.

Chapter 7, Building an NPC That Allows Us to Travel, helps you create a nice, animated **Non-Player Character** (**NPC**) who will provide us with a travel service between worlds.

Chapter 8, Acquiring a House, teaches you how to program a script that allows players to obtain a house. We will connect the owner with their building purchased on Firestore.

Chapter 9, Turning Our World into a Multiplayer Room, gives us the necessary knowledge to turn our metaverse into a multiplayer world, using Photon SDK to do so.

Chapter 10, Adding Text and a Voice Chat to the Room, to perfectly complement a multiplayer world, covers how to add a text chat and a voice chat, which will allow players to communicate in real time.

Chapter 11, Creating an NPC that Allows Us to Change Our Appearance, will provide the knowledge necessary to create a new NPC that will allow the player to change their avatar.

Chapter 12, Streaming Video like a Cinema, teaches you how to play streaming videos on a cinema-style screen.

Chapter 13, Adding Compatibility for the Meta Quest 2, as metaverses and VR are concepts that go hand in hand, covers how to convert the project to run on your Meta Quest 2 glasses.

Chapter 14, Distributing, will teach you how to compile our project for platforms such as Windows, Linux, Mac, Android, and iOS, as well as extra tricks for further performance optimization.

To get the most out of this book

Throughout this book, we will work within the Unity 3D game engine, which you can download from `https://unity.com/download`. The projects were created using Unity 2021.3.14f1, but minimal changes should be required if you're using future versions of the engine.

If there is a new version out and you would like to download the exact version used in this book, you can visit Unity's download archive at `https://unity3d.com/get-unity/download/archive`. You can also find the system requirements for Unity at `https://docs.unity3d.com/2022.1/Documentation/Manual/system-requirements.html` in the Unity Editor system requirements section.

To deploy your project, you will need an Android or iOS device. For the sake of simplicity, we will assume that you are working on a Windows-powered computer when developing for Android and a Macintosh computer when developing for iOS.

Software/hardware covered in the book	Operating system requirements
Unity 2021.3.14f1	Windows, macOS, or Linux
Unity Hub 3.3.1	Windows, macOS, or Linux
Meta Quest 2 (optional)	Android

If you are using the digital version of this book, we advise you to type the code yourself or access the code from the book's GitHub repository (a link is available in the next section). Doing so will help you avoid any potential errors related to the copying and pasting of code.

Download the example code files

You can download the example code files for this book from GitHub at `https://github.com/PacktPublishing/Build-Your-Own-Metaverse-with-Unity`. If there's an update to the code, it will be updated in the GitHub repository.

We also have other code bundles from our rich catalog of books and videos available at `https://github.com/PacktPublishing/`. Check them out!

Conventions used

There are a number of text conventions used throughout this book. Code in text: Indicates code words in text, database table names, folder names, filenames, file extensions, pathnames, dummy URLs, user input, and Twitter handles. Here is an example: "This gives us the code needed – in particular, the `GameNotificationManager` class – to be added to our script."

A block of code is set as follows:

```
public void ShowNotification(string title, string body,
                             DateTime deliveryTime)
{
    IGameNotification notification =
    notificationsManager.CreateNotification();
    if (notification != null)
```

When we wish to draw your attention to a particular part of a code block, the relevant lines or items are set in bold:

```
    {
        notification.Title = title;
        notification.Body = body;
        notification.DeliveryTime = deliveryTime;
        notification.SmallIcon = "icon_0";
        notification.LargeIcon = "icon_1";
```

Bold: Indicates a new term, an important word, or words that you see onscreen. For instance, words in menus or dialog boxes appear in **bold**. Here is an example: "Once open, click the **Continue** button and follow the initial setup process."

> **Tips or important notes**
> Appear like this.

Get in touch

Feedback from our readers is always welcome.

General feedback: If you have questions about any aspect of this book, email us at customercare@packtpub.com and mention the book title in the subject of your message.

Errata: Although we have taken every care to ensure the accuracy of our content, mistakes do happen. If you have found a mistake in this book, we would be grateful if you would report this to us. Please visit www.packtpub.com/support/errata and fill in the form.

Piracy: If you come across any illegal copies of our works in any form on the internet, we would be grateful if you would provide us with the location address or website name. Please contact us at copyright@packtpub.com with a link to the material.

If you are interested in becoming an author: If there is a topic that you have expertise in and you are interested in either writing or contributing to a book, please visit authors.packtpub.com.

Share Your Thoughts

Once you've read *Build Your Own Metaverse with Unity*, we'd love to hear your thoughts! Scan the QR code below to go straight to the Amazon review page for this book and share your feedback.

`https://www.amazon.in/review/create-review/error?asin=1837631735`

Your review is important to us and the tech community and will help us make sure we're delivering excellent quality content.

Download a free PDF copy of this book

Thanks for purchasing this book!

Do you like to read on the go but are unable to carry your print books everywhere?

Is your eBook purchase not compatible with the device of your choice?

Don't worry, now with every Packt book you get a DRM-free PDF version of that book at no cost.

Read anywhere, any place, on any device. Search, copy, and paste code from your favorite technical books directly into your application.

The perks don't stop there, you can get exclusive access to discounts, newsletters, and great free content in your inbox daily

Follow these simple steps to get the benefits:

1. Scan the QR code or visit the link below

`https://packt.link/free-ebook/9781837631735`

2. Submit your proof of purchase
3. That's it! We'll send your free PDF and other benefits to your email directly

Part 1: Getting Started

In this part of the book, we will start exploring the Unity 3D editor, from its installation to the creation of a first scene, as well as getting to know each part that makes up this impressive video game engine. We will also learn how to create a player controller, which will allow us to move in the three-dimensional space of any scene in our project.

We will design and create virtual worlds, full of houses, pavements, and trees, and optimize performance for a smooth experience. Throughout all the chapters that make up this part of the book, we will learn in depth about all the services that make up Firebase. In addition, we will practice how to integrate the Firestore database into our project.

This part has the following chapters:

- *Chapter 1, Getting Started with Unity and Firebase*
- *Chapter 2, Preparing Our Player*
- *Chapter 3, Preparing Our Home Sweet Home: Part 1*
- *Chapter 4, Preparing Our Home Sweet Home: Part 2*
- *Chapter 5, Preparing a New World for Travel*

1

Getting Started with Unity and Firebase

The **metaverse** and the future are a combination of increasingly united concepts, a new formula that was born to stay and lay the foundations for new forms of communication in the coming years. A metaverse is a shared, three-dimensional virtual world where people can interact with each other and with digital objects in real time. Its utility lies in providing an expansive space for creativity, collaboration, and entertainment.

It allows users to explore immersive virtual environments, engage in social, commercial, and educational activities, and experience new forms of expression and experiences. In addition, the metaverse can serve as a platform for developing innovative solutions in a variety of areas, such as remote work, distance learning, product design, and simulation, offering endless possibilities for human interaction and imagination.

In the coming years, the way we attend work meetings, concerts, training courses, and even dating to find a partner may occur in the metaverse. This book will help you get started with programming a metaverse by using the **Unity 3D** video game engine, the **Google Firebase** suite, and other amazing tools.

In this first chapter, you will learn the first steps of configuring your computer so that you can start developing virtual worlds in Unity 3D. We will start by laying down the basics about the Unity 3D video game engine, introducing the services offered by Firebase and explaining how they will help us in our project, and covering what other tools we will need.

Once we have reviewed all the concepts and technologies we will use, we will install Unity 3D on our computers. This chapter's mission is to ensure we completely configure our work environment for satisfactory project development.

We will cover the following topics:

- Installing Unity
- Organizing your project assets
- Choosing an input handler
- Understanding Firebase services

> **Note for advanced readers**
> If you have previously worked with Unity, you will know how to install it correctly on your system. If you already have it installed, you can skip the following steps – just keep the version of Unity that we will use as a base in this project in mind to avoid incompatibilities in future chapters.

Technical requirements

For the Unity 3D game engine, you will need a Windows or Mac computer. Normally, Unity can be run on most computers nowadays, but if you want to have a look at the minimum requirements for a computer to run Unity, go to `https://docs.unity3d.com/Manual/system-requirements.html`.

If you plan to compile your metaverse for iOS or Mac, you must have a Mac, with the latest version of macOS and also a current Apple developer account. If you only want to compile for Android, Windows, or Linux, you can use a Windows or Mac computer indifferently. In this book, the screenshots provided have been taken from a Mac, but this should not affect you in any way, even if you have Windows.

Fortunately, our project can be tested by running the scene directly from Unity. This will make it much easier for you to progressively see small changes with just one click. Only when we get closer to the end of this book will we need an Android and/or iOS device to test compilations on mobile devices, but the latter is optional if you want to compile for mobile devices.

You must follow this first chapter in detail to make sure that your computer uses the same version of Unity and other dependent plugins that we will show throughout.

Almost certainly, the project that we will be creating throughout this book will be compatible with future versions of Unity 3D. So, if, for any reason, you need to use another version of Unity that's different from the one recommended here, it will not be a problem.

This book's GitHub repository contains the complete project that we will work on: `https://github.com/PacktPublishing/Build-Your-Own-Metaverse-with-Unity/tree/main/UnityProject`.

Installing Unity Hub

To develop our metaverse, just like a video game, we will need to install Unity on our computers. In this section, we will follow all the necessary steps to do so. We will install Unity Hub, which will allow us to manage multiple versions of Unity in a simple and orderly way, create a project from scratch, and explore the fundamental parts of the editor to get familiar with the graphical interface.

Unity Hub is an application that allows us to manage and install multiple versions of the Unity editor on our computer. Also, for each version, it allows us to install different build platforms and tools in an isolated way. It is very common that throughout our professional careers developing video games, we need different versions of Unity to support different projects, which, for whatever reason, cannot be updated or do not work quite right in later versions of Unity.

This is where Unity Hub comes into play – it allows us to easily install any version of Unity in isolation without it conflicting with other versions. It also allows us to load existing Unity projects on our computers and configure which version of Unity it should use when opening.

To install this tool, execute the following steps:

1. Please download and install the **Unity Hub** program from `https://unity3d.com/get-unity/download`. It is recommended that you use **Unity Hub** to manage the Unity versions on your computer instead of downloading and installing the Unity editor directly.

   The screenshots in this book were taken in Unity Hub version **3.0.0**. The functions and buttons in earlier or future versions should not change.

2. You will need to activate a license for Unity if you have not already done so. The Unity User license has free plans for Students and Community; you can decide if you need to upgrade your plan to Plus or Pro in the future. For our project, a free plan is sufficient since all Unity features are open in the free license.

3. Press the **Download For Mac or Download For Windows** button, as shown in the following screenshot:

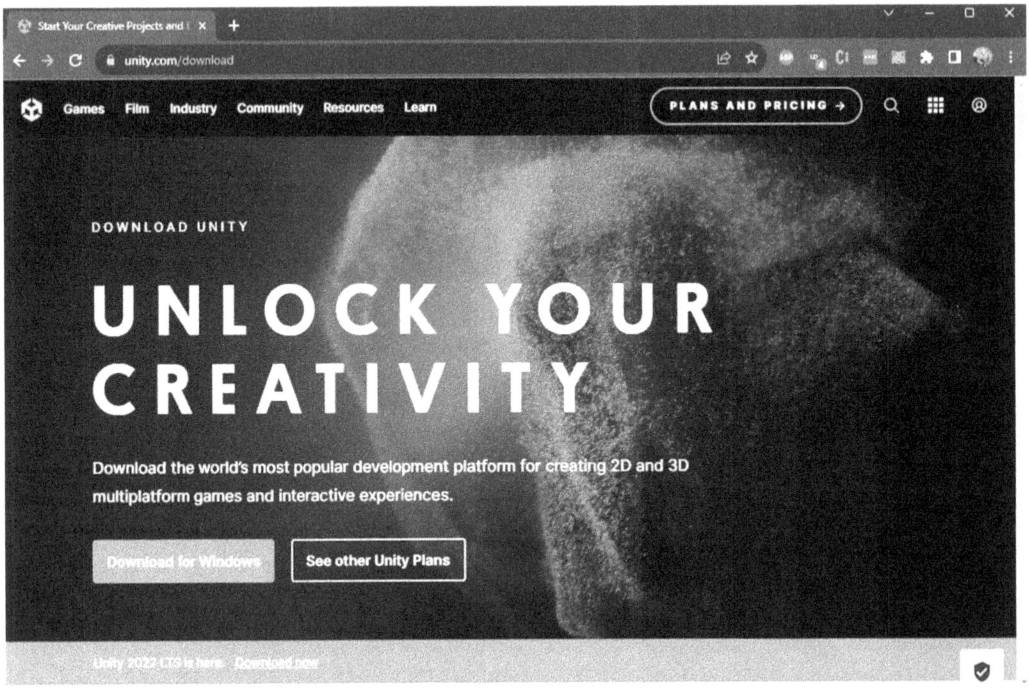

Figure 1.1 – Installing Unity Hub instead of downloading Unity directly

4. If this is your first experience with Unity or you want to expand your knowledge, I recommend that you take a look at the **Learn and Community** menus that you will find in Unity Hub when you open it.
5. If this is the first time you are using Unity Hub, you will need to create a new account for free. You can also log in with your Google, Facebook, or Apple account and accept the terms and conditions of use.

Once we have downloaded Unity Hub, we can proceed to install it on our computers. Unity Hub allows us to manage multiple installations of different versions of the Unity editor on our computers. This will allow us to maintain compatibility with all previous projects that we have or those that we will create in the future. Next, I will guide you through the process of installing a new instance of the Unity editor step by step.

Installing the Unity editor

As a recommendation, whenever you start a new project in Unity, it is interesting to download the latest version available. That is why, in our metaverse, we will do it this way.

When starting a new engine installation, Unity will show us different options, such as the recommended LTS version, other recent versions, and a history of previous versions, all of which are very useful when you have a legacy project that needs a specific version to work.

To start a new installation, execute the following steps:

1. On the **Unity Hub** main screen, select the **Installs** tab and click on the **Install Editor** button:

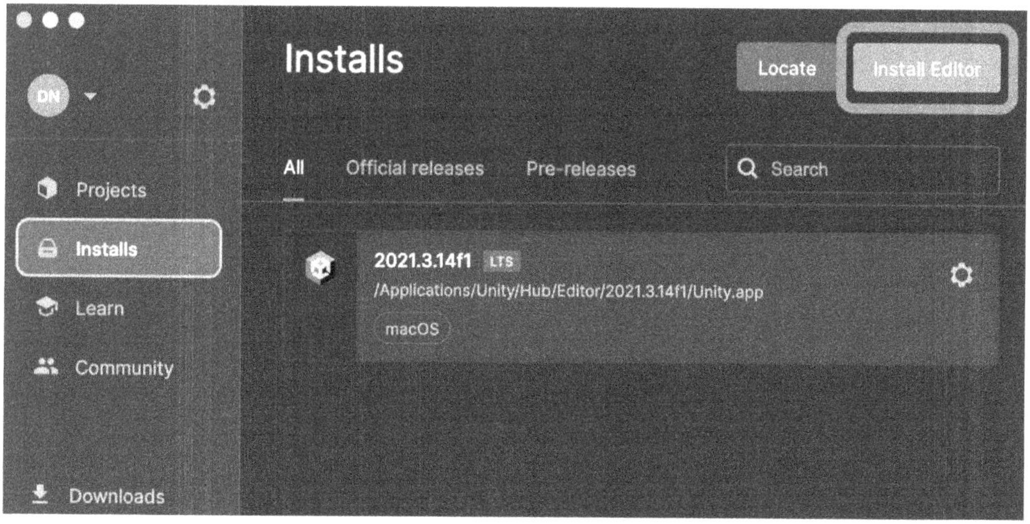

Figure 1.2 – The Install Editor button in Unity Hub

2. Click on the **Archive** tab and then click on the **download archive** link, a tab will open in your browser that will show the Unity version catalog. If you wish, you can also go to the url `https://unity.com/releases/editor/archive` in your browser:

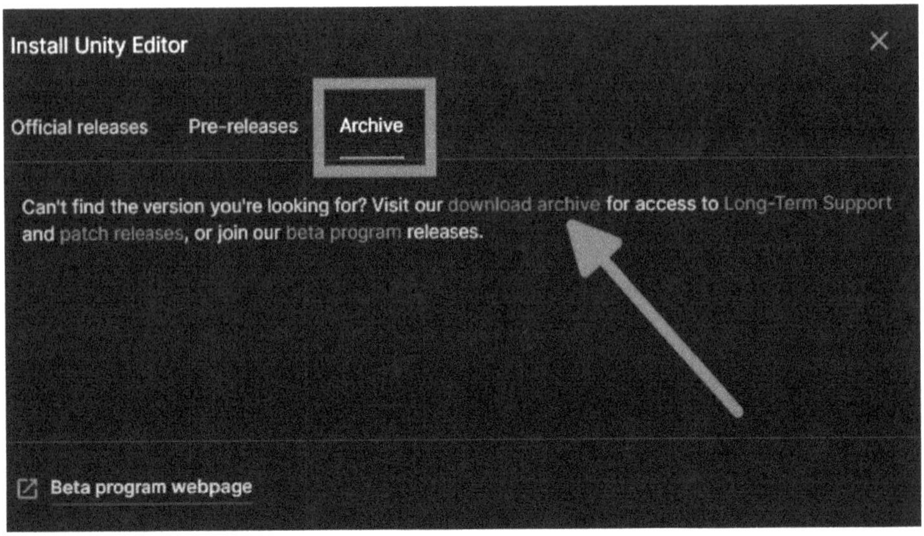

Figure 1.3 – Archive tab in Unity Hub

3. This project has been built from version **2021.1.14f1**. Once you are on the **Unity** version catalog page, you can search for the desired version. Click on the **Unity 2021.X** tab to find the **2021.1.14f version**, finally click on the **Unity Hub** button, this action will launch the download in **Unity Hub**.

> **About the Unity version**
>
> In this book, we have built the project on version 2021.3.14f1. This version will most likely be lower than the version Unity recommends at the time of installation. We cannot guarantee that the project will work in future versions of Unity. To use a particular version of Unity or if the version you are looking for is not listed in Unity Hub, you can access the version history at `https://unity.com/releases/editor/archive`.

4. The next step is very important: to configure the installation, we must select the components that will be installed initially. Do not worry if you have skipped this step – you will be able to reinstall other components in the future. For the correct course of our project, we will need to check the following:

    I. If you are working on Mac, you will have to select Visual Studio for Mac, whereas if you are on Windows, the Microsoft Visual Studio Community 20xx option will appear.

    II. In *Chapter 14, Distributing*, we will learn how to compile and export our project on different platforms. To successfully carry out this part, at this point (or later if you wish) we need to install the necessary components to compile for Android, iOS, and Windows.

III. Check the following components:

- Android Build Support, with OpenJDK and Android SDK and NDK Tools
- iOS Build Support
- Linux Build Support (IL2CPP)
- Mac Build Support (IL2CPP) (Mono) if you are using Windows
- Windows Build Support (Mono) (IL2CPP) If you are using Windows

Some of the components are heavy and may take a while to download and install; it's a good time for a coffee!

Changing Unity's installation location

Using the Unity Hub gear icon, you can change the location where the Unity version you have selected will be installed. You can change the path from the **Installs** tab:

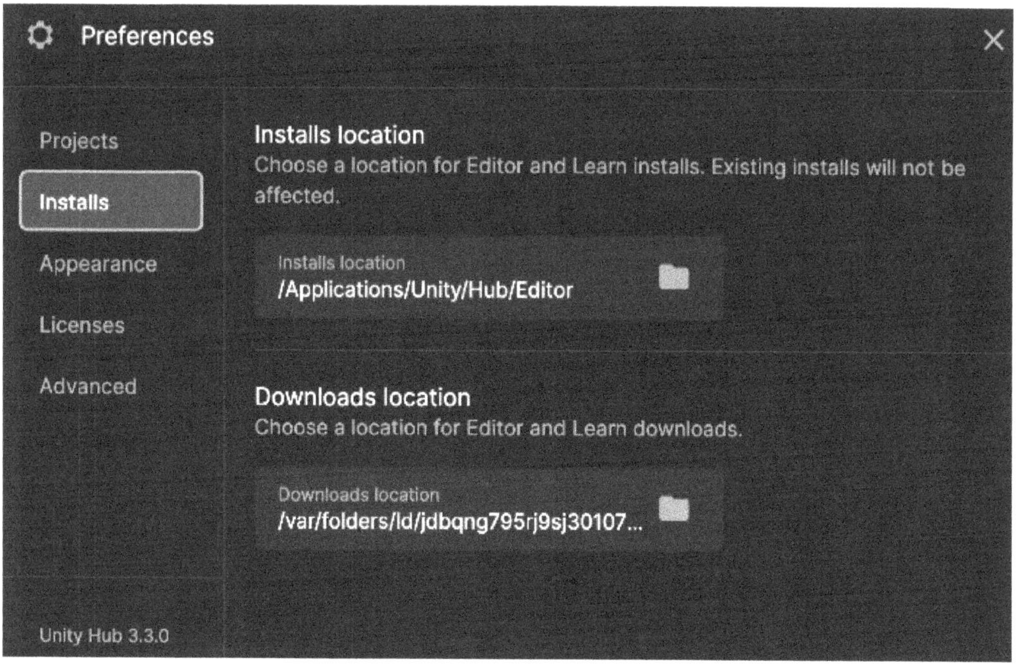

Figure 1.4 – Setting the default installation folder

Well, after knowing the main requirements to be able to work with Unity3D without problems, we will see in detail how to create new projects from Unity Hub.

Creating and managing Unity projects

The **Unity Hub** tool also serves to manage our projects, regardless of the version of Unity in which they were conceived. We can have an infinite number of projects, old and new, and each of them can be configured to be edited in a different version of Unity. You can easily add the projects you have in some directory on your computer using the **Open** button.

Don't worry if you have loaded a project that uses a version you don't have installed in Unity Hub. As you can see in *Figure 1.5*, a warning icon will appear to warn you that the project uses a version of the engine that you do not have installed and allows you to install that particular version by clicking on the button specifying the version number:

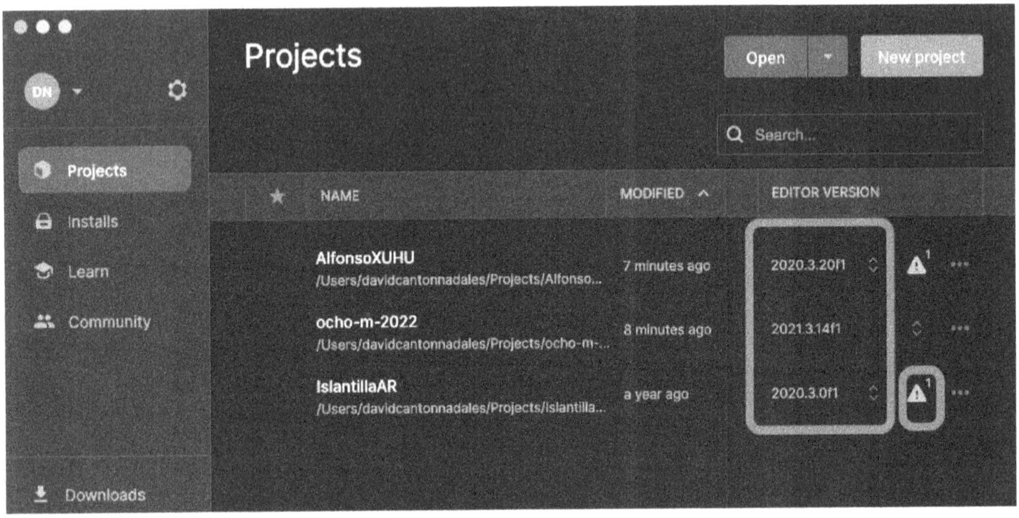

Figure 1.5 – Managing multiple projects with different versions of Unity

To begin, we will need to create a new project. Each project that we'll create from Unity Hub will be created in a specific folder that we create. Don't worry about this – the project creation wizard will ask us each time where we want to create it.

If you always want to create them in the same directory and you want to avoid configuring this every time you create a new project, you can configure it at a general level in the same window that appears in *Figure 1.4*.

In the **Projects** tab, you will see which directory is used by default for each new project. Just click and select the new destination and click **Accept**.

Now, to create our new project, execute the following steps:

1. With Unity Hub open with the **Projects** tab active, click on the **New Project** button, located at the top right. Once the new project creation wizard opens, it will ask us to update the initial configuration we want to have (**template**), the name of our project, and the location where it will be created:

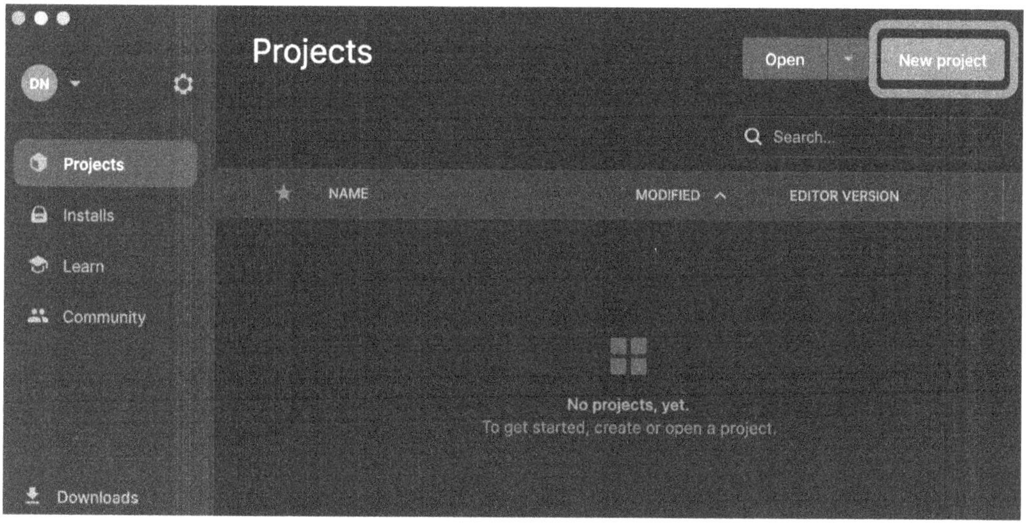

Figure 1.6 – The New project button in Unity Hub

2. Scroll down and select the **3D (URP)** template. We will see why we have selected this item and not another one shortly. It is possible that you may not have used the **3D (URP)** template before and you need to download it; in this case, click on the **Download template** button:

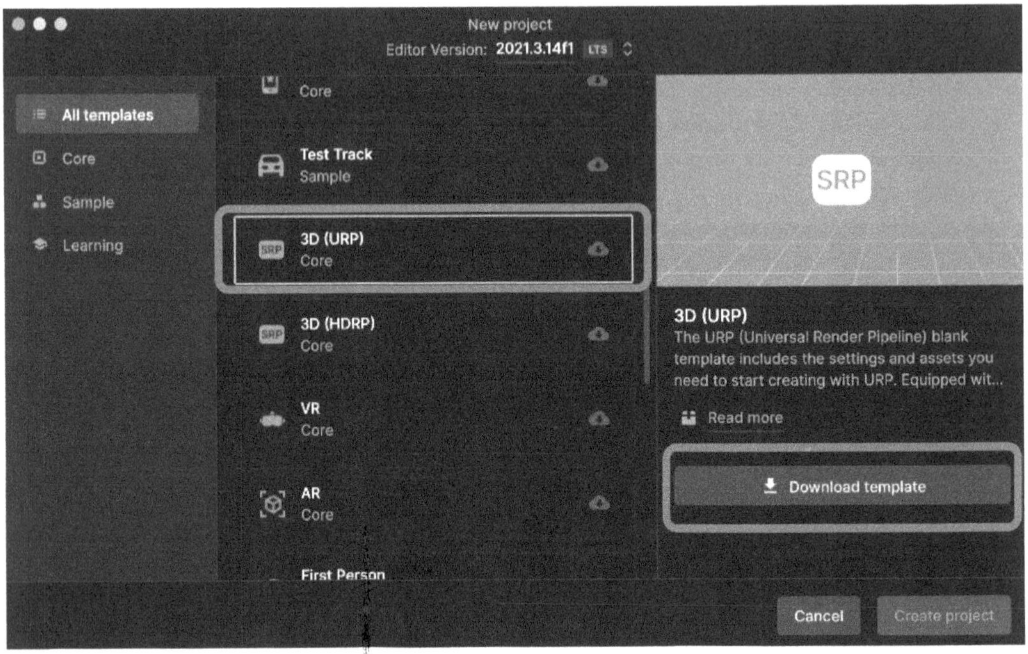

Figure 1.7 – Creating a new project

3. Once you have downloaded the template, you can fill in the title of your new project and its location. In my case, I am going to call this project `MyMetaverse` and I will save it in my default directory.
4. Remember that you can choose the directory you prefer as this will not influence the development of the project. In my case and for your reference, I have used the `/Users/davidcantonnadales/Projects/MyMetaverse` path. If you're using Windows, it can be something like `C:Projects/MyMetaverse`:

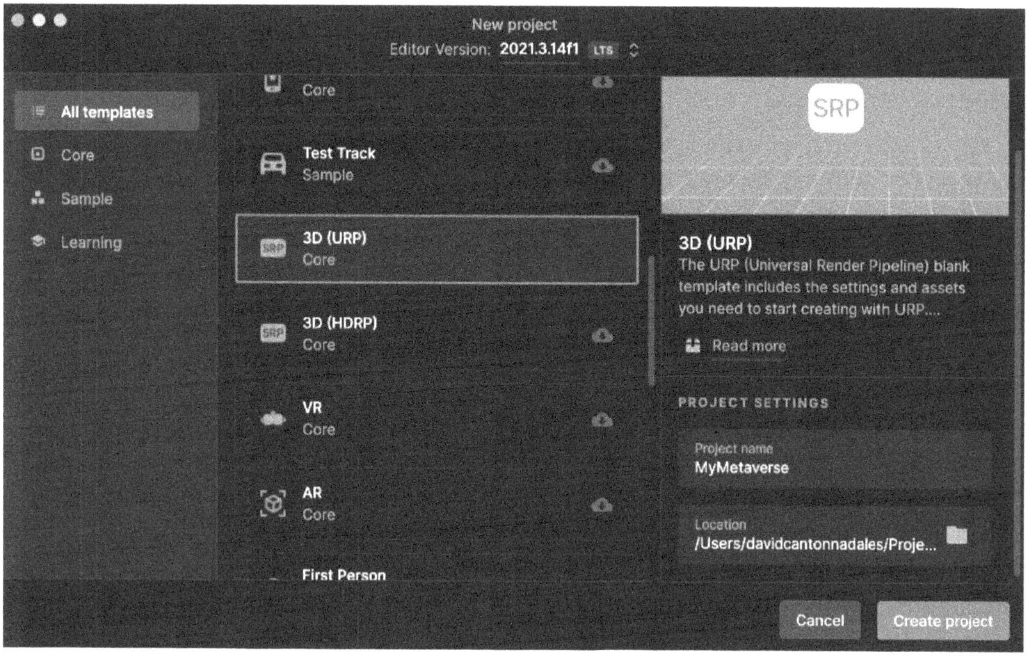

Figure 1.8 – Setting up the name and location of the new project

> **Things to consider when choosing your project name**
>
> In some projects I have done, I have had problems with the spaces in the project creation directory. Keep in mind that Unity will create a folder with the same name as the project you have chosen. For example, if we had chosen `My Metaverse` instead of `MyMetaverse`, it would create a directory with a blank space. To avoid possible problems later on, I recommend that you avoid blank spaces.

You may have some doubt in your mind as to why we have used the URP rendering method instead of the default one. I have specified the technical benefits we have obtained by selecting the URP rendering method for our project in the following note.

> **Why did we choose the 3D template (URP)?**
>
> Unity allows you to create **Universal Render Pipeline** (**URP**) or **High Definition Render Pipeline** (**HDRP**) projects so that you can focus on performance or quality rendering.
>
> URP offers better graphics quality than the default **Standard Render Pipeline** (**SRP**) built-in system in Unity. URP is designed for casual games for mobile and PC/Mac, offering great compatibility between different devices. On the other hand, there is a superior render pipeline called **High Definition Render Pipeline** (**HDRP**) that's oriented to AAA games for consoles and PC with great features.
>
> If you want to learn detailed technical information about the differences between URP and HDRP in Unity, I recommend that you check out this blog post: `https://vionixstudio.com/2022/02/12/urp-vs-hdrp-in-unity`.

1. Once you have clicked the **Create** button, Unity will take a few seconds to configure and create the new project. It will also load the default plugins. Once everything has finished loading, the editor will open with a default scene. When this whole process is finished, you will see something similar to *Figure 1.9*:

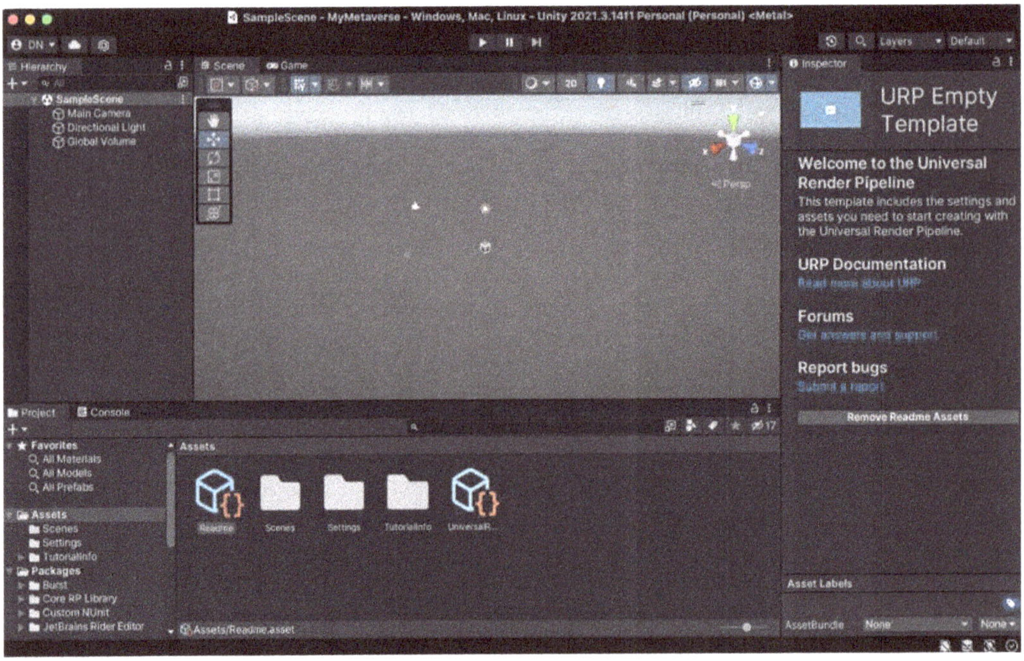

Figure 1.9 – The appearance of the Unity editor after creating a new project

With that, we have created our first project. As you can see, Unity has created our first blank scene called **SampleScene**. Before we move on to the next point, let me briefly explain something about the target platforms that Unity offers.

As I commented at the beginning of this book, our project will be compatible with multiple platforms – that is, it can be compiled and exported for Android, iOS, Windows, macOS, and Linux. At the end of this book, we will learn how to do this for each of the platforms.

In Unity, when we create a new project, by default, it is configured to be exported for Mac/Windows. You can view the target platform's settings by opening the **Build Settings** menu. On Mac and Windows, you can access this menu by clicking on **File** in the toolbar and then on the **Build Settings** menu:

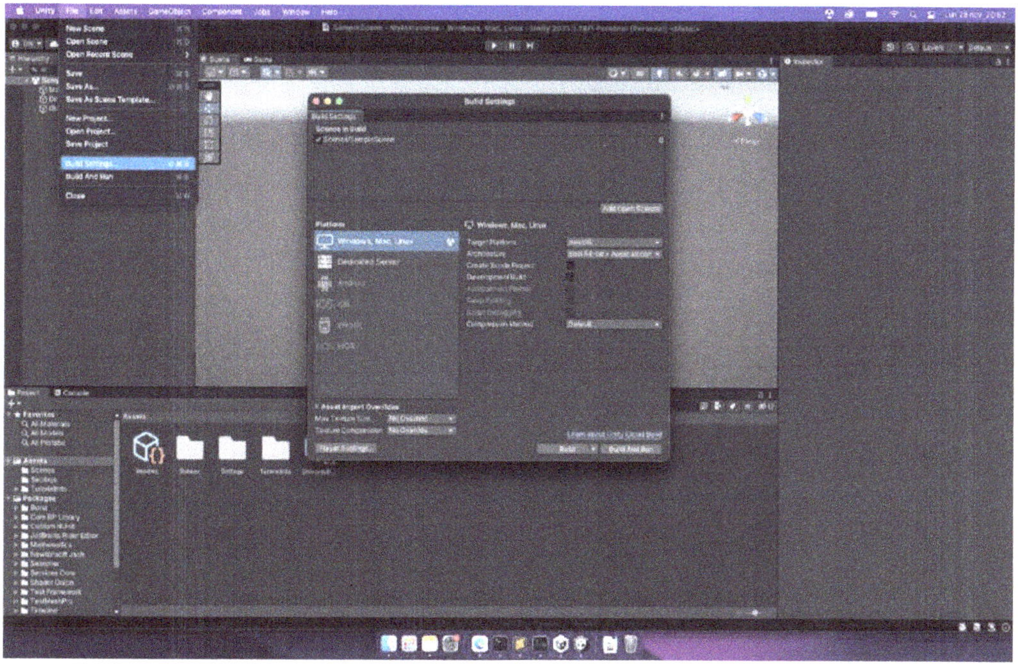

Figure 1.10 – The Build Settings window

For now, and until we finish most of the chapters and get to the **Cross Platform** compilation part, we will not need to change these settings. We will dig into this later.

Basics of using the Unity editor

If this is the first time you're using Unity, do not skip this section as we will briefly review all the parts that comprise Unity. We will explore the most important menus and begin to learn the basic actions that will lay the foundations of the knowledge necessary for you to advance to the following chapters.

The first and most basic thing to know about Unity is its organization. By default, when we create a new project, we see the Unity editor with five clear divisions called **layouts**. There are several predefined layouts that we will see later. In addition, we can customize this aspect to our taste and save the configuration.

If we pay attention to the Unity editor, we can distinguish mainly five sections in the layout that load by default. To learn about other layouts, take a look at **Windows | Layouts**, as shown in the following figure:

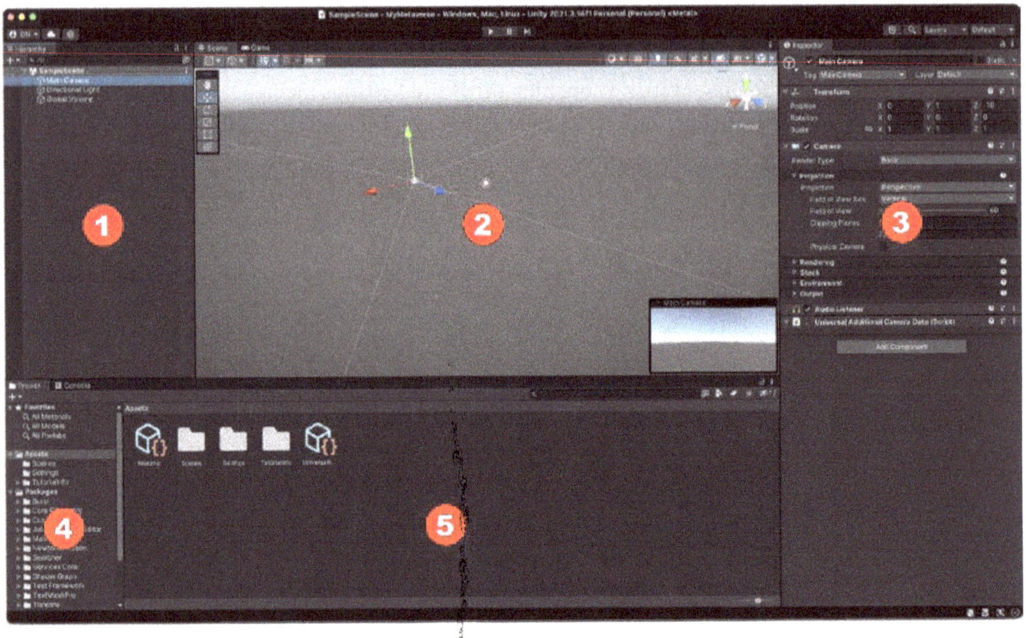

Figure 1.11 – The default layout in the Unity editor

Let's review what the divisions we have numbered in the preceding screenshot are for:

1. **Hierarchy window**: All the objects that appear in our scene (GameObjects) are organized as a tree, and they can be relocated and grouped in parent-child mode. There is a root node with the name of the scene; the rest will hang from here. It is here where we will select the object we are going to work with so that we can move it, enlarge it, edit it, and so on.
2. **Scene window**: This is the three-dimensional representation of the world we are building. This is where the objects that we have in the **Hierarchy** window will be represented graphically. When you select an object of the hierarchy, it will be marked in the scene, as shown in the preceding screenshot with the GameObject called **Main Camera**.
3. **Inspector window**: This section displays information about the position, rotation, and scale properties of the selected object in the **Hierarchy** window. The changes we make to these values will affect the visual representation of the **Scene** window. Also, in this **Inspector** window, we can manage the components of an object.

   By way of summary, it can be said that components are pieces with unique functionality, which are attached to a GameObject and cause it to acquire a new functionality or feature. These components can also be scripts that we make in C#. We will start programming our own scripts in *Chapter 5, Preparing a New World for Travel*.

4. **Project window**: Here, you will find the folders of the added resources and the assets or plugins that have been installed in our project. For example, here, we will find the images, audio, or 3D objects that we will be using in some of the scenes.
5. **Project window in detail**: As a master detail view, this window shows the contents of the folders that were selected in the **Project** window. The assets that we select here will have their information displayed in the **Inspector** window.

We already know how a layout is composed in Unity; inside each window of the layout, we can find other functionalities that we can access by changing tabs, as shown in the following figure:

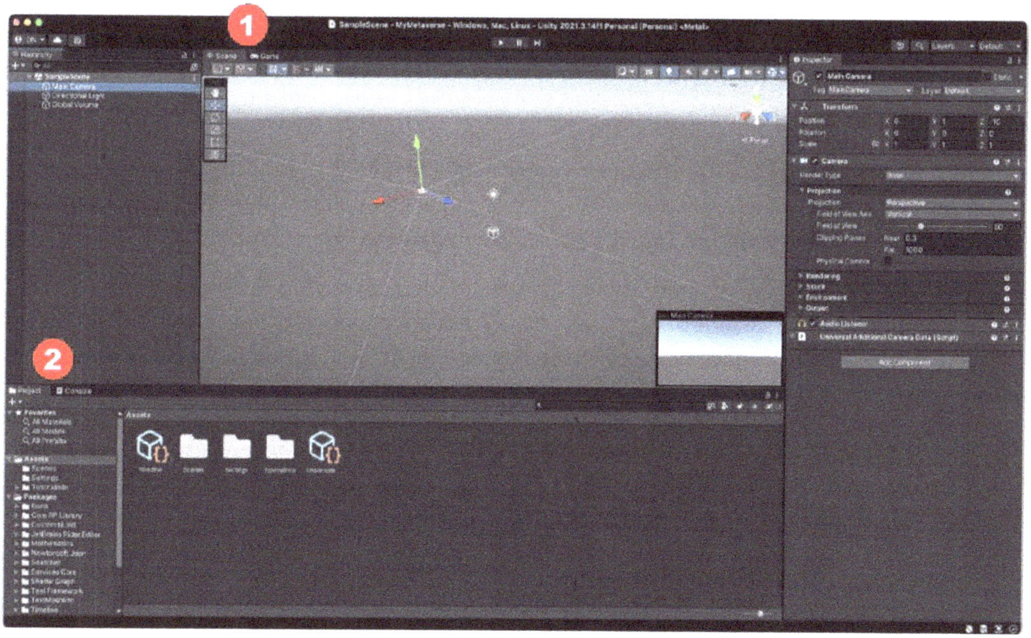

Figure 1.12 – Tabs with other functions in the layout

If we look at the **Scene** window in *Figure 1.12*, we will see that there are two tabs called **Scene** and **Game** (**1**). The **Scene** tab, which is checked by default, as we mentioned previously, shows the 3D representation of the objects placed in the **Hierarchy** window. Here, we can move freely with the mouse and keyboard to explore the scene.

In the **Game** tab, we will see what is focusing and rendering the camera in our scene – in this case, the **Main Camera** GameObject.

In the **Project** window, there are two tabs – **Project** and **Console** (**2**). In **Project**, as we have seen previously, the folders of the resources and assets that we have in our project are shown. In the **Console** tab, we will see information about errors and logs when we run our project.

On the other hand, we have the main menu bar of Unity, in which we can find the following options:

Figure 1.13 – The Unity editor's main menu bar

Let's take a closer look:

- **File**: Here, you can find the options to create a new scene, save the changes of the current scene, or open another one. You also have options to open or close a project. Finally, in **Build Settings**, you can change the **Target Platform** settings, as we saw previously.
- **Edit**: Here, we can see different actions, such as ordering or copying and pasting, that we can use for the GameObject of the current scene. Another very important menu that you can access from here is **Project Settings**, which contains all the settings applicable to our project.
- **Assets**: This menu allows you to search, import, and manage your project assets.
- **GameObject**: This menu allows you to add new items to your scene. It also gives you advanced options to create them as parents or children of others. There is a list of GameObjects that comes with Unity by default, such as **lights**, **particle systems**, **cameras**, and others.
- **Component**: This menu shows a list of built-in components that you can add to your selected GameObject.
- **Window**: This menu includes very important options, such as layout management, which we talked about previously. There are also actions to organize the editor panels.

  A very important menu that we will see later is called **Package Manager**. This is an asset manager. When you acquire a new asset or when you want to acquire a new asset, this is where you can find it and add it to your project.

Last but not least, we have the main tools for manipulating the GameObject in the scene, as shown in *Figure 1.14*:

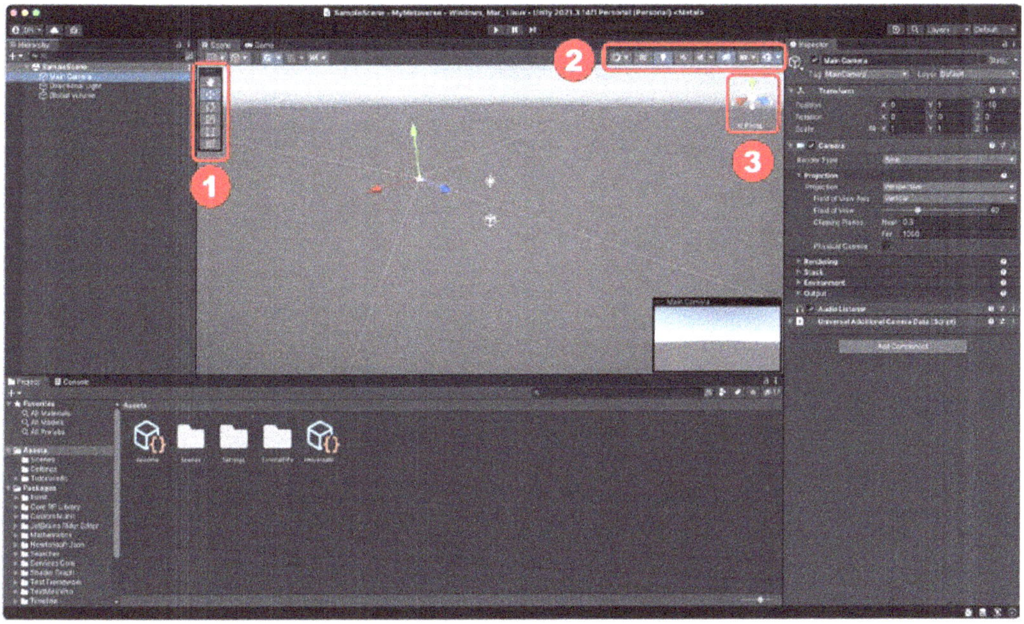

Figure 1.14 – The main toolbars in the Unity editor

Let's take a look:

1. **Tools**: These are the buttons you will use the most in any project you create in Unity. These five buttons are used to rotate, move, transform, and scale the currently selected GameObject.
2. **View options**: This bar has display options that affect the current scene, such as enabling or disabling the lighting, changing the camera rendering to 2D/3D, disabling the sound, and so on.
3. **Gizmo**: This cross is used to change the viewing perspective in the scene. Here, you can view the scene from above, left, right, or below... just click on the different orientations.

Creating a new scene

Now that we've covered a lot of the theory regarding Unity and its interface, it's time to get down to business. For this, we are going to create our first scene. For now, this scene will only have a floor, but it will serve as a base for the following chapters.

Follow these steps to create a new scene from scratch:

1. Create a new scene from the **File** menu by selecting **New Scene**.
2. The **New Scene** creation window will appear. Since our project is being created with URP, select the **Standard (URP)** option. This will create the scene with the appropriate settings:

Figure 1.15 – Creating a new scene with URP configuration

3. Upon clicking the **Create** button, you will need to choose the name of your new scene. The scene we are creating will be our main scene, so you could call it `MainScene`, for example.

   The **Where** option allows you to specify the directory inside the project where you will create the scene. For the moment, you can leave it as-is. We will learn more about how assets are organized in the project later:

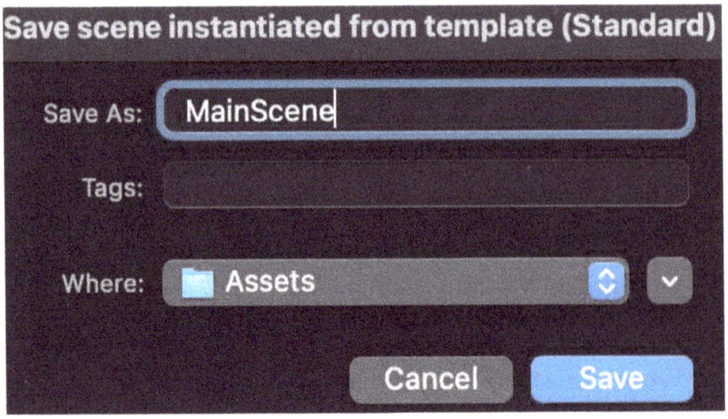

Figure 1.16 – Choosing a name for the new scene

4. After clicking the **Save** button, the new scene will be created and opened. By default, Unity adds a **Camera** object, a **Light** object, and a GameObject called **Global Volume**, which is in charge of adding post-processing settings to our camera and will make everything look much better.
5. Now, to create our floor, we will use a **Plane** object. Click **GameObject** | **3D Object** | **Plane**.
6. We already have a floor; you can see that it has been created in the **Hierarchy** window. Now, you can change the name of the newly created object. For example, you can call it `Floor` instead of **Plane**.
7. Make sure that the ground is at the lowest part of the scene and in the center. For this, modify the position values by clicking on the GameObject from the **Hierarchy** panel and changing the **X**, **Y**, and **Z** positions to `0`, `0`, and `0`. This can be seen in *Figure 1.17*:

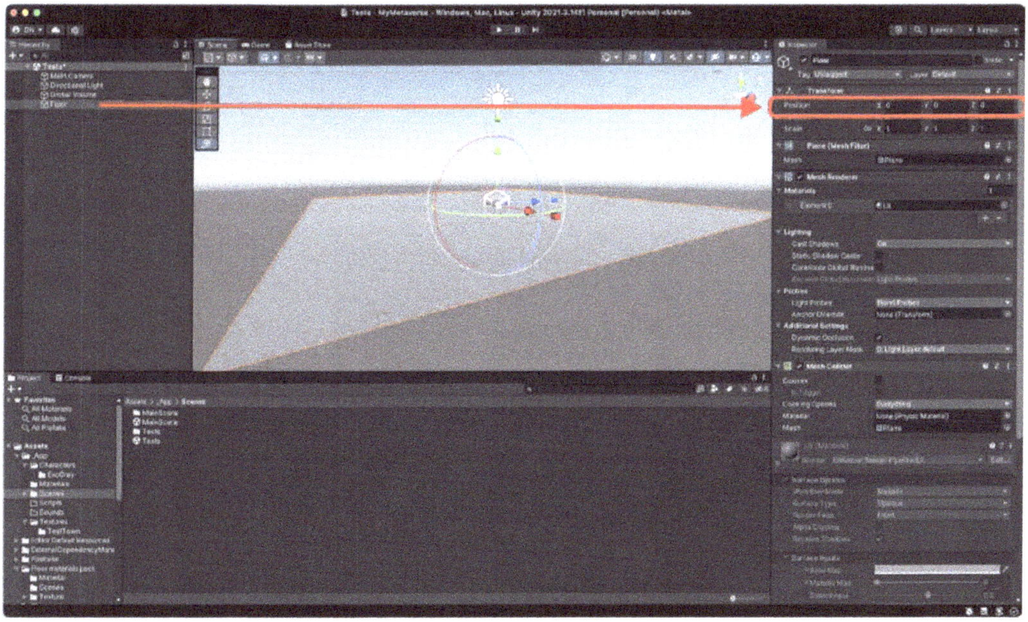

Figure 1.17 – Changing the position of the floor

8. The floor we've created is too small – we would like it to be wider so that the character we create later can walk several steps before falling into the void. You can modify the floor's size by manipulating the **Scale** values in the **Inspector** panel.

   By default, the panel is created with `1`, `1`, `1` values; we will change these to `10`, `10`, `10`. As you change these values, you will see how the floor is enlarged.

9. Finally, save the changes you've made to the scene via the **File** | **Save** menu. The scene will now look like this:

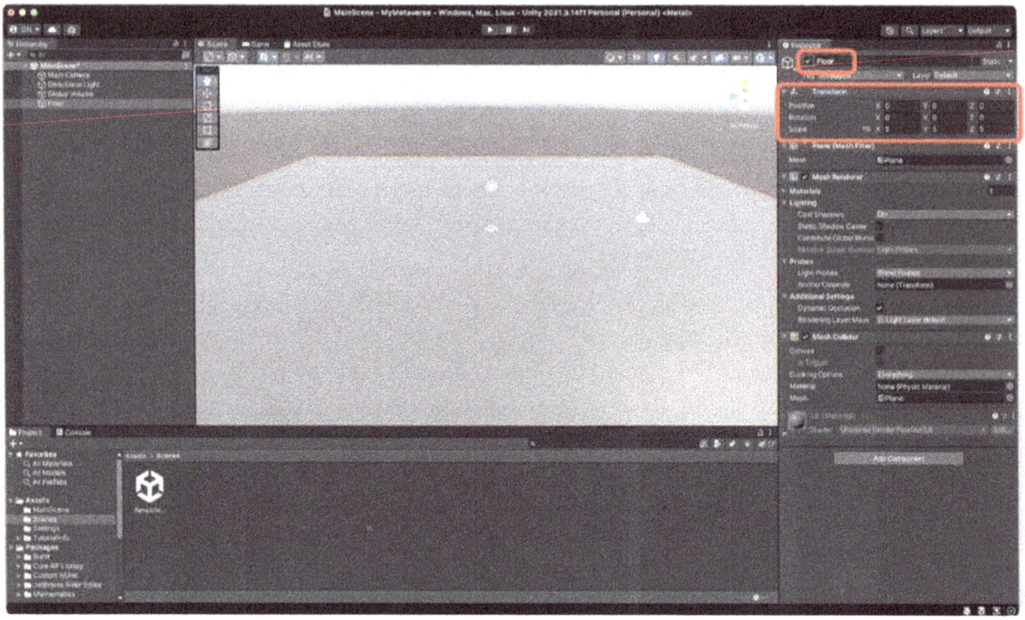

Figure 1.18 – The new scene with a floor

Getting familiar with Scene View controls

At this point, we already know the basics of the Unity interface, including how to create a new scene and change the **Position** and **Scale** properties of a GameObject. At this point, I recommend that you explore the controls to navigate the scene with the mouse and keyboard.

For example, using right-click to pivot the view, press *Alt* + left-click (*Option* + left-click on Mac) to orbit the view around the center of the view and center-click the mouse to move the view.

You can also use *Alt* + right-click (*Option* + right-click on Mac) or the mouse scroll wheel to move closer or further out. For more detailed information about keyboard and mouse combinations, you can visit `https://docs.unity3d.com/Manual/SceneViewNavigation.html`.

> **Advanced object positioning documentation**
>
> You can find extended information on how to change the position, rotation, and scale properties of a GameObject at `https://docs.unity3d.com/Manual/PositioningGameObjects.html`.

Now that you have learned everything you need to install Unity to create new projects and scenes, you know the most important points of the user interface, as well as how to place objects and modify their properties in a scene. Next, we'll look at a very important technique that will lay the foundation for efficiently organizing files, folders, and assets in your project.

Organizing your project assets

When we talk about organizing assets and resources in our project, we are referring to how we want the folder structure in the **Project** panel to look. There is not a single established way to do this - it depends on how we like and feel more comfortable working. Unity recommends a series of points you should take into account when doing this, as well as to avoid file problems.

The following are recommendations for organizing our project:

- Establish a folder and file naming guide before starting.
- Don't lose track of the folder naming throughout the project. On many occasions, we will add files and resources and forget to rename them.
- Do not use blank spaces in folders or individual files.
- Separate the files to be used in testing from the production files by creating another folder dedicated to them.
- Always use the `Assets` root folder; everything you create should hang from this folder. Avoid creating other folders at the same level as the `Assets` folder.
- Keep your private files separate from downloaded assets or Unity's resources.

  As we have said before, there is no standard way to organize projects in Unity. I use a technique that helps me have good order and be more efficient when working. *Figure 1.19* is the default structure that can be seen when creating a project:

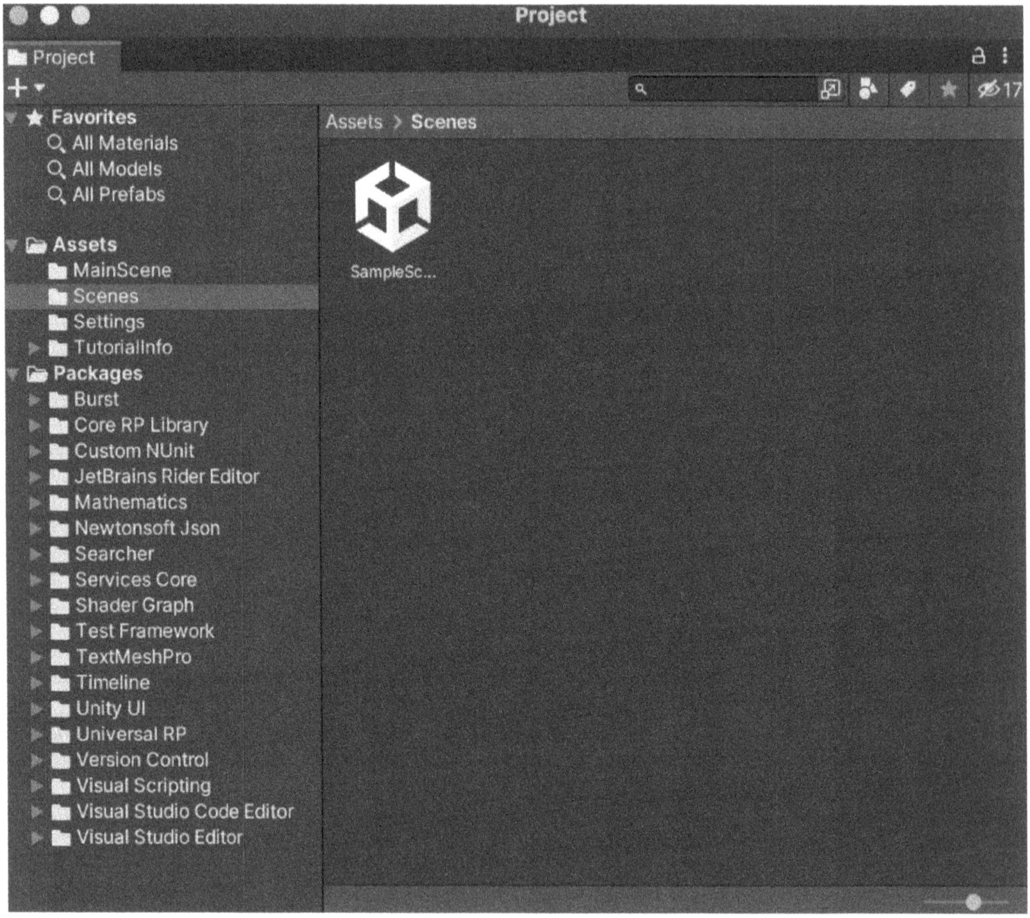

Figure 1.19 – Detailed view of the Project panel

As shown in the preceding figure, under the root `Assets` folder, we have `MainScene`, `Scenes`, `Settings`, and `TutorialInfo`. At first sight, this seems like a good form of organization, but in reality, it is not quite like that. Let me explain why.

As the project grows, you will be downloading assets and installing them in the project. This will cause an infinite number of new folders that hang from the `Assets` folder to be created. These folders are sorted alphabetically, and this can become chaotic if you try to locate your own folders, mixed in among the folders of third parties.

A widespread technique among Unity programmers is to identify your folders with the underscore symbol (_) so that they always appear at the top, as shown in *Figure 1.20*:

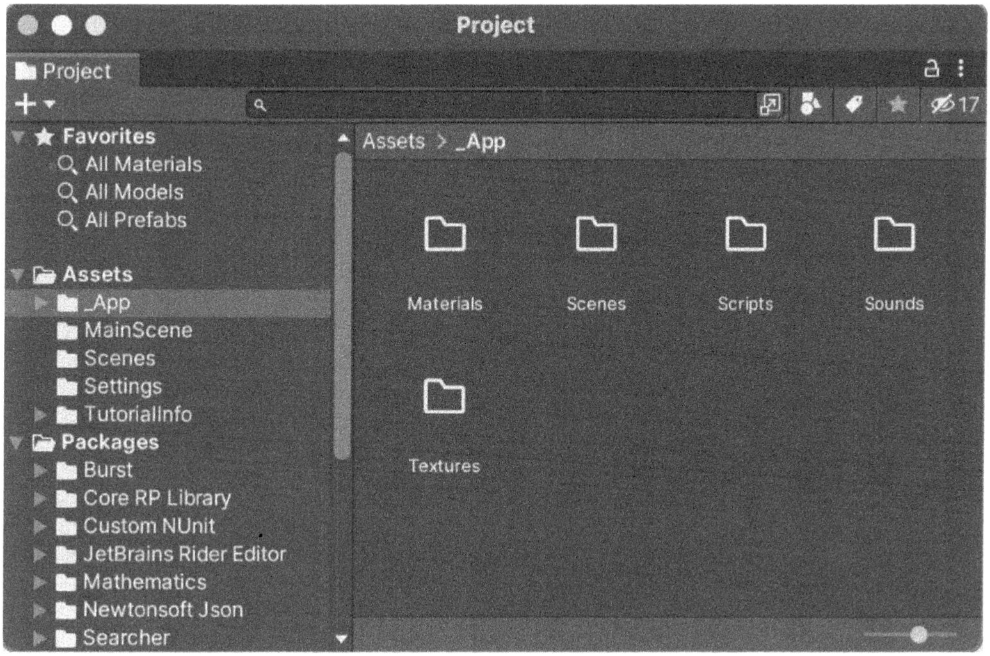

Figure 1.20 – Custom organization

This way, you will no longer care that third-party plugins can create an infinite number of folders in your `Assets` root folder since you will always have yours at the top of your `_App` folder.

Within the `_App` folder, you can create the folder relationship that suits you best. From my experience, most assets can be perfectly organized into materials, scenes, scripts, sounds, or textures.

To finish organizing the project as a whole, we will move the scene we have created to its new location and delete what Unity has created:

1. Move the `MainScene` scene file, along with its configuration folder, inside `_App` | `Scenes`:

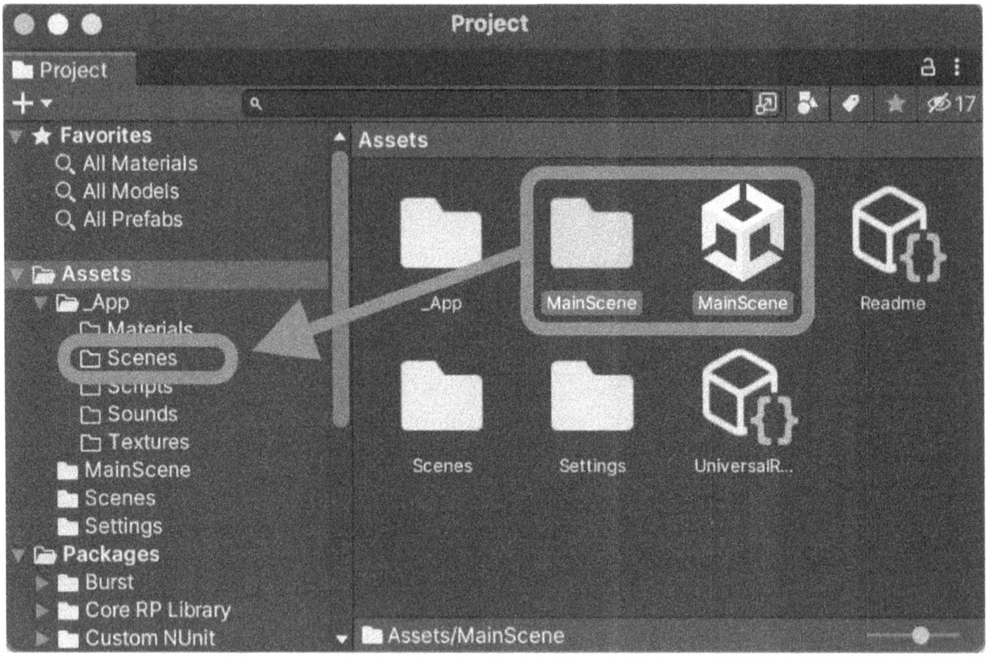

Figure 1.21 – Reorganizing our assets

2. Delete the `TutorialInfo` folder from the root `Assets` folder.
3. Delete the old `Scenes` folder located in the root `Assets` folder. Inside this is the `SampleScene` example scene, but don't worry, we don't need that either.

Once we have reorganized our assets and removed the sample material, our **Project** panel will look something like *Figure 1.22*:

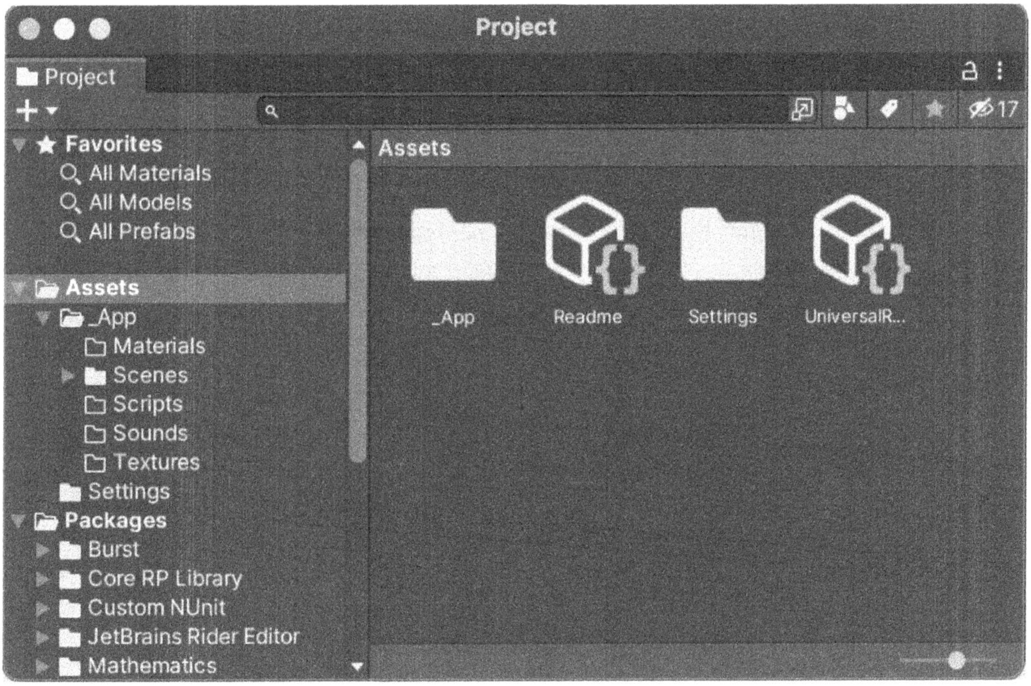

Figure 1.22 – Screenshot of what the new organization should look like

> **Note**
>
> If you want to dive deeper into this matter, I recommend `https://unity.com/how-to/organizing-your-project` and `https://docs.unity3d.com/Manual/SpecialFolders.html`.

Good job! With that, you have learned everything you need to have your work well organized. Over time and when you have worked on multiple projects, you will notice the difference if you have your files well organized. As projects grow, if we don't apply some technique to keep everything organized, we will lose a lot of time looking for the assets we need. Next, we will cover a very easy and interesting point regarding how to configure our project so that later, in *Chapter 2, Preparing Our Player*, we can move our character with the keyboard and mouse.

Choosing an input handler

The Input Manager in Unity is the area where you define the keys or axes of the keyboard and mouse that will be detected when the user plays the game. These keys and axes can be related to a name and then accessed via code. For example, in one of our scripts, we can ask if the user has made a jump because they have pressed the spacebar.

A few years ago, Unity used an Input Manager system that is now obsolete due to the arrival of a new system called the **new Input System**.

Unity's new Input System was created to unify the controls on PC, PlayStation, Xbox, and other systems.

Let's start by adding the new Input System to our project:

1. Open **Package Manager**. This can be found under **Window** | **Package Manager**:

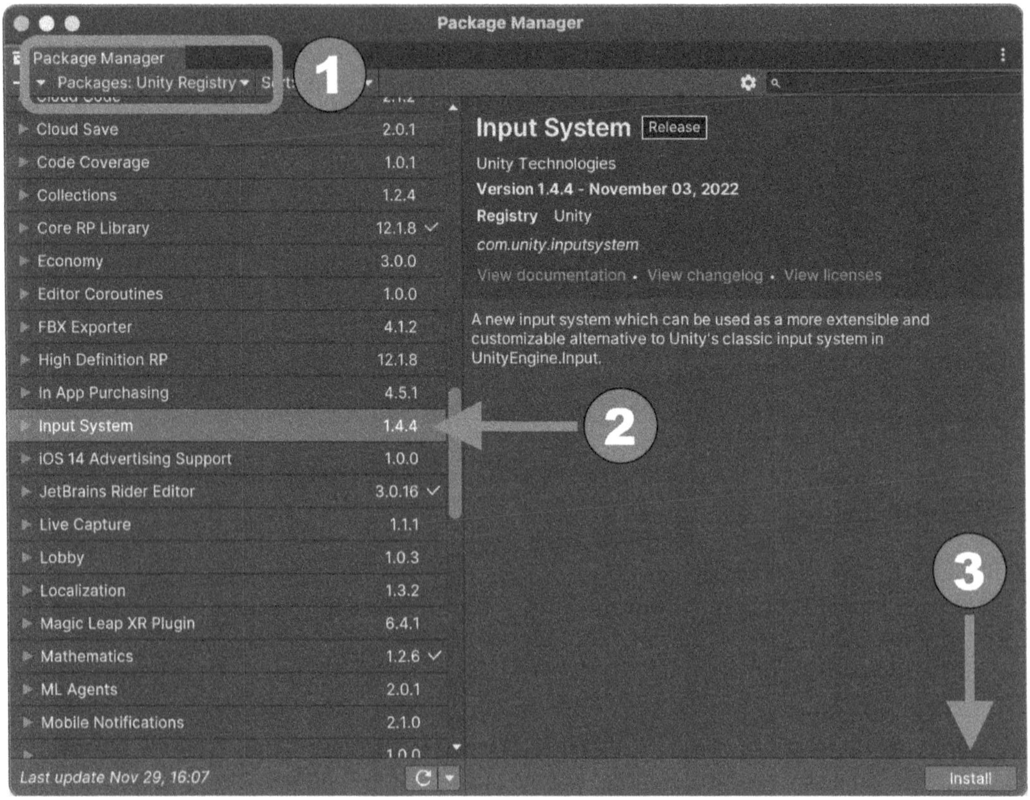

Figure 1.23 – Locating the new Input System plugin in Package Manager

2. Change the source of **Packages: In Project** to **Packages: Unity Registry** (**1**).
3. Scroll down the list and look for **Input System** (**2**) and click the **Install** button (**3**).
4. Unity will display a warning that for the changes to activate the new Input System, the project must be restarted. Click the **Yes** button to restart the editor.

If you have unsaved changes in the scene, you will see a popup, warning you of this so that you can save it before the editor closes:

Figure 1.24 – Unity notice about applying changes to the project

> **Note**
>
> You can find detailed information about the differences between the old and the new system at `https://blog.unity.com/technology/introducing-the-new-input-system`. This is recommended reading.

Perfect! Our project has been configured to work. Now, it's time to get to know Firebase and learn about all the services that it offers.

Understanding the different Firebase services

Firebase is a cloud platform for web and mobile application development that's owned by Google. It has SDKs for a wide range of platforms and technologies, which help you use it in virtually any environment, whether it is on the web with JavaScript, a desktop application written in .NET, a native mobile app on Android or iOS, and so on. Most importantly, it has a huge community, which makes this suite a great ally in any project.

It was created in 2011 and Google acquired it in 2014. Initially, it was only a real-time database. Due to its success, more and more features were added, which, in part, allows us to group the SDKs of Google products for different purposes, making it easier to use.

What is Firebase for?

Its essential function is to simplify the creation of web applications and mobile, desktop, and even REST API projects and their development, allowing for high productivity or helping to make the work faster, but without sacrificing quality.

Its tools cover practically any technical challenge that a complete application must overcome, considering that its grouping simplifies the management tasks to the same platform. The functionalities of these tools can be divided into four groups:

- Compilation
- Launch and monitoring
- Analytics
- Participation

It is especially interesting because it helps developers not need to spend so much time on the backend, both in terms of development and maintenance issues.

Firebase functionalities

This chapter will help you understand and love Firebase from these very pages. During the journey of our project, we will be connecting features and functionalities of our metaverse in this suite.

Compilation

The **Compilation** section is where Firebase includes the most important functions that you will use the most. Services such as user identification, file hosting, and databases can be accessed from here. The following is a more detailed review of the most important services in this section:

- **Authentication**: Identifying the users of an app is necessary in most cases if they want to access all its features. Firebase offers an authentication system that allows both registration (via email and password) and access using profiles from other external platforms (such as Facebook, Google, Apple, or Twitter), which is a very convenient alternative for users who are reluctant to complete the process.

  Due to this, these types of tasks are simplified. This is also because access is managed from Google servers and greater security and data protection can be achieved. It should be mentioned that Firebase can securely store login data in the cloud, which prevents someone from having to identify themselves every time they open the application.

- **App Check**: This offers security techniques that protect your token-based applications.

- **Firestore Database**: One of Firebase's most prominent and essential tools is real-time databases. These are hosted in the cloud, are non-SQL, and store data as JSON. They allow the data and information of the application to be hosted and available in real time, keeping them updated even if the user does not perform any action.

  Firebase automatically sends events to applications when data changes, storing the new data on disk. Even if a user is offline, their data is available to everyone else, and the changes that are made are synchronized once the connection is restored.

  Nowadays, Firebase maintains two real-time databases – Realtime Database and Firestore Database. The latter is a very powerful evolution of the former and allows you to manage indexes and make and apply more complex filters from the SDK. We will use the latter in our project.

- **Extensions**: These are prefabricated actions that facilitate automated tasks on your database, such as sending an email when a new item is created, optimizing a photo when a user uploads it, translating text from a database field, and so on.

- **Storage**: Firebase has a storage system where developers can store their application files (and link them with references to a file tree to improve app performance) and synchronize them. Like most Firebase tools, it can be customized through certain rules. This storage is a great aid for handling user files (photos they have uploaded), which can be served more quickly and easily. It also makes downloading file references more secure.

- **Hosting**: Firebase also offers a server to host the apps quickly and easily – that is, static and secure hosting. It provides SSL and HTTP2 security certificates automatically and free of charge for each domain. It works by placing them on Firebase's **Content Delivery Network** (**CDN**), a network that receives the uploaded files and delivers the content.

- **Functions**: This option allows you to deploy functions hosted in Firebase as a backend. These functions can be of the REST type (GET, PUT, DELETE, and POST) or of Trigger type – that is, they are launched automatically when something happens in the database, such as when a new document is added, a value changes in a specific field, or a document is deleted.

  Finally, you can also create schedule functions. These are equivalent to cron tasks and can be programmed to be executed in a time range – for example, every 5 minutes, every 24 hours, and so on.

- **Machine Learning**: These are pre-built functions that make it easier for you to perform tasks in mobile applications, such as recognizing text in a photo, detecting a face, and tracking objects using the camera.

- **Remote Config**: This allows you to dynamically configure your project based on *constants* that you generate in the cloud. These constants can be numeric values, text, or even store a whole tree in JSON. This tool is very powerful if you want to dynamize the look and feel of your app without needing to recompile and deploy. For example, in these constants, we can store a discount of our shop, a Boolean variable that decides whether to display a banner or not.

Launch and monitoring

This section is where Firebase groups the services related to quality and analysis. Important services such as monitoring and managing bugs in our software with Crashlytics, obtaining performance metrics with Performance, robotically testing app features, and distributing binaries to testers can be accessed from here. The following is a more detailed review of the most important services in this section:

- **Crashlytics**: To maintain and improve the quality of the app, special attention must be paid to bugs, so bug tracking (and also the overall performance of the app) is key to being able to act and fix them. For this reason, Firebase offers crash reporting, which detects and helps solve the app's problems, obtaining a very detailed error report (with data such as the device or the situation in which the exception occurs) and organizing them as it groups them by similarity and classifies them by severity.
- **Performance**: This provides statistics on the load times of a mobile app or website. Here, you can find metrics calls to other REST APIs that occur within your code.
- **Test Lab**: The Test Lab allows you to test your app on virtual Android devices based on the parameters you configure. This way, it is much easier to detect possible errors before launching the application.
- **App Distribution**: This allows you to distribute an APK or AAB to private testers. This makes it much easier to manage new versions. Firebase automatically sends notifications to users when you upload a new version for testing.

Analytics

Have you met or heard about Google Analytics before? Firebase Analytics is practically the same but oriented to mobile apps. With this service, you can get powerful information about the users of your app.

Analyzing data and results is key to making consistent and informed decisions for your project and associated marketing strategy. With Firebase Analytics, you can monitor various parameters and obtain a variety of metrics from a single dashboard for free. It is compatible with iOS, Android, C++, and Unity and, among other functions, allows you to do the following:

- Obtain measurements and analysis of the events taking place in the application. You can receive unlimited reports with up to 25 attributes.
- Check the performance of events, notifications, and advertising campaigns in networks, based on user behavior.
- Get to know the user with segmented information. For example, you can obtain statistical data on language, access devices, age, gender, location, and so on. Insights on usage and loyalty to the app can also be obtained.

Participation

Participation is where Firebase groups the services related to communication or interaction with other users of our app. Push messaging communication, the creation and consumption of *Magic Links* with Dynamic Links, and A/B testing are some of the most interesting services that we will find here. Let's look at these in more detail:

- **A/B testing**: This allows you to use Remote Config to measure the success of your project on two different variants.
- **Messaging**: This is useful for sending notifications and messages to various users in real time and through various platforms.
- **Dynamic Links**: These are *smart* links, which allow the user to be redirected to specific areas or contents of the application, depending on the objective to be achieved and the personalization given to various parameters of this URL.

  Thus, the operation of these links is directed as we want and ensures a pleasant experience for the user on various platforms. They are particularly useful for directing content to certain segments of users, whether they're current or potential, in which case they may receive a recommendation to install our app.
- **AdMob**: This allows you to monetize your mobile apps easily and intuitively; with a few lines of code, you can display banners and ads automatically.

Advantages and disadvantages of using Firebase

You will probably agree with me that this tool has many benefits for developers who use it. We will add below an interesting link with official documentation and examples, ideal to broaden your knowledge of the services mentioned above.

Here is a list of the benefits of Firebase:

- Highly recommended for applications that need to share data in real time.
- Its functionalities complement each other very well and can be easily managed from a single panel. Moreover, it is not necessary to use all these options for the application, and you can choose only those that interest you most.
- They have official SDKs for client applications developed in iOS, Android, JavaScript, C++, and Unity. They also have libraries to integrate into an API for the NodeJS, Java, Python, and GoLang languages. Finally, although they do not have official support, there are libraries for frameworks such as Angular, Ember, Flutter, React, React Native, Rx, and Vue that are recommended on their website.
- It allows you to send notifications. They are very simple to implement and manage, and they are also extremely useful for maintaining communication with users.

- Allows monetization – that is, from Firebase itself, advertising can be added to the app, allowing it to be easily monetized.
- Includes Analytics, which specializes in certain mobile application metrics and is integrated into the Firebase central panel with a very intuitive operation.
- Google offers abundant introductory and informative documentation (in great depth) to make diving into Firebase much easier.
- Free email support, regardless of whether the developer uses the free or paid version.
- It is scalable – that is, the startup is free but allows it to adapt to the needs of the application with different payment plans.
- Offers security to users with SSL certificates.
- Allows developers to take their attention away from the backend and complex infrastructures to focus entirely on other aspects.

The most mentioned disadvantage is Firebase's price. We have already discussed the scalability of Firebase, where starting with the Spark plan is free. However, it has limitations (mainly regarding the number of concurrent users and storage space), so it may be necessary to purchase a paid version. The paid plans are Flame ($25 per month) and Blaze (pay-as-you-go).

As a side note, in the many years that I have been using Firebase, I have deployed more than 50 projects, some small, some medium, and some larger, and I can say that I have received invoices of a few cents or on occasion less than $10. This is an insignificant amount, in my opinion, for the stability, quality, and monitoring of the services offered.

You can always set up budget alerts and decide to stop the service when your budget is exceeded – you will receive an email when you are approaching your limit.

> **Note**
> You can find technical documentation and tutorials by browsing to `https://firebase.google.com/docs`.

At this point, you know what Firebase can offer us. As you have seen, it is a suite full of fascinating services. Can you imagine how all this will help us in our impressive metaverse? From *Chapter 3, Preparing Our Home Sweet Home: Part 1*, we will perform various actions and integrate our project with Firebase services, step by step and progressively. Now, let's get down to business and create our project in the Firebase console.

Configuring a new Firebase project

Firebase will be our traveling companion throughout this book. It is like a Swiss Army knife that will provide us with solutions to technical problems, such as data persistence in the cloud, dynamic configuration, hosting user-generated files, identifying users in our system, and much more.

To explore the tools offered by this Google suite, we must create an account for free:

1. Navigate to `https://console.firebase.google.com`.
2. Since Firebase is a Google service, you will need to use your Gmail account to log in. If you are already logged in with Gmail, you will be taken directly to the welcome screen; otherwise, you will be prompted to log in. Once logged in, you will see the following screen:

Figure 1.25 – The welcome page in the Firebase console

3. Click on the **Create a project** button. At this point, we will be asked for the name of the project. We can use the same name we used for our project in Unity:

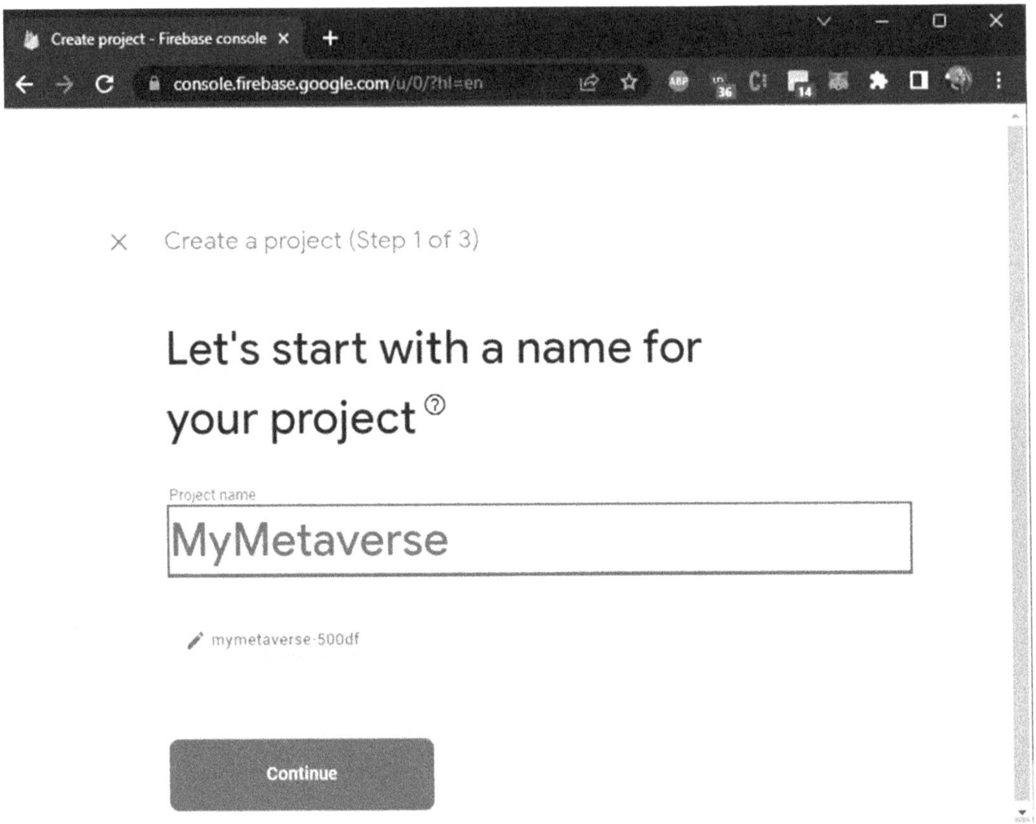

Figure 1.26 – Setting up the name of your project in Firebase

4. After you have entered the name of your project, you can click the **Continue** button.
5. In **step 2 of 3**, the wizard will ask you if you want to activate **Google Analytics**. I recommend enabling the **Enable Google Analytics for this project** option as it will be very useful in the future.
6. When you click on **Continue**, it will ask you the location that your Analytics account will have. **Google Analytics** will ask you to select an active account or create a new one, if you do not have one, you can select the **Create new account** option, in this case, after choosing an account name, accept the Google Analytics terms of use and check the **Use the default settings to share Google Analytics data** box to make the account creation easy and click the **Continue** button.
7. Firebase will take a few seconds to finish configuring the project, so be patient. Once the process has finished... congratulations! You have successfully created your first Firebase project:

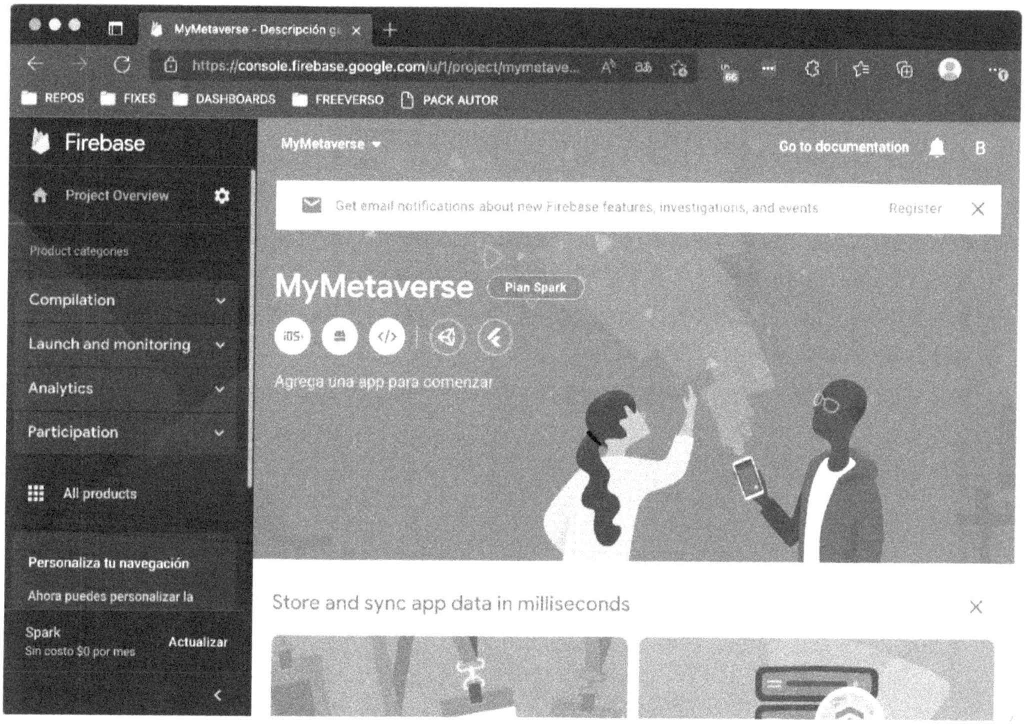

Figure 1.27 – The home page of the new Firebase project

Great! With that, you know all the theory behind Firebase's services and created your first project in the console. Now, let's go one step further and download the Firebase SDK for Unity. This is a very important first step so that in future chapters, where we need to program our scripts to access Firebase, we have done all the necessary work already and we won't have to stop.

Installing the Firebase SDK

We have reached the last part of this chapter; after this step, we will focus more on practice and less on theory. We have already seen all the benefits that Firebase offers; after this section, you will understand the reason you should use it in our project.

Next, we are going to follow a few simple steps to download and install the **Firebase SDK** in our project. To do this, execute the following steps:

1. Navigate to `https://firebase.google.com/download/unity`.
2. The Firebase SDK will automatically start downloading. If this does not happen, look for a hyperlink that looks like what's shown in *Figure 1.28*:

Figure 1.28 – Manual download button

3. It may take some time to download as it is approximately 940 MB. In the folder where your downloads land, you will find a ZIP file named `firebase_unity_sdk_X.X.X.zip`. In my case, I have downloaded the `firebase_unity_sdk_10.1.1.zip` version.
4. When unzipping, a new folder called `firebase_unity_sdk` will appear, the contents of which are shown in *Figure 1.29*:

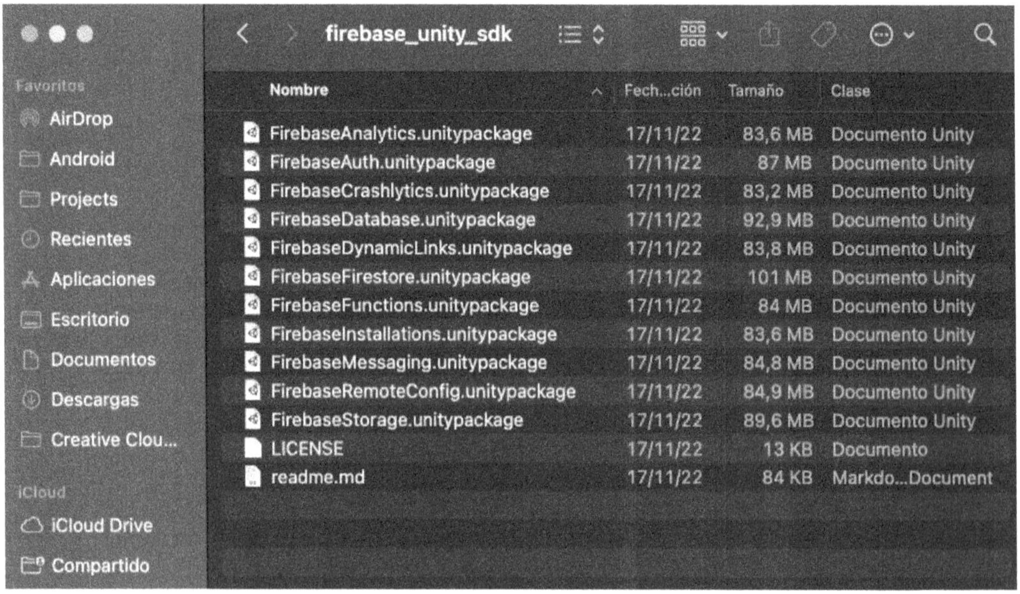

Figure 1.29 – Firebase SDK files for Unity

5. This is fantastic! As you can see, the SDK comes fragmented by services. This will help us integrate only what we are going to need and expand in the future if we need new functionality that requires it. We will avoid having plugins and unnecessary code in our project.
6. To get started in the following chapters, we will need to install `FirebaseAuth.unitypackage`, `FirebaseFirestore.unitypackage`, and `FirebaseStorage.unitypackage`. When we need more functionality in our project, we will go back to this folder and install what we need.
7. Double-click on `FirebaseAuth.unitypackage` to install it. A modal window will open in Unity, warning us about the import and showing us which files are going to be created. Click on the **Import** button:

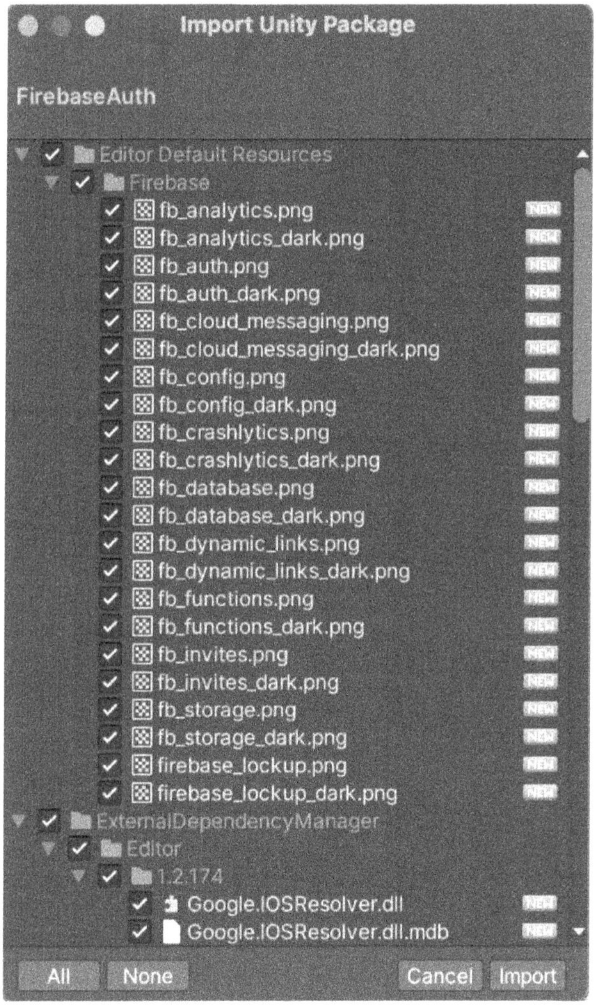

Figure 1.30 – Unity's new asset import dialog

8. When the progress bar finishes, do the same with the `FirebaseFirestore.unitypackage` and `FirebaseStorage.unitypackage` files.

   Once you've finished installing these, you will see that a new folder has been created in your project called `Firebase`, which contains the internal libraries needed to use it in the code we will program later:

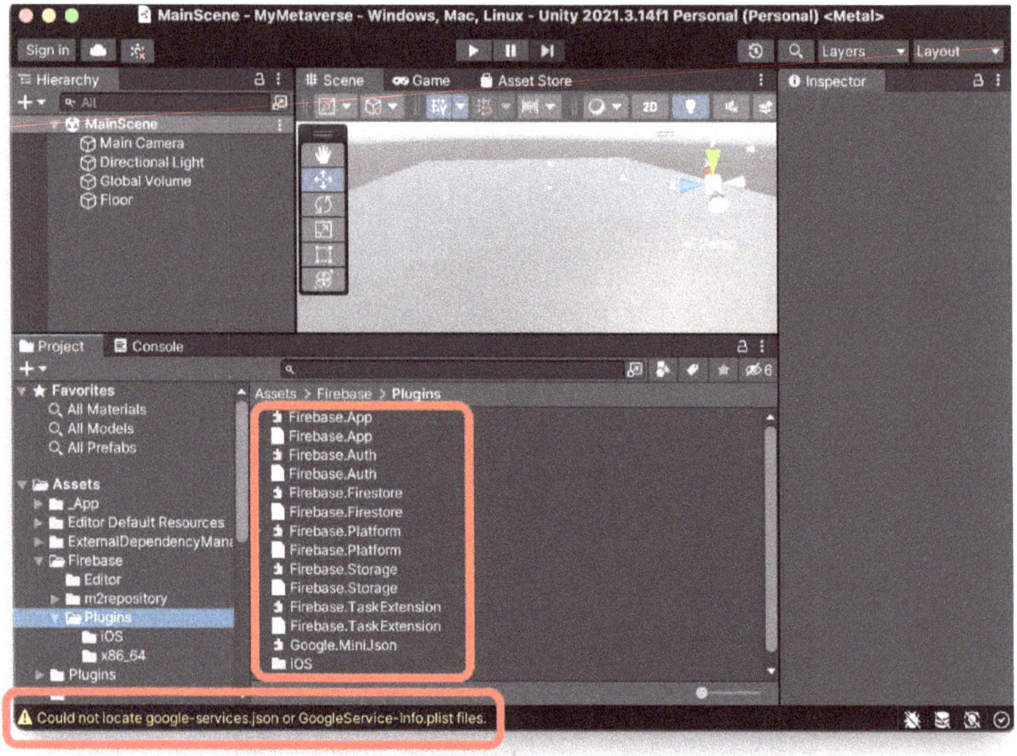

Figure 1.31 – New Firebase folder and configuration warning

As you may have seen, a yellow warning about a missing configuration will appear. You might be wondering, *Once I have imported the SDK and I have created my project in the Firebase console, how can I connect the Unity project with the Firebase console?* This is done with the `google-services.json` file for Android and the `GoogleService-Info.plist` file for iOS. For Windows, Mac, and Linux, Firebase will take the `google-services.json` mobile file and automatically generate one called `google-services-desktop.json` the first time you run the project after importing the files.

9. To generate these files, go to the Firebase console at `https://console.firebase.google.com`.
10. On the main screen, click on the Unity icon. You can also access it from the gear button and click on the **Project settings** option:

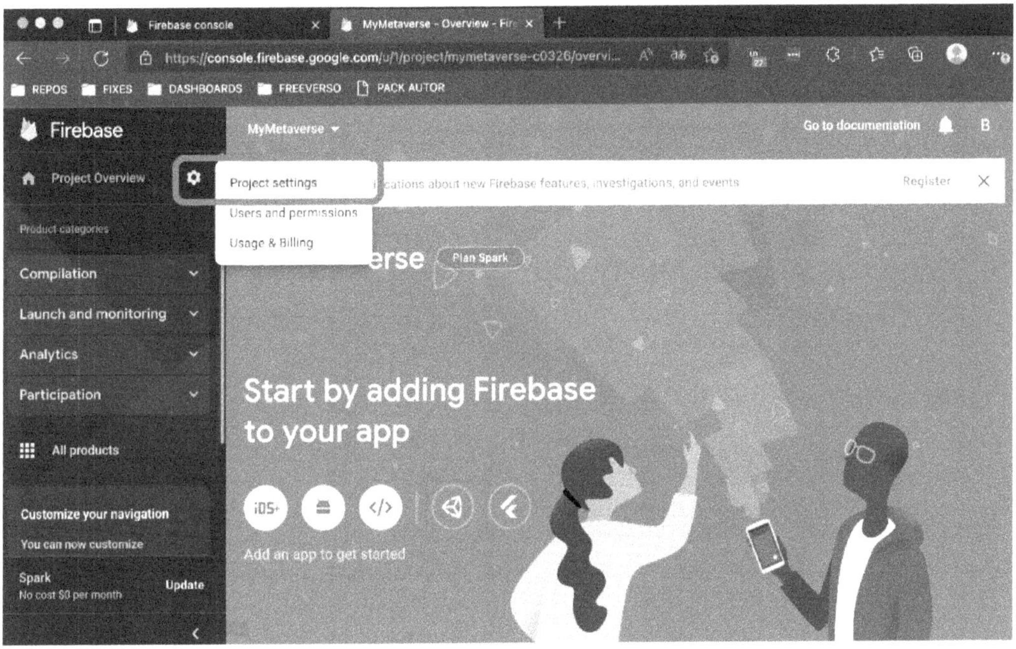

Figure 1.32 – Preparing Unity's configuration in the Firebase console

11. You will be presented with a form that contains simple information that you must fill in to continue.
12. Check the **Register as an Apple app** check box and enter an **Apple Package ID** property of your choice. Here, I have entered `com.mymetaverse.app` (it must be in *inverted domain* format, which is comprised of three parts – *xxx.xxx.xxx*).
13. Check the **Register as Android app** check box and enter an **Android Package Name** property. You can use the same one you entered for iOS.
14. Click on the **Register App** button.

15. Once this process has finished, you will be prompted to download the two files that we had previously named. The remaining steps of the wizard do not require further action, so you can click on the **Continue** button until it is finished:

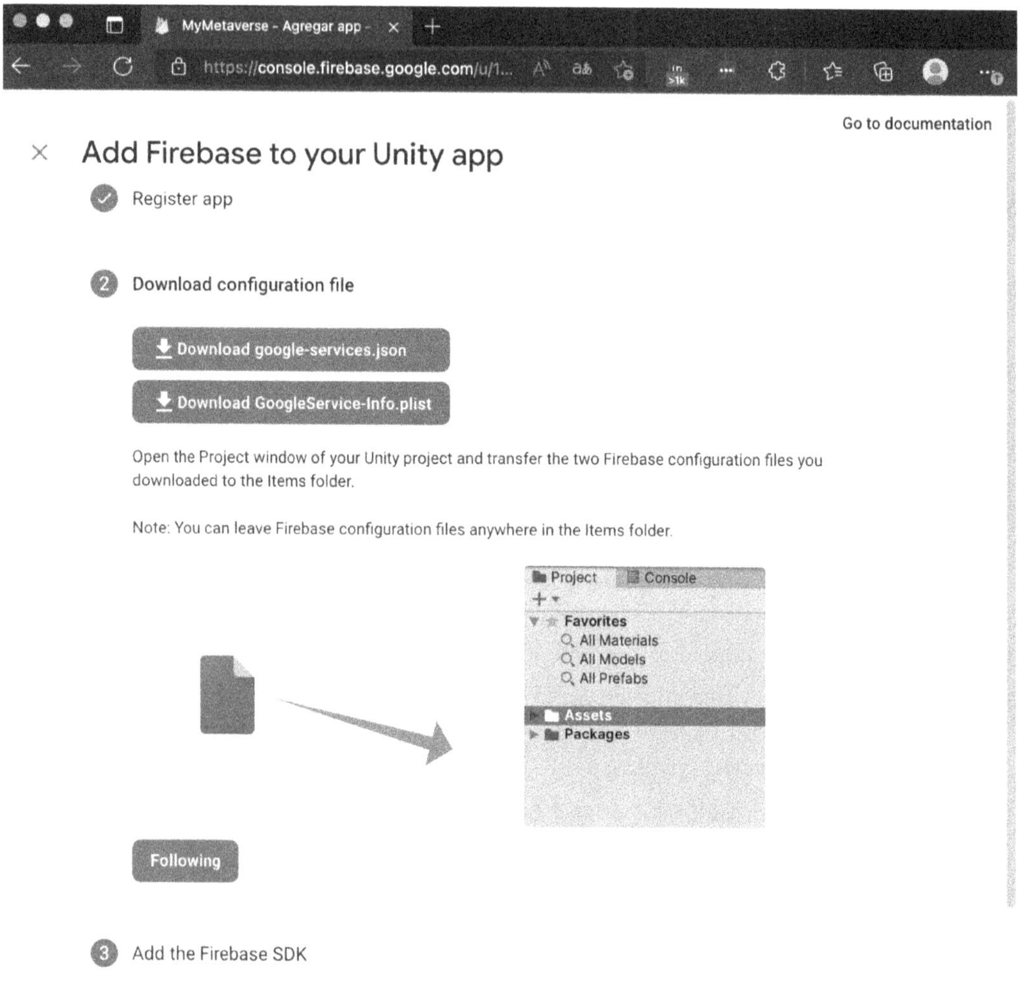

Figure 1.33 – Downloading the configuration files for the Unity project

16. Finally, drag the two files (`google-services.json` and `GoogleService-Info.plist`) to the root of your Unity project, just inside the `Assets` folder:

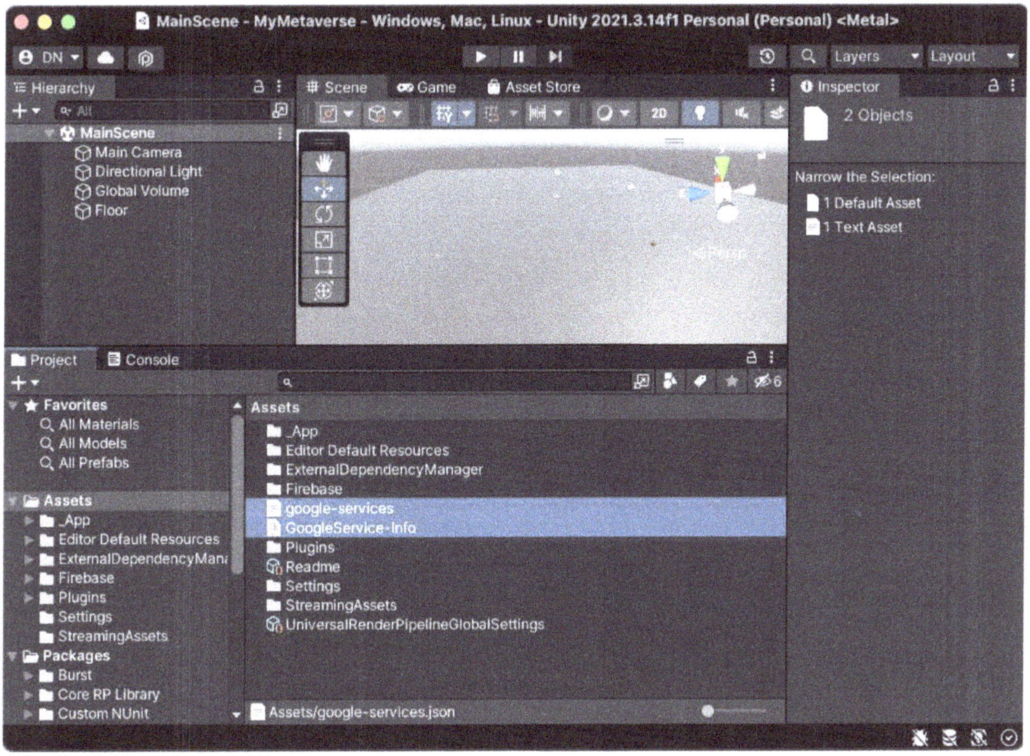

Figure 1.34 – Adding the configuration files to the Assets folder

Congratulations! With this last step, we have perfectly configured our project with Firebase. Finally, I would like to warn you about the yellow warning that appeared before. This will not disappear on its own – you have to press the **Clear** button that is located in the **Console** panel.

Congratulations! You have finished configuring Firebase and connected it to our project. I know that this part may seem tedious because it is very theoretical and focuses on configuration tasks, but it is tremendously necessary to be able to advance without obstacles in the following chapters.

Having connected the project in the Firebase console to our Unity project, we will be able to start programming functionalities that *attack* our Firebase database directly, for example. We will be able to integrate a sign-in/sign-up system for our users, make queries to the database, and much more.

Summary

In this chapter, we learned the basics of working with Unity, from downloading and installing it to understanding and putting the basic concepts into practice. We already know about the main parts of the Unity interface, and we will learn about them more in the future. We will expand our knowledge of the game engine by exploring new features, menus, and tricks.

At this point, we have our project with URP created and configured so that we can advance correctly in the following chapters. On the other hand, we have learned about what Firebase is and how its services can help us in multiple ways.

In addition, we learned how to configure a project from scratch in the Firebase console, which allowed us to connect it to our Firebase project, as well as how to download and install the SDK.

In the next chapter, we will learn how to create an animated 3D character that we will be able to make move, jump, and run by pressing certain keys, as well as look around by moving the mouse. Fascinating, isn't it? See you in *Chapter 2, Preparing Our Player*.

2
Preparing Our Player

After a long introductory chapter full of theory, we finally land on the fun part of the book. During this new chapter, we will learn how to create the main, most basic, and most necessary thing: the **Player**. Our player will be a 3D character able to move around the world we build, following our orders.

Using the keyboard and mouse, our character will be able to move forward, backward, and sideways, turn with the movement of the mouse, jump, run, and so on.

We will cover the following topics:

- Creating a third-person controller
- Working with colliders
- Changing the default avatar

Technical requirements

This chapter does not have any special technical requirements; just an internet connection to browse and download an asset from Unity Asset Store. We will continue with the project we created in *Chapter 1, Getting Started with Unity and Firebase.*

Remember that we have the GitHub repository, `https://github.com/PacktPublishing/Build-Your-Own-Metaverse-with-Unity/tree/main/UnityProject`, which will contain the complete project that we will work on here.

Creating a third-person controller

One of the many advantages of using Unity is its large asset marketplace. Unity Asset Store is a place where you can find content created by third parties and also assets created by Unity, free and paid.

Unity has a multitude of assets called **Starter Assets** that greatly facilitate and accelerate the development of your project. In our case, we will use **Starter Assets - Third Person Character Controller** for our player. This player controller is already adapted to the new input system with which we previously configured the project.

To download and install **Starter Assets - Third Person Character Controller**, follow these steps:

1. Go to `https://assetstore.unity.com` in your browser.
2. In the search engine, type `Starter Assets - Third Person Character Controller`.
3. Click on the single item that will appear in the search results.
4. Click on the **Add to My Assets** button.

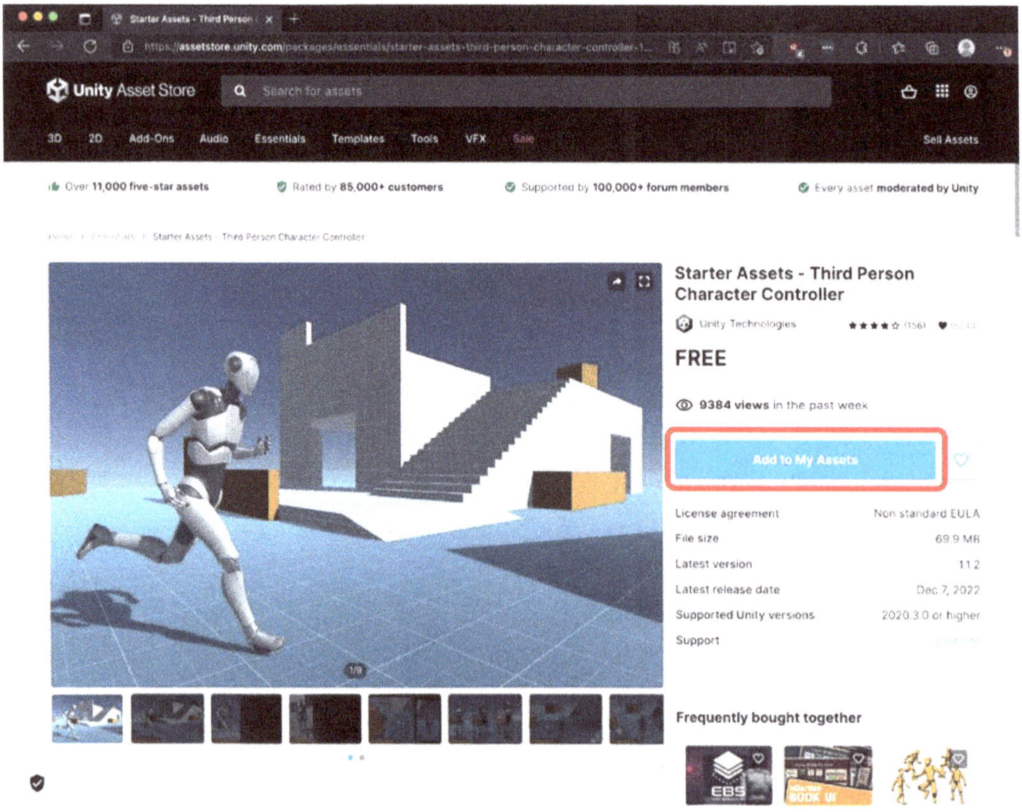

Figure 2.1 – Downloading Starter Assets - Third Person Character Controller

5. If you are not logged in, you will be redirected to the **Sign In** page. If you are already logged in, you will return to the same page but you will now see a button called **Open in Unity**. Click on it.

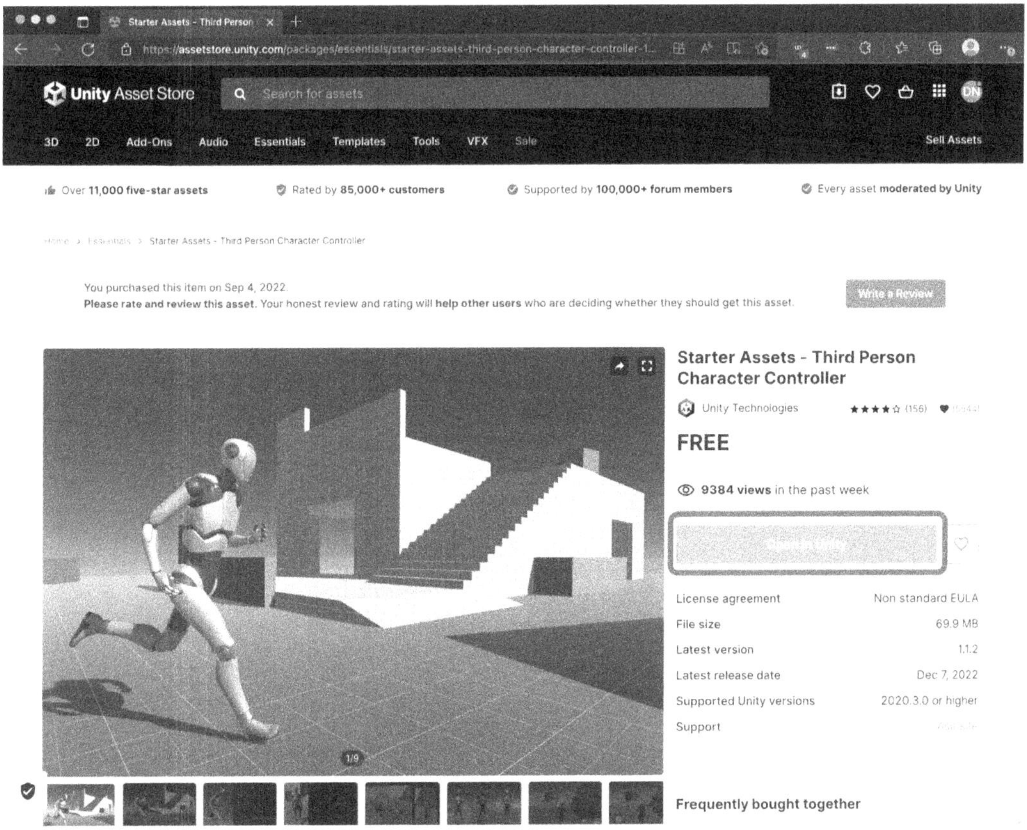

Figure 2.2 – Add Starter Assets - Third Person Character Controller to Unity editor

6. The **Open in Unity** button will open the editor and display the **Package Manager** window with the newly downloaded asset marked. Click on the **Download** button.

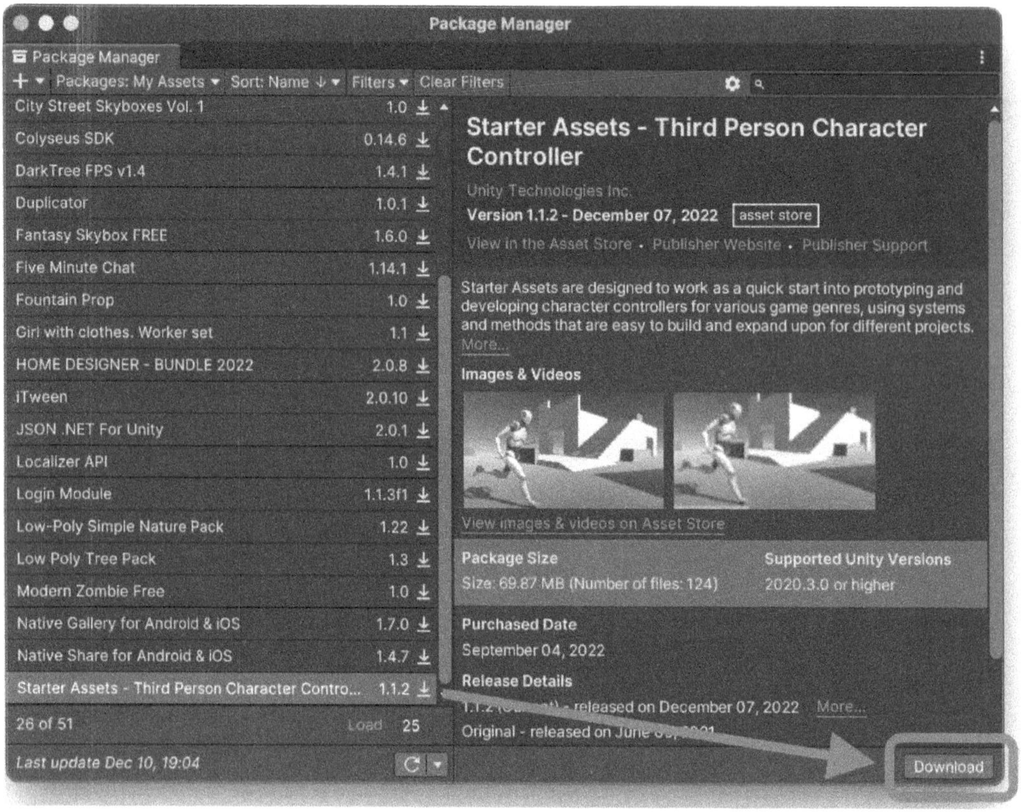

Figure 2.3 – Adding Starter Assets - Third Person Character Controller to the Unity editor in Package Manager

7. When the download is finished, an **Import** button will appear in the same place. Click on it.
8. It is possible that when you click on the **Import** button, you will see a warning about updates and dependencies. Click on the **Install/Upgrade** button.

Figure 2.4 – Warning about upgrade dependencies

9. Finally, as we have seen before when a new asset is added, a confirmation window appears informing you about the content to be added to the project. Click on the **Import** button to finish the process.

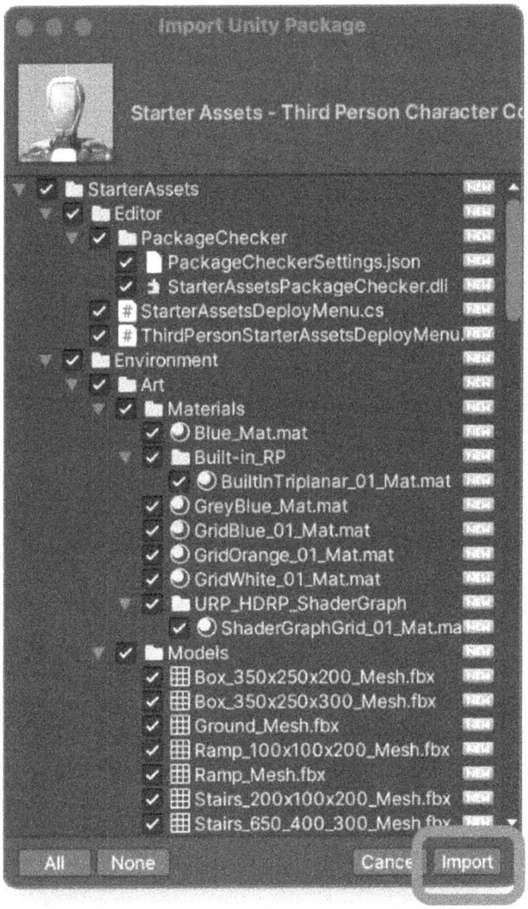

Figure 2.5 – Import confirmation popup

We have already added to our project everything we need to start using our player. It's time to test it. To do so, follow these steps:

1. Open **MainScene**.
2. In the **Project** window, go to **Assets** | **Starter Assets** | **ThirdPersonController** | **Prefabs**.

3. Drag and drop the **PlayerArmature** Prefab into the scene. Your scene will look something like this:

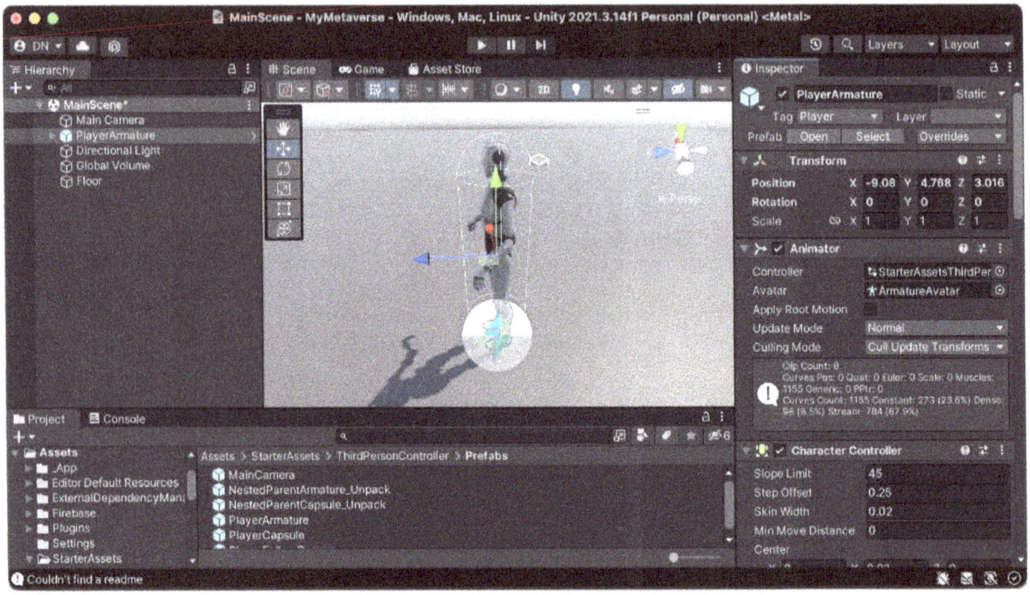

Figure 2.6 – The look of our scene after importing the player's Prefab

Testing our player

If you click on the **Play** button now and use the *W*, *A*, *S*, *D* keys or the arrow keys to move the character, you will see that the character moves, but the camera remains fixed where it was first created. This happens because we have not finished configuring the player's Prefab correctly.

To correct this problem, in the main menu bar, click on **Tools** | **Starter Assets** | **Reset Third Person Controller Armature**.

As you can see, a new GameObject called **PlayerFollowCamera** has been created that will make the active camera in our scene follow the character wherever they move.

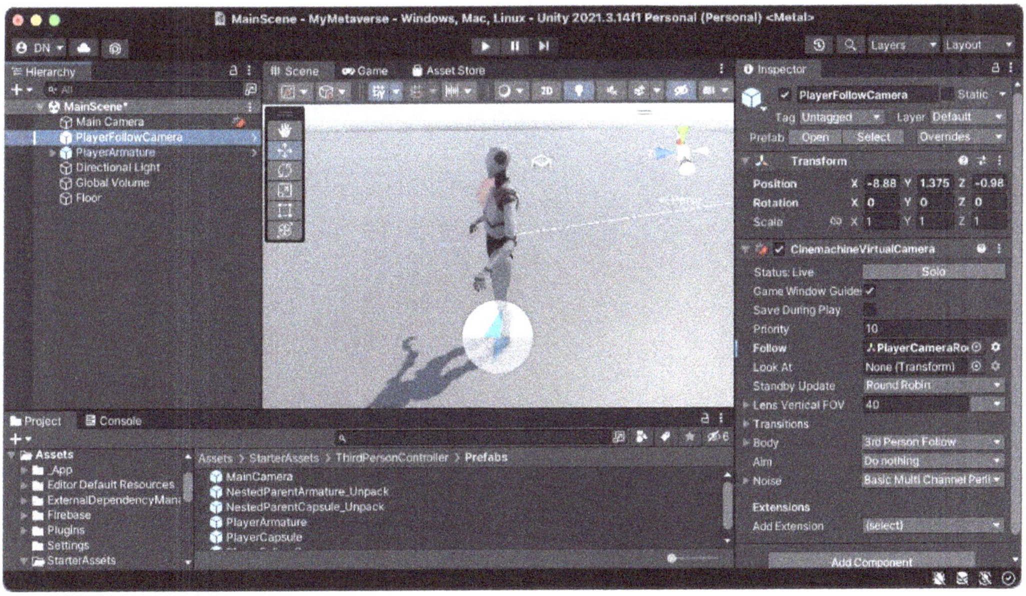

Figure 2.7 – The appearance of our scene after resetting the player's controller

Now, if you click again on the **Play** button and move the character with the *W*, *A*, *S*, *D* keys, use the *spacebar* to jump, and move the mouse, you will see that the camera always focuses on the player when they are moving.

Incredible, isn't it? You already have in your project a controllable player, as in any game you know.

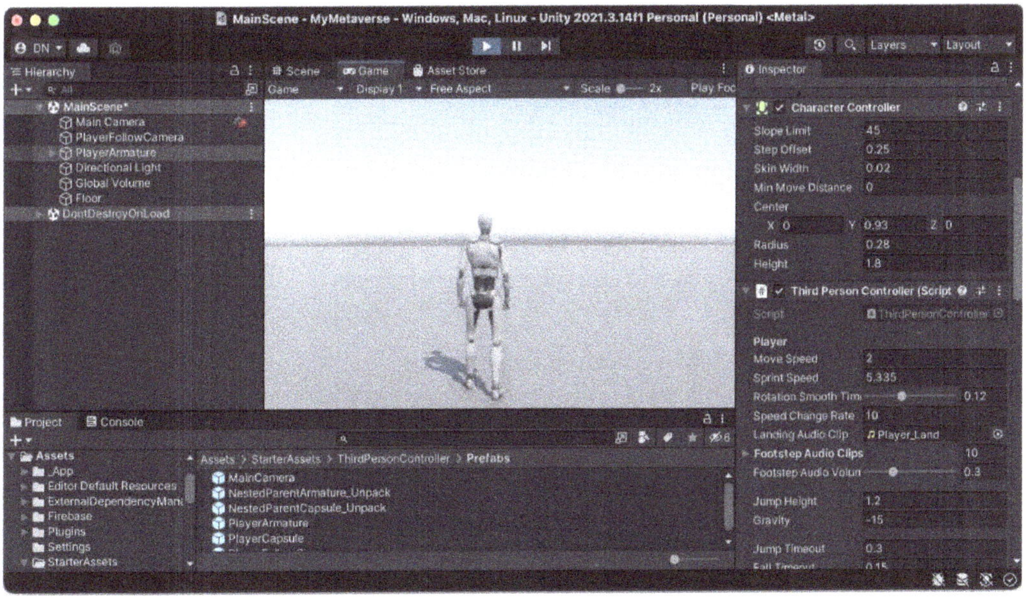

Figure 2.8 – Testing how the camera follows the player

If you look at the components on top of the **PlayerArmature** GameObject, you will notice one in particular called **Third Person Controller**.

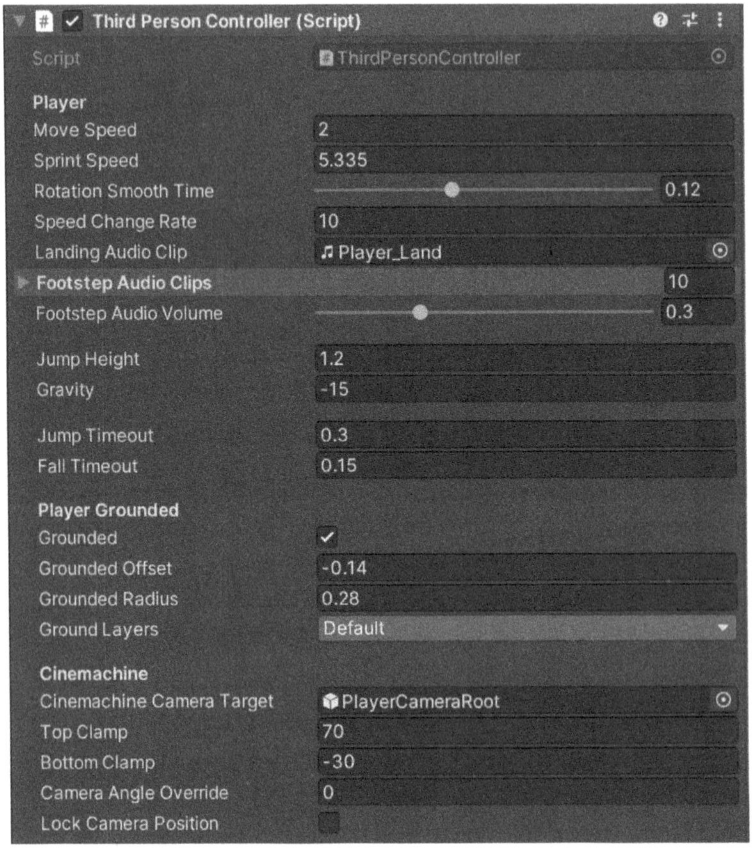

Figure 2.9 – Third Person Controller properties

Here, we see the constants that determine the speed, gravity, jump height, and other information that, if you change it, the player's behavior will be affected. For the moment, we are satisfied with the default values, but if you want to play and experiment, I encourage you to change the values and click **Play** to test the new results.

> **Tip: changing values in a component**
>
> If you want to change the values of a component to test different behaviors, you can do it after clicking the **Play** button, so the new values will take effect immediately. You have to keep in mind that when you finish and exit the **Run** mode, the new values you entered will disappear and the default values will be loaded again. If you changed the values before pressing **Play** and you want to return to the default values, click on the button with three dots to the right of the component name and press the **Reset** button.

Working with colliders

As you may have noticed, the player walks on the floor that we placed in *Chapter 1, Getting Started with Unity and Firebase*. If you are a little naughty, you may also have noticed that if your player leaves the floor, he will fall into the void. This is explained by the fact that the floor that we placed is a plane. By default, this type of geometry comes with a component called **Mesh Collider**, which serves as a physical barrier so that any object that is placed on top does not fall into the void.

In order for one object not to fall through another or not to be passed through, both objects must each have a collider component. If you noticed earlier, when we created our player controller, a green capsule appears around the character. This capsule is nothing more and nothing less than a type of collider (Capsule Collider). By default, Unity includes one for us in our player.

Let's do another test to learn more about how colliders work:

1. Place a square in front of our player by selecting **GameObject** | **3D Objects** | **Cube** in the main menu bar and you will see a cube has been created in the scene. This cube has a collider called **Box Collider**, which does the same function as the Mesh Collider – the plane we use as a floor.
2. Move the cube that has just been created and place it a little bit ahead of the player.

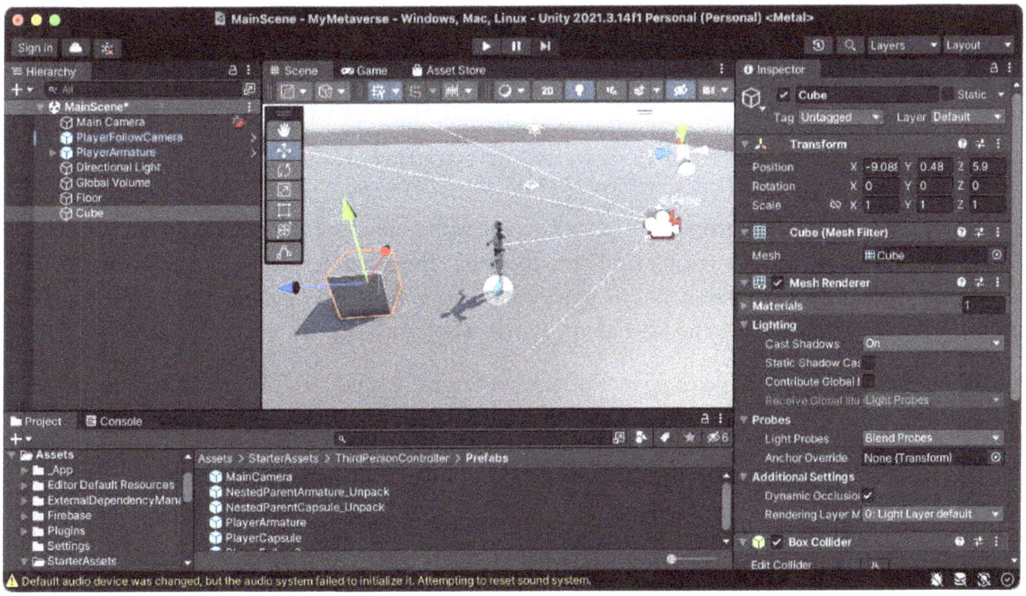

Figure 2.10 – The cube we created and placed in front of the player

3. Click on **Play** and try to go through the cube with the character. You will see that it is impossible. Even if you jump over the cube, you will only be able to get on top.

This is how colliders work in a basic way. Later, we will see in more detail how to use them effectively and on more complex objects. Finally, we will do another fun test with colliders, and this time we will see how they work together with physics.

4. Add a new GameObject to the scene. This time, add a sphere by clicking on the main menu bar and selecting **GameObject | 3D Objects | Sphere** to make a sphere appear in the scene. Move it to separate it from the character. Your scene will look similar to this one:

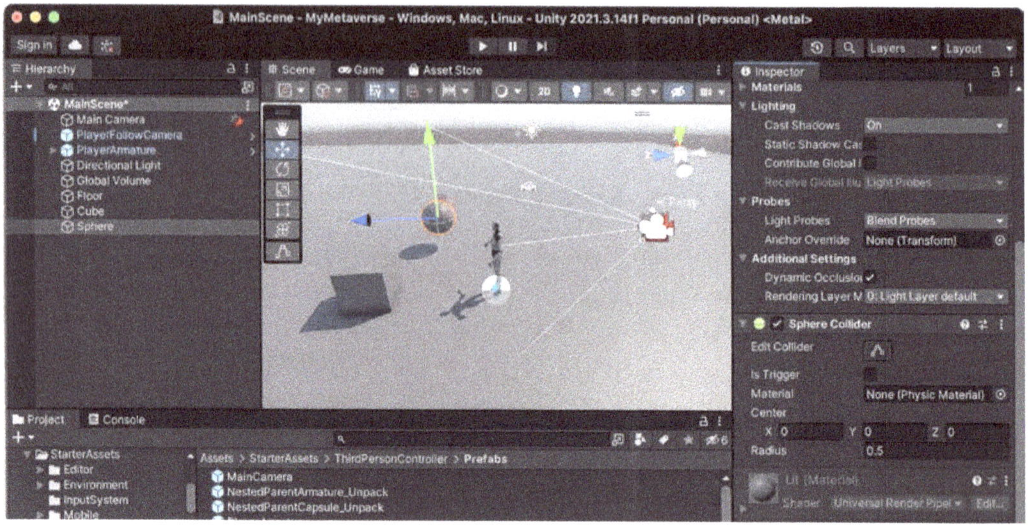

Figure 2.11 – The sphere we created and placed in front of the player

Well, we already know that the geometries in Unity come with a collider component by default. In fact, if we play and test it, we will see a result similar to the cube. The character will collide with the sphere and will be unable to pass through it, but we don't want that to happen. When the player collides with the sphere, we want it to react to the hit and move with a behavior similar to a soccer ball.

To make this happen, we have to do some more setup, but before we finish, let's create a ball bounce effect:

1. With **Sphere** selected in the **Hierarchy** panel, scroll down the list of components that appear in the **Inspector** panel and click on the **Add Component** button.

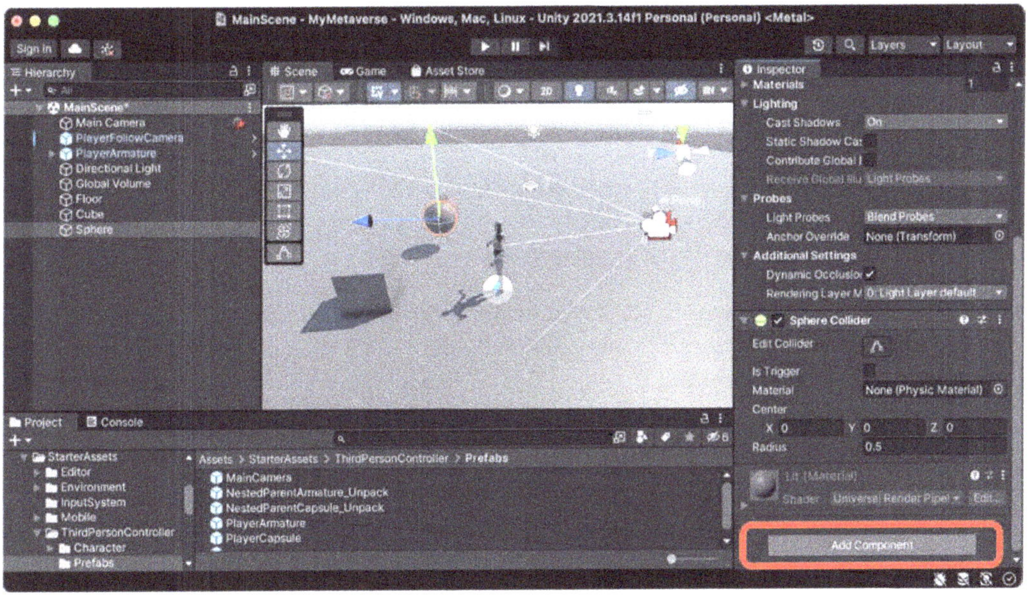

Figure 2.12 – Adding a new component to the Sphere GameObject

2. Type `Rigidbody` in the search box that appears and select it. A new component will be added to the list.

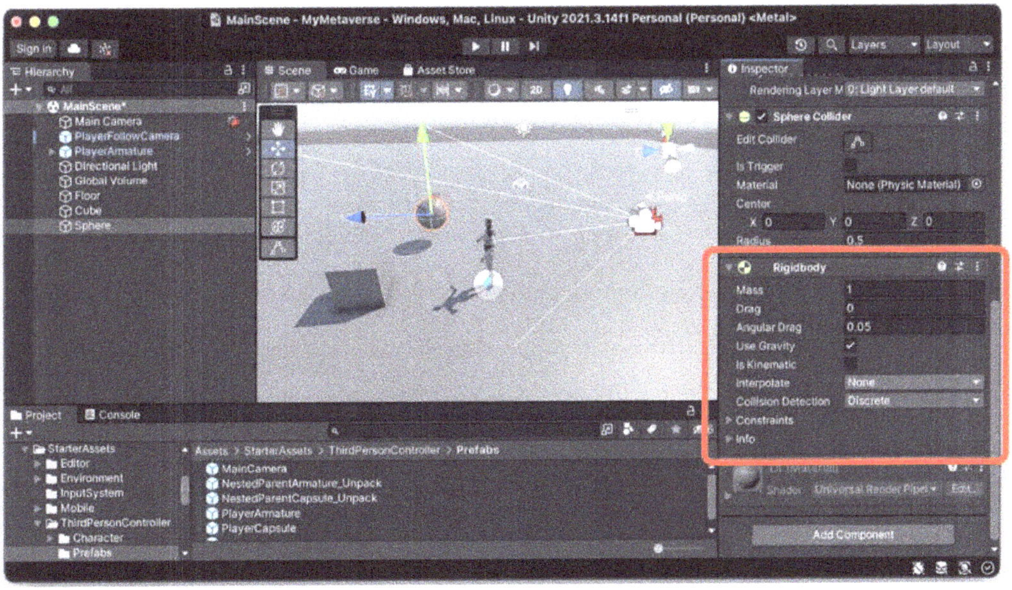

Figure 2.13 – Recently added Rigidbody component

If you click **Play** now, you will see that the sphere falls under its own weight and remains fixed on the ground. If we approach it and collide with it, it does not react to our hit and does not move. With the **Rigidbody** component that we have just added, we give it some physical properties so that the object has a weight and a mass, but something else is missing before it will behave like a soccer ball.

Although we are not going to work with the **useGravity** and **isKinematic** properties in this chapter, it is interesting to know about them. The **useGravity** and **isKinematic** properties are two important features used to control the behavior of **Rigidbody** objects in Unity:

- **useGravity**: The **useGravity** property is a boolean found on the **Rigidbody** component of a GameObject in Unity. When set to true, the **Rigidbody** object is affected by the gravity of Unity's physics engine. This means that the object will fall downward on the *Y* axis (downward in the scene) due to the influence of gravity. This is useful for objects that need to respond to the force of gravity in a natural way, such as characters, falling objects, or physical elements in the game.
- **isKinematic**: The **isKinematic** property is also a boolean found in the **Rigidbody** component of a GameObject. When set, the **Rigidbody** object will no longer respond to physical forces applied by the physics engine.

  This means that the object will not be affected by gravity or collisions with other objects. However, you can still move the object manually through its transformation. This is useful for objects that need to be controlled by scripts rather than relying entirely on the laws of physics, such as moving platforms controlled by the player or kinematic elements controlled by animations.

We can summarize the behavior of both properties: **useGravity** is used to allow a **Rigidbody** object to respond to gravity in a realistic way, while **isKinematic** is used to disable physical forces and allow a **Rigidbody** object to be controlled by code or animations. These properties are fundamental to the simulation of realistic physics and behavior on objects within the Unity engine.

Advanced information on all the properties and the operation of the **Rigidbody** component can be found on the following page of the documentation: `https://docs.unity3d.com/2021.2/Documentation/Manual/class-Rigidbody.html`

We are going to add physics materials to our sphere to turn it into a ball. To do this, follow these steps:

1. In the **Project** panel go to the **Assets** | **_App** | **Materials** folder.
2. Right-click on the **Project** panel and select **Create** | **Physic Material**.

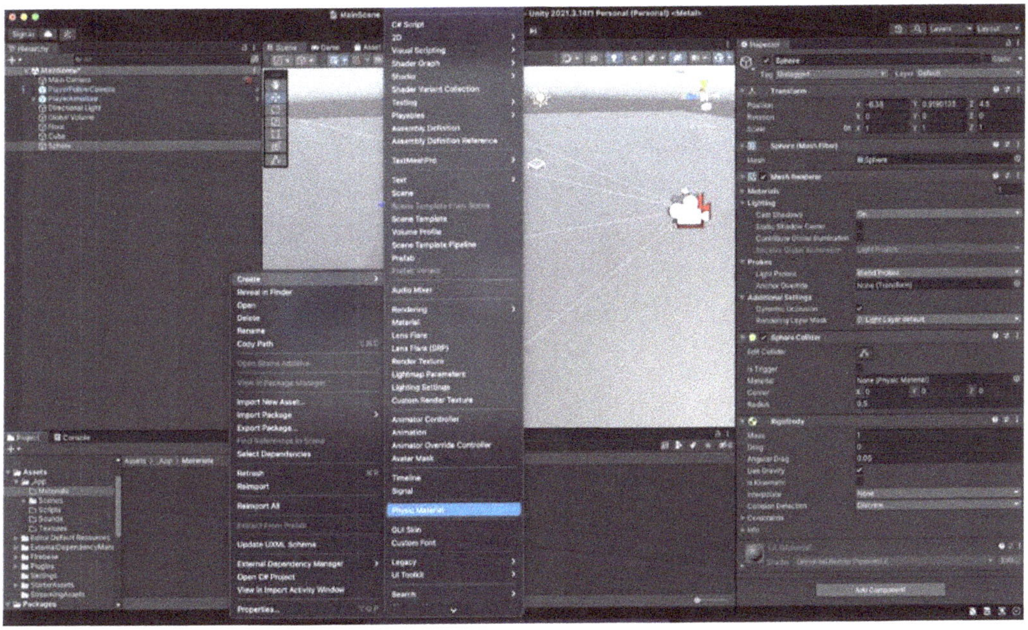

Figure 2.14 – Creating a Physic Material

3. A new asset will be created in the `Materials` folder called `New Physics Material`. Rename it to `MyBall`.

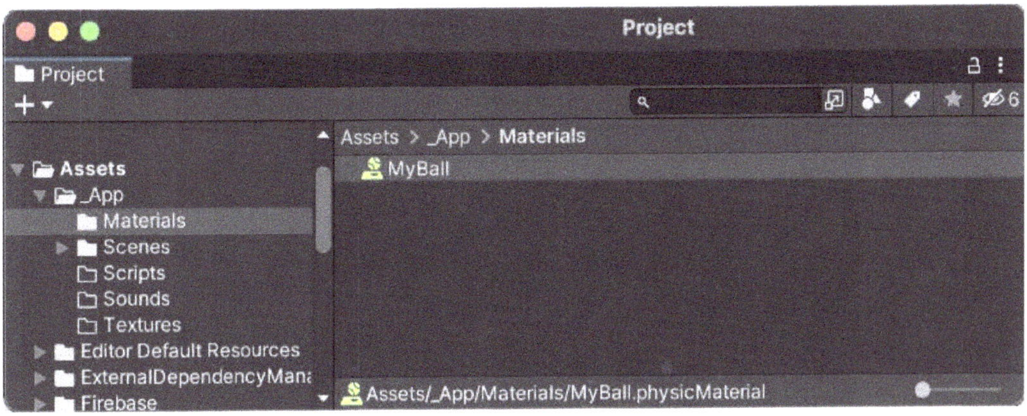

Figure 2.15 – Asset renamed to MyBall

4. Drag the **MyBall** material to the **Material** slot, which is located in the **Sphere Collider** component.

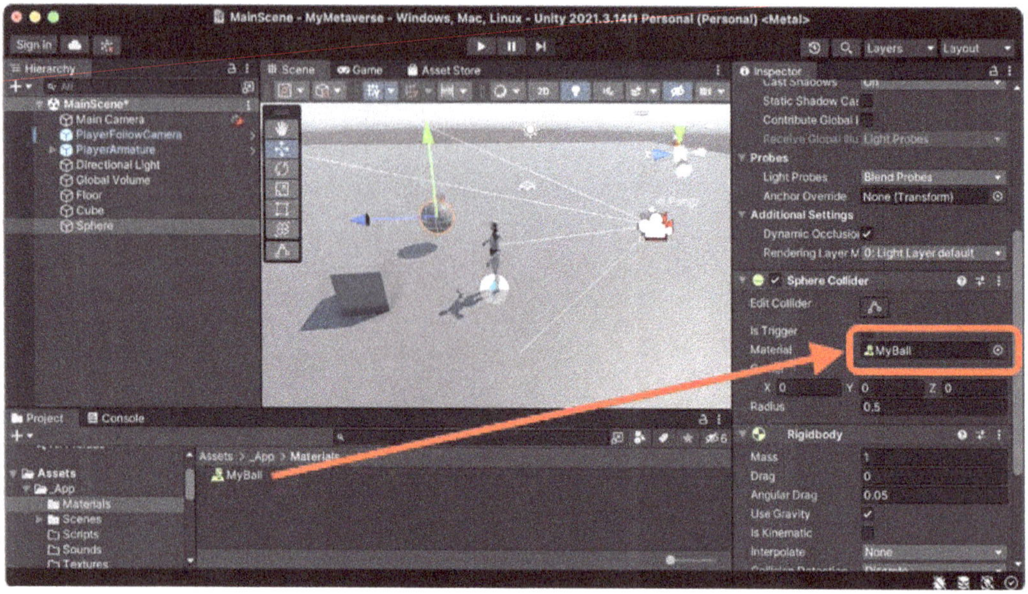

Figure 2.16 – Adding the material to the component

5. If you click on the **Play** button, you will see that the behavior has not changed much when adding the physics material. This is because the parameters configured in **MyBall** do not differ much from the default behavior without this material. To see the result better, click on the **MyBall** material and, in the **Inspector** panel, change the values to these:

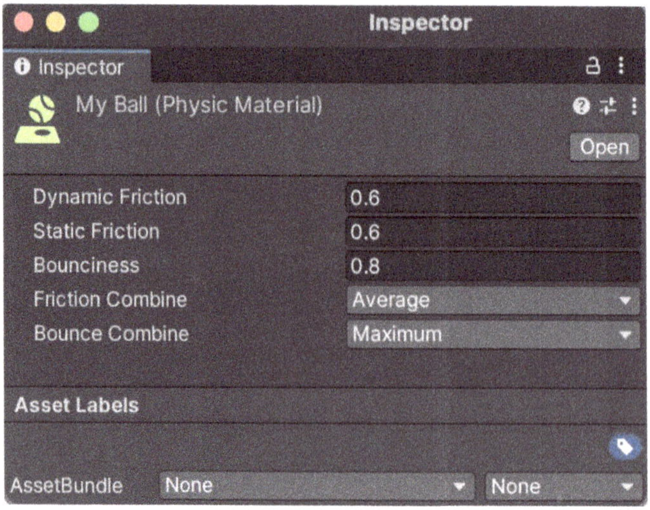

Figure 2.17 – Setting up the Physic Material

Change the **Bounciness** property to `0.8`, since with the value `0` it will not cause a bounce when colliding. We will also change the **Bounce Combine** property to **Maximum**.

> **Tip: advanced information on physics material properties**
>
> If you want to know more about what each property of a physics material is used for, I recommend the following website: `https://docs.unity3d.com/Manual/class-PhysicMaterial.html`.

Well, now, if we click on the **Play** button, we will see that the behavior of the sphere has changed. Now when it falls to the ground, it bounces several times. If you approach it and collide with it, it still does not move. This happens because, by default, our player is configured not to exert forces when colliding with another GameObject that has an associated Rigidbody component:

1. If we look at the components in our **PlayerArmature** GameObject, we will see one called **Basic Rigid Body Push**. This component defines the behavior that other objects in the scene that have a **Rigidbody** should have when our character makes contact.

   By default, it is configured as follows:

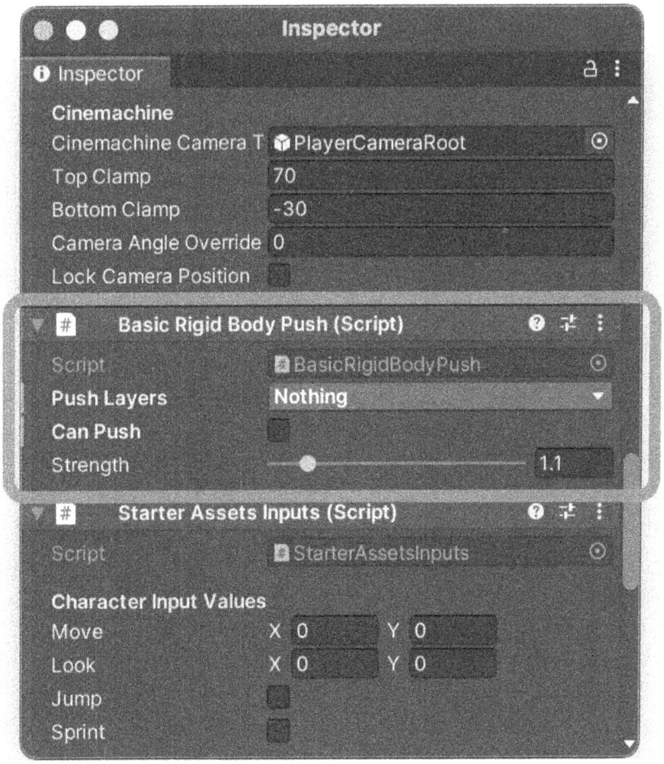

Figure 2.18 – Default configuration of the Basic Rigid Body Push component

2. The Boolean **Can Push** property is disabled, so it will not be able to exert physical impulses in contact with other objects. We also see a property called **Push Layers**, which defines a filter to apply physics forces only to objects belonging to a layer. Finally, we have the **Strength** property, which is the amount of force to exert when colliding.
3. If we change **Push Layers** to **Everything**, the **Can Push** checkbox is checked, and we click **Play** to collide with the sphere, we will see the new behavior, as we thought.

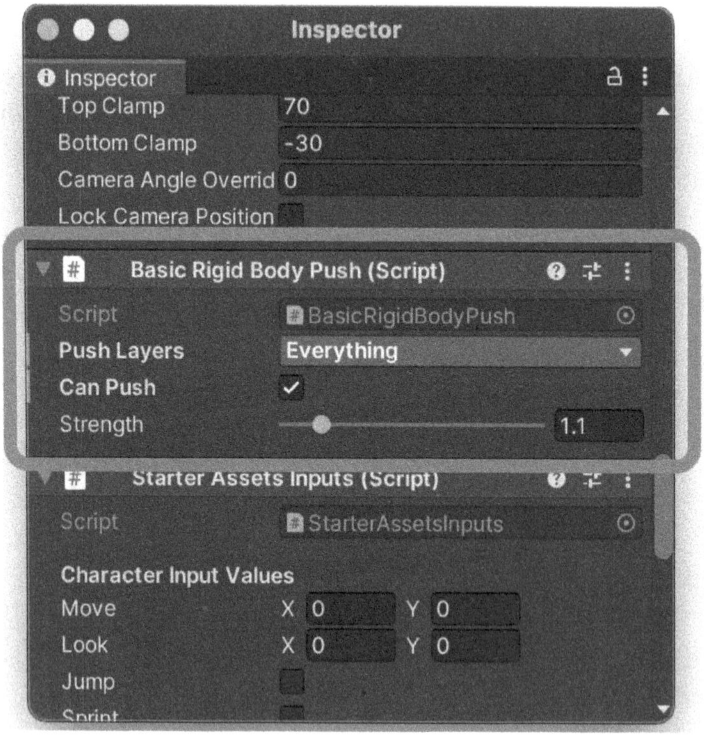

Figure 2.19 – New configuration in the Basic Rigid Body Push component

> **Tip: more information on layers and tags**
>
> **Layers** and **Tags** in Unity are useful tools for organizing and managing objects in the scene. Layers allow you to classify objects into logical groups, such as players, enemies, scenery objects, and so on, making it easier to control and manipulate specific objects through code or the collision system. Tags, on the other hand, are labels assigned to objects to quickly identify them according to a common purpose or function, such as **Player**, **Collectible**, or **Enemy**. These tags allow objects to be accessed and perform specific actions in a more efficient and readable way in the code, improving organization and facilitating development and interaction between objects in the game. Advanced information can be found in the official Unity documentation at `https://docs.unity3d.com/Manual/class-TagManager.html`.

Fun, isn't it? We already have our player ready and fully configured. As you can see, Unity offers us a very complete Starter Asset that comes ready with basic animations and actions such as moving, jumping, and running. We have also learned about some basic interactions with colliders, such as the Mesh Collider on our floor, the Box Collider for the cube, and the Sphere Collider for our ball, which we have also added physics to.

Changing the default avatar

By default, Unity creates our character to look like a robot – possibly not what we'd like the most. In future chapters, we will see how to change the appearance of our player at runtime, as we play, as we would do in any video game. Now, as a preview, we will learn how to change the initial appearance of the player in a very easy way.

There is a web platform called **Mixamo**. It is a huge catalog that offers hundreds of 3D characters and animations ready to use in your Unity project.

As a mandatory step, visit `https://www.mixamo.com` and register for free or log in if you already have an account.

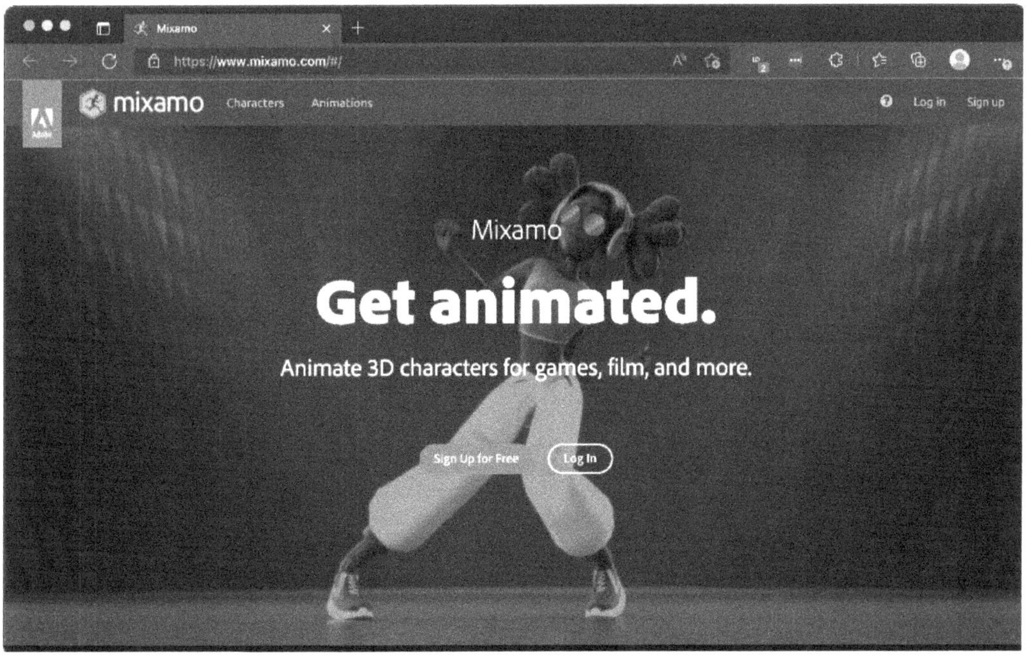

Figure 2.20 – Mixamo.com home page

Once identified, let's look for a character in the catalog to replace the robot avatar. To do so, follow these steps:

1. Click on the **Characters** link at the top of the web page.

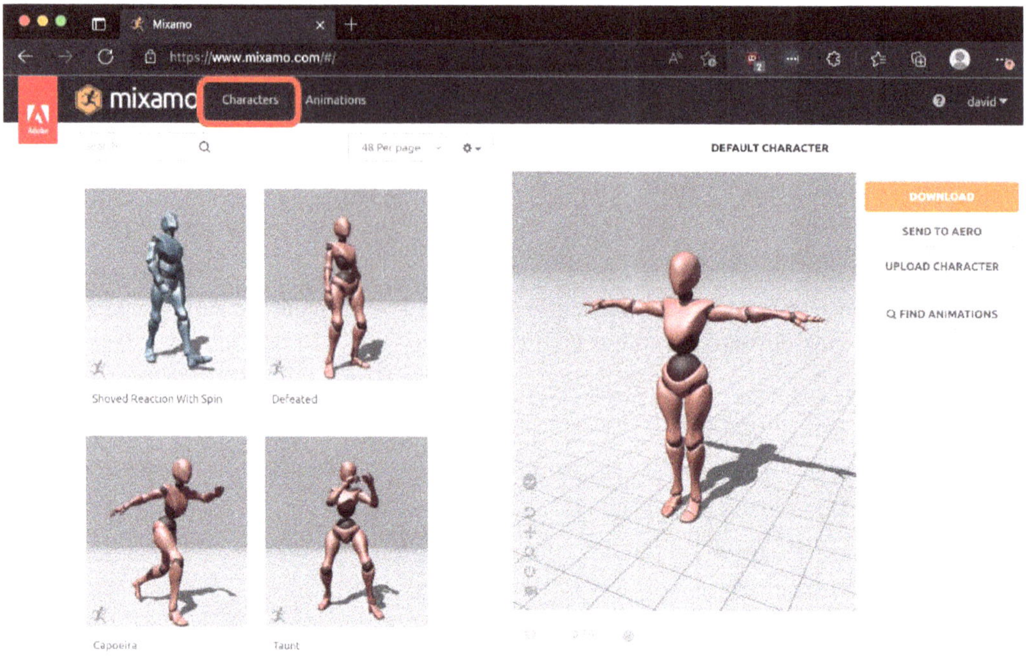

Figure 2.21 – Mixamo Characters link

2. Voila! On the first page appears a futuristic character perfect for our metaverse – **Exo Gray**. Click on it to see it in more detail in the viewer on the right and then click on the **DOWNLOAD** button.

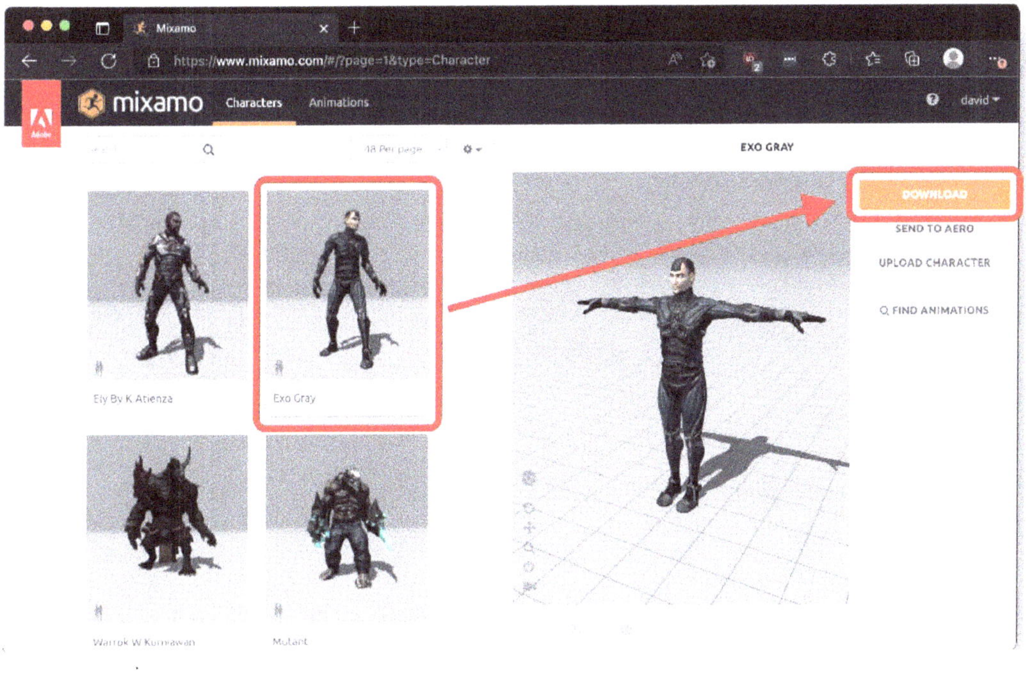

Figure 2.22 – Selecting and downloading the new character

3. A popup will open, allowing us to select some settings for the 3D file we are going to download. Select **FBX for Unity(.fbx)** for **Format** and **T-pose** for **Pose**, and click again on the **DOWNLOAD** button. A file named `Exo Gray.fbx` will then be downloaded to your download folder.

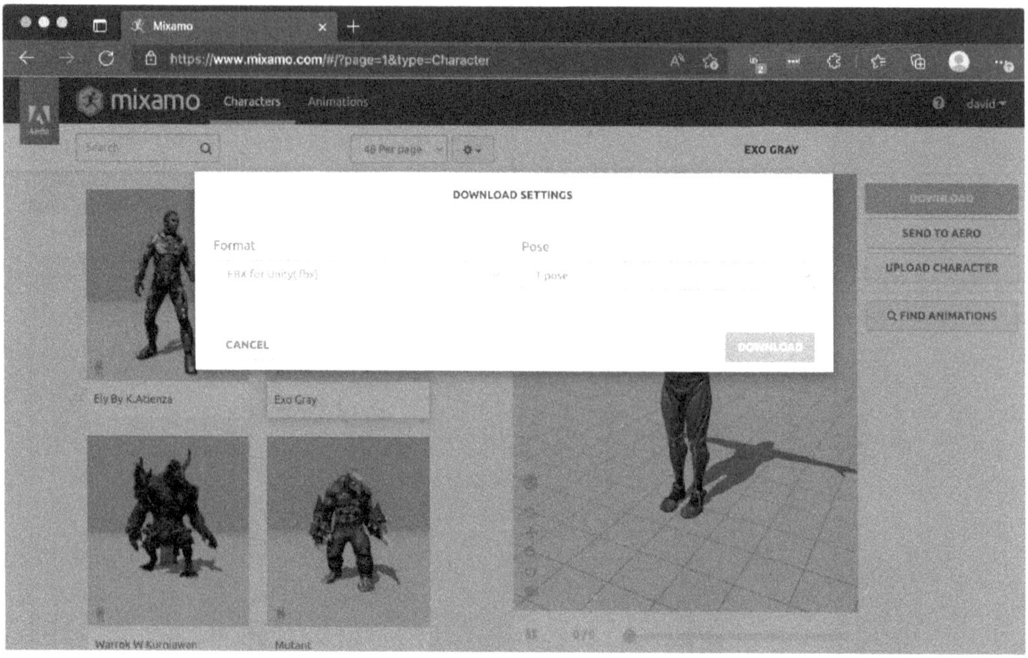

Figure 2.23 – Configuring the FBX options

4. Go to the Unity **Project** panel and create a folder called `Characters` inside **Assets** | **_App**. Then create a new folder inside `Characters` called `ExoGray` and then drag the `Exo Gray.fbx` file into it. The `Exo Gray.fbx` file will be in the **Assets** | **_App** | **Characters** | **ExoGray** path.

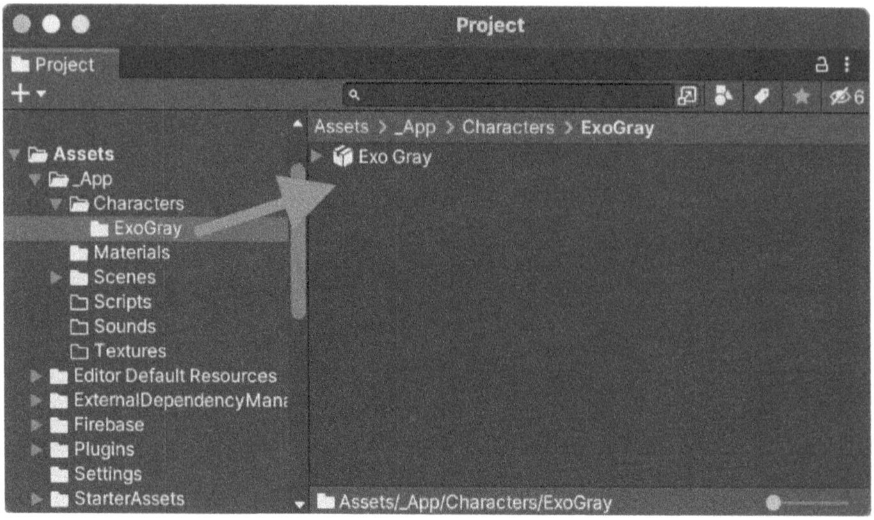

Figure 2.24 – Placing the Exo Gray file inside the folder

5. To finish configuring our new model for the player, we need to change the **Rig** type. To do this, click on the **Exo Gray** asset and look at its properties in **Inspector**. Click on the **Rig** tab and change **Animation Type** to **Humanoid** and **Avatar Definition** to **Create From This Model**, and click on the **Apply** button.

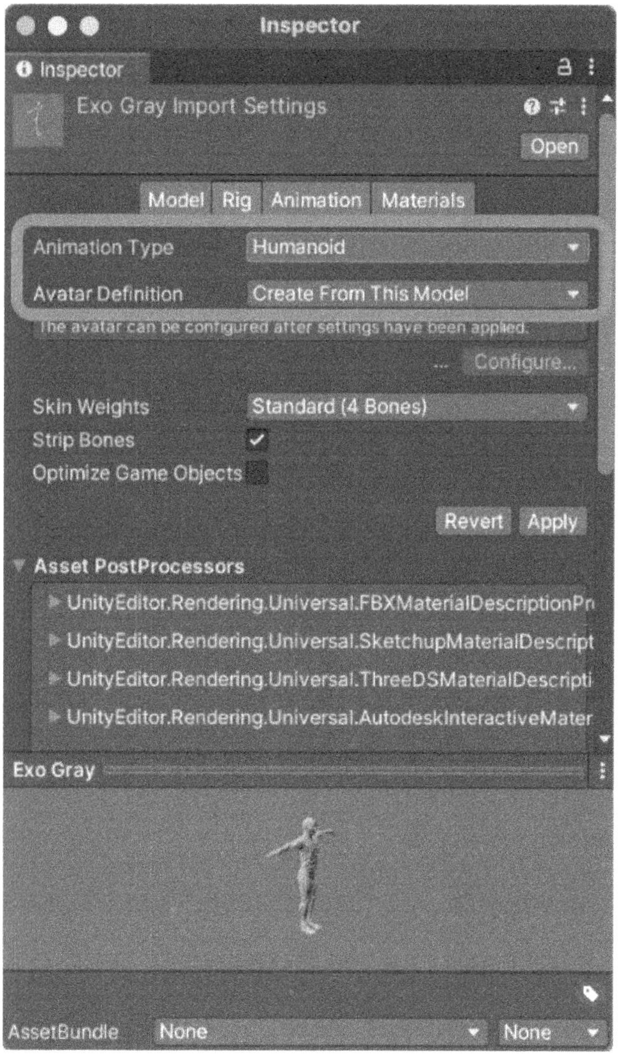

Figure 2.25 – Setting up the model Rig type

6. If you now look at the **Exo Gray** asset and open it by clicking the arrow button, you will see that an asset called **Exo GrayAvatar** has been created, which will be essential to replace the robot with the new character.

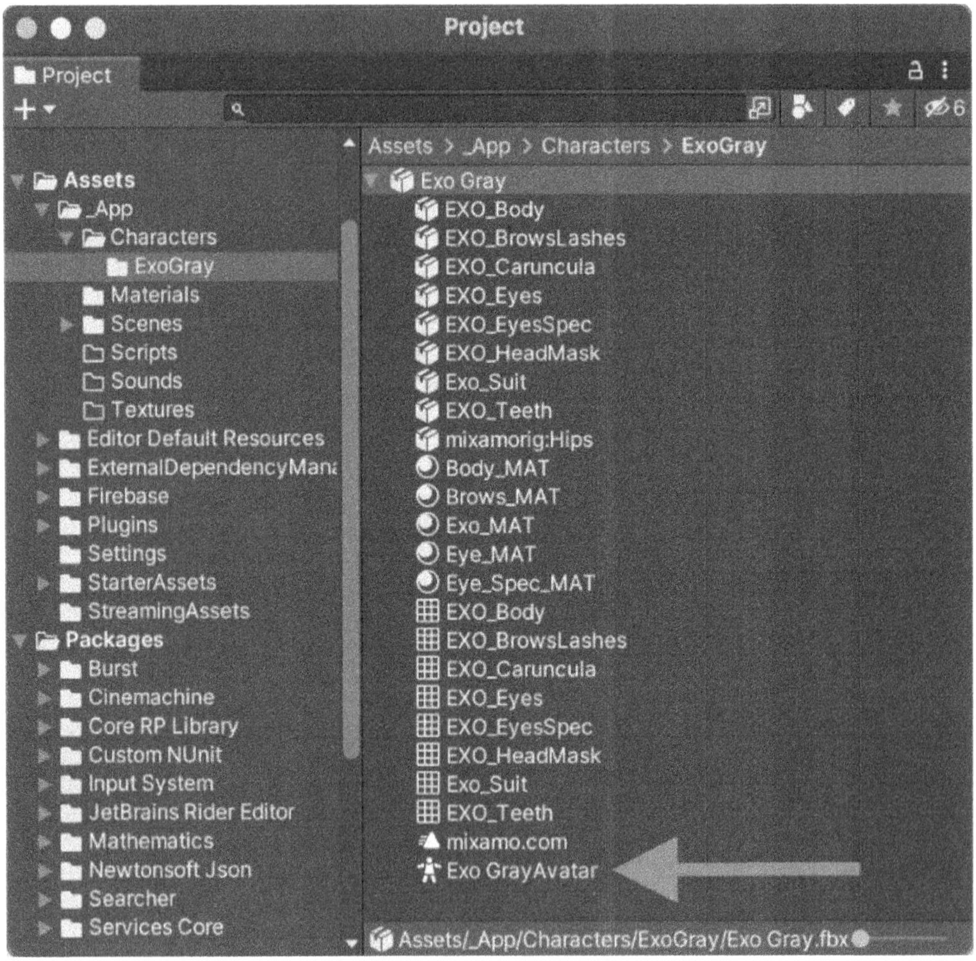

Figure 2.26 – New avatar created

7. Okay, let's proceed to replace the robot. Look at the **Hierarchy** panel on the left. Select the **PlayerArmature** item. The item's text is blue. It is blue because it is a Prefab. We want to remove the robot. This cannot happen as long as it is still a Prefab. However, you can unzip a Prefab. Therefore, right-click on **PlayerArmature**. In the menu that appears, select **Prefab | Unpack Completely**.

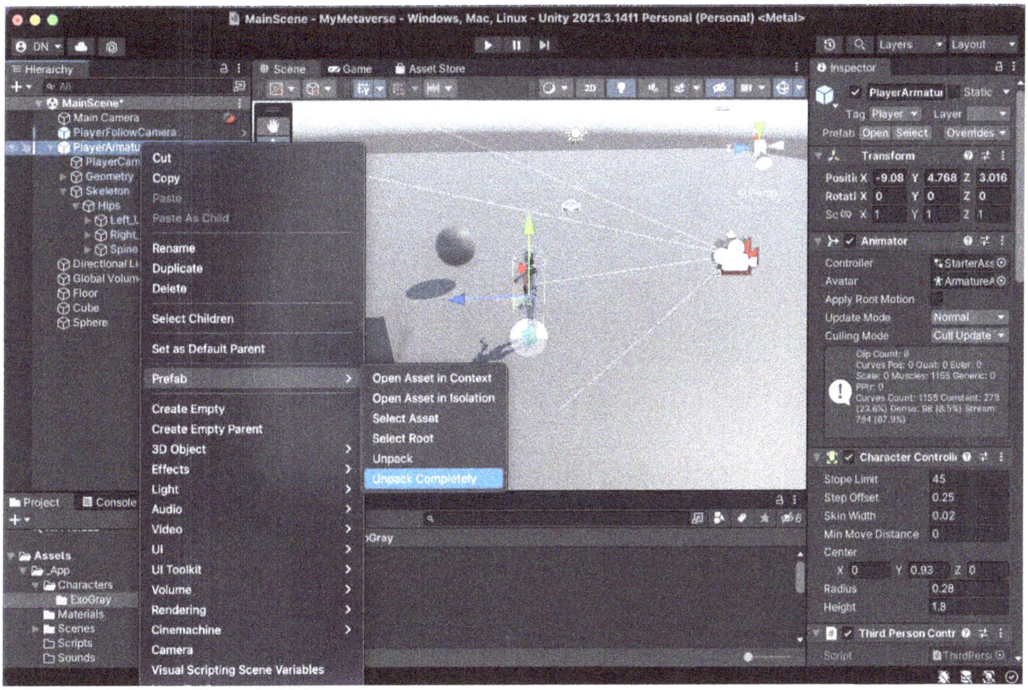

Figure 2.27 – Unpacking the PlayerArmature Prefab

8. You may have noticed that the **PlayerArmature** GameObject has lost its blue color. Now it is no longer a Prefab, and we can modify it. We unbundle the GameObject and look for a child object called **Armature_Mesh**, which is inside the **PlayerArmature | Geometry | Armature_Mesh** hierarchy, and click on it.

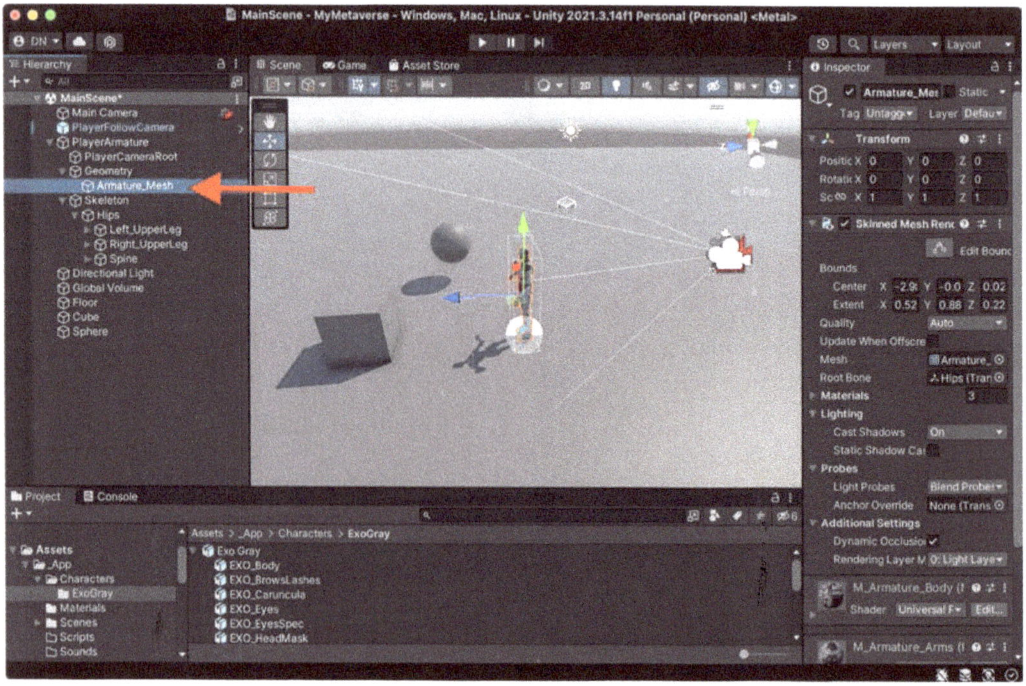

Figure 2.28 – Selecting the Armature_Mesh GameObject

9. Now, with the **Armature_Mesh** GameObject selected, we delete it by pressing the *Delete* key or by right-clicking and clicking the **Delete** option (without fear).

10. Drag the **Exo Gray** asset into the **Geometry** hierarchy, in the same place we just deleted the **Armature_Mesh** GameObject from, and you will see something similar to this:

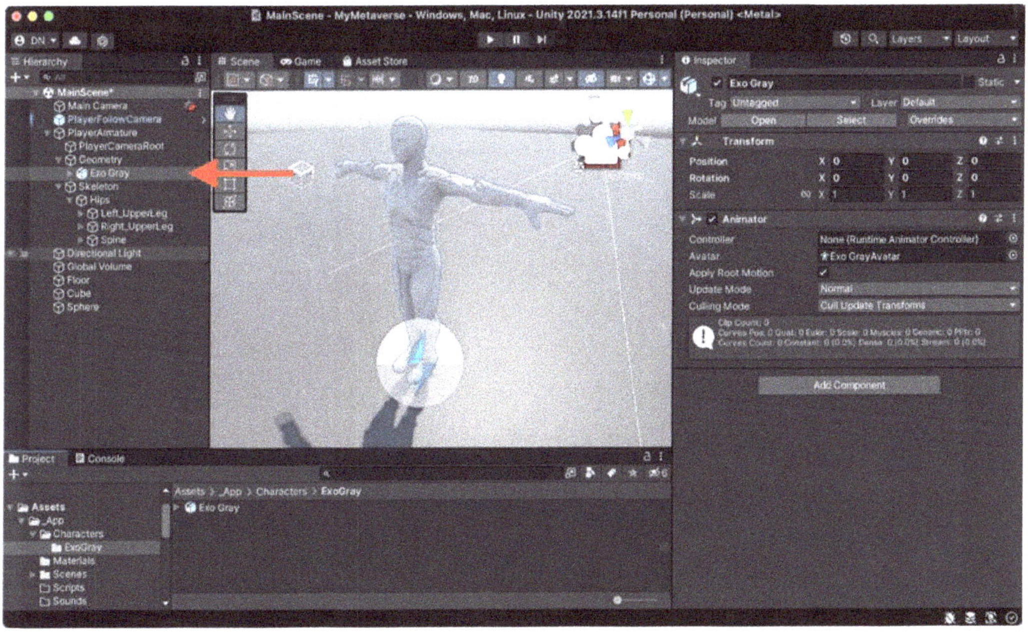

Figure 2.29 – Replacing Armature_Mesh with the Exo Gray asset

11. Almost there. We only need to assign the avatar of the new character to the animation engine. To do it, click on the **PlayerArmature** GameObject and assign the avatar named **Exo GrayAvatar** to the **Avatar** box, in the **Animator** component, clicking on the button with the circular icon to select it.

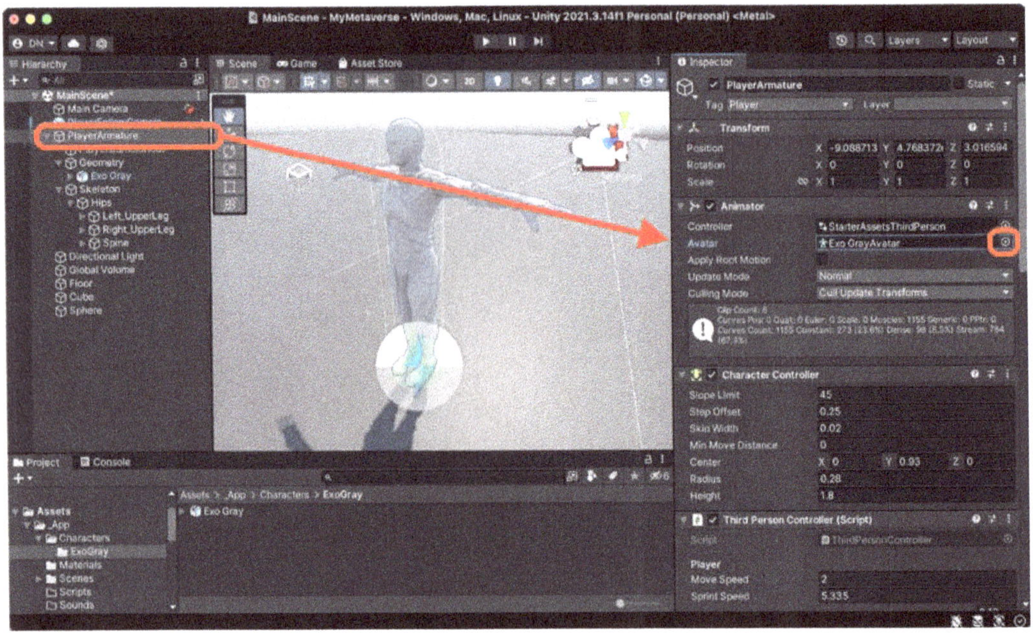

Figure 2.30 – Replacing the avatar in the Animator component

12. That is it... Oh, wait… My new character is white. Where are the textures? Sometimes Unity goes crazy when it comes to using textures that are embedded in the FBX model, but it's an easy fix. In the **Project** panel, select the **Exo Gray** asset and click on it.

13. In the **Inspector** panel, select the tab called **Materials** and click on the **Extract Textures...** button.

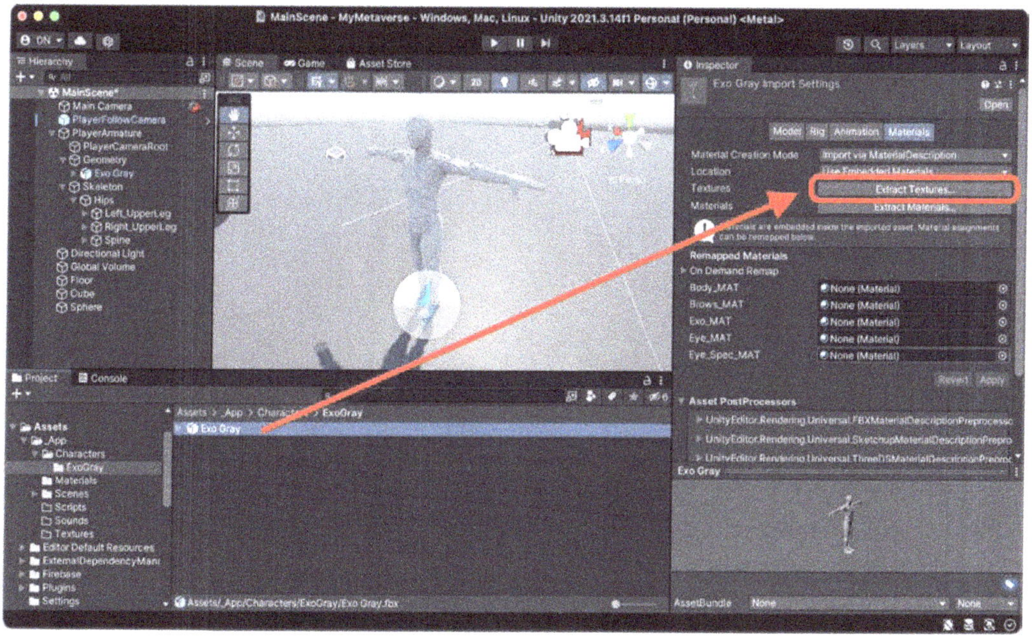

Figure 2.31 – Extracting model textures

14. Unity will ask you where you want to extract the textures. By default, it will have selected the folder where you have the FBX model. That's okay – you can click on the **Choose** or **Accept** button in Windows.

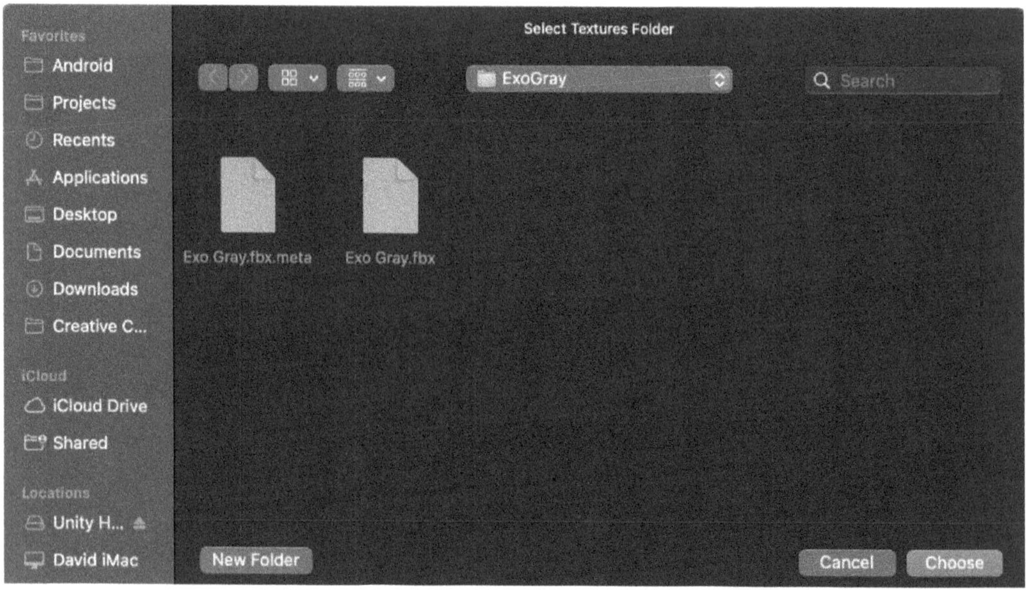

Figure 2.32 – Choosing the target folder for textures

15. Finally, when Unity finds a bug or a problem in a texture, it will try to fix it. That's why you may see a pop-up window like the one that follows. Whenever you see it, click on the **Fix now** button to apply corrections.

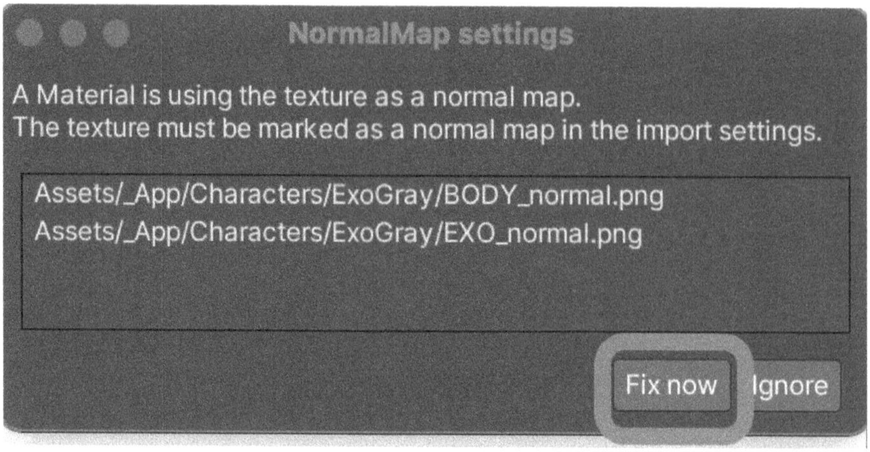

Figure 2.33 – Automatically fixing textures

Fantastic! We've done it – we've replaced the ugly (or not so ugly) default robot. Now, we have a splendid futuristic character in its place. Go on – click the **Play** button and try it out.

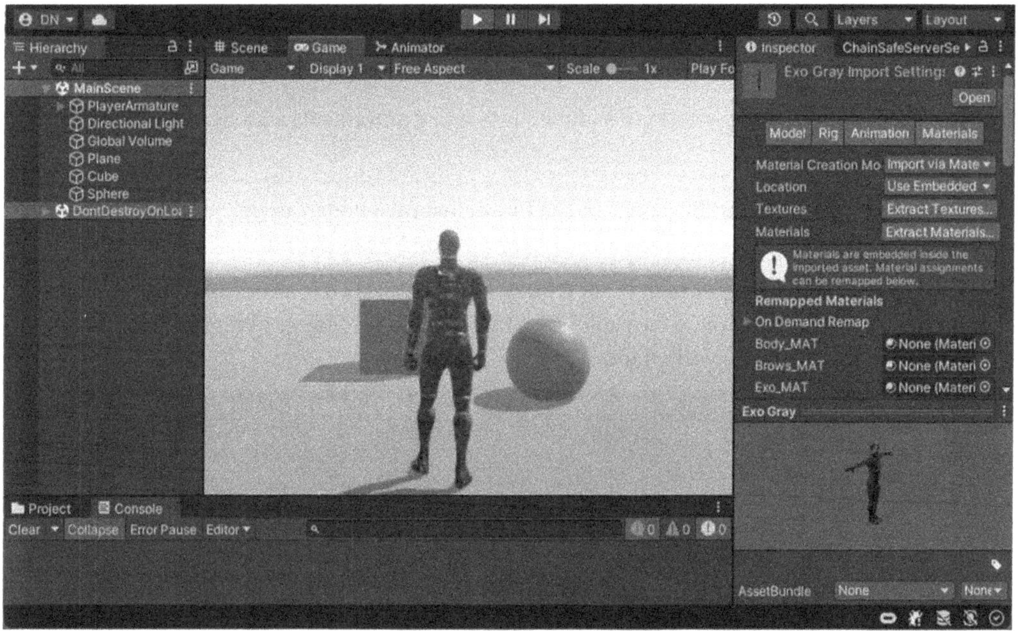

Figure 2.34 – Trying the new character in Play mode

You will have seen how easy and fun it has been to change the appearance of our character. Now that you know Mixamo, think of the infinite possibilities it can offer us in our project. In future chapters, we will draw from the Mixamo catalog to give our users the possibility to choose a more personalized appearance. Don't forget to save the project.

Tip: more info about Prefabs

Prefabs are a type of *container* object. They can be created to facilitate cloning. One of the great benefits of using Prefabs is that you can apply modifications and extend these changes to all the clones that exist in a project. Another great advantage is that you can compose a multitude of objects in a single Prefab, for example, you can create a Prefab containing a house, its doors, windows, and details in a single object, which greatly facilitates working with them. You can learn more about Prefabs by visiting `https://docs.unity3d.com/es/530/Manual/Prefabs.html`.

Adding Mobile Support

So far, our character is able to move around with PC controls, that is with the *W*, *A*, *S*, *D* keys, *Spacebar*, mouse clicks, and so on. But you may ask, what about on a mobile device? Here we solve it. Fortunately, Unity offers us an easy solution to solve the character movement with touch controls.

Next, we will add a step by step system that allows us to make movements with our fingers. To do so, we will follow these steps:

1. We must make some modifications in the **PlayerInstance** Prefab, double click on the Prefab located in **Assets** | **Photon** | **PhotonUnityNetworking** | **Resources** to edit it.
2. Unity offers a Prefab that practically brings this system, it is called **UI_Canvas_StarterAssetsInputs_Joysticks** and we look for it in the path **Assets** | **StarterAssets** | **Mobile** | **Prefabs** | **CanvasInputs** and we drag it to the **PlayerInstance** Prefab.

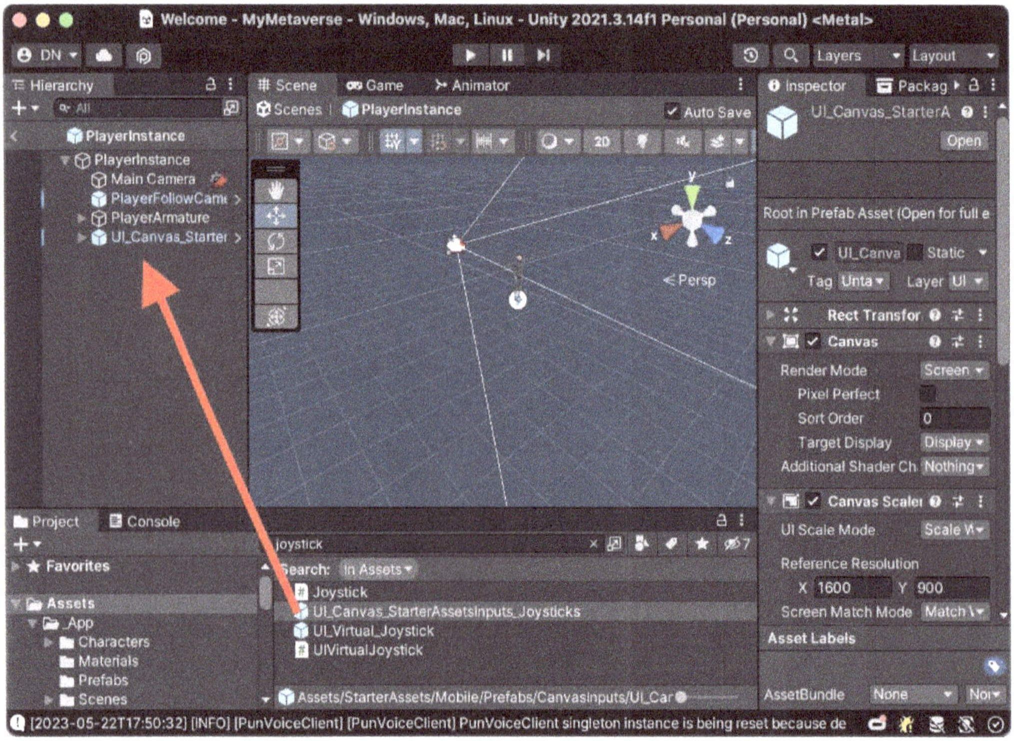

Figure 2.35 – Adding Mobile Controls Prefab

3. Now, we select the **UI_Canvas_StarterAssetsInputs_Joysticks** GameObject that we have dragged in the **Hierarchy** panel and in the **Inspector** panel. We have a look at the **Starter Assets Inputs** properties of the **UI Canvas Controller Input** component and **Player Input** of the **Mobile Disable Auto Switch Controls** component. Drag the **PlayerArmature** GameObject to the empty slots to finish the configuration.

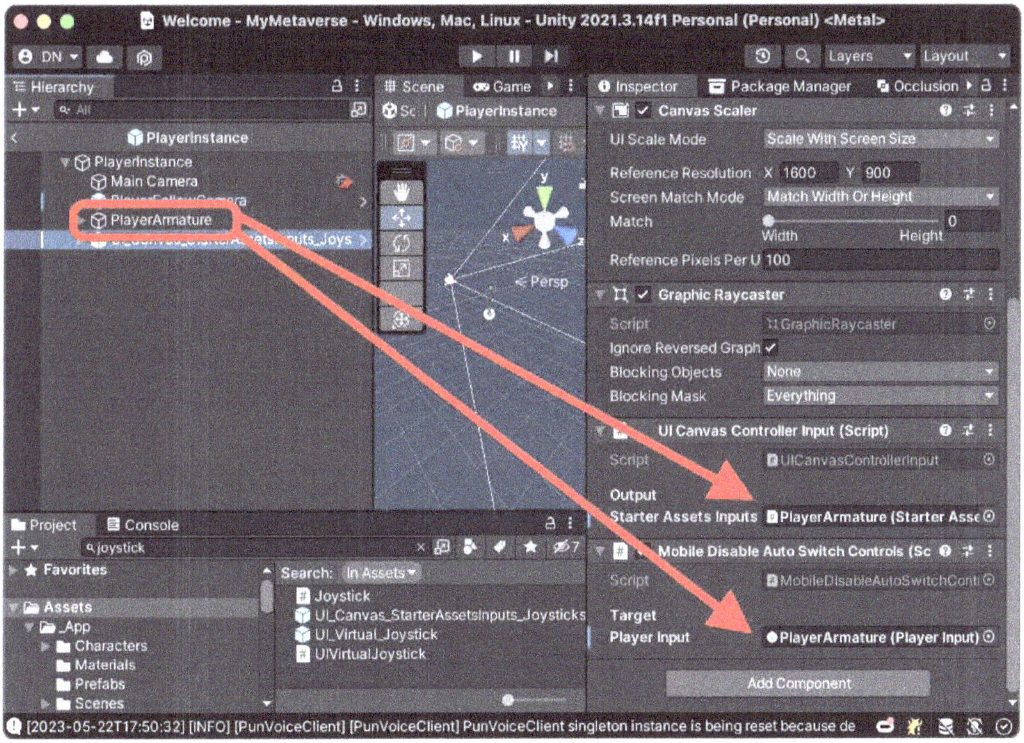

Figure 2.36 – Assigning PlayerArmature to empty slots

This Prefab that Unity offers us to manage mobile controllers has its own system that will show or hide depending on whether or not we are on a mobile device. This task is performed automatically by the **Mobile Disable Auto Switch Controls** component.

If we run our project now, and we also have the **Android Platform** active in the **Build Settings** menu, we will be able to see that Unity loads a mobile controller for our character.

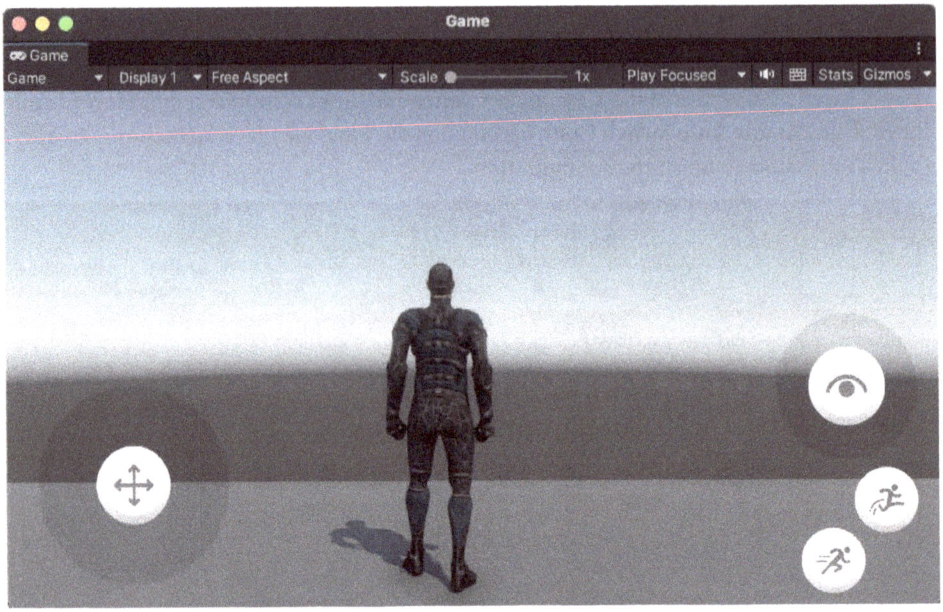

Figure 2.37 – Final result Adding Mobile Support

Good job, it was easy, wasn't it? But we still need to make a few small corrections to make sure that everything works perfectly.

We have just implemented a control system for mobile users. We now offer support for both desktop and mobile users. These mobile controls offer buttons to move, turn your head to look, walk and run. Everything you need at the moment.

Summary

Throughout this chapter, we have learned how to create a complete controller for our player. Now we have a character capable of being controlled by the keyboard and mouse, with totally professional effects. We discovered how Unity facilitates development with Starter Assets that offer compact solutions to complicated tasks such as creating a controllable player.

We practiced collisions by placing a cube and a sphere in the scene with the physics of a soccer ball, which we are able to move upon collision. To enhance the player we created, Mixamo surprised us with its huge catalog of free characters ready to use in our Unity project.

In the next chapter we will learn how to design a scene from scratch, with buildings, decorative elements such as trees, pavements and even a fountain! We will create a wonderful little village that will serve as the main scene for our players to land in as soon as they connect to our metaverse. We will also make our first foray into performance optimisation tactics in the game. Finally, we will create our first collection in Firestore.

3

Preparing Our Home Sweet Home: Part 1

In many online games and in the existing metaverses, it is very common to see a cozy place where you always appear when you log in. This place is usually called **hall** or **meeting point**. In our project, we will do the same – we will offer our users an amazing place to show up, a place to meet friends, find information about things you can do, and interact with NPCs.

Our Meeting Point will look like it is in a town square; it will be bounded with closed perimeters to make it intuitive and easy to understand so that the user does not have to travel long distances to reach their goal, with all services offered centrally.

This square or Hall will be the first place that your users will see when they enter the metaverse, so we must ensure that it is pleasing to the eye, easy to navigate, and fun.

We will cover the following topics:

- Understanding the hub concept
- Designing the scene

Technical requirements

This chapter does not require any special technical requirements, just an internet connection to browse and download an asset from Unity Asset Store.

We will continue on the project we created during *Chapter 1, Getting Started with Unity and Firebase*. Remember that we have the GitHub repository, `https://github.com/PacktPublishing/Build-Your-Own-Metaverse-with-Unity/tree/main/UnityProject`, which will contain the complete project that we will work on here.

Understanding the Hub concept

Before we start designing our Hub, let's understand what features it should offer to the user. A good Meeting Point or Hub is a pleasant place where the user lands when they log in or enter our metaverse for the first time. That is why we need it to be a pleasant place in every way.

We will work on designing a wide space, without complications in terms of navigation; we do not want it to be a labyrinth but instead what the user to find everything quickly. If we put ourselves in the shoes of the newcomer, we think that they will want to find out what they can do in our metaverse without having to *walk* a lot or go through virtual streets.

For this task I can't think of a better place than the square of a small town, you know, a wide space but delimited in its perimeter, so that with a brief turn of the head, we can see what we can interact with.

If we were to summarize our to-do list of things to set up for the Hub, it would look something like this:

1. A closed square with buildings, oval or circular in shape.
2. A small water fountain in the center, which serves as a reference for users who are friends to meet, and for meeting others. You know... We met in the town square.
3. Environmental sounds that transmit serenity and harmony, such as birds singing, murmurs of people talking, and water from the fountain.
4. Natural and pleasant light, such as a sunset.
5. NPCs scattered around the square, each with their own functionality.

To populate our Hub scene with buildings, we're going to get them totally free in Unity Asset Store. After this chapter is finished, your scene will look similar to the following figure. Impressive, right? We won't delay any longer discussing the theory and will get into the practice.

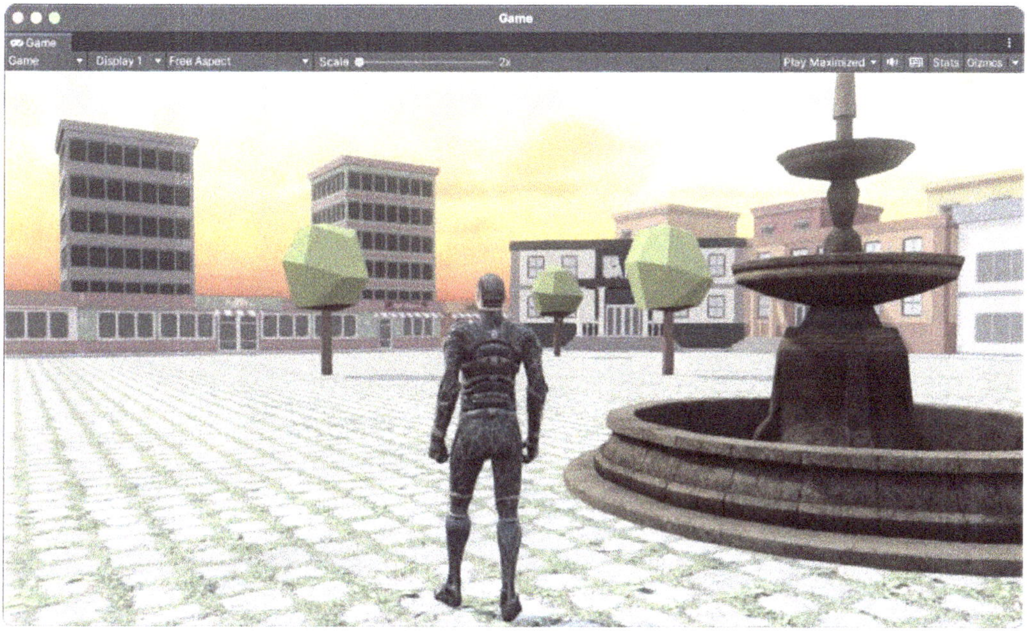

Figure 3.1 – Final result of the main scene

Designing the scene

We have come to one of the most beautiful and interesting parts of this book; we will design the main scene, the sweet home of all our users, a friendly, quiet place where we will give a warm welcome to all visitors to our metaverse.

Customizing the floor

The first thing we will do is customize the floor of our scene. Let's work on the main scene that we have created in previous chapters, the **MainScene** scene:

1. To do this, if you don't have it open in the Unity editor, double-click on **_App** | **Scenes** | **MainScene**.

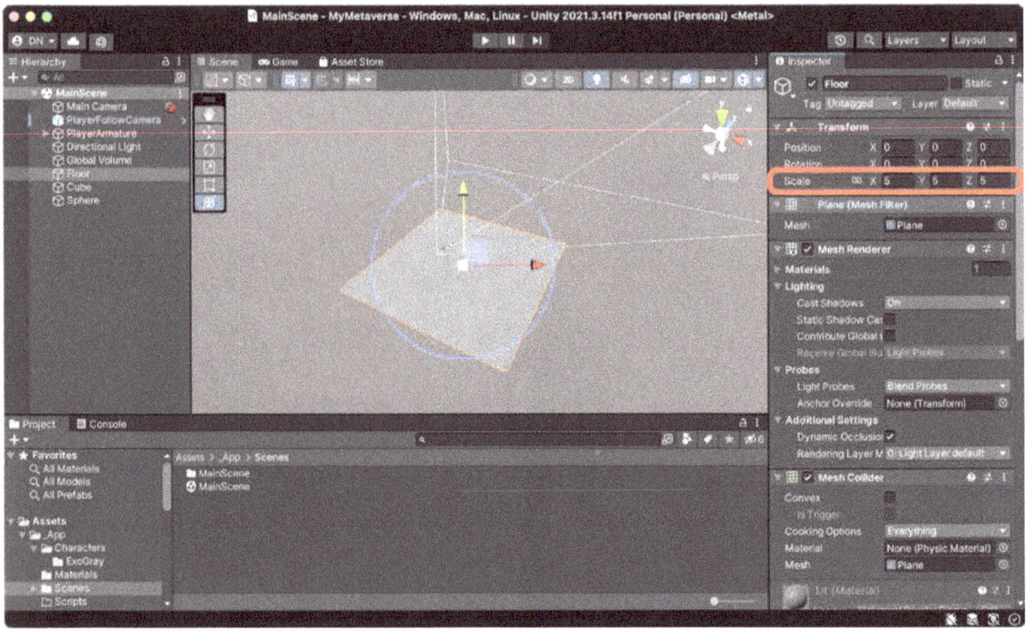

Figure 3.2 – Resizing the floor

Our **Floor** GameObject is currently a bit small; we are going to expand it so that when we place the buildings, there is still ample space in its central area. We will simply change the **Scale X/Y/Z** values `5`, `5`, and `5` to `10`, `10`, and `10`.

The next thing we will do with the floor is to decorate it a little. We are going to change the current gray texture for a more realistic, more beautiful floor.

We will need a texture, we can find it for free in Unity Asset Store.

2. To get it, go to `https://assetstore.unity.com` in your browser.
3. Search for `floor texture` in the search engine and sort the result by **Price (Low to High)**, from lowest to highest, so we can see the free assets first.

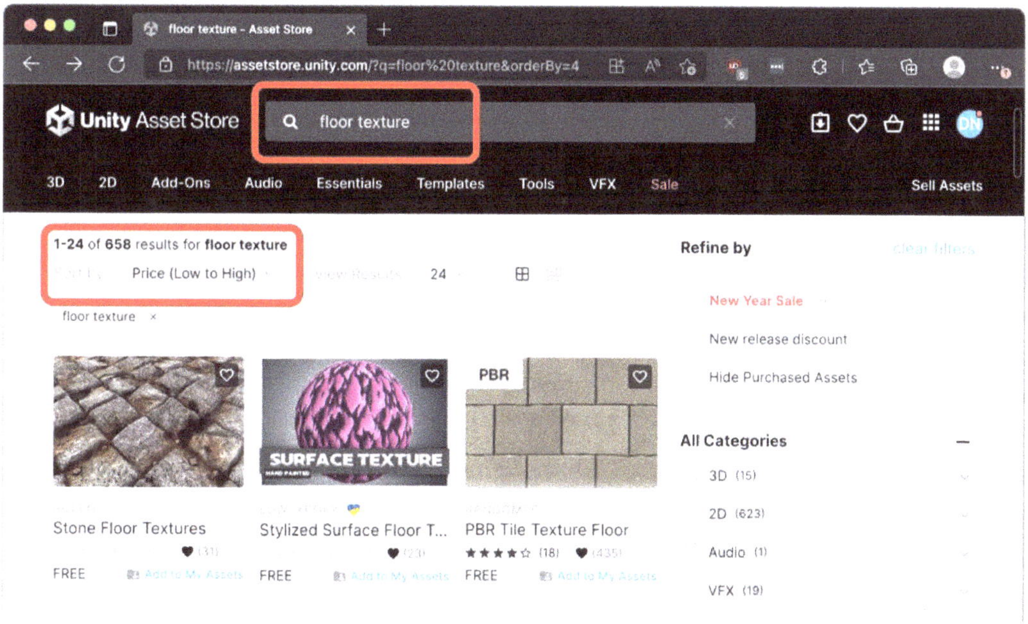

Figure 3.3 – Searching for a floor texture in Unity Asset Store

4. Scrolling down, I personally have chosen this asset since it's *free*, with more variety, which will be useful later, and a nice rustic-type plot. Click on the marked asset to go to the download page.

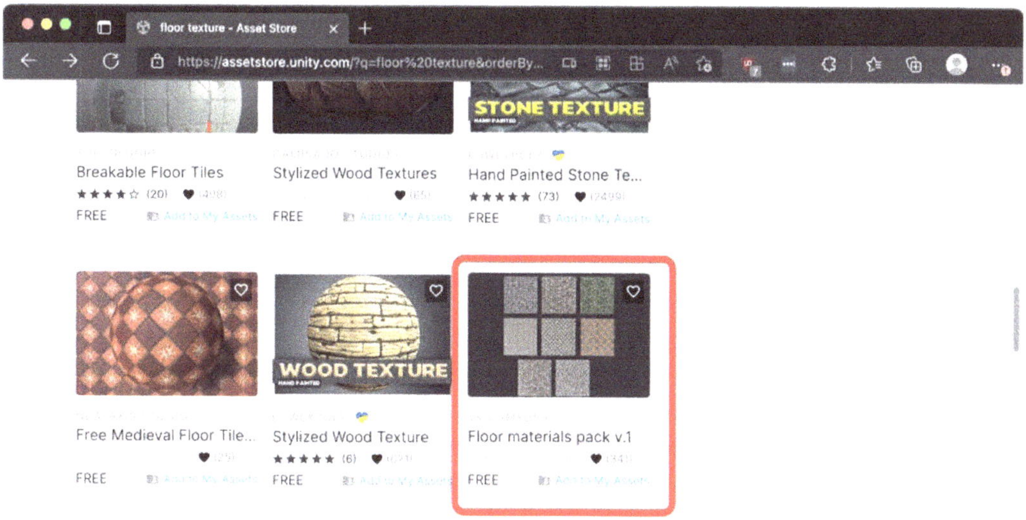

Figure 3.4 – Selected material for our floor

5. Click on the **Add to My Assets** button.
6. Then, you will see a button at the top called **Open in Unity**; click on it.

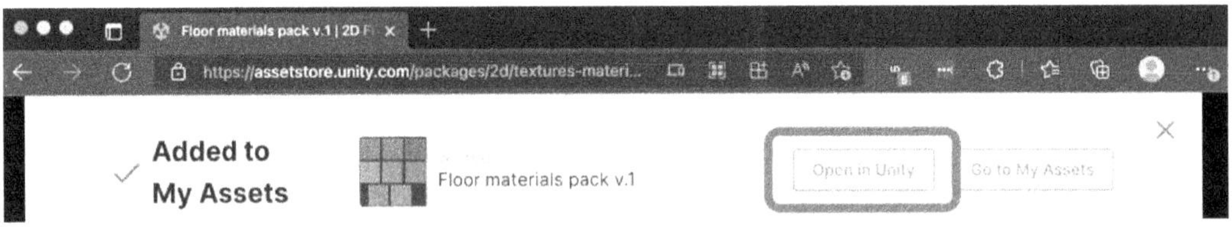

Figure 3.5 – Open in Unity button for the newly acquired asset

Then, the asset will be loaded into our Unity Package Manager, and a modal window will be displayed, as we have seen on previous occasions with the new asset.

7. Click on the **Download** button and click on the **Import** button when the download is finished.

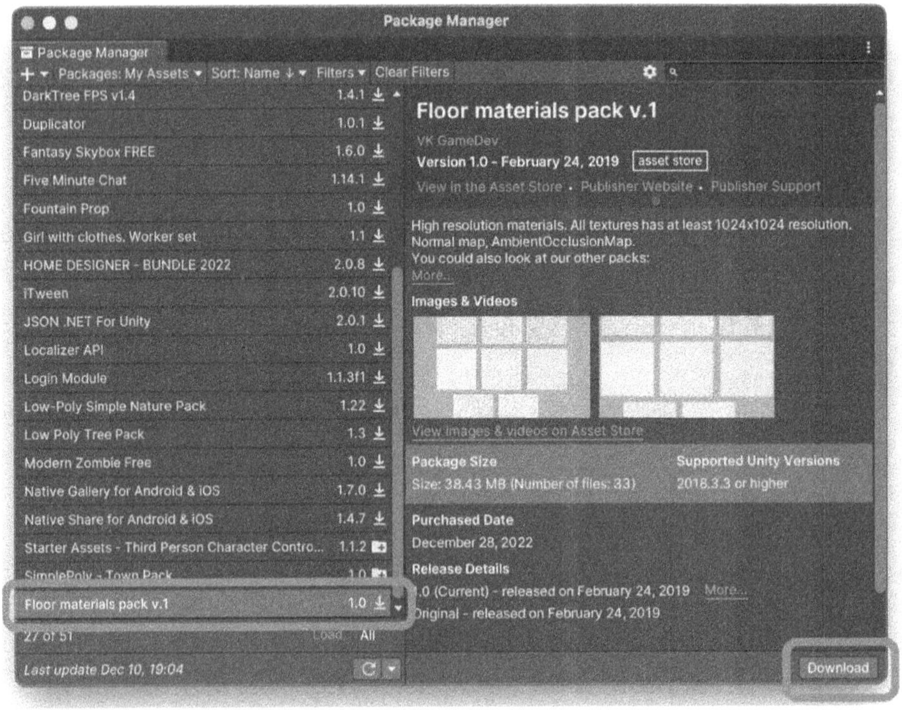

Figure 3.6 – Downloaded Floor materials pack asset

8. We already have a new asset in our project; as you can see, a folder called `Floor materials pack` has been created.

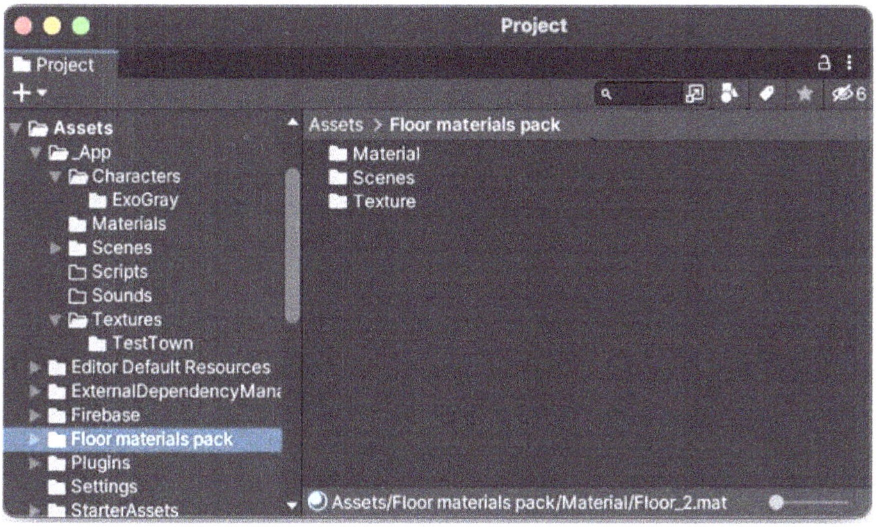

Figure 3.7 – Floor materials pack folder

It is important that we check the compatibility of the newly imported materials with the URP configuration of our project. Generally, in Unity Asset Store, authors often put a warning if their asset is already optimized for the **Universal Render Pipeline** (**URP**). Otherwise, we will have to make an adjustment to convert it. But don't worry, it's a simple process.

9. To check that the floor materials we have downloaded are compatible, we go to **Assets** | **Floor materials pack** | **Material** and click on any that we find inside the folder. If the preview of the material looks fuchsia or pink, you have compatibility issues.

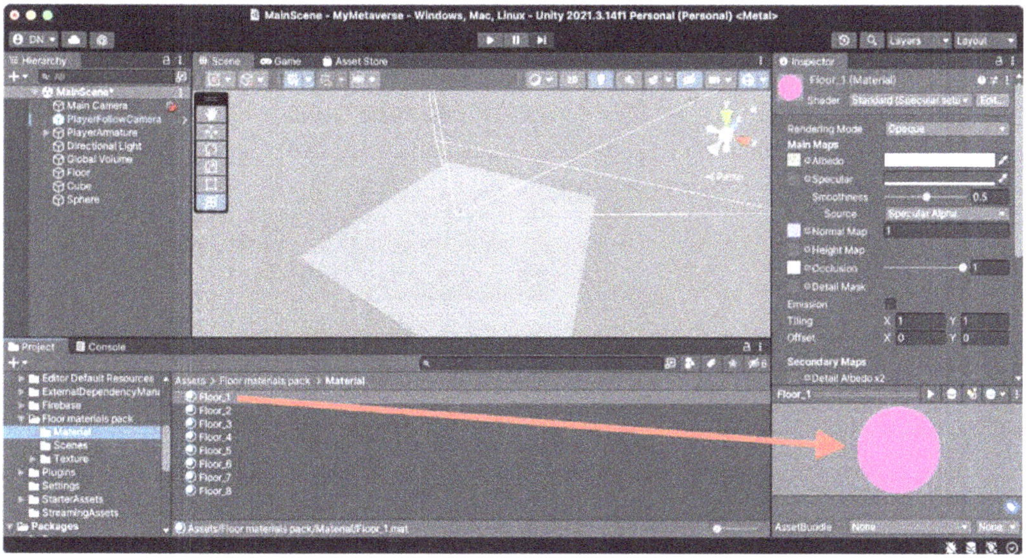

Figure 3.8 – No URP material previsualization

10. To correct the materials and convert them for the URP render system, select all the materials in the folder, and in the main menu bar, click on **Edit** | **Rendering** | **Materials** | **Convert Selected Built-in Materials to URP** and confirm by clicking the **Proceed** button in the confirmation window that will appear immediately after. With that, we already have the material ready to use.

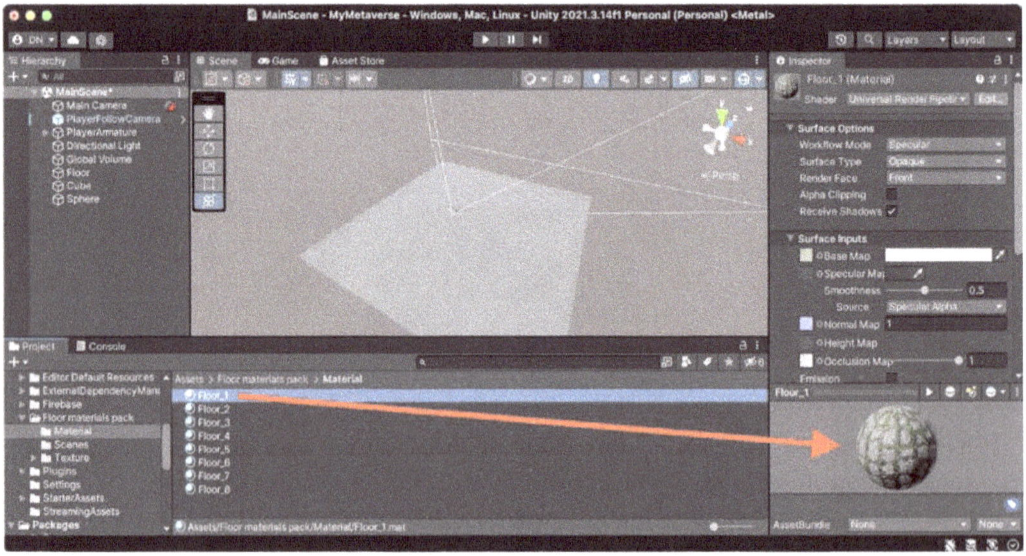

Figure 3.9 – Fixed material

11. With the **Floor** GameObject selected, we drag the `Floor_1` material into the **Materials** box, which is located in the **Mesh Renderer** component. As you may have noticed, the floor now has the new texture that we have downloaded, but the proportion of the floor is wrong; the stones are immense.

    Normally, this proportion is already corrected in prefabricated models that we will find online, but being a native Unity GameObject, it does not know what size to apply to the texture to make it correct. We just have to modify some values to correct this.

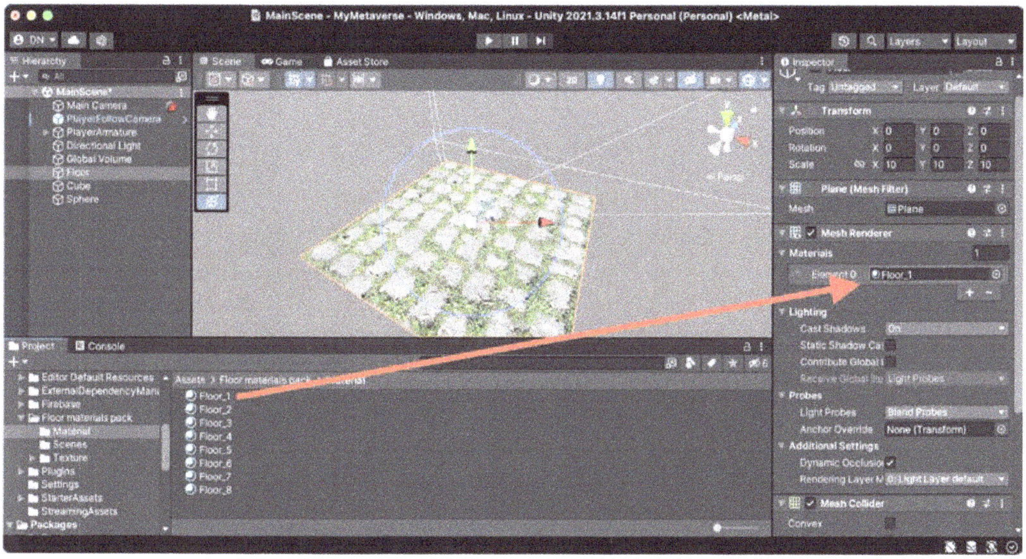

Figure 3.10 – Changing the material in the Floor GameObject

12. Finally, in the **Inspector** panel of the **Floor** GameObject, we can find some properties called **Tiling**, which by default are set to `1`. This means that the texture will be repeated only once in relation to the size of the object where it is applied.

    If we change these values to `30`, we will see that the stones are now more proportional; this means that the texture will repeat 30 times across the object to which it is applied. You can play with these values to make the floor to your liking.

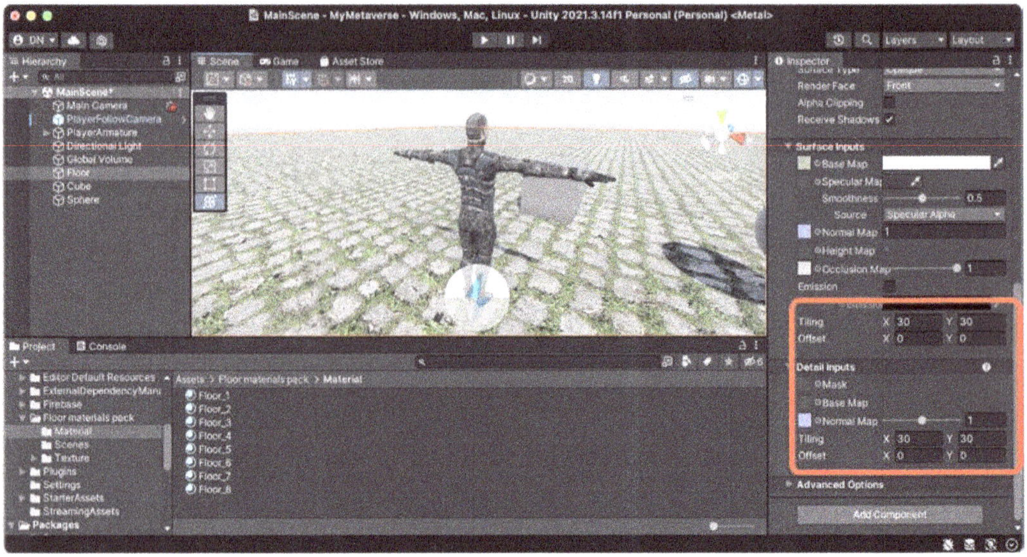

Figure 3.11 – Setting up Tiling values

Good work! You have designed a beautiful floor for our scene. Now that you have acquired the knowledge that allows you to add a new GameObject, download, and apply a texture correctly, let's find out how to populate our scene with nice buildings that we will download to emulate a small village.

Downloading the buildings

The floor has turned out nice, hasn't it? Our scene is taking shape, but it is nothing compared to how it will be when we have the buildings ready. As we have done with the floor texture, we are going to look for some ready-to-use models from Unity Asset Store.

Looking through the store, I found some low-poly buildings that will be perfect for our Hall. Not just any building will do; we always have to make sure that it is low-poly so as not to harm the performance.

> **Tip: what is low-poly?**
>
> **Low-poly** is an expression that brings to mind several ideas about what it could mean. It deals with designs with a low number of polygons, that is, geometric elements that have few faces or vertices, such as spheres, cylinders, or cubes. Low-poly allows a lower resolution in its compositions, which facilitates one of its main advantages: a higher rendering speed.

Let's proceed to download the building kit that I have selected for this book. To do so, follow these steps:

1. Go to Unity Asset Store at `https://assetstore.unity.com`.
2. Write `SimplePoly - Town Pack` in the text box of the search engine.
3. Click on the single result that appears to go to the download page.

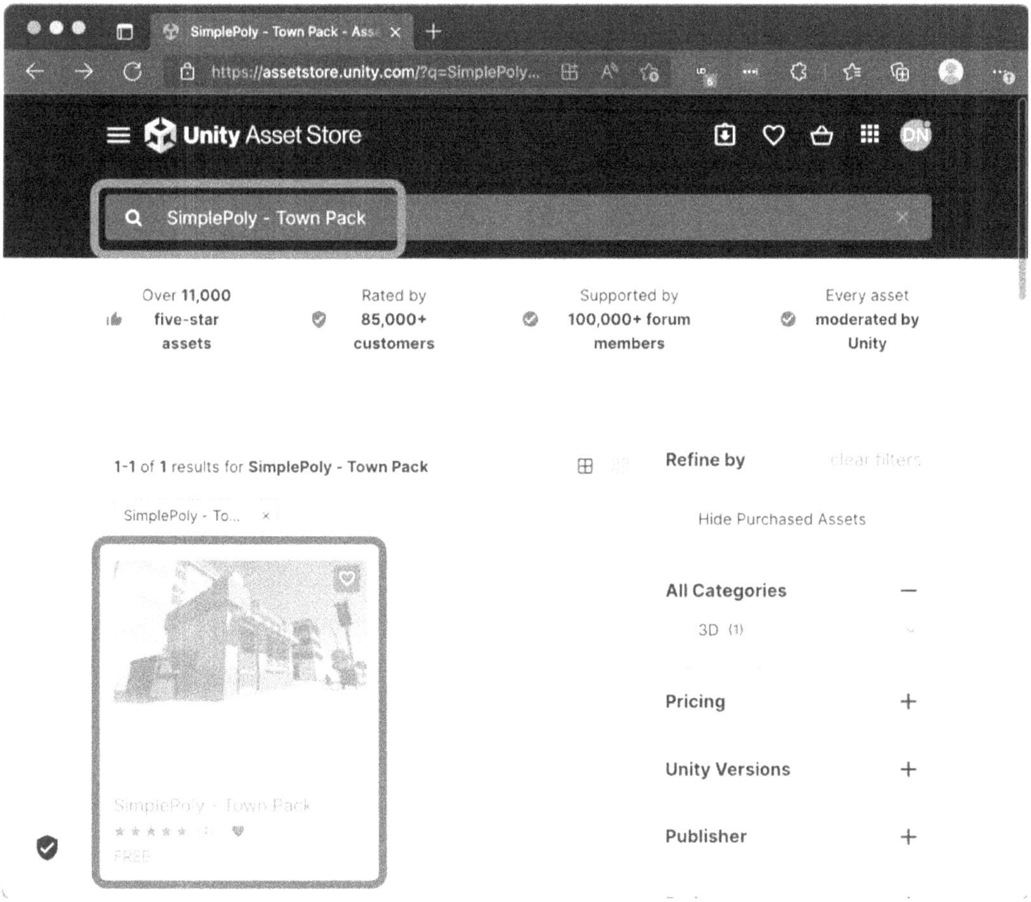

Figure 3.12 – SimplePoly - Town Pack asset

4. Click on the **Open in Unity** button.

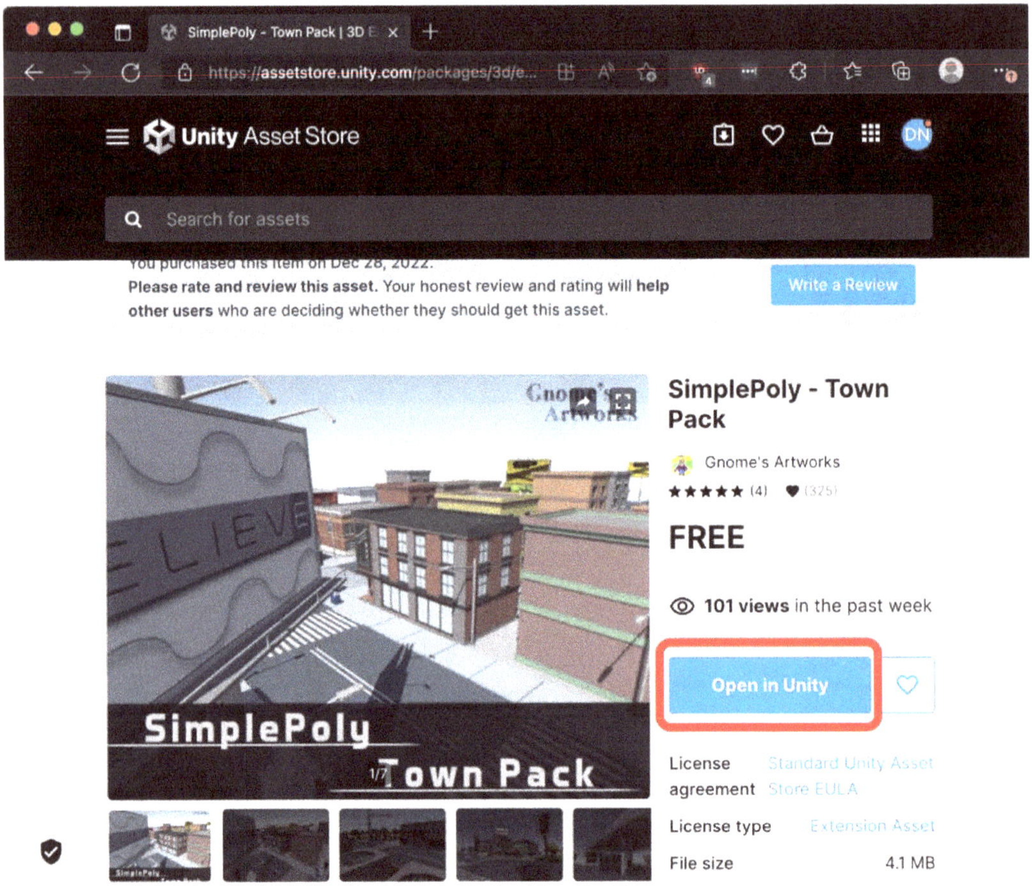

Figure 3.13 – SimplePoly - Town Pack download page

5. The **Package Manager** window will open with the newly downloaded asset selected. Click on the **Download** button.

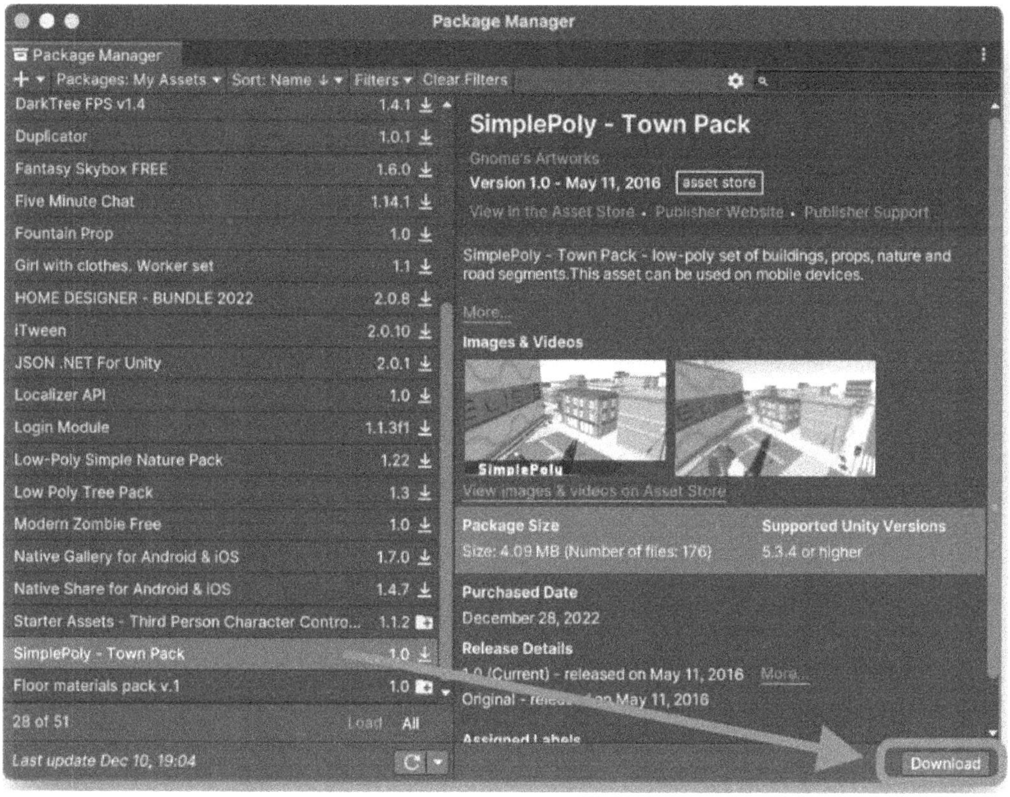

Figure 3.14 – SimplePoly - Town Pack in Package Manager

6. Well, we have already downloaded the asset; now we just need to add it to the project. Click the **Import** button.

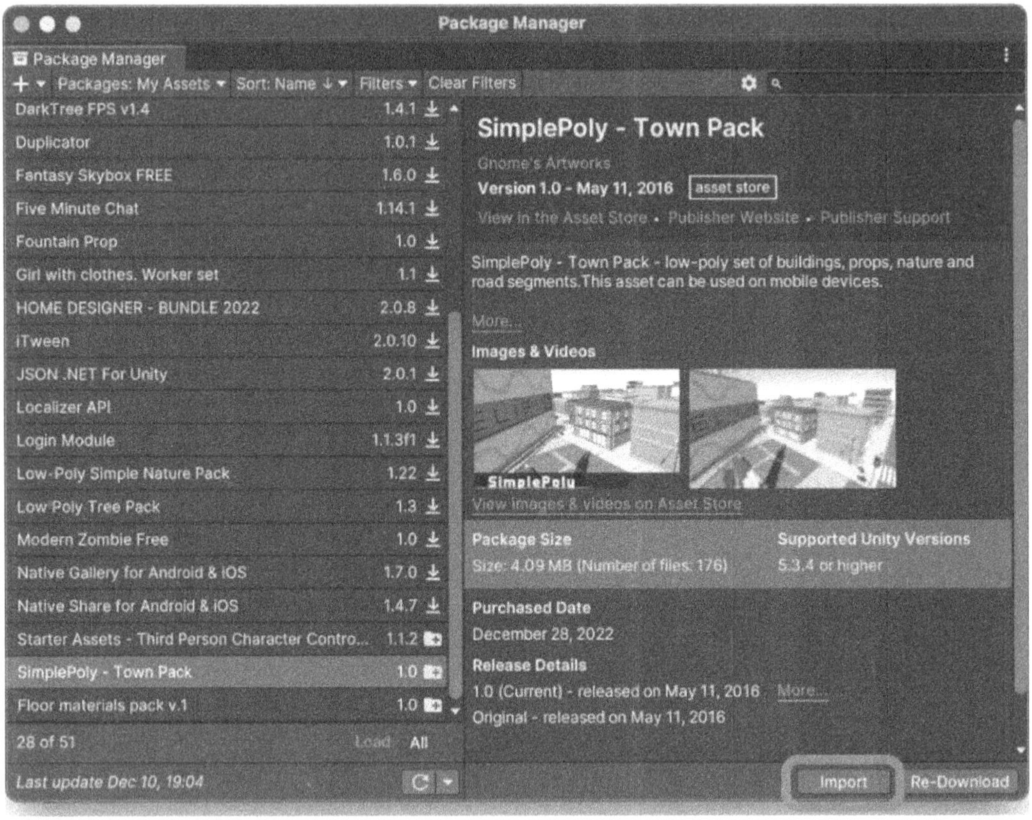

Figure 3.15 – SimplePoly - Town Pack freshly downloaded and ready to import

7. You may receive a pop-up warning that some imported files need automatic correction to conform to the new Unity API. Click on the **Yes, for these and other files that might be found later** button.

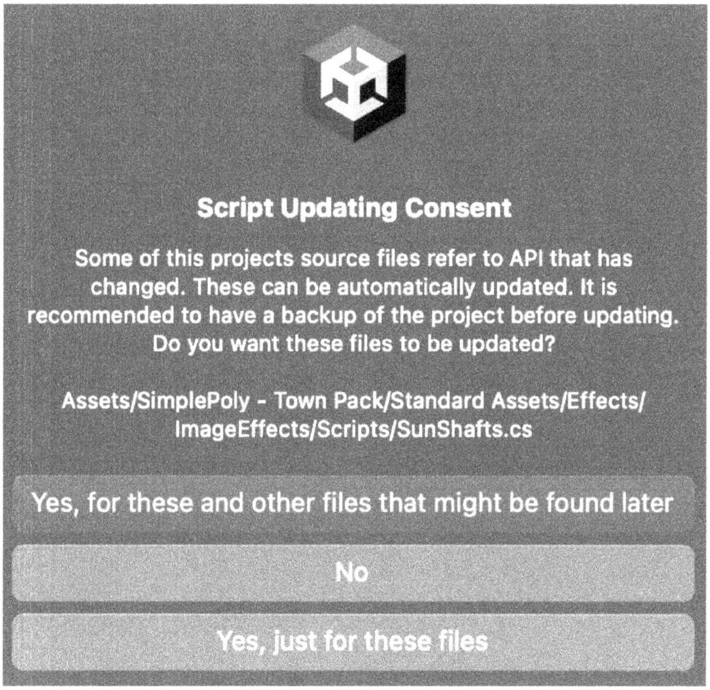

Figure 3.16 – Confirmation window for scripts upgrade

Okay, we have managed to download and install the buildings we are going to need for our little town. Next, we are going to start assembling buildings in the scene. Keep in mind that I'm going to place buildings according to my point of view (I know I have very bad taste).

The position, rotation, and scale of the buildings are not fixed and can be varied if you wish. Principally, they must be placed so that the player cannot leave the scene and fall into the void.

Assembling the buildings

As you may have noticed, Unity has created a new folder in our **Project** panel called `SimplePoly - Town Pack`.

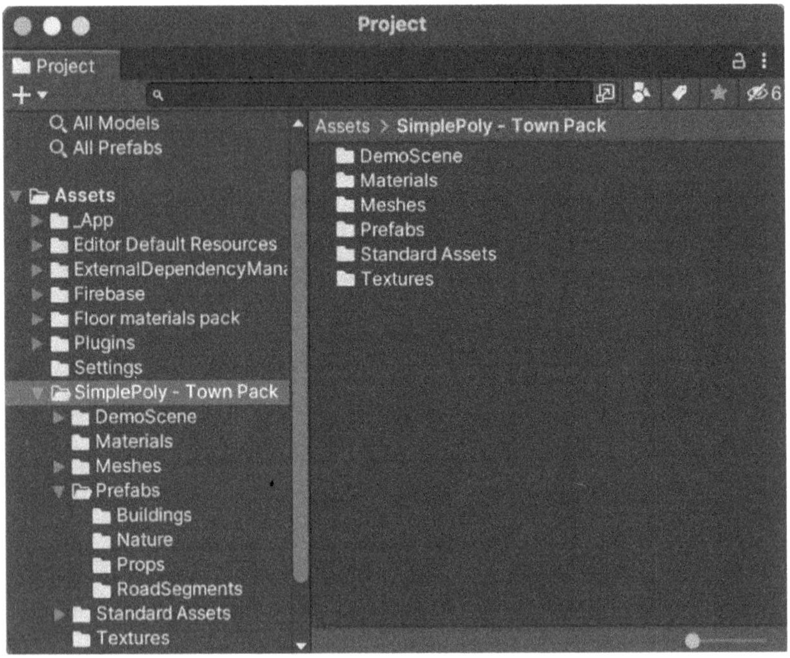

Figure 3.17 – SimplePoly - Town Pack folder

Inside, we have everything you need; the first thing we are going to do is see whether the materials are already prepared for URP rendering or need correction. To do this, view the content of **Assets** | **SimplePoly - TownPack** | **Materials**.

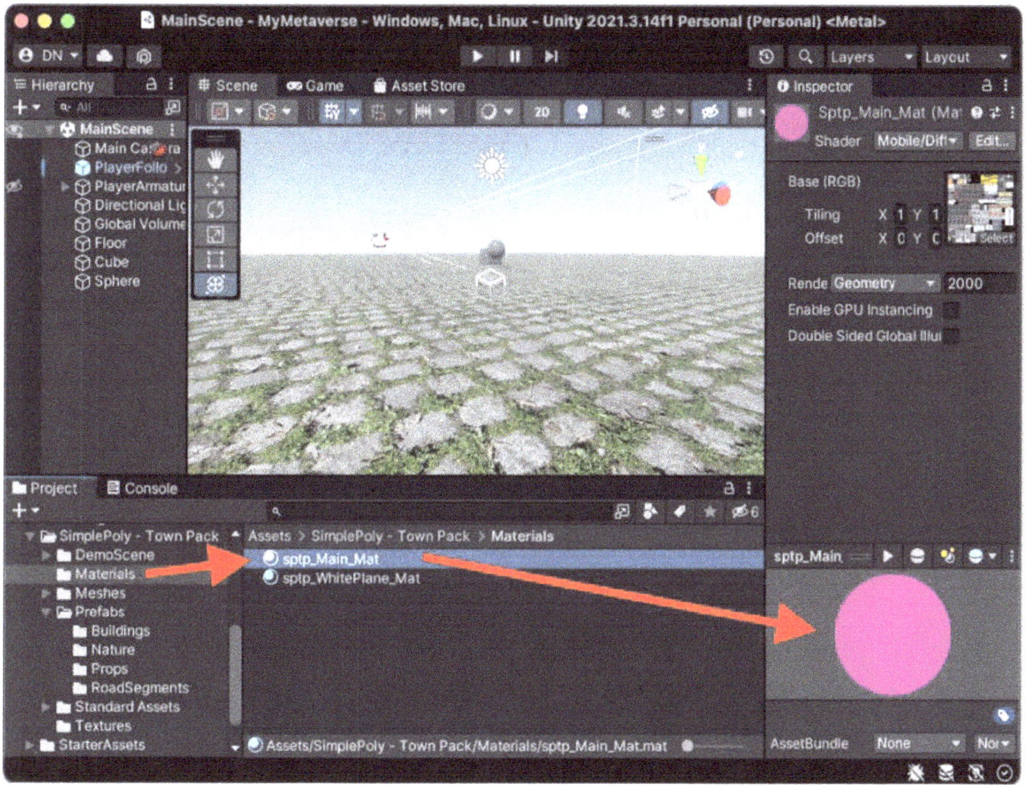

Figure 3.18 – SimplePoly - Town Pack | Materials folder

We see again that the materials are previewed in fuchsia or pink, which is no problem; we proceed to apply a correction as we did previously. To do this, select the two materials from the `Materials` folder, and in the main menu bar, click on **Edit** | **Rendering** | **Materials** | **Convert Selected Built-in Materials to URP** and confirm by clicking the **Proceed** button.

Well, if we return again to the contents of the `SimplePoly - Town Pack` folder we see that there is a subfolder called `Prefabs`, and inside, in a very organized way, the author of this asset has grouped the different assets that we can use into folders with very intuitive names.

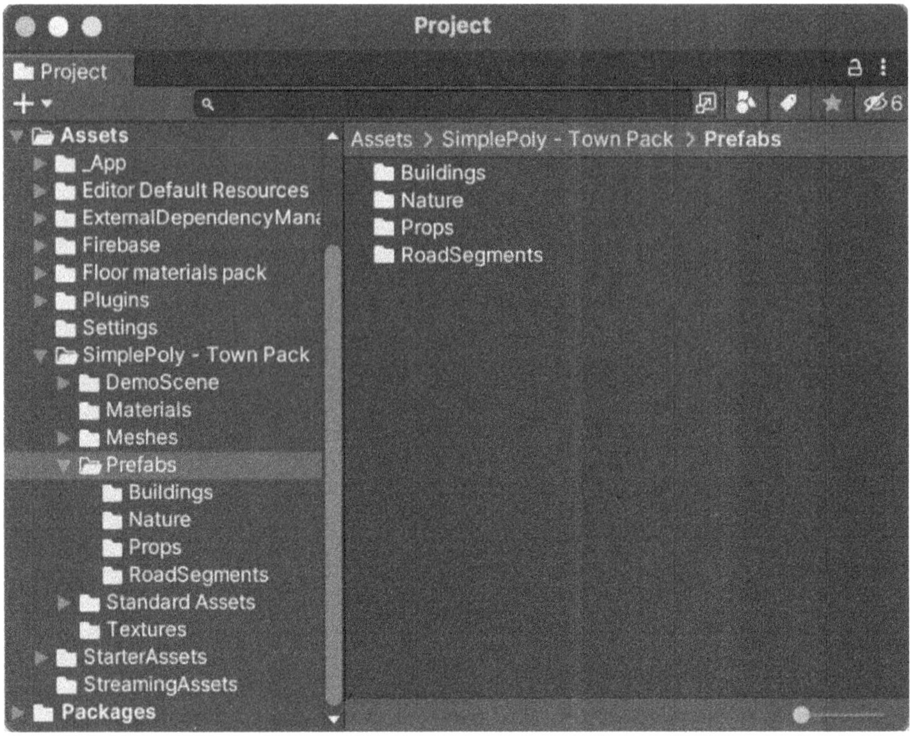

Figure 3.19 – SimplePoly - Town Pack | Prefabs folder

Before we continue, let's take a look at the structure of the organized folders and what types of Prefabs can be found inside them.

- `Buildings`: Inside this folder, we will find all the available buildings; they are already prepared in Prefab format, which means that they are ready to be dragged directly into the scene and enjoyed without the need for extra configuration
- `Nature`: A small collection of trees and stones in Prefab format
- `Props`: Prefabs of objects that we can find on the street, such as bins, traffic lights, street lamps, and so on
- `RoadSegments`: Pieces of roads in Prefab that can be combined with each other to compose a complete asphalt road

Now that you know a little more about this magnificent (and free) asset, let's proceed to place buildings on the scene. I remind you that the buildings you will see in the figures do not have to be exactly in the same position in your project. Also, if you prefer, you can place another model if you do not like one.

Let's start placing buildings, follow these steps:

1. We go to **Assets | SimplePoly - Town Pack | Prefabs | Buildings**.
2. I have decided to start with the first building that appears in the list. Select it with a left click, and without releasing the mouse, drag and drop it more or less in the position as seen in the following figure. As you can see, if you select a Prefab, the preview will load to the right of the editor.

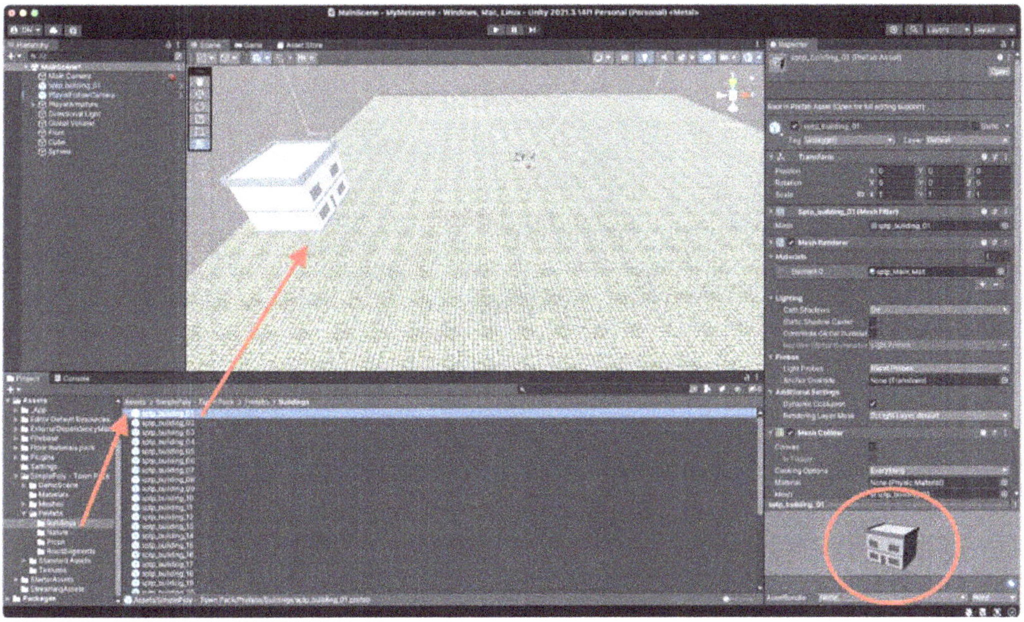

Figure 3.20 – The first building newly added to the scene

3. You can repeat *step 2* with the other buildings, placing them close enough together that the player cannot throw himself into the void.
4. When we reached the corner at which we must change the direction of the buildings, I decided to rotate the building a little to give a more oval appearance to the square.

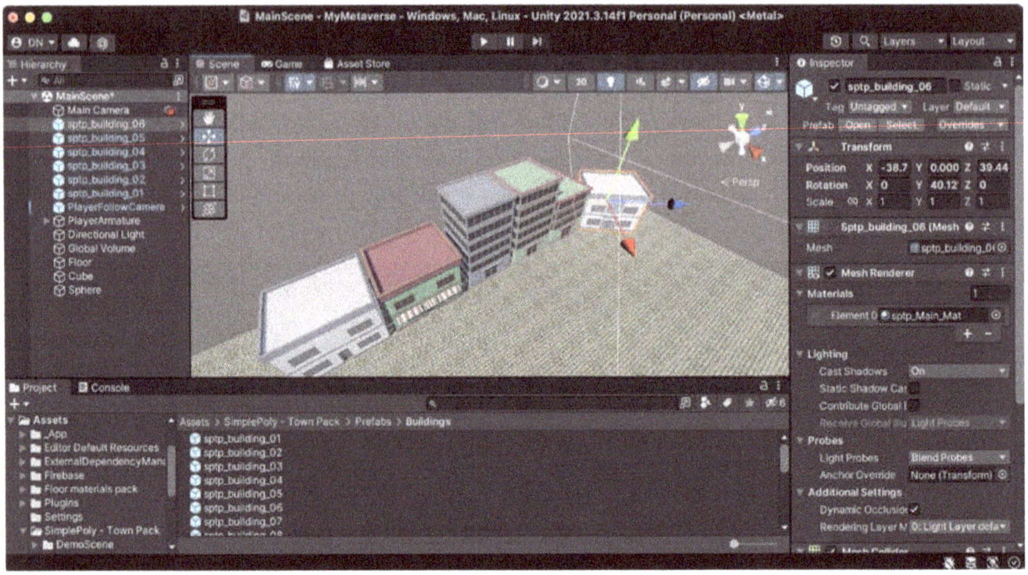

Figure 3.21 – Adding a building on the first corner

5. To rotate the building a little, first, place it as you have done with the previous ones. Then, click on the **Rotate** tool, and with a left click of the mouse, move the green vector and turn it to your liking.

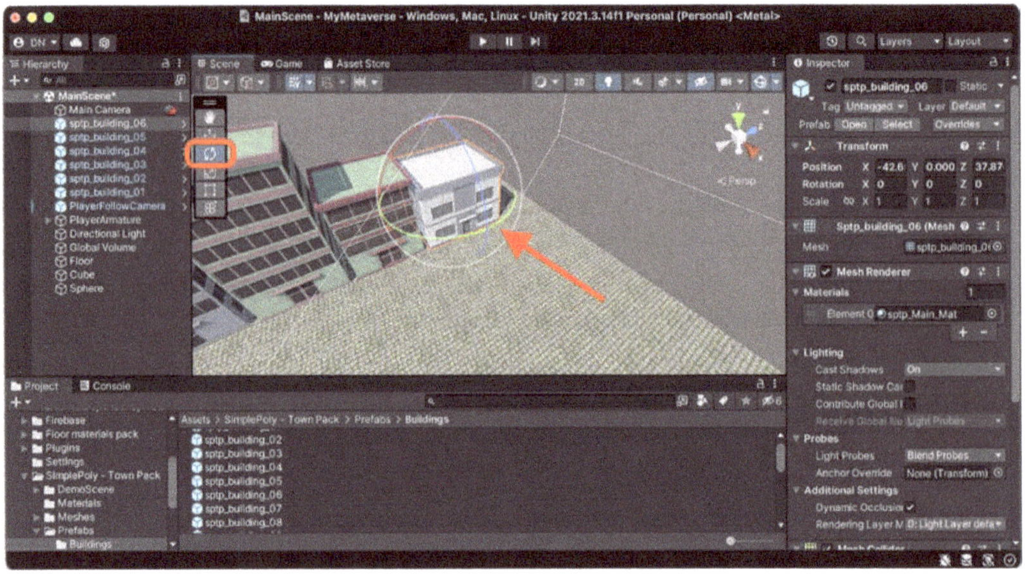

Figure 3.22 – Using the Rotate tool on a building

6. Once you have turned the building, click on the **Move** tool; you can drag the blue and red arrows to place it where you like.

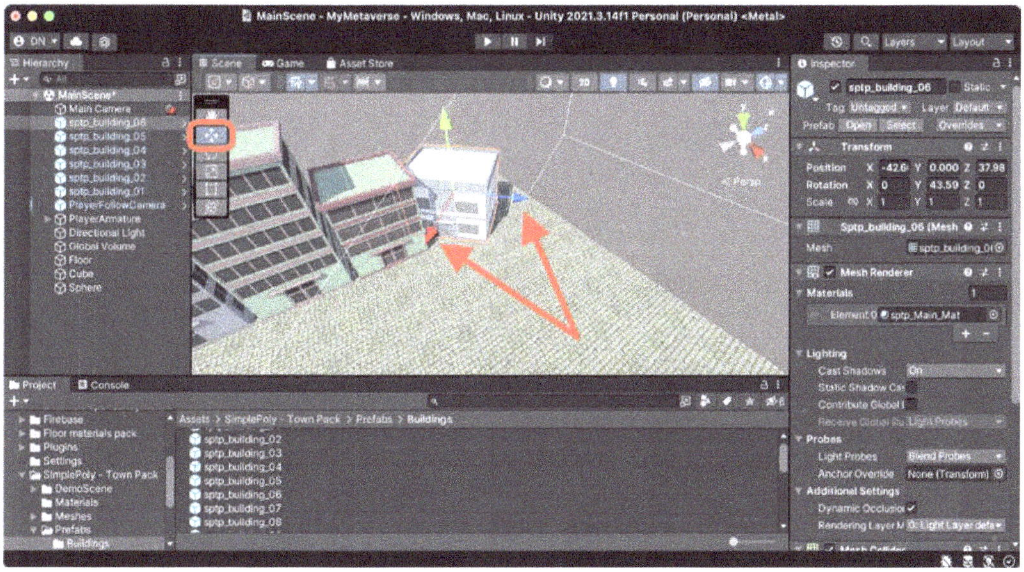

Figure 3.23 – Using the Move tool on a building

7. Once you have placed the corner building, you can continue adding buildings to the next flank. As you can see, we are creating an enclosed space. You will have to rotate the buildings for this flank since when dragging them into the scene, they are placed facing to the right. No problem; follow the same directions as *steps 5* and *6* to make it look something like this.

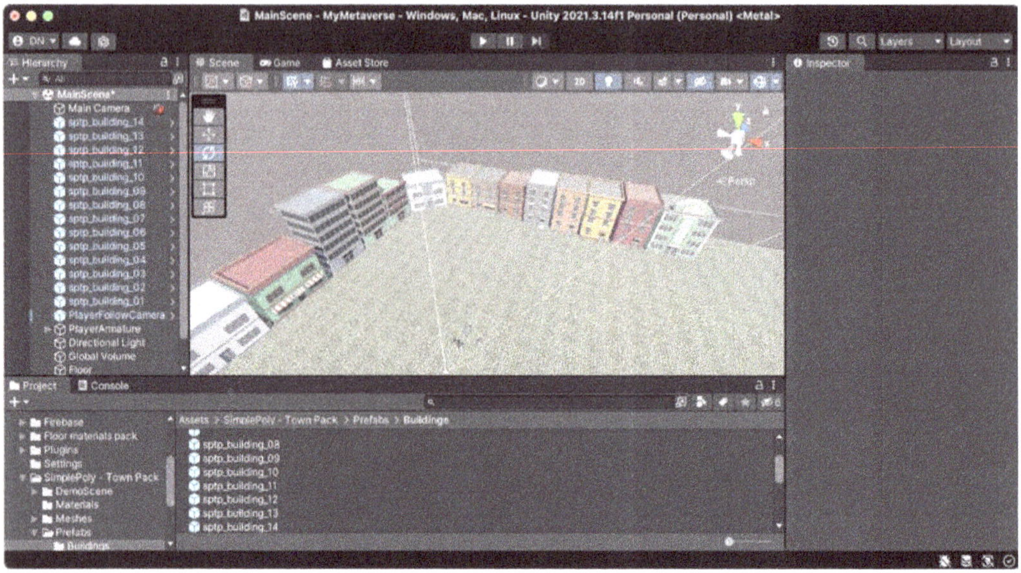

Figure 3.24 – Adding more buildings to the second flank

8. When we reach the second corner, repeat *steps 5* and *6* with the next building so that it is somewhat rotated.

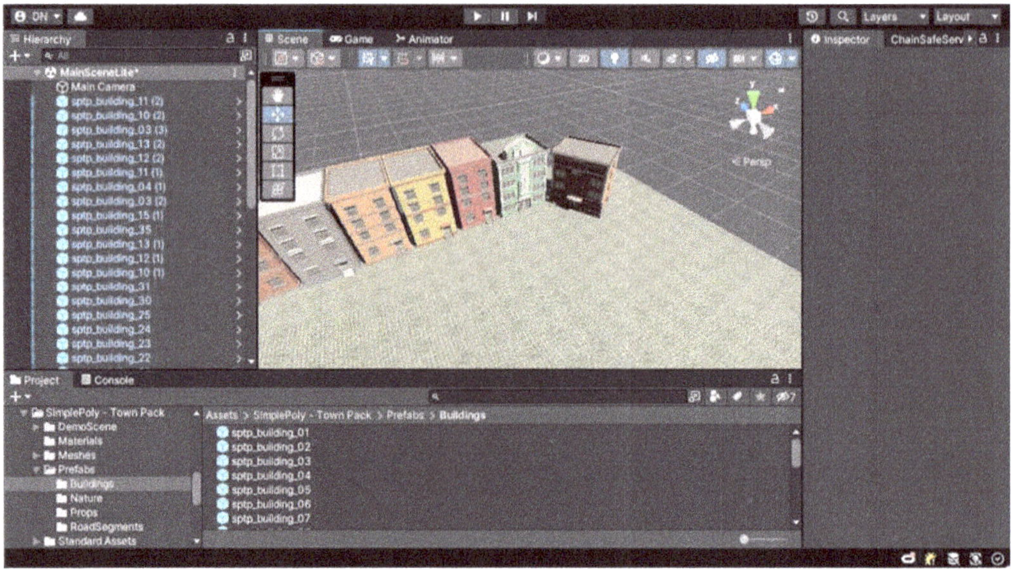

Figure 3.25 – Adding a new building to the second corner

9. We continue adding buildings to the third flank of our square. If you have followed the same order of buildings, you will have noticed that there is a gas station in this row of buildings. I have decided to place it as seen in the following figure.

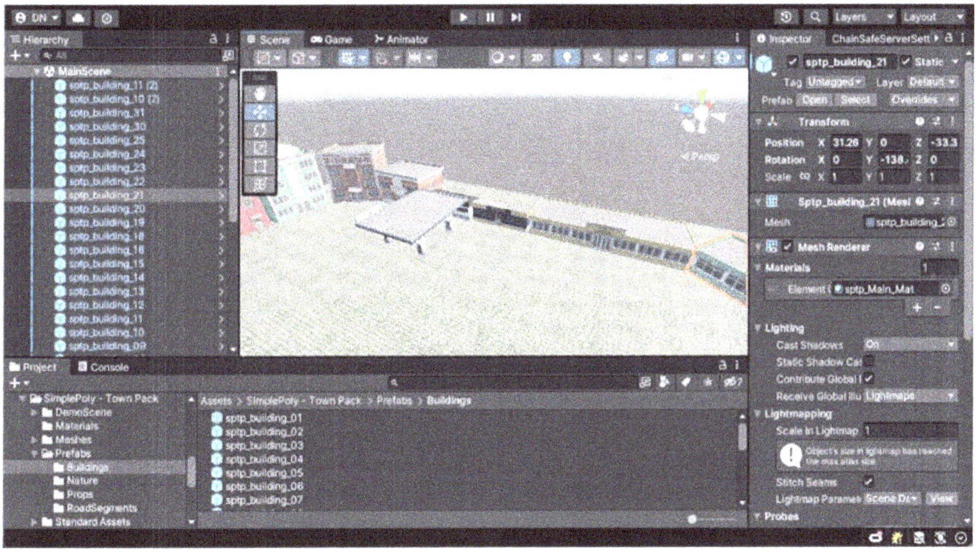

Figure 3.26 – Third flank infill of new buildings

10. We arrive at the third corner of the square. I have decided to place a supermarket in this corner; you can apply the same *steps 5* and *6* to place it.

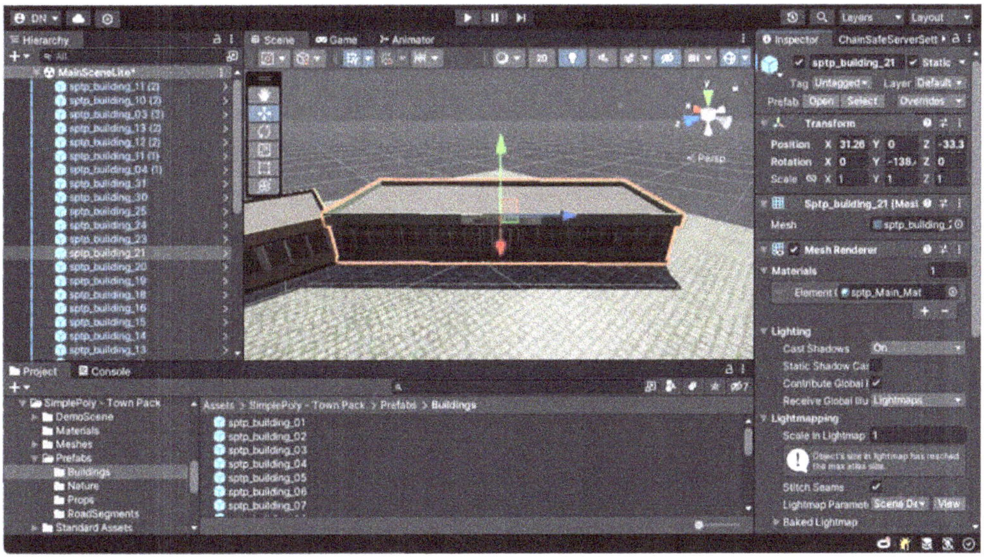

Figure 3.27 – Adding a new building to the fourth corner

We are almost finished; we continue to place buildings on this flank until we reach the last corner.

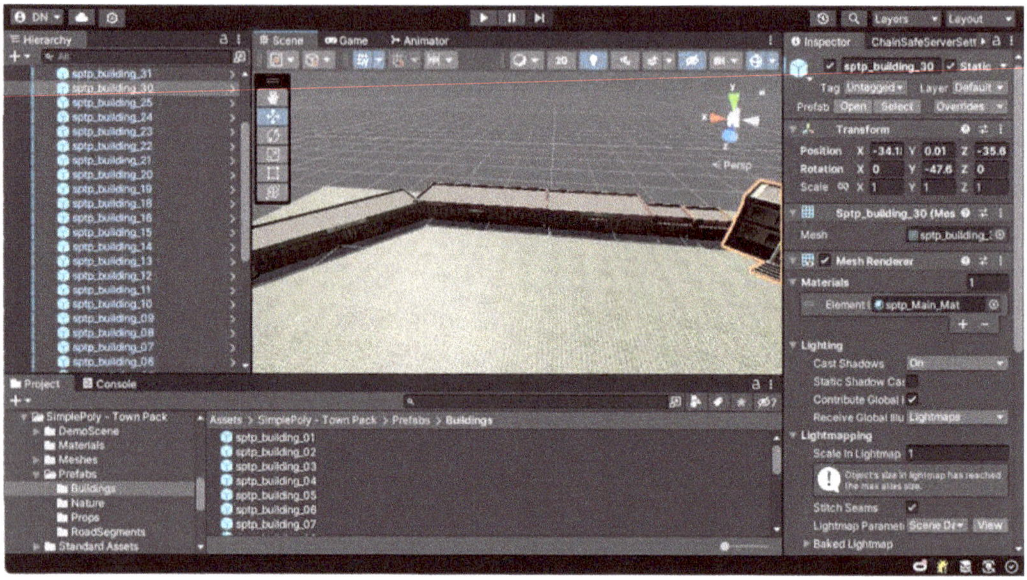

Figure 3.28 – Closing the last flank with new buildings

11. To finish the last corner, I have decided to place the buildings `sptp_building_30` and `sptp_building_31`, but you are free to place the ones you want. In my case, the last corner has been set up like this:

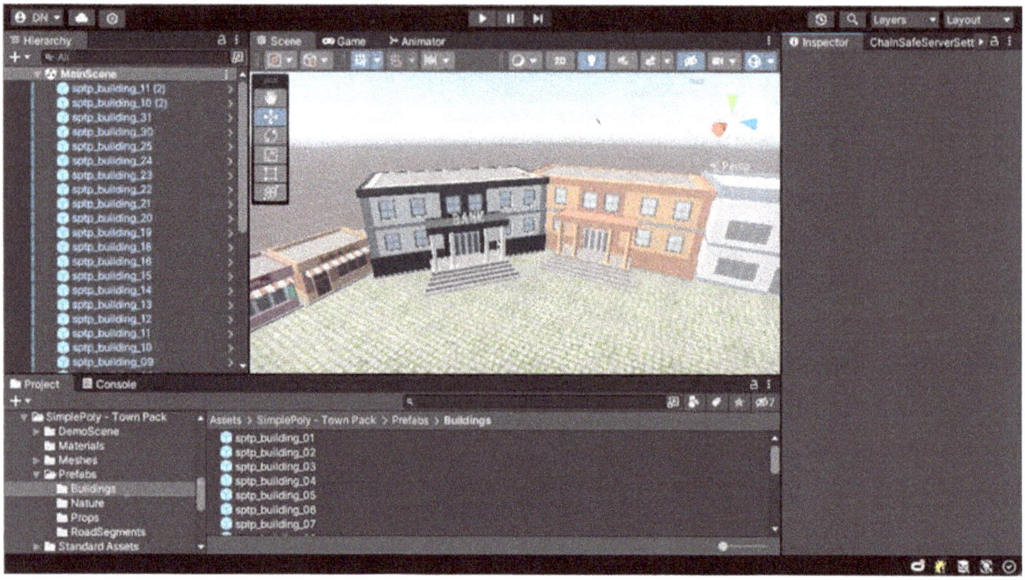

Figure 3.29 – Adjusting the last corner with a rotated building

Congratulations – you have finished placing the buildings in our beloved village. Now it's time to try out how it looks in first person; you can click the **Play** button and move with your player throughout the scene. Don't forget to save scene changes often.

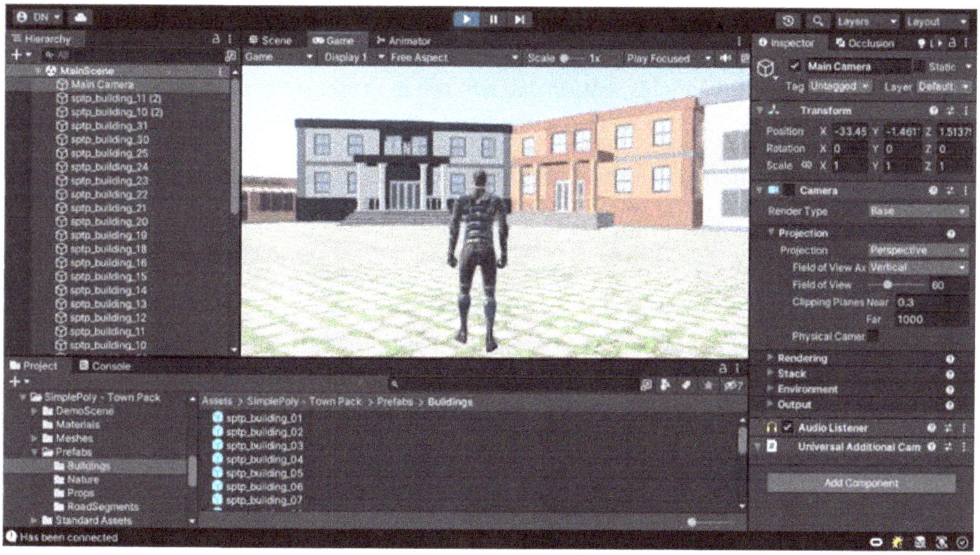

Figure 3.30 – Previsualization in Game Mode

As you may have noticed, the player is unable to go through the buildings. This is because the buildings we have used have a **Mesh Collider** component. If you click on one of them in the scene and look at its **Inspector** panel, you will see it:

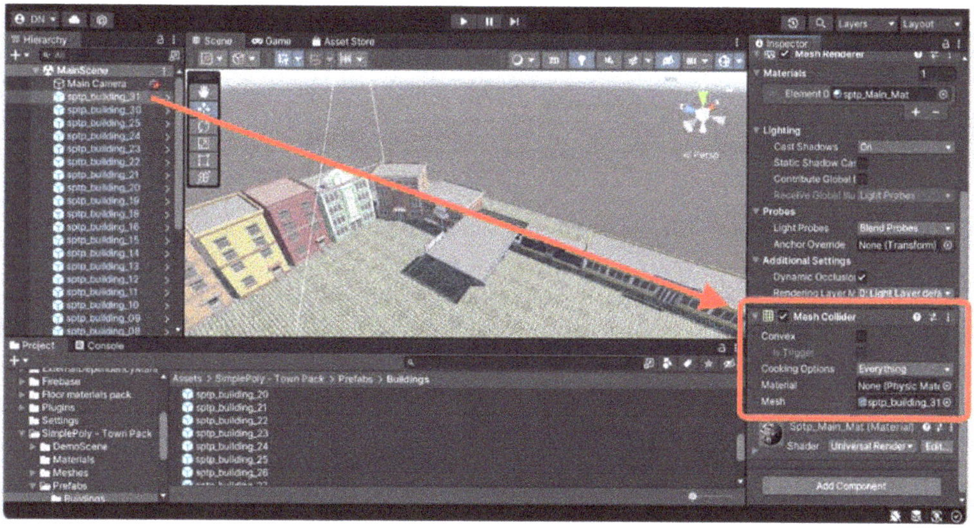

Figure 3.31 – The Mesh Collider component

Mesh Collider is a type of collider that takes the same shape as the object itself; this helps any polygon detect the collision. For example, if a building has columns and stairs, the player can automatically climb the stairs thanks to **Mesh Collider**, which will wrap the object in its entirety, offering a complex collision.

This type of collider is not optimal in large scenes since it needs more processors to calculate collisions in complex objects but, to begin with, it is fine. In *Chapter 4, Preparing Our Home Sweet Home: Part 2*, in the *Understanding optimizations* section, we will see in more depth what kind of alternative optimizations can be applied to improve the performance of our metaverse in this scene.

Now, we can finish the creation of the scene, but I think that there is room for improvement. We can make the scene even more beautiful and then we will see what other details we can add.

Customizing sidewalks

Observing the result, it has certainly been very good. The player can move and observe some beautiful buildings, and our little village looks great. However, the **SimplePoly - Town Pack** asset that we have downloaded offers more interesting objects that we can use. Let's add a special touch to the scene.

I would like to add a sidewalk under the buildings all along. Fortunately, we have a variety of Prefabs for this. Open the folder in **Assets | SimplePoly - Town Pack | Prefabs | RoadSegments.**

If you currently see the assets in vertical list format, slide the slider to the right, as shown in the following figure. With this, you will make the icons larger and you can easily preview the Prefab.

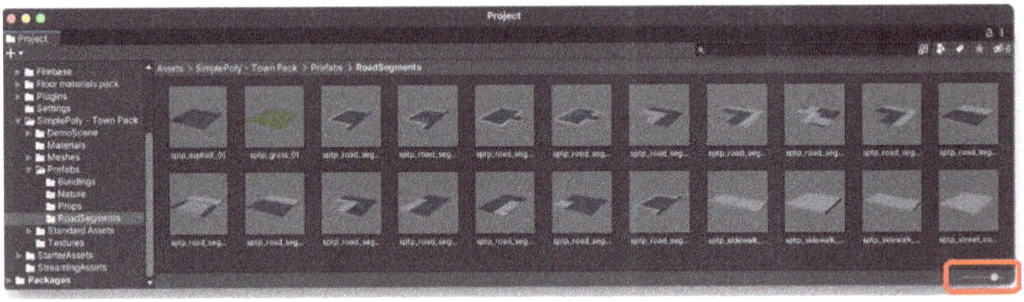

Figure 3.32 – Changing icons' size in the Project panel

As we said before, you are free to use the objects that you like the most to decorate the scene. This will not influence the behavior we are looking for in this chapter. I will personally use the Prefab called `sptp_sidewalk_segment_03`; it is a basic sidewalk.

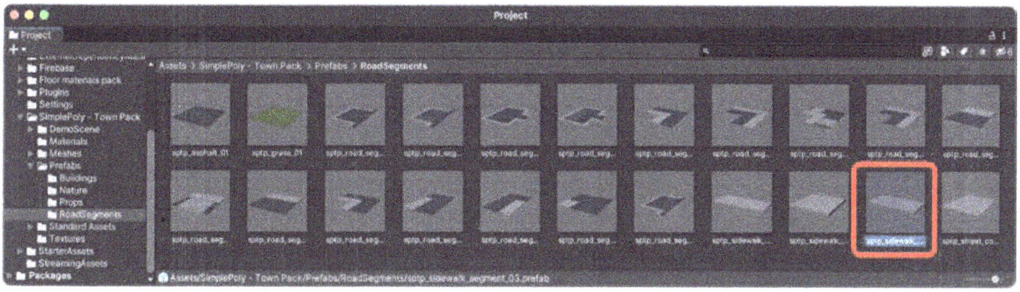

Figure 3.33 – Selected sidewalk Prefab

To add the sidewalks, the procedure is that used previously to add buildings to the scene:

1. We drag the `sidewalk` Prefab named `sptp_sidewalk_segment_03` (or the one you chose) into the scene.
2. We use the **Move** tool to place it as follows:

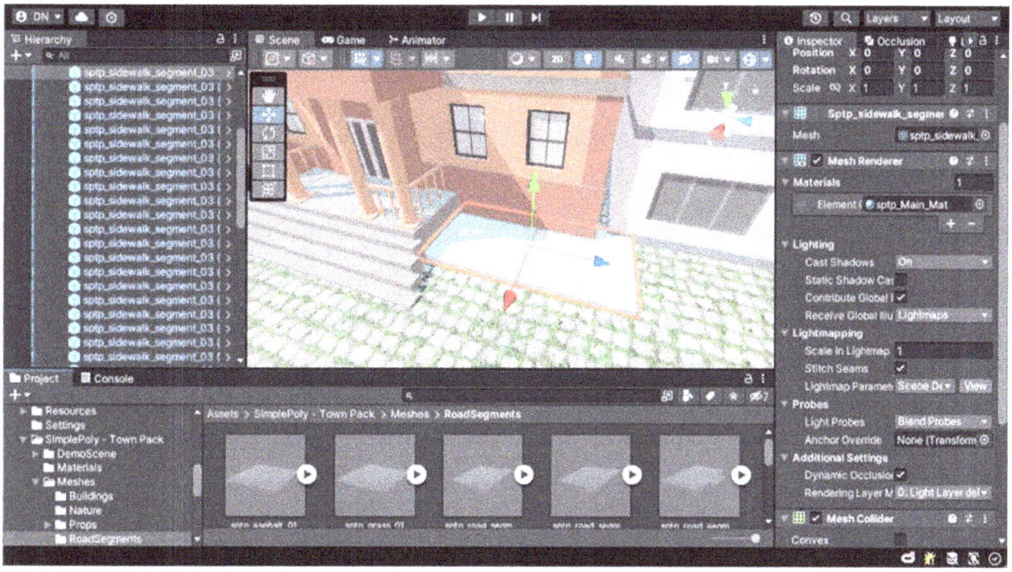

Figure 3.34 – The first sidewalk Prefab in the scene

3. To duplicate the same object and maintain its position, rotation ,and scale, you can right-click on the object in the **Hierarchy** panel and select the **Duplicate** option. This will create an exact clone.

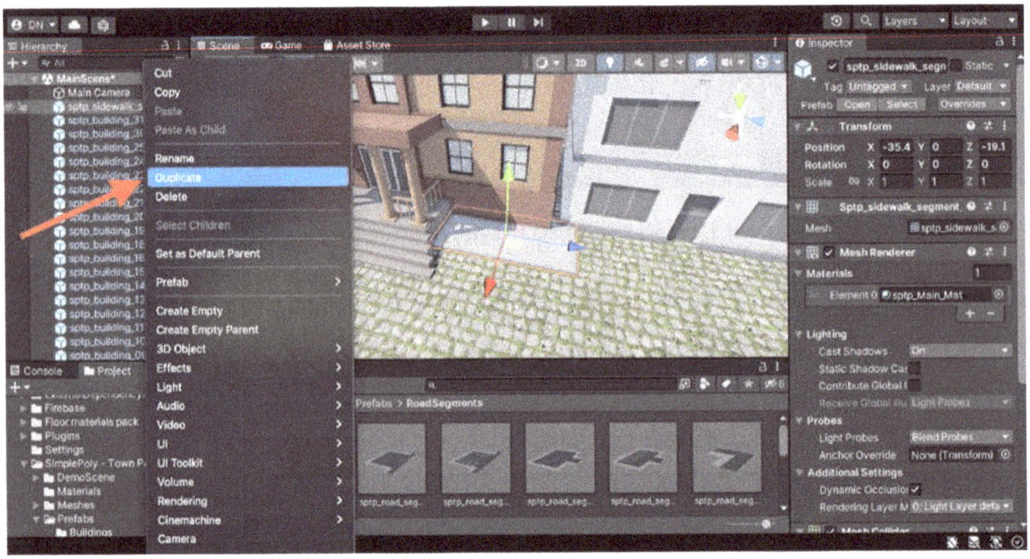

Figure 3.35 – Duplicating Prefab

> **Tip: focusing on the GameObject**
>
> Sometimes, when you want to zoom in on an object in the scene with the mouse wheel, you'll notice that the zoom doesn't advance the amount you want. If you press the *F* key, you will make a quick focus on the object you have selected in the **Hierarchy** panel.

4. A new object will be created in the **Hierarchy** panel named `sptp_sidewalk_segment_03 (1)`. With this GameObject selected, use the **Move** tool to place it next to the first one. You can move it to the right using the blue arrow.

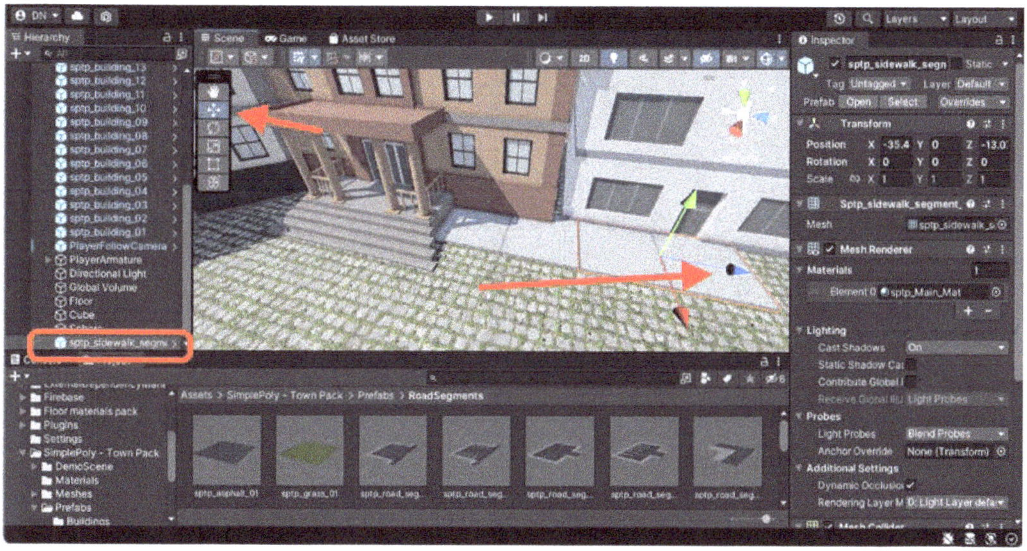

Figure 3.36 – Moving a duplicated Prefab

Tip: duplicating multiple GameObjects

The task of duplicating objects can be tedious, but there is a very simple trick. You can select multiple objects from the **Hierarchy** panel. For this, click the first, then press the *Shift* key, and, without releasing it, click the second, and so on until you select all the items you want to duplicate.

Finally, when you have all of them selected, right-click on any of the selected objects and select the **Duplicate** option. All selected objects will automatically be duplicated. When the new objects appear in the **Hierarchy** panel, keep in mind that the new ones will be selected. Then, you can use the **Move** tool.

If you try, for example, to move something to the right with the **Blue** arrow, you will notice, as shown in *Figure 3.37*, that everything will move all at once. This is fantastic and will save you a lot of time.

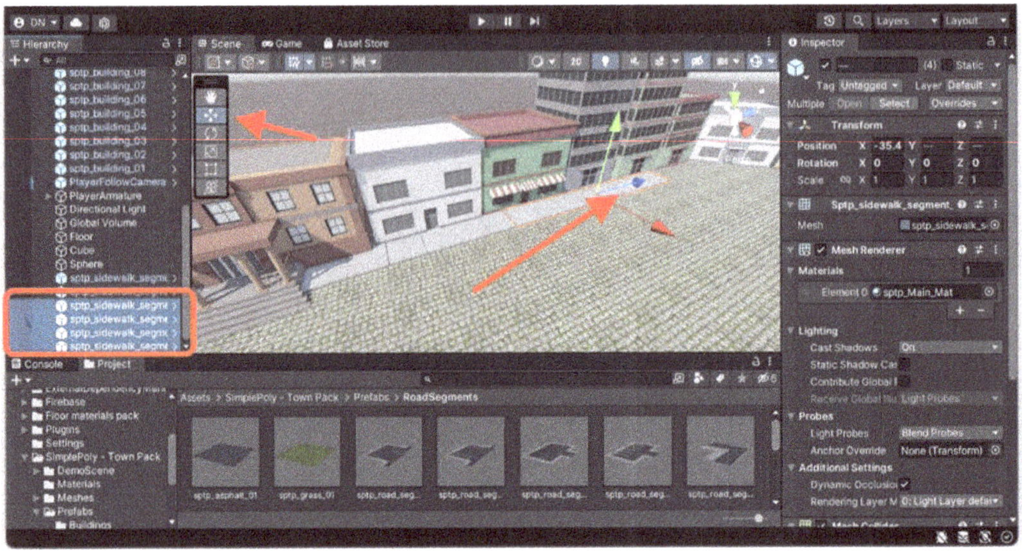

Figure 3.37 – Multiple GameObject selection

5. Repeat the previous steps until you have the sidewalk placed along this entire flank. When you are finished, you will have something similar to this:

Figure 3.38 – First flank with sidewalks

When we reach the first corner, we can perform the same action as with the buildings; we can rotate the piece.

6. We select the last Prefab that we have added, the one we will use for the corner. Use the **Rotate** and **Move** tools until you get the desired position, whatever you like. In my case, it was adjusted as shown here:

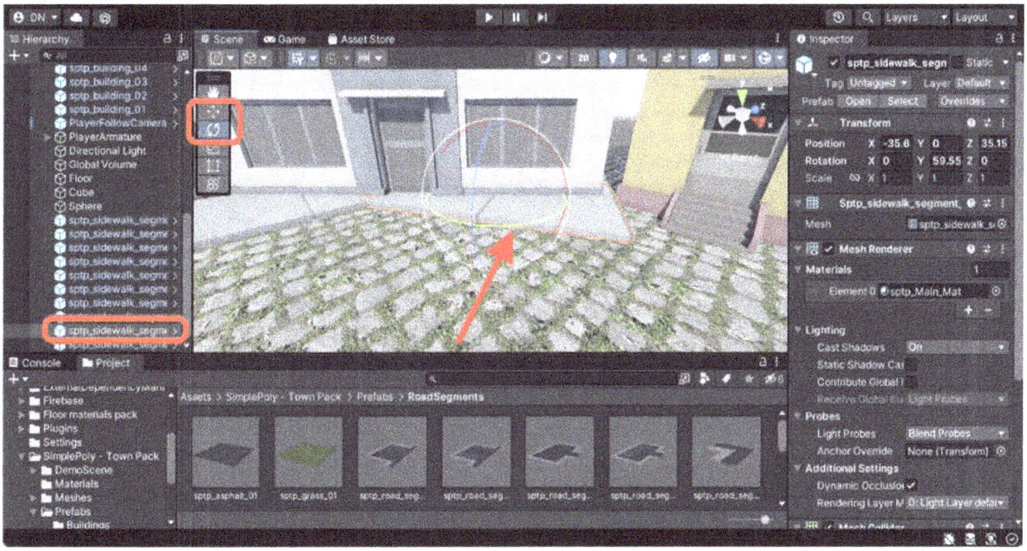

Figure 3.39 – Rotating the sidewalk in the corner

7. We start adding the sidewalk to the second flank, which you can do in two ways – you can add a new Prefab of the `sptp_sidewalk_segment_03` object we have used so far, or you can also duplicate the previous object and, using the **Rotate** tool, place it as follows to start covering the second flank.

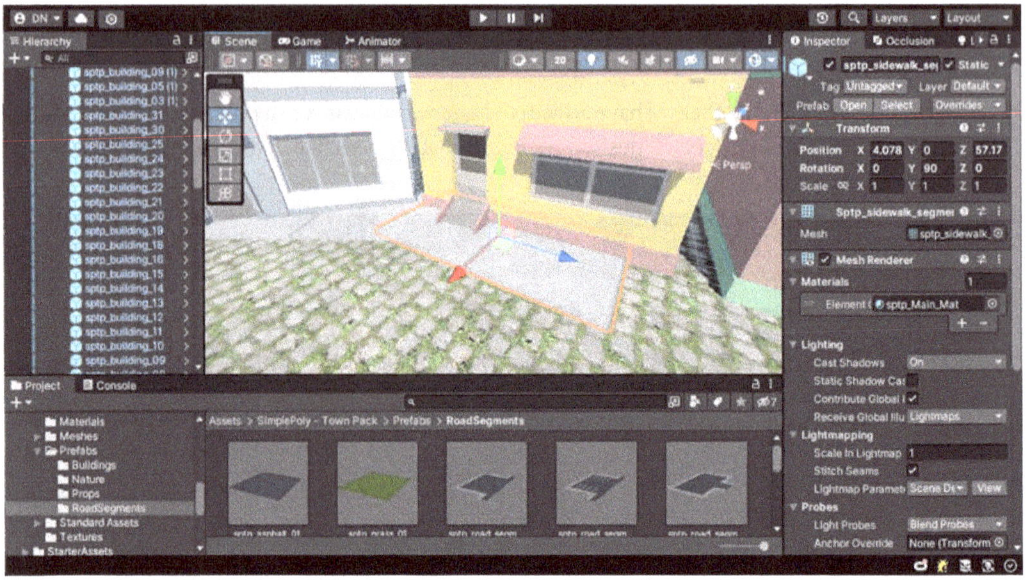

Figure 3.40 – Adding more sidewalk Prefabs

8. We continue to duplicate the first sidewalk of this second flank and, using the **Move** tool, we place one after the other. When you reach the end, you will have something similar to this:

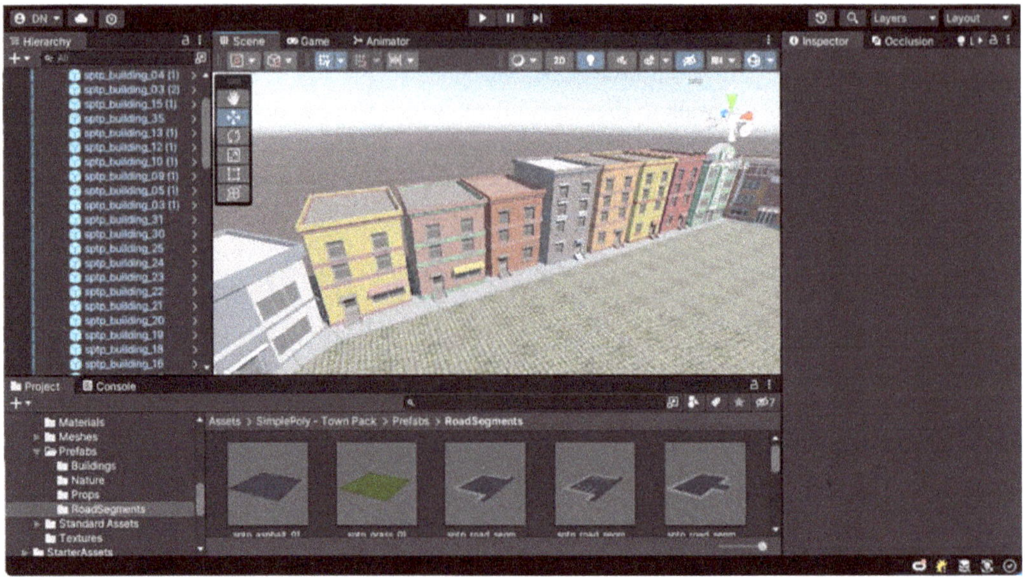

Figure 3.41 – Second flank with sidewalks

9. We reach the second corner and apply the same process; repeat *steps 6* and *7* to rotate the pieces. In my design, I will need two pieces since the corner is longer than the previous ones:

Figure 3.42 – Rotating the sidewalk Prefab in the corner

10. We start on our third flank. Easy, right? Now, you just have to repeat *steps 8* and *9* to continue adding pieces to this side.

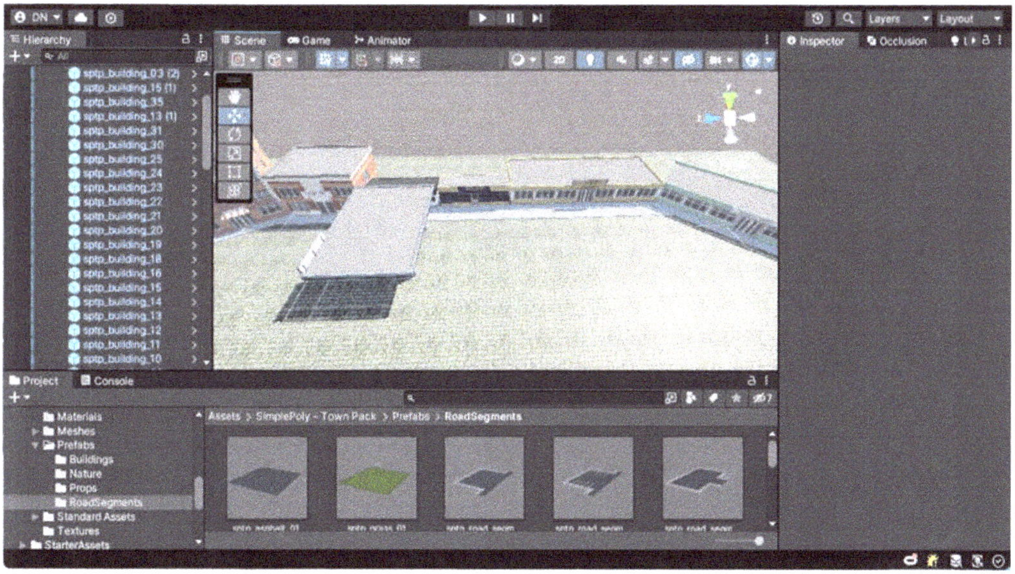

Figure 3.43 – Third flank with sidewalks

11. When you reach the next corner, repeat *steps 6* and *7*, add a new sidewalk Prefab, and apply some rotation. This corner is something special because the building is very long, so you must place more pieces than in previous corners.

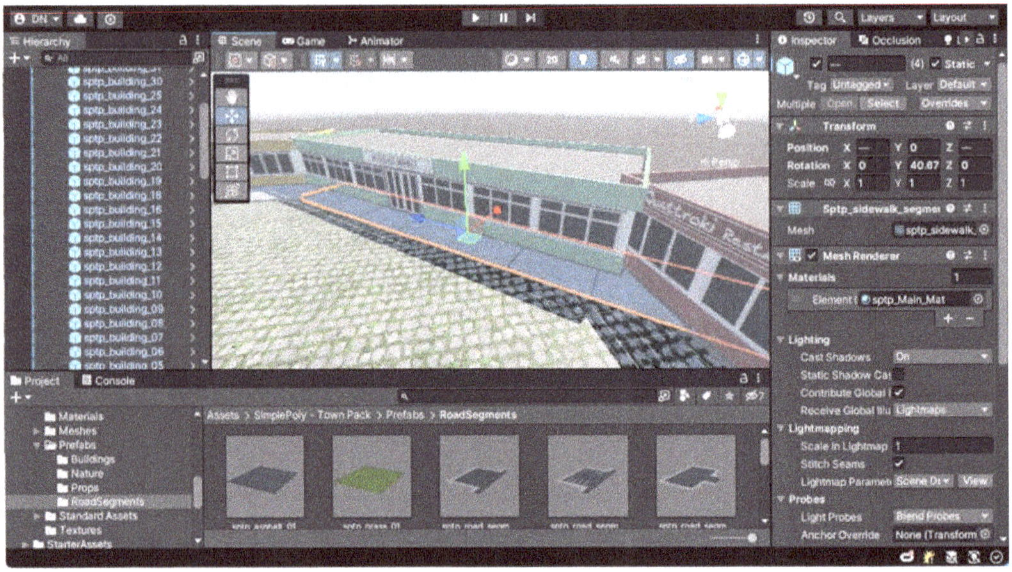

Figure 3.44 – Adding sidewalks to the corner

12. We repeat *steps 8* and *9* to continue filling this flank with sidewalk pieces until we reach the end.

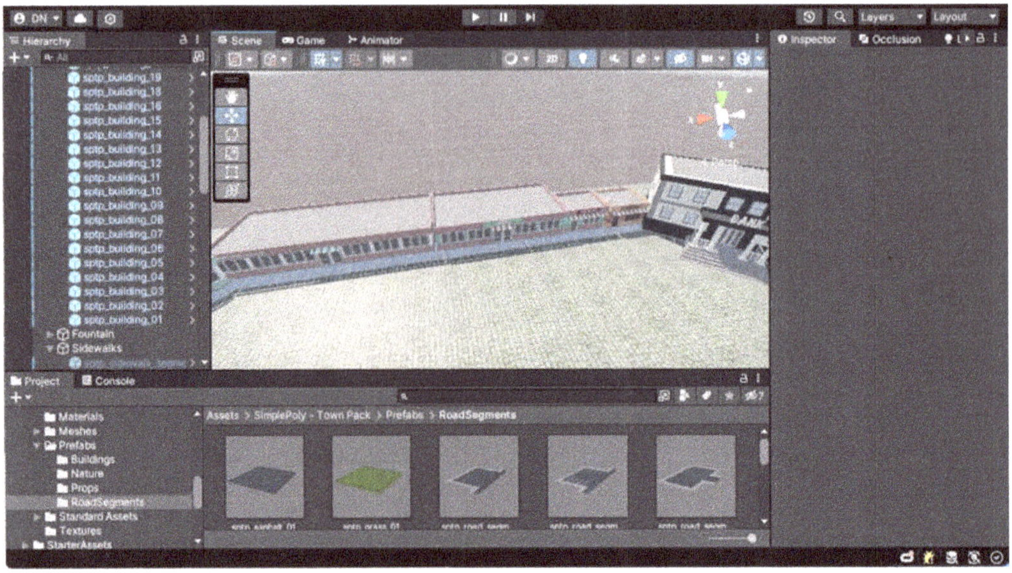

Figure 3.45 – Last flank with sidewalks

13. We are going to place the last pieces in the last corner, add a new Prefab, and apply a rotation as we have done in *steps 6* and *7*. This corner is long, so you must duplicate some more pieces to cover this part. You will have something similar to this:

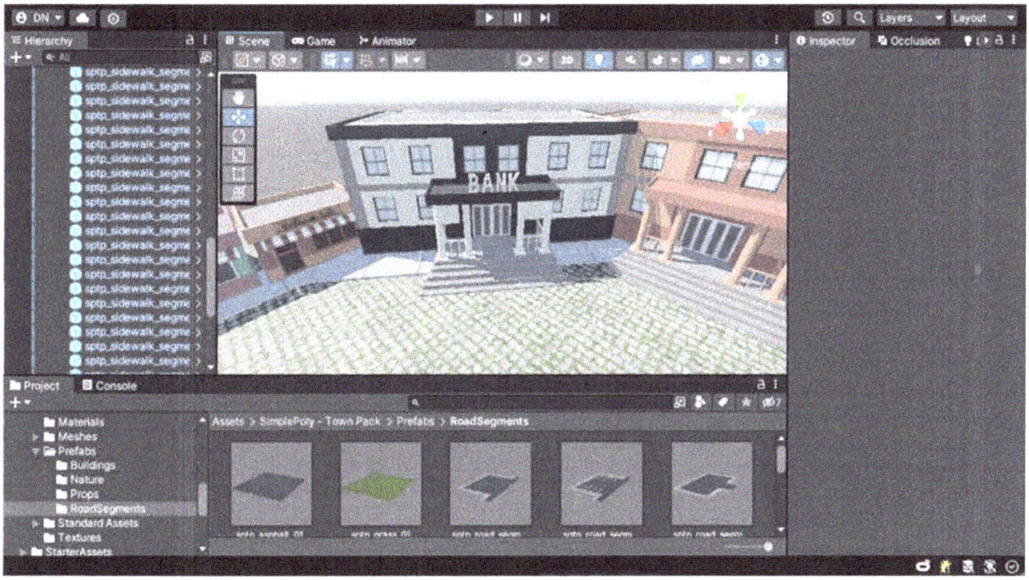

Figure 3.46 – Adding sidewalks to the corner

Congratulations! What a scene we are left with; you have completed the placement of sidewalks and it has been great. Now, you can click the **Play** button and see how it looks from the player's view. Don't forget to save your scene changes.

> **Tip: improving the design**
>
> As you place new objects in the scene, in the latter case with the sidewalks, you may notice some design problems in the buildings – for example, you may notice that they need to be moved to make them look better. There is no problem if you need to reposition the objects or even swap them for others that you like more. This task we are performing is purely visual and any variation will not affect the desired performance. I encourage you to explore your creative side and decorate the scene as you see fit.

Now that we have a scene full of buildings with a great sidewalk, we are going to give a magical touch to the sky. We like the one that comes by default, but it can be improved. Next, we will get a perfect skybox from Unity Asset Store for our virtual world.

Customizing the sky

Before continuing, we need to settle the knowledge of a new term – Skybox. We will read it many times throughout the book. Surely, you have seen beautiful skies that decorate the scene in hundreds of games throughout your life; this is called a **Skybox**. I will explain how it works.

A **Skybox** is a technique used in game graphics that consists of creating a cube with six faces. This cube is placed behind all the graphical elements of the game and is rendered around the entire scene. The main purpose of the Skybox is to give the illusion that the game has complex scenery on the horizon, which creates a sense of immersion for the player.

It is important to note that Skyboxes are rendered after all opaque objects in the game scene. This means that the main elements of the game, such as characters and solid objects, are drawn first and then the Skybox is placed as a background.

To create the Skybox, two types of meshes are used: a box with six textures, where each face of the cube is assigned a different image representing the sky and horizon from different perspectives, or a tessellated sphere, which is divided into multiple triangles to form a sphere and is also assigned corresponding textures.

What we see as a spherical sky is composed as follows:

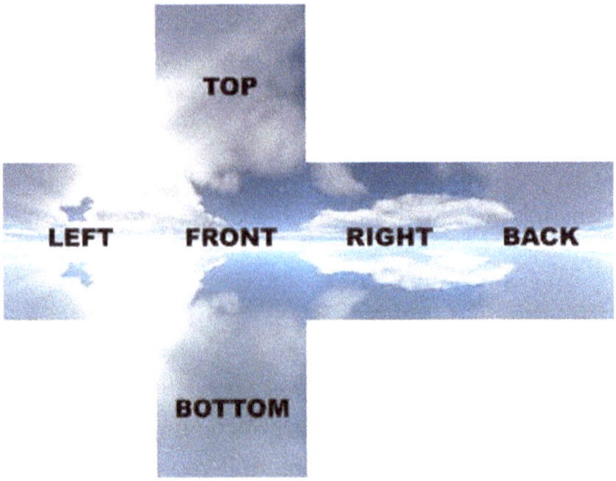

Figure 3.47 – Cubemap composition example

This six-sided cube, when closed and wrapped around our scene, will give the visual aspect of the natural sky.

Next, we will download a more impressive one that comes by default and apply it to the scene:

1. Go to Unity Asset Store by selecting `https://assetstore.unity.com`.
2. Type `Skybox` in the search engine and do not forget to sort the results by price, from lowest to highest, so that the free assets appear first. Select the asset called **Fantasy Skybox FREE**.

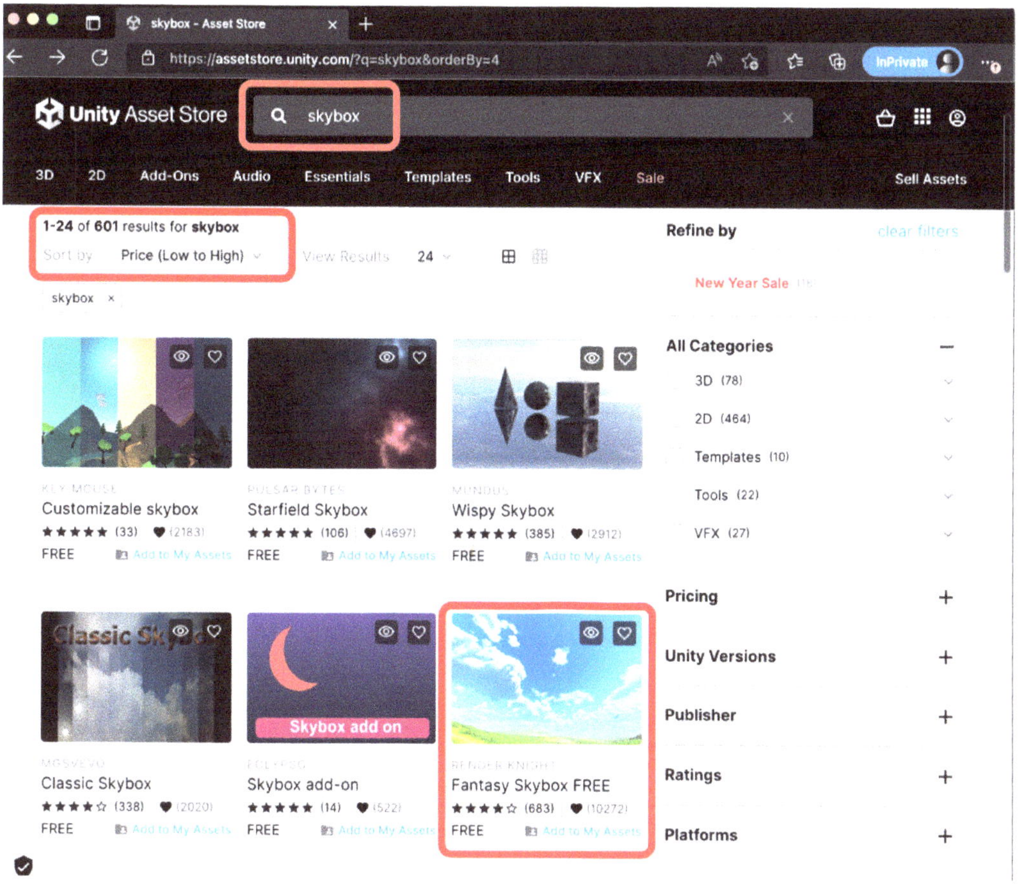

Figure 3.48 – Skybox search in Unity Asset Store

3. Click on the **Add to My Assets** button and then click the **Open in Unity** button that will appear on the same page.

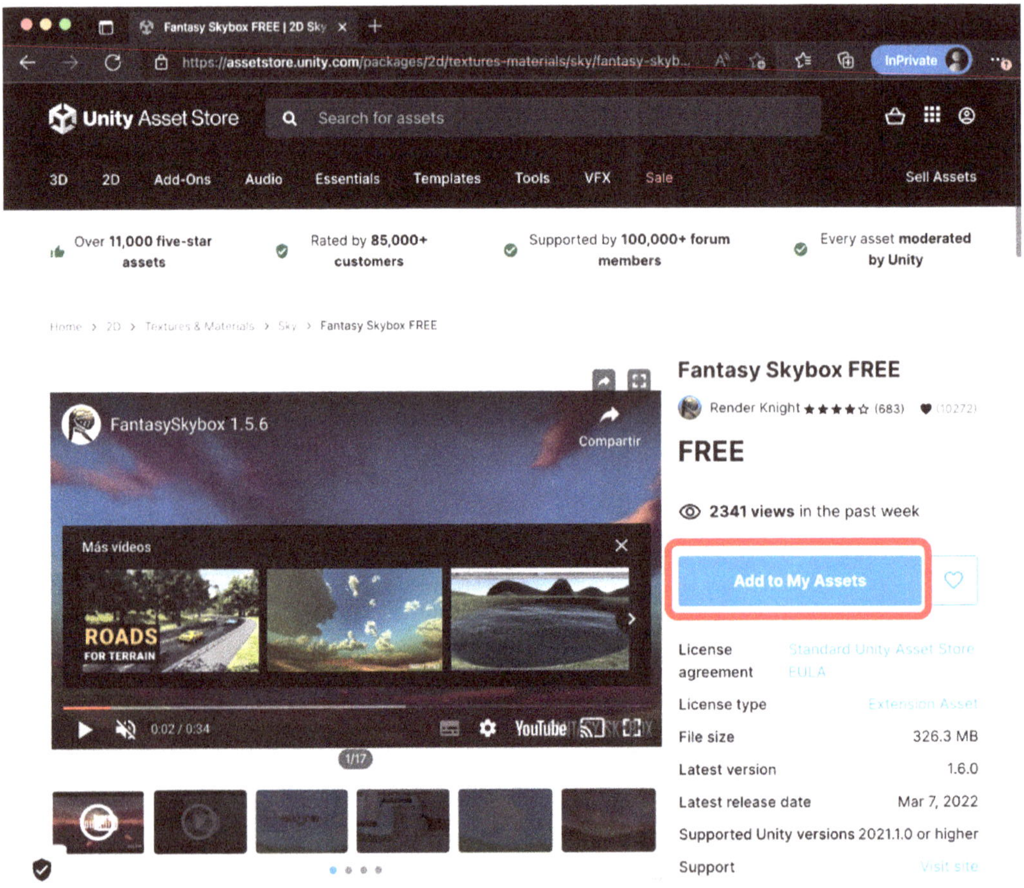

Figure 3.49 – Fantasy Skybox FREE download page

4. Next, the **Package Manager** window in Unity will open, and the downloaded asset already selected will appear. Click on the **Download** button. Once the download is finished, in the same place, the **Import** button will appear; click on it.

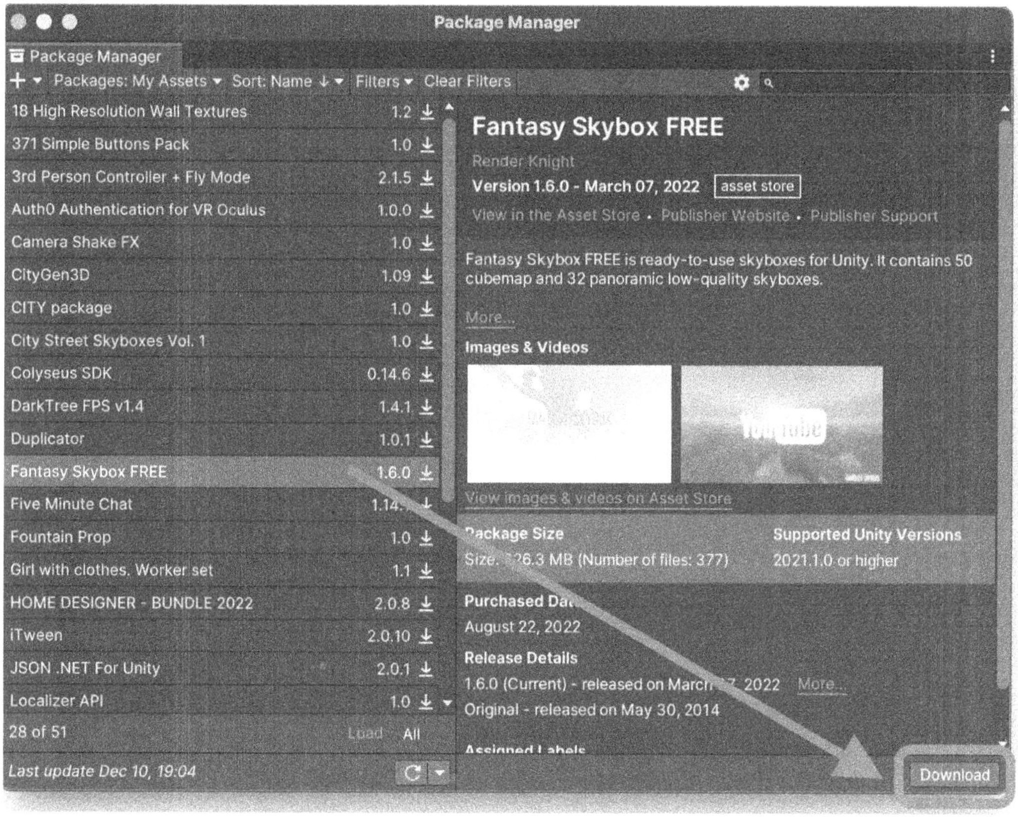

Figure 3.50 – Fantasy Skybox FREE in Package Manager

5. Well, once downloaded and imported, we will see the new folder that has been created in the **Project** panel, called `Fantasy Skybox FREE`.

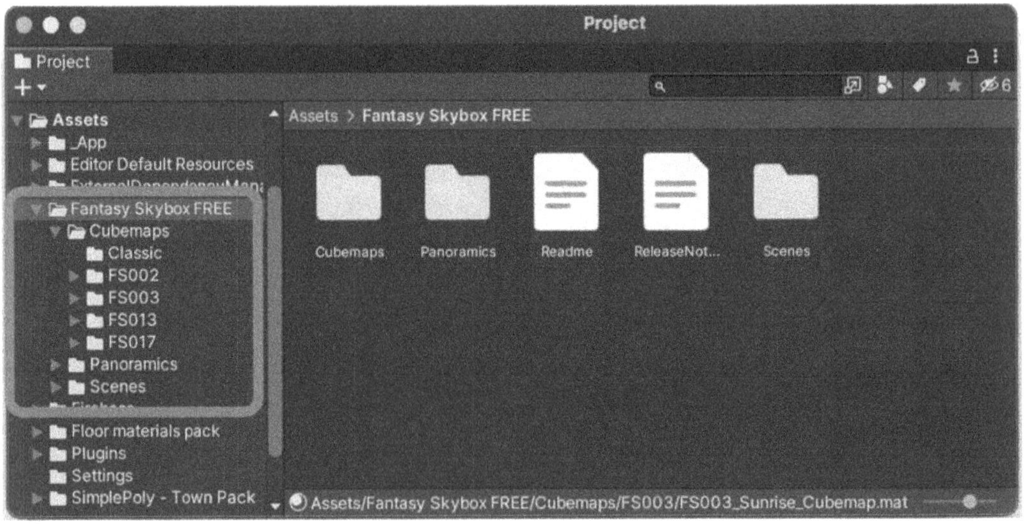

Figure 3.51 – The Fantasy Skybox FREE folder

6. Navigate to look inside **Assets** | **Fantasy Skybox FREE** | **Cubemaps**; we will see something similar to this:

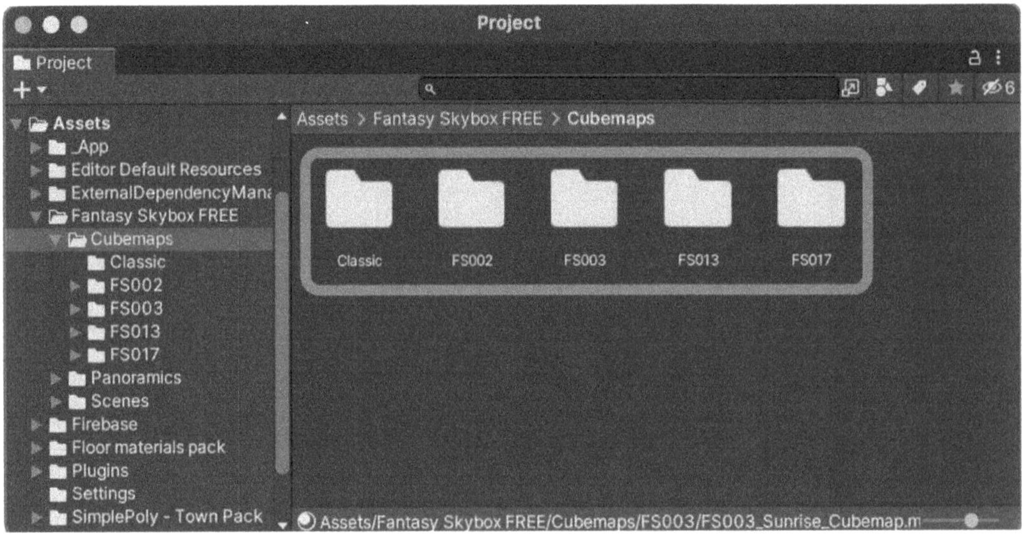

Figure 3.52 – Fantasy Skybox FREE | Cubemaps folder

This asset comes with five different sky versions. You can click on each folder and see the appearance inside – for example, we, for this scene, are going to use the `FS003_Sunrise_Cubemap` Skybox found in the `FS003` folder.

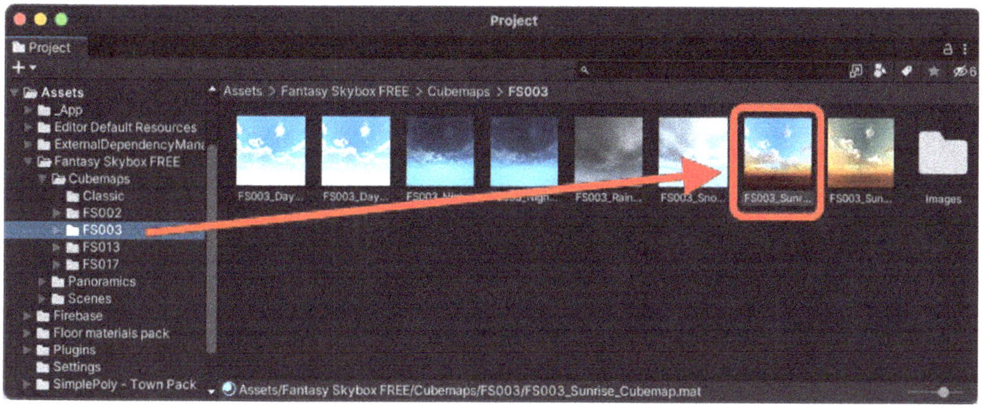

Figure 3.53 – Fantasy Skybox FREE selected asset

> **Tip: your favorite Skybox**
>
> If you like another Skybox of those included in this asset, you are totally free to choose it and it will not negatively affect the operation of the project. I encourage you to choose your favorite; just follow the same instructions to configure it.

7. Finally, select the `FS003_Sunrise_Cubemap` asset with a left click of the mouse, and without releasing it, drag and drop it into the sky in the scene.

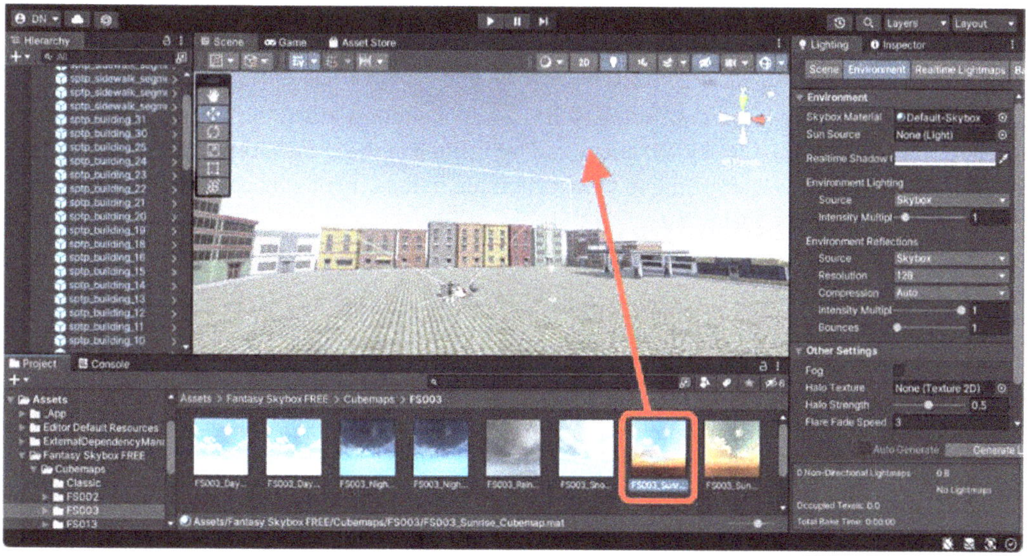

Figure 3.54 – Replacing the Skybox

Our scene has become a beautiful little village experiencing a beautiful sunset, with just a few simple steps.

Figure 3.55 – Previsualization of the replaced Skybox

You may have noticed that when setting the new sky, the tone of the environment has changed as well; this is because Unity uses the colors in the tones of the sky to infuse them into the ambient lights in the scene. Don't forget to save the scene.

You are doing great. Now, we have our scene with beautiful buildings, sidewalks that decorate it a lot, and a harmonious and relaxing sky. What else can we include in the scene? As we mentioned at the beginning of the chapter, we are going to place a fountain in the center, which serves as a reference place to meet friends in the metaverse.

The village fountain

When I think of a town, I always imagine it with a fountain, and our virtual town is no different. Particularly, I find it very appropriate to have a fountain in the center of the scene since, throughout my life, I have played many online games including MMOs, and it was very common to meet friends at a fountain, or in the center of the village. This kind of setup feels very medieval-style. There, you could see people gathered, waiting to get into a game.

We're going back to Unity Asset Store, this time to get a free fountain. To do this, follow these steps:

1. Go to Unity Asset Store by selecting `https://assetstore.unity.com`.
2. Type `Fountain` into the search engine and do not forget to sort the results by price, from lowest to highest, so that the free assets appear first. Select the asset called **Fountain Prop**.

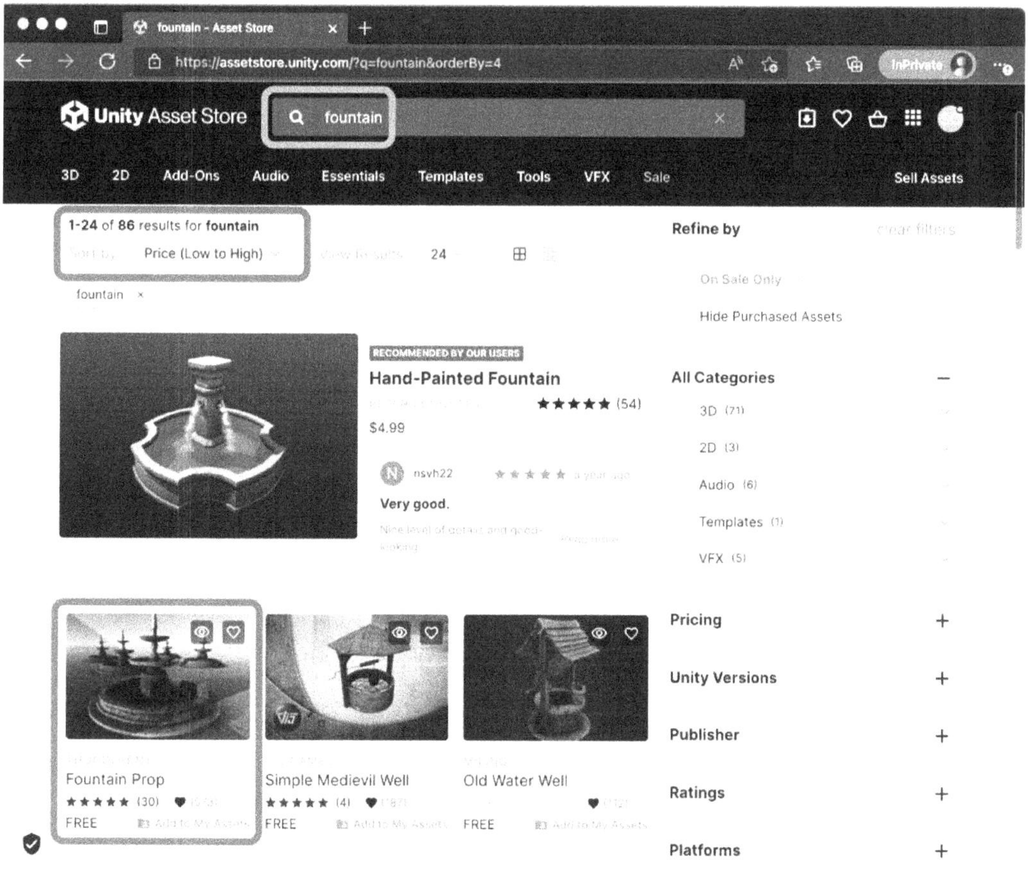

Figure 3.56 – Searching for fountains in Unity Asset Store

3. As we previously did for our other assets, click the **Add to My Assets** button and then click on the **Open in Unity** button.
4. Once the **Package Manager** window is open, click on the **Download** button and finally on **Import**.
5. We should already have a new folder in our project called `Stone Fountain`. If we take a look inside, we see again that the 3D object is previewed as fuchsia or pink. We must apply a correction to its material to adapt it for URP rendering.

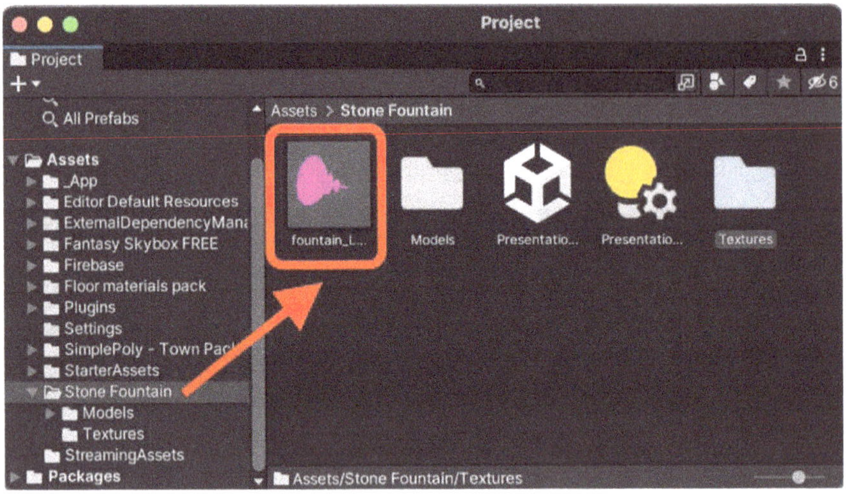

Figure 3.57 – Non-URP Material previsualization

6. The materials of this asset can be found in **Assets** | **Stone Fountain** | **Models** | **Materials**.

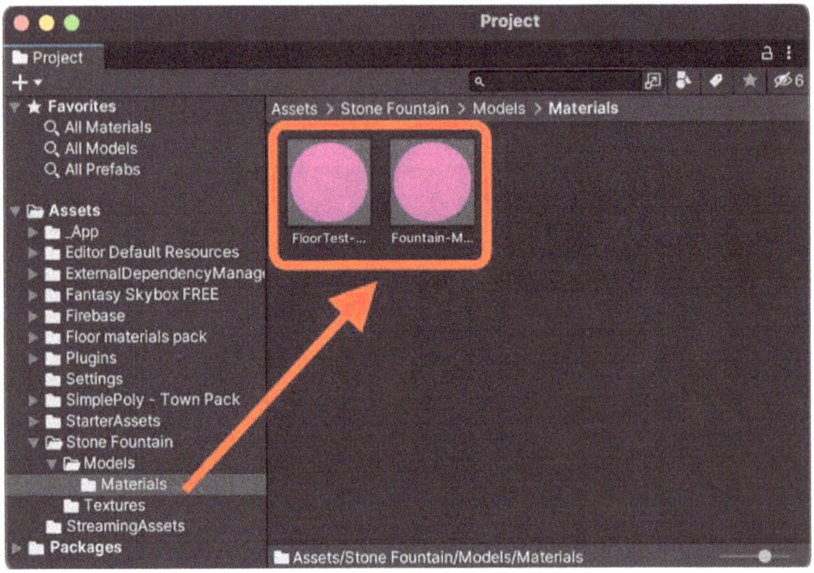

Figure 3.58 – Selecting materials to apply a fix

7. Select the two existing materials and click on **Edit** | **Rendering** | **Materials** | **Convert Selected Built-in Materials to URP** from the main menu bar and confirm. They will be corrected and show their correct appearance.

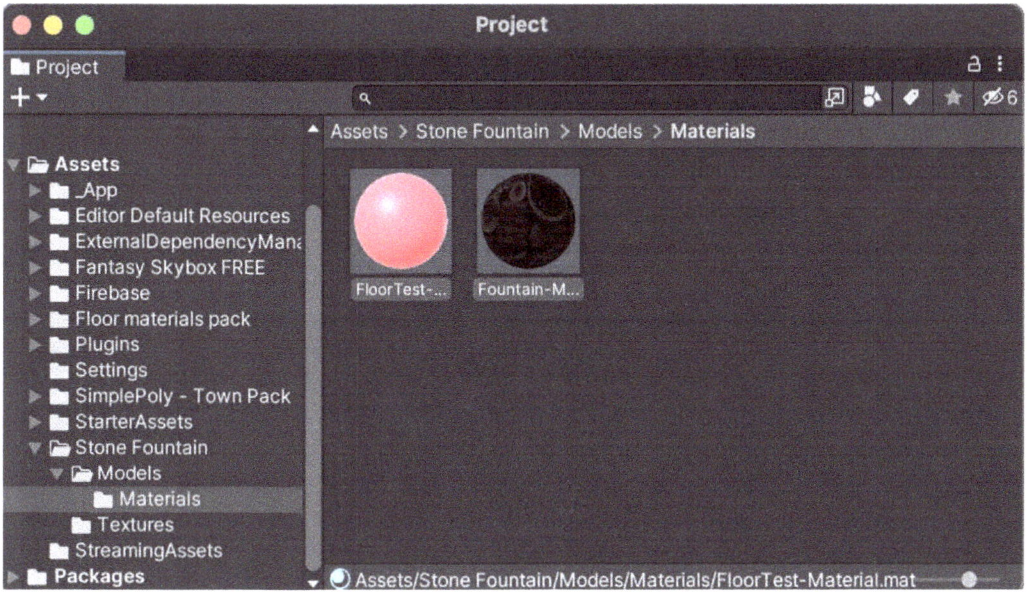

Figure 3.59 – Fixed materials for the URP

8. The font Prefab is located in **Assets** | **Stone Fountain** and is called `fountain_LOD0.prefab`. We select it and, without releasing our left-click, we drag and drop it into the scene.

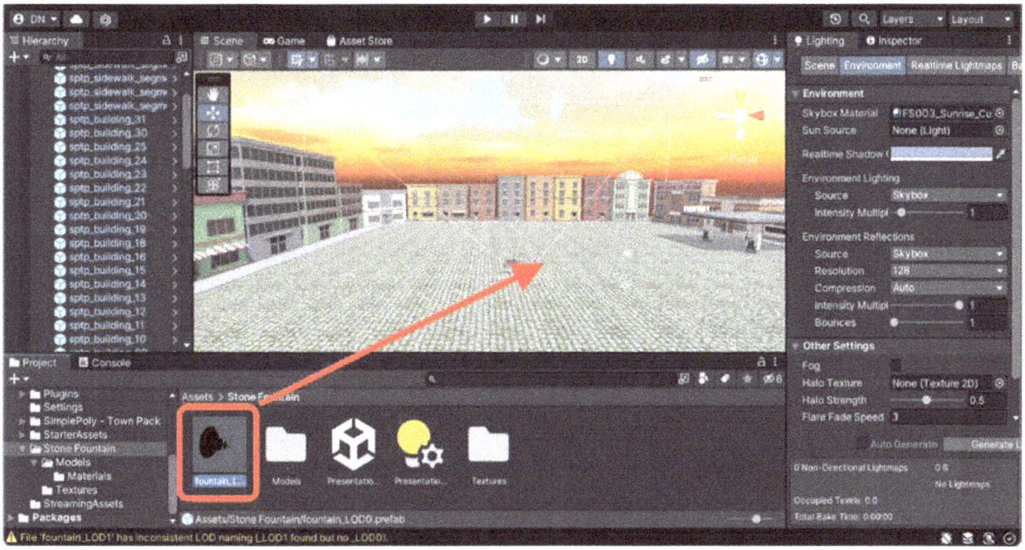

Figure 3.60 – Placing the fountain Prefab into the scene

> **Tip: Prefab remains fuchsia**
>
> Sometimes, after converting materials into a URP format as we did in *step 7*, the Prefab may still appear fuchsia or pink; don't worry, everything went well. Save the scene by clicking **File** | **Save** in the main menu bar or by pressing the key combination *Ctrl* + *S* in Windows or *Cmd* + *S* in Mac. It will automatically refresh and appear successfully.

9. You can use the **Move** tool and place the font where you prefer; I have decided to place it as centrally as possible.

Figure 3.61 – Moving the fountain Prefab within the scene

10. For my taste, the fountain is very small by default compared to our player. You can use the **Scale** tool to enlarge it. To make it grow proportionally along all its vectors, select the square that joins the three vertices and move it until you get the desired size:

Figure 3.62 – Scaling the fountain Prefab in the scene

11. Note that by scaling the fountain, it will grow on all sides. After resizing, you will have to use the **Move** tool to put it back on the ground. You can also change the position manually in the **Inspector** panel, replacing the current value of **Y** under **Position** to `0`; this way, it will be placed right on the ground.

Figure 3.63 – Changing the position of the fountain Prefab in the scene

12. We already have a wonderful fountain in our village. It's time to test how the scene plays out. Click the **Play** button and move with the player to the source.

Figure 3.64 – Player inside the fountain

13. Our player is able to pass through the source. This is because this Prefab does not include any collider. To add one, with the **Fountain** GameObject selected in the **Hardware** panel, scroll down all the way in the **Inspector** panel and click on the **Add Component** button.

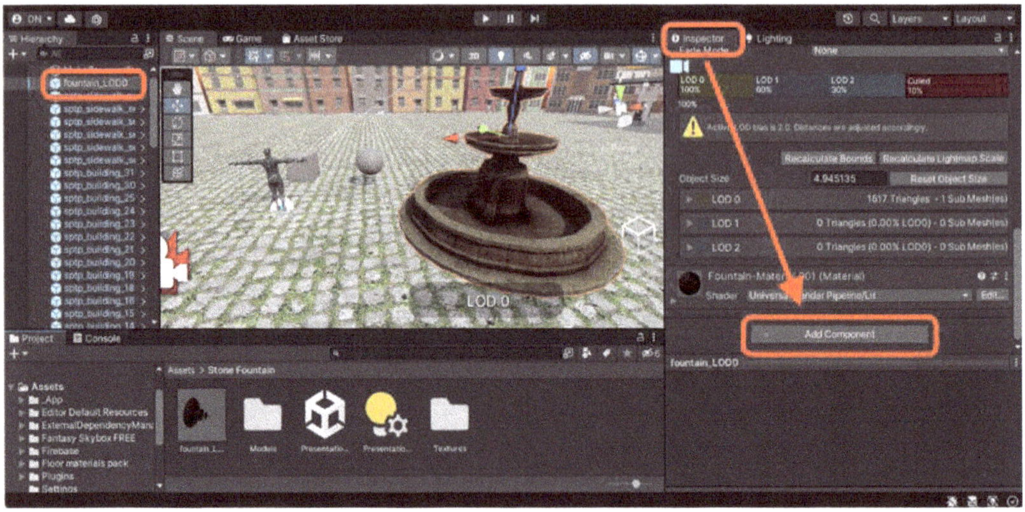

Figure 3.65 – Adding a new component to the fountain

14. Type `Mesh Collider` into the search engine in the small modal window that appears and select `Mesh Collider` from the list.

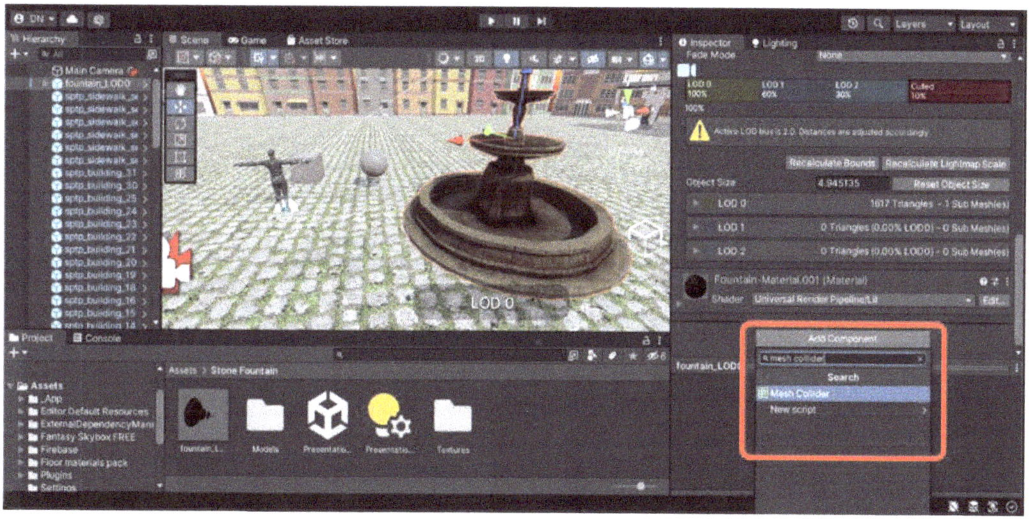

Figure 3.66 – Adding a new Mesh Collider to the fountain

15. Finally, if you now click the **Play** button again and direct the player to the source, you will see that he is not able to pass through it.

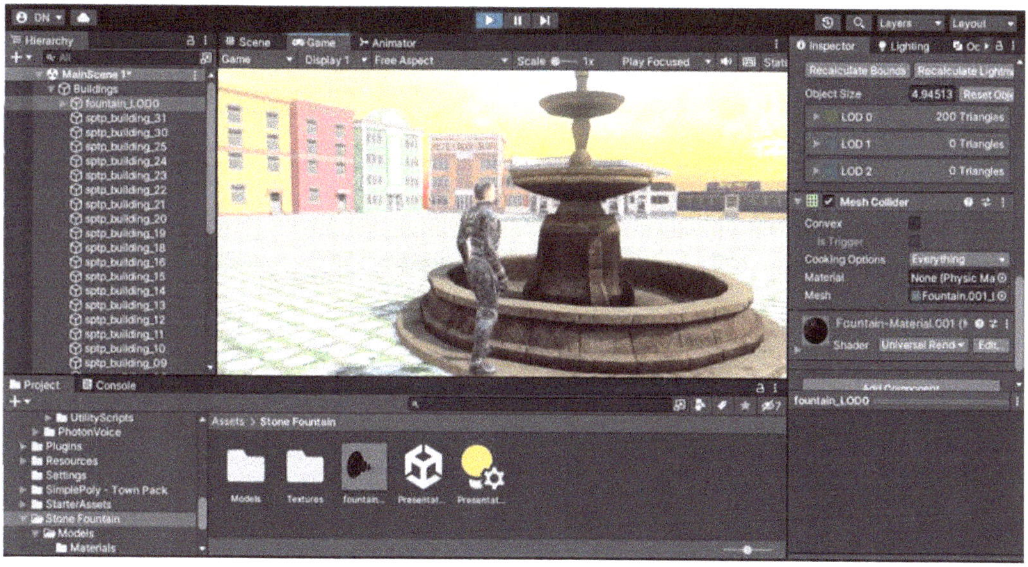

Figure 3.67 – Trying Player collision in the fountain

Great – we are coming to the end of decorating our scene. At home, you can play with all the buildings and objects that we have downloaded during this chapter and decorate the scene as you like.

Finally, we are going to add some buildings to the second line to give the look of a more complex horizon, which makes it seem that we are in the center of a larger town.

Expanding the horizon

Perhaps, if you are as fussy as I am, you will have noticed that beyond the buildings we have placed, except for the Sun and the sky, nothing can be seen.

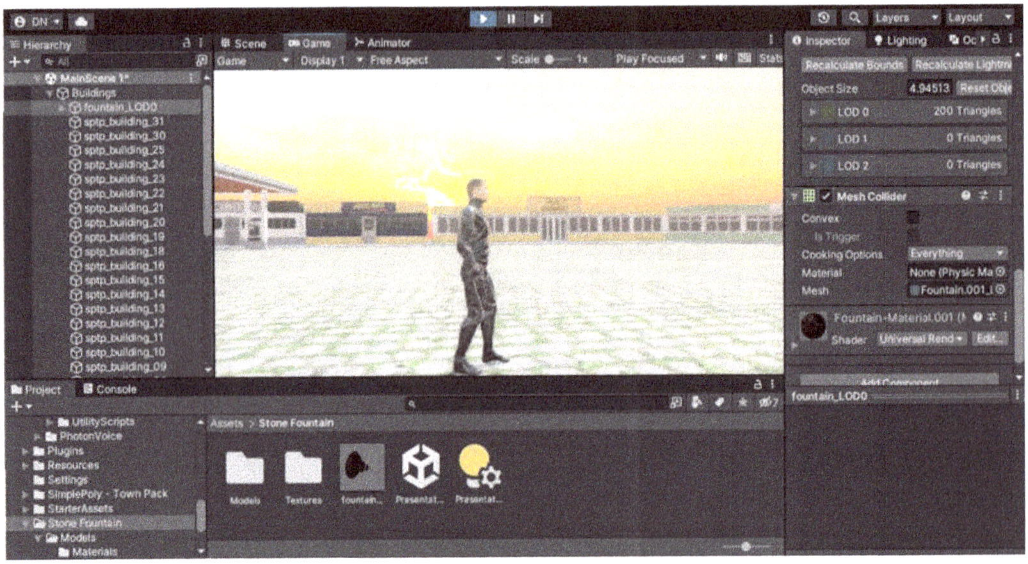

Figure 3.68 – Previsualization of the scene's horizon

This step is optional. We will add taller buildings from behind; this will give a feeling of being in a larger town, and the horizon will be more complex and visually beautiful.

Simply choose the tallest buildings from the **Assets** | **SimplePoly - Town Pack** | **Prefabs** | **Buildings** path.

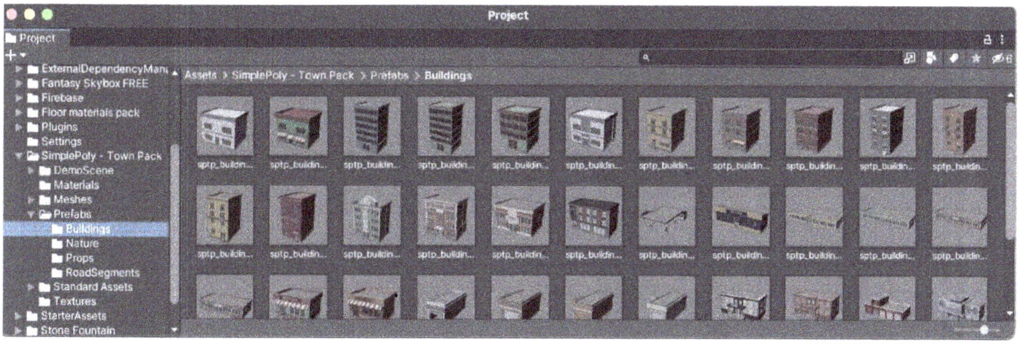

Figure 3.69 – SimplePoly – Town Pack buildings folder

Add a few buildings, organized as you like, scattered around the back of the buildings we initially placed.

Figure 3.70 – Adding more buildings to the scene

Repeat the process until there is a second ring of buildings. They do not need to be organized. Just place them around the scene and we will soon see the result. In my case, my scene looks as follows:

Figure 3.71 – Scene previsualization

Now it's time to survey our artwork. Click the **Play** button and take a walk with your character. Incredible, right?

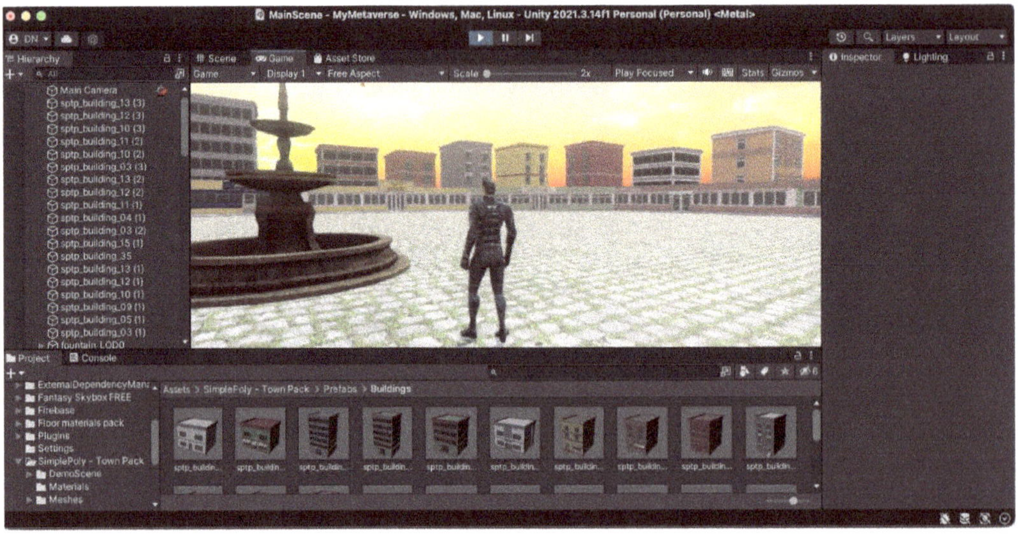

Figure 3.72 – Scene test in Game Mode

Now our town seems bigger. Being in the center is overwhelming! We have done a good job; congratulations!

Finally, we are going to give an ecological touch to our square. We are going to take advantage of the vegetation Prefab that comes as part of **SimplePoly - Town Pack** to give a green touch to our town.

A greener town

A village without trees – really? Next, we will plant a few trees in the square of our town. We will give it a more natural touch.

To do this, as we have done previously with buildings, we use the Prefabs that are in the **Assets** | **SimplePoly - Town Pack** | **Prefabs** | **Nature** path.

The process is simple. Just like we have done before, we will use the Prefabs called `sptp_tree_01` and `sptp_tree_02`. We select one of them and, without releasing the left click, drag the object and drop it into the scene. You can repeat the process as many times as you want until you plant all the trees you want. In my scene, it looks like this:

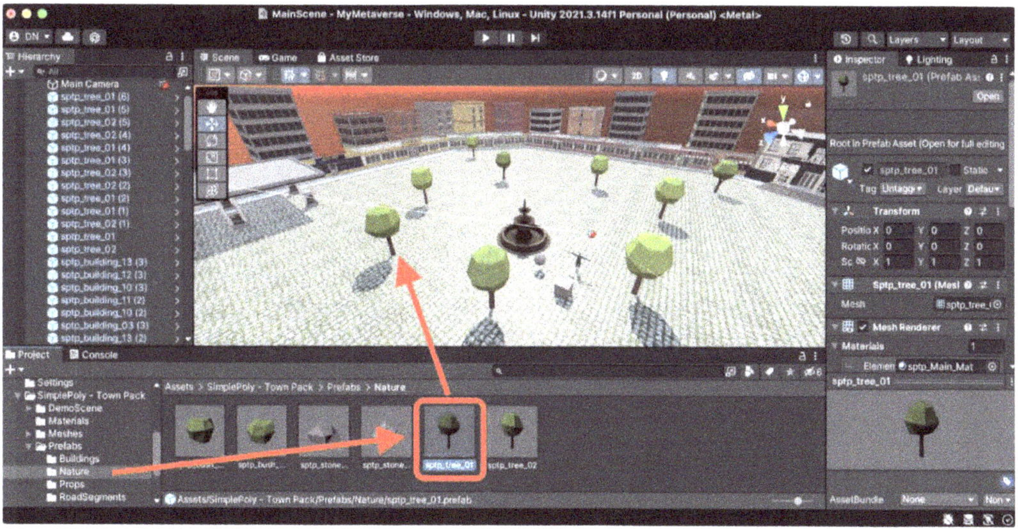

Figure 3.73 – Placing trees in the scene

Excellent! We could continue to decorate the scene forever, but I think we have already learned a lot in this chapter about how to decorate our initial scene. Feel free to add or remove buildings or design the scene to your liking.

Last but not least, we will apply some order and cleanliness to the scene. We have built a lot today. Now, we have to leave the scene ready for *ourselves* in the future, make sure we know what we have done, and make it easy to continue working on this main scene.

Order and cleanliness

As you may have seen, we left the **Hierarchy** panel very messy, full of assets of all kinds. A good practice to have a well-organized scene is to group things; you can group similar elements within the parent GameObjects.

To carry out this task of arranging and cleaning, we will execute the following steps:

1. Let's start with the trees, which are the first GameObjects we found here. We select them all. A technique to select a range of objects is to first click and select the first one, hold down the *Shift* key, and then click on the last. This way, we will select a range between the start and end.

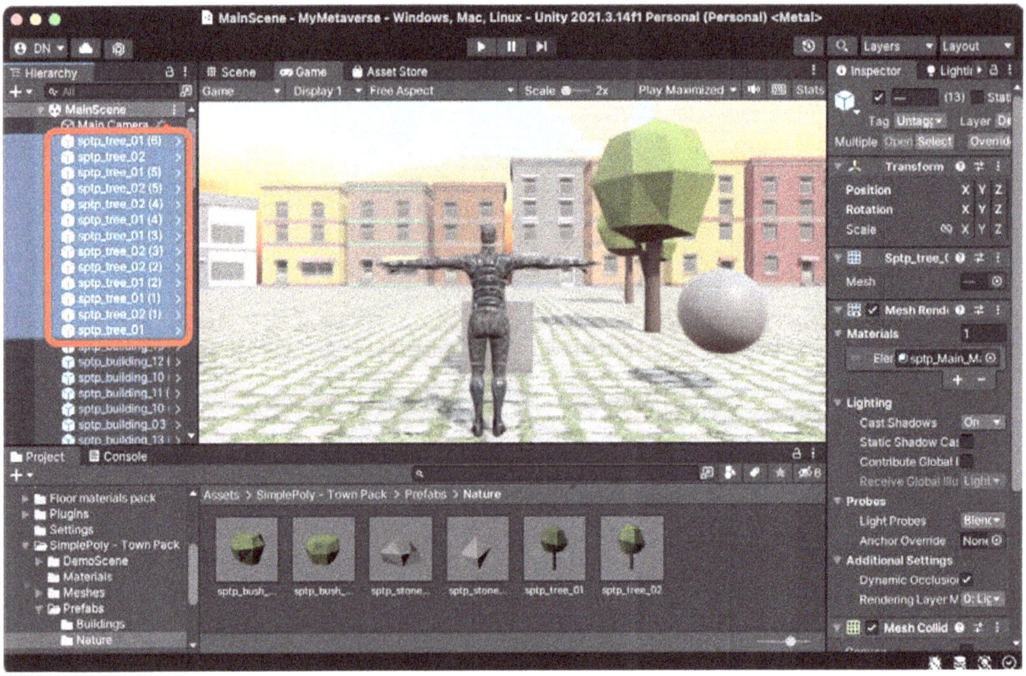

Figure 3.74 – Selecting GameObjects

2. For any selected item, we right-click and click on the **Create Empty Parent** option.

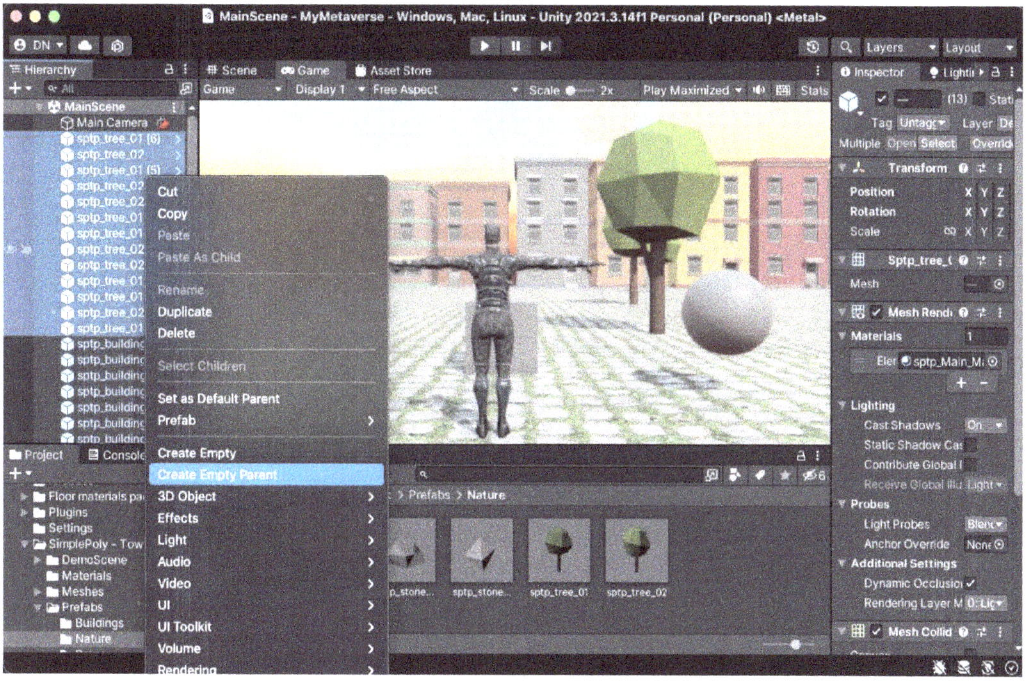

Figure 3.75 – Creating a parent for GameObjects

3. This action that we have just performed groups the selected GameObjects under a *parent*. As you may have noticed, a parent object called **GameObject** has been created, and inside, hanging from it, are the trees.

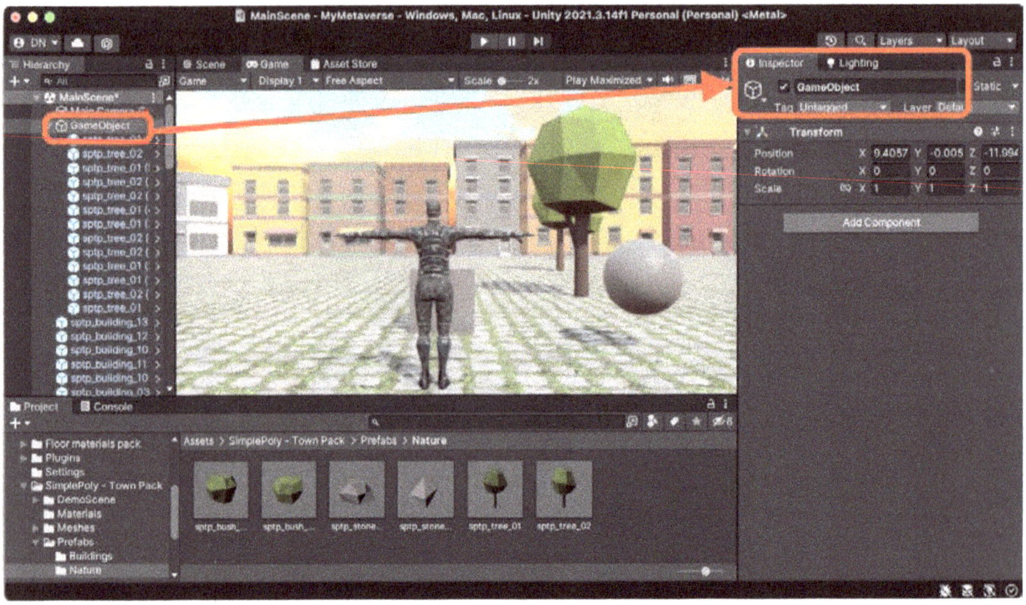

Figure 3.76 – Renaming the parent GameObject

4. Click on this parent object called **GameObject** and, to its right, in the **Inspector** panel, you can change the name by editing the text box. For example, type `Trees` and press the *Enter* key to apply the changes. You'll see that you can fold the parent by pressing the small arrow-like button.

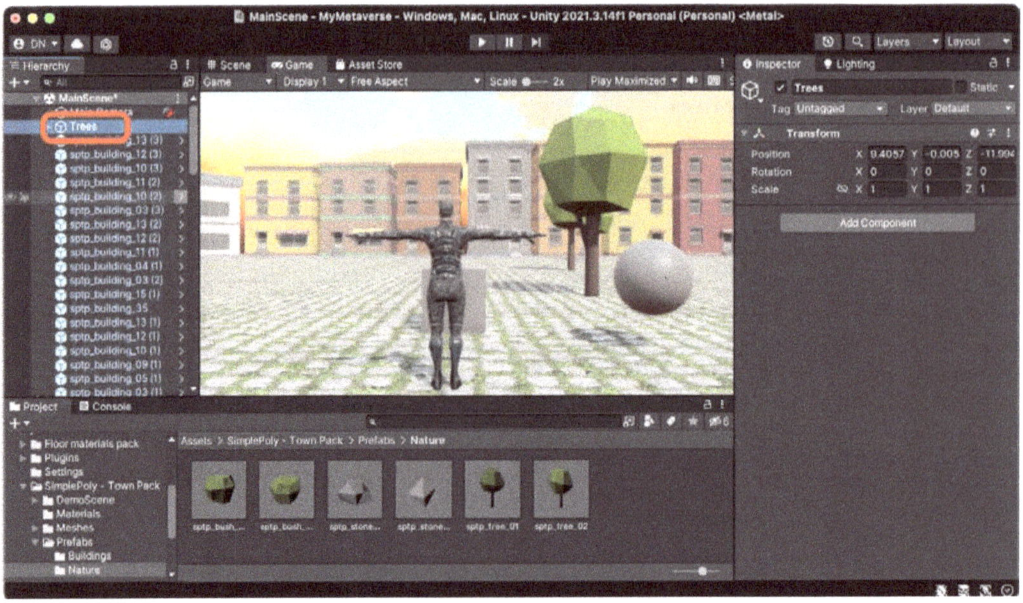

Figure 3.77 – Collapsing the parent GameObject

5. We repeat the process this time with the **Building** Prefabs. Select as many as you can, right-click on one of them, and click the **Create Empty Parent** button again.
6. We renamed the parent GameObject `Buildings` and our **Hardware** panel should look something like this:

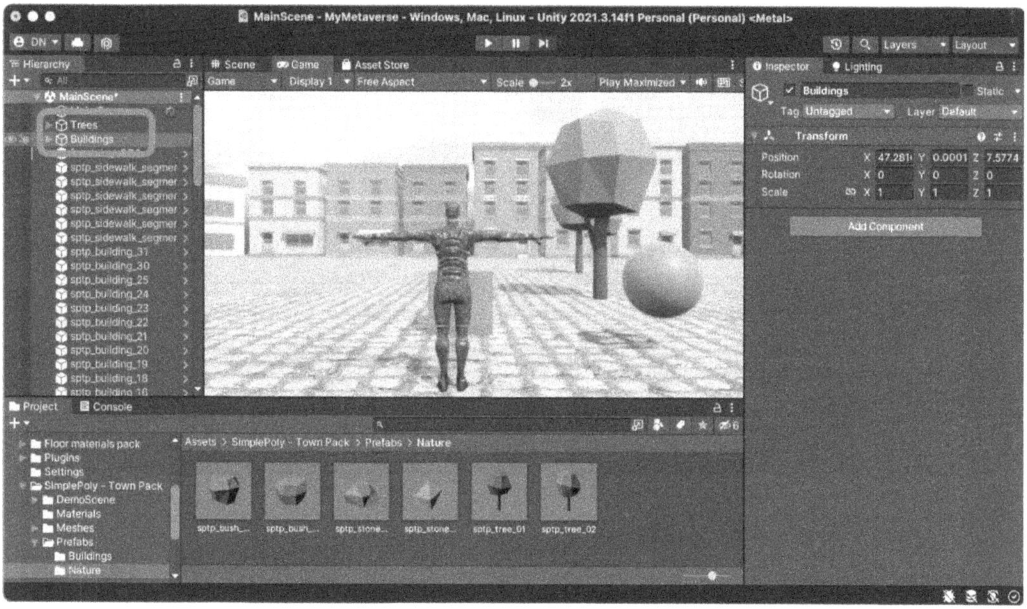

Figure 3.78 – Collapsing the parent GameObject

7. Don't worry if building Prefabs have been left out of the folder. You may not have been able to select them all at once. To move them to their corresponding parent folder, select the ones you want and drag and drop them into the parent GameObject, just as you usually do to move files to folders on a desktop.
8. We will perform the same action this time with the sidewalks. Select all sidewalks GameObjects, right-click to select the **Create Empty Parent** option again, and call it `Sidewalks`.
9. Finally, even if we only have one GameObject, it is advisable to create a parent to the source. This way, if in the future, we want to add more objects to the source, they will be organized into a folder. For example, if we want to add another water object, right-click on the fountain GameObject, select **Create Empty Parent** again, and rename it to `Fountain`.

Fantastic – much cleaner, right? Now, our **Hierarchy** panel will look something like this:

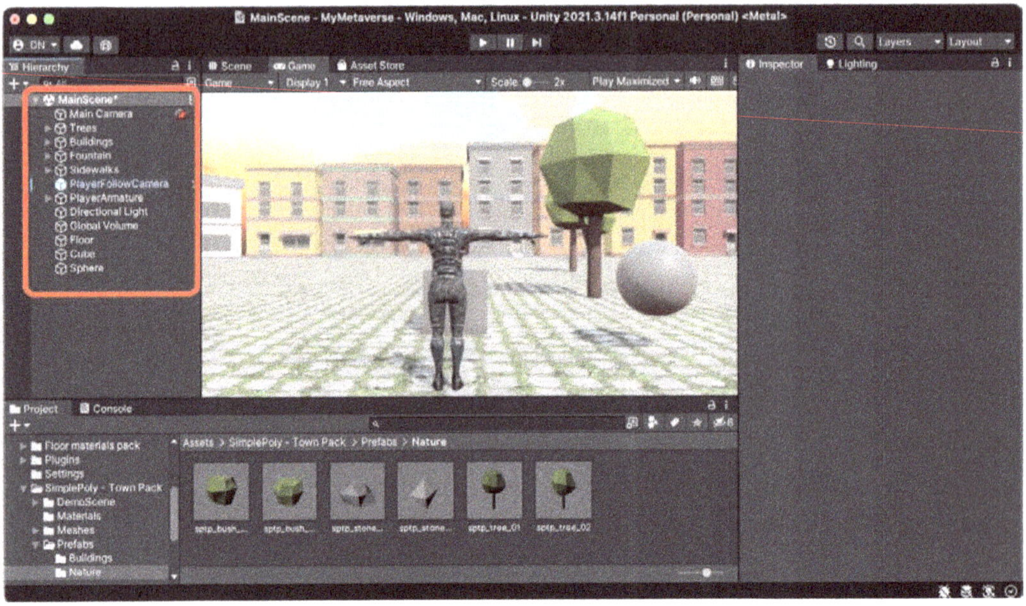

Figure 3.79 – Cleaned scene result

Finally, we are going to delete the objects that we used to test at the beginning of the book. We are going to proceed to eliminate the **Cube** and **Sphere** GameObjects. For this, right-click on each of them and click the **Delete** option. Once you've finished this, don't forget to save the scene.

Congratulations on getting here. At this point in the book, we have done a lot together. Now you know how to design a scene in Unity in a professional way, and we have placed a floor, buildings, sidewalks, a fountain, and trees in it. I hope and wish that you are as satisfied with the scene as I am, but this is only the beginning. We are creating a wonderful metaverse from the beginning.

Now we must dedicate a few pages to understanding how we can optimize the scene for a more fluid performance. Possibly, when you have clicked on the **Play** button, everything went perfectly for you, but it is possible that in the future, a user of your metaverse who has a lower-powered system will experience slowness or technical problems.

Summary

Throughout this chapter, we came to understand the concept of the Hub, Meeting Point, or Lobby in our metaverse project. We laid the foundations in terms of designing a scene and placing, moving, and rotating GameObjects. We discovered Unity Asset Store and how to download and import new assets into our project.

We also saw how to change the look of the sky, added new textures to objects, and finally, we walked with our player through a world created by us from scratch.

In the following chapter, we will learn about the primary techniques for optimizing the performance of our project. We will also take our first dive into the world of Firestore, Firebase's real-time database.

4
Preparing Our Home Sweet Home: Part 2

Optimization in the world of video games is fundamental. If we do not do this task, the user experience will be terrible, even if we have good hardware. In the previous chapter, we learned how to design a scene with all the details: the sky, the buildings, the sidewalks, the ground, the trees, and even a fountain – everything we designed from scratch.

Throughout the following pages, we will learn what techniques Unity offers us to improve performance. You will be surprised to see how with a few simple processes, we gain FPS and rendering speed. Then, we will do our first write operation in the Firestore database.

We will cover the following topics:

- Case study on the need for optimization
- Understanding optimizations
- Adding the first world to the Firebase database

Technical requirements

This chapter does not have any special technical requirements, just an internet connection to browse and download an asset from the Unity Asset Store.

We will continue with the project we created in *Chapter 1, Getting Started with Unity and Firebase*. Remember that we have a GitHub repository, `https://github.com/PacktPublishing/Build-Your-Own-Metaverse-with-Unity/tree/main/UnityProject`, which contains the complete project that we will work on here.

Case study on the need for optimization

Imagine you are developing a game for mobile devices, such as a 3D car racing game, with detailed graphics and impressive visual effects. The game scene contains a long and complex track with many objects, trees, buildings, traffic signs, lampposts, and animated spectators. However, when you test the game on a mobile device, you notice that the performance is extremely low, with a very low frame rate and constant stuttering.

In this case, optimization in Unity becomes crucial for the game to run smoothly and attractively on mobile devices. Some of the optimization techniques you could apply are as follows:

- **Static batching**: The game track probably has many static objects, such as buildings and rocks, that do not change position. By marking these objects as static, you can take advantage of baking lights to pre-calculate static lighting at design time. This will significantly reduce real-time lighting processing time during gameplay, which is especially useful on resource-constrained mobile devices.
- **Dynamic batching**: Ensure that objects in the scene meet the criteria for dynamic batching, such as sharing the same material and having similar render settings, to reduce the number of render calls.
- **Occlusion culling**: Use occlusion culling to avoid rendering objects that are completely or partially hidden by other objects, which will decrease the number of objects sent to the GPU.
- **Texture optimization**: Adjusts the size and compression of textures so that they take up less memory and reduce the load on the GPU. This will be covered in *Chapter 14, Distributing*.

By applying these optimization techniques, you can ensure that your game runs smoothly on mobile devices, providing a more enjoyable gaming experience for players and avoiding performance issues that can negatively affect gameplay.

Remember that optimization in Unity is an ongoing process and can involve a trade-off between performance and visual quality. It is important to find the sweet spot that delivers a smooth and engaging gameplay experience for players, without compromising too much on visual quality and gameplay.

Now that we know in which cases we should optimize our game and which techniques we can apply, let's go deeper. Next, we are going to see in a practical example what factors change when we apply optimizations. This will help you to visualize the result.

Understanding optimizations

Without a doubt, optimization is the most important part of a video game. If we don't pay attention to this point, the game will run slow, with interruption peaks, and offer a lousy user experience.

Unity offers very simple tools to see performance-related information at a glance, from the **Game** panel itself. We will see as follows how to view this information and how to interpret the data it provides.

When we click the **Play** button in our Unity editor and launch our metaverse, there is a button at the top of the **Game** panel called **Stats**.

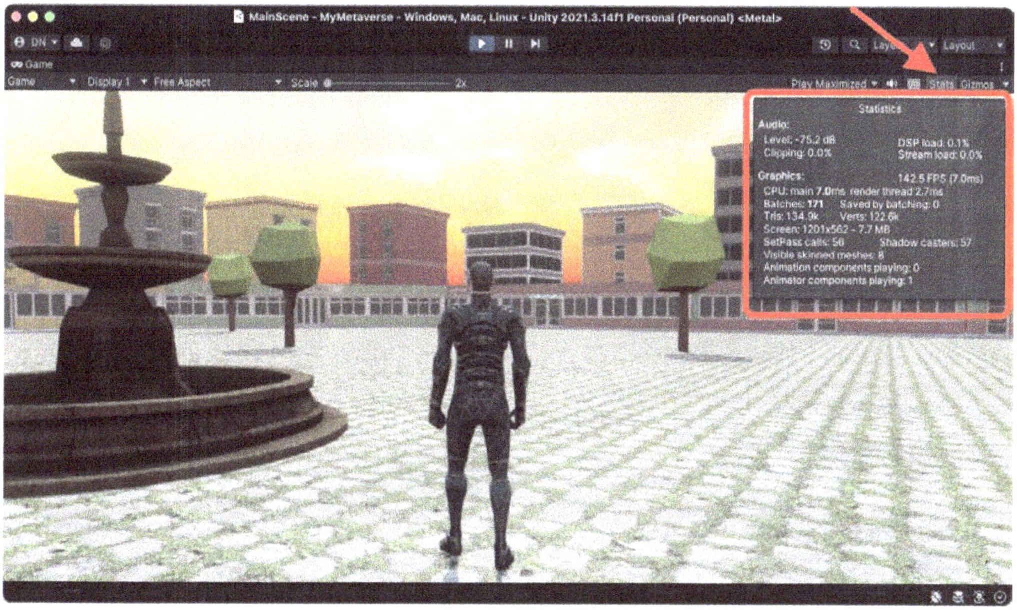

Figure 4.1 – Performance stats

If we click on it, Unity will display a small modal window with very useful real-time information about the performance of the scene. We can see here the most relevant information:

- **FPS**: Standing for **frames per second**, the higher the value, the smoother the gaming experience. This value represents the speed at which the information the camera is observing is refreshed.
- **CPU: main**: If the value we see in **FPS** means how many frames the game is capable of displaying per second, the **CPU: main** value calculates how long the computer takes to process one frame.
- **Render thread**: The time it takes for the computer to render a frame on the screen.
- **Batches**: Represents the number of calls to the graphics APIs that are made in order to satisfy the rendering of each frame.
- **Tris**: Undoubtedly one of the most important values to take into account, each 3D object is composed of triangles. The more triangles, the more complex and heavy it is to render. That is why on mobile devices, 3D objects tend to be less realistic and more polygonal. On older game consoles, where there was little rendering power, the models were tremendously primitive compared to today's game consoles.

- **Verts**: Also a value to take into account when optimizing the game, it represents the number of total vertices.

> **Tip: Extended information about rendering stats**
>
> If you want more information about what is displayed in the **Stats** panel, you can navigate to `https://docs.unity3d.com/Manual/RenderingStatistics.html`.

All the optimizations that we are going to learn about here are extensible to all the scenes that we will create in the future throughout this book, and in future scenes that you create in your own project.

The main optimizations that we are going to work on and that are very effective will be as follows:

- **Static/dynamic GameObjects**
- **Camera Occlusion (Culling)**
- **Light Baking**
- **Colliders**

Let's start with the basics, knowing how to recognize which objects in our scene should be marked as static to improve performance.

Static/dynamic GameObjects

There are two types of GameObject in Unity, dynamic and static. Dynamic GameObjects are those that are susceptible to movement during gameplay, while static GameObjects are usually just that, static. Both types of GameObject have a special precompute method. Unity understands that every object can be moved during gameplay. That's why it offers a check so that we can tell it which one is static and, thus, we can optimize.

By indicating that a GameObject is static, we help Unity to save computation time in processing this kind of object, because if they are never going to move, it will perform computation operations at startup and reuse the computation processes during the whole game, which will drastically improve the performance of our game.

Static batching

Now that we have understood the difference between static and dynamic GameObjects, we know that all our buildings, sidewalks, ground, fountain, and trees are static. To mark static objects, you just have to mark the **Static** checkbox as active. This checkbox is located in the **Inspector** panel, to the right of the GameObject name:

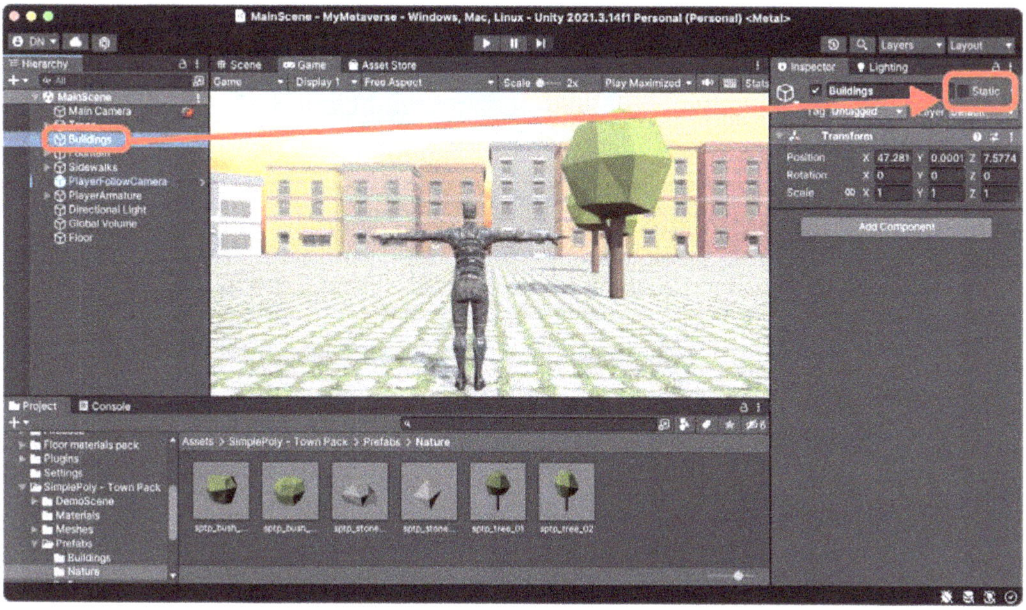

Figure 4.2 – Marking a GameObject as static

Note that you can check the checkbox for individual GameObjects or also parent GameObjects. In the latter case, Unity will ask you whether you want to apply the change to the elements inside it. For example, you will see the following message if you set the **Static** checkbox to active for the parent **Buildings** GameObject:

Figure 4.3 – Apply to children confirmation window

Then, click the **Yes, change children** button, which will mark all child GameObjects as static. Perform the same action with the parent **Trees**, **Fountain**, **Sidewalks**, and **Floor** GameObjects and save the scene.

Now that we have marked many GameObjects as static, we can click the **Play** button and re-analyze the statistics:

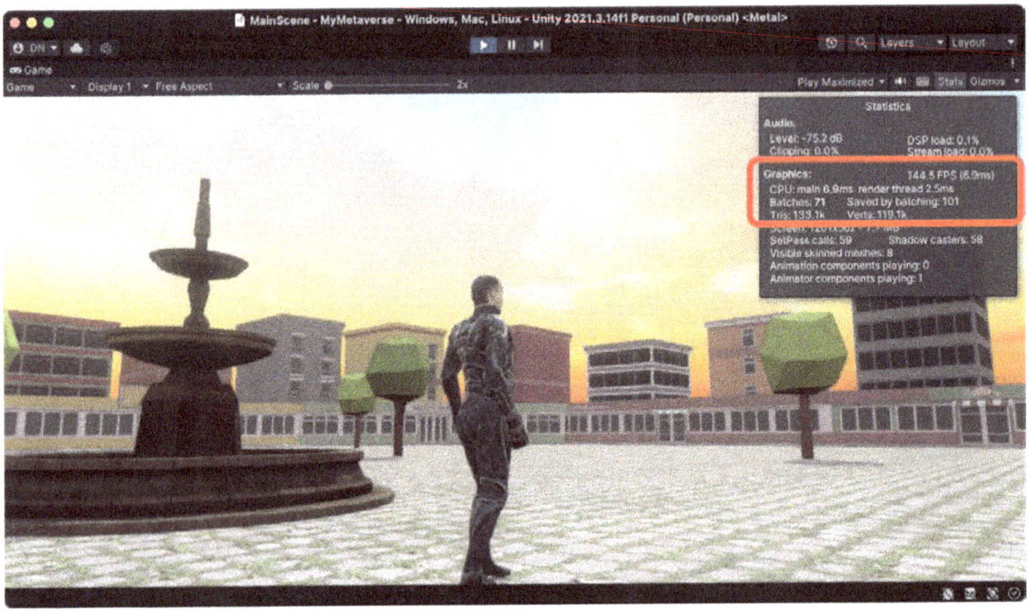

Figure 4.4 – Performance stats

Now, we can observe that the value of **Saved batching** is `101`, which means that Unity is optimizing the rendering of the scene, applying a cost-saving algorithm to the general rendering. Thanks to static objects, we have a lighter scene.

Dynamic batching

Dynamic batching optimization in Unity is a technique that allows you to combine multiple small objects into a single mesh (batch) during the real-time rendering phase. This is especially useful when you have many similar objects, such as particles, decorative objects, or repetitive elements in the scene.

When objects are dynamic (that is, they change their position, rotation, or scale at runtime), they cannot normally be batched to optimize performance. However, Unity uses dynamic batching to combine objects with the same material and render settings in small batches in real time, even if they are dynamic objects.

By reducing the number of rendering batches the engine has to process, performance is improved by reducing GPU calls and minimizing CPU overhead. However, it is essential to note that dynamic batching has a limit to the number of triangles it can efficiently process, and there may be situations where other optimization techniques, such as static batching or GPU instancing, are more suitable for improving performance in complex scenes.

To enable dynamic batching in your Unity project, follow these steps:

1. In the main menu bar, click **Edit** | **Project Settings**. Then, in the **Project Settings** window, select the **Player** tab.
2. In the **Other Settings** section, make sure the **Rendering** submenu is displayed.
3. Look for the option called **Dynamic batching** and check the box next to it to enable it. This will allow Unity to combine dynamic objects that share the same material and render settings in batches during the real-time rendering process.
4. Once you have enabled **Dynamic batching**, be sure to save your changes to your project settings.

Remember that dynamic batching will only work with objects that meet certain criteria, such as having the same material and render settings. Also, there is a limit to the number of triangles that Unity can batch dynamically, so make sure you optimize your objects and settings to get the maximum benefit from dynamic batching.

Checklist for using dynamic batching

In order for Unity to be able to apply dynamic batching to our scene, it will consider the following:

- The objects you want to batch dynamically batch use the same material. Unity can only combine objects with the same material in a batch, so try to reuse materials wherever possible.
- Objects that you want to batch dynamically should have similar render settings, such as the same textures, shadows, and lighting. Avoid having objects with very different settings if you want Unity to combine them in a batch.
- The fewer triangles your objects have, the more efficient dynamic batching will be. Try to use simple objects, with a mesh of a few polygons.
- Objects that have complex movement, such as deformations or complex animations, cannot be batched dynamically. If possible, avoid it.
- If you have objects that do not change position, rotation, or scale during the game, you can mark them as static and use static batching (which we have seen previously), which combines objects in batches during the build phase, before the game starts running. This can further improve performance and reduce runtime overhead.

Remember that while dynamic batching is a powerful technique for improving performance, it is not always the best option in all situations. It is important to test and evaluate different optimization techniques to find the right combination for your specific game and scenes.

The next optimization is regarding the camera. We will optimize the performance to avoid adding objects that are not visible.

Camera occlusion (culling)

Occlusion culling in Unity is an optimization technique that helps improve game performance by avoiding rendering objects that are completely or partially obstructed by other objects in the scene and therefore not visible to the camera.

During the calculation phase, Unity determines which objects are hidden or out of the field of view and excludes them from the rendering process, which significantly reduces the graphics load and improves real-time performance. This technique is especially beneficial in complex scenes with numerous objects, allowing for the more efficient use of resources and a smoother gaming experience.

To understand it better, we will look at an example image based on the main scene that we just finished decorating:

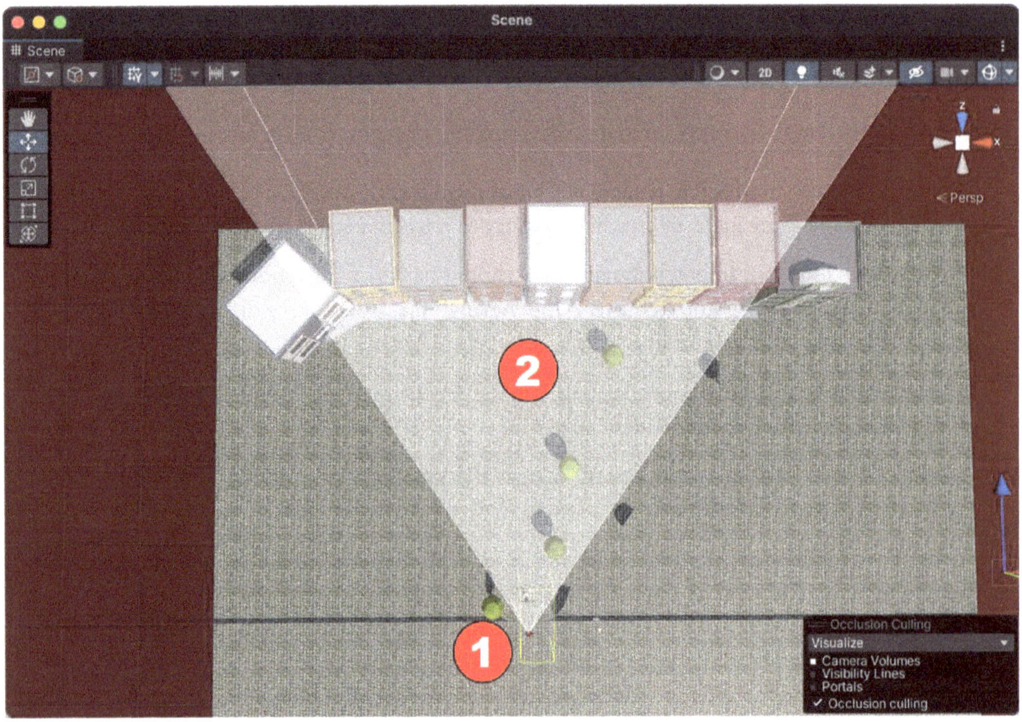

Figure 4.5 – Camera occlusion demo

As you can see, it is a capture of the scene taken from above. After applying a camera occlusion technique, the result is as shown in the preceding figure. We can see that point **1** is where the camera is located. The triangle at point **2** is the camera's angle of view, which is what the player sees.

Everything that is not inside the viewing area (**2**) disappears from the scene. As the player turns their head, the camera also turns, so the viewing angle also moves. Therefore, dynamically and automatically, objects that are inside the viewing area (**2**) will appear in the scene, and those that are outside will disappear.

> **Tip: The importance of using a culling mask**
>
> By applying culling and baking in Unity, a significant improvement in the rendering of triangles in the game scene is achieved. Culling avoids rendering non-visible objects, reducing graphics overhead; baking pre-calculates static lighting; and reducing the number of real-time computations optimizes render calls, saving resources and improving overall performance. These techniques combined reduce the number of triangles processed, allowing for smoother and more efficient gameplay, especially on resource-constrained devices.

As you can see, this is a technique that must always be applied since, as we can see in the preceding screenshot, Unity saves the rendering of all buildings, trees, sidewalks, and other objects that are out of the camera's vision.

Now, how is this optimization applied? Very easily; we will do it together by following these steps:

1. The first thing we are going to do is open the **Occlusion** panel. To do this, in the main menu bar, click on **Window** | **Rendering** | **Occlusion Culling**.

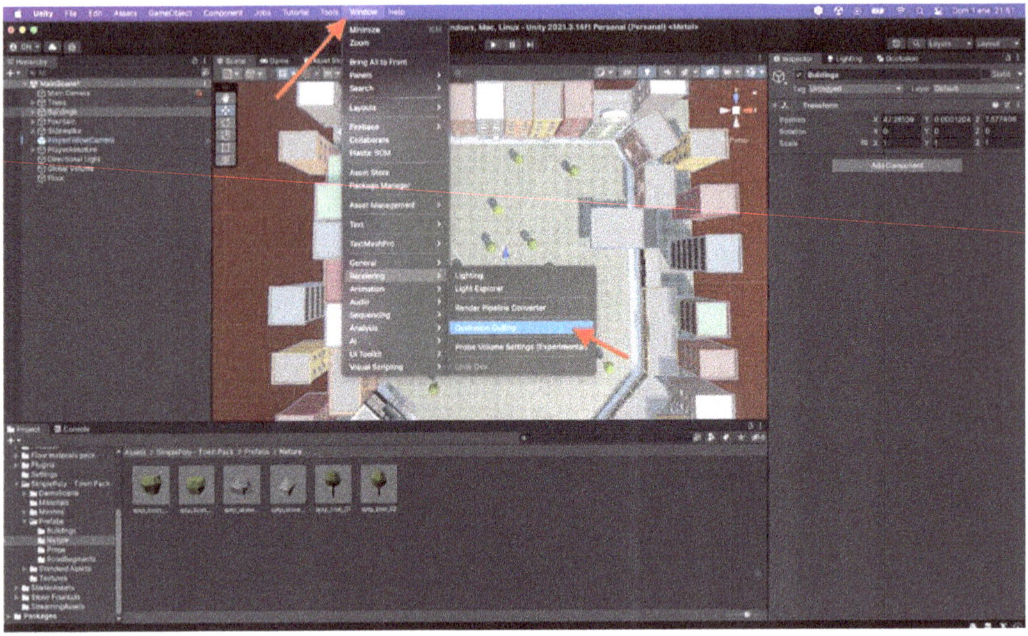

Figure 4.6 – Occlusion Culling menu

Once the **Occlusion** panel is opened, we will see three tabs:

- **Object**: This allows you to explore objects in the scene to view and edit whether they are static or not
- **Bake**: This tab is where the camera occlusion configuration is launched
- **Visualization**: This simulates the result by displaying the viewing angle of the active camera

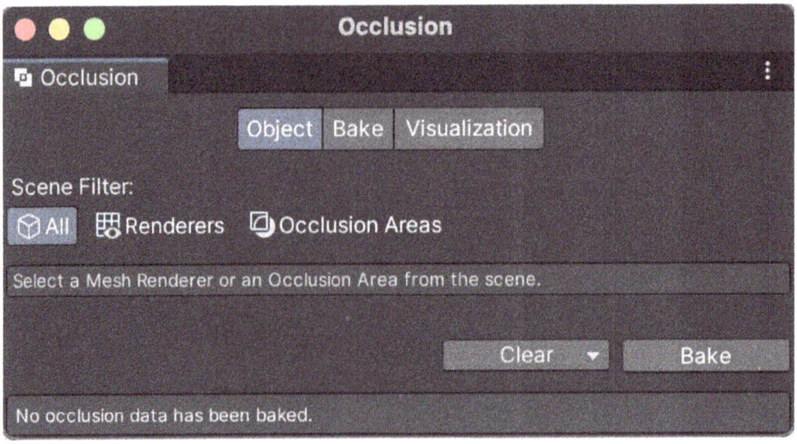

Figure 4.7 – Occlusion panel

2. Click on the **Bake** tab. The **Smallest Occluder** and **Smallest Hole** values serve to configure the granularity of the objects we want to hide. Change the value of **Smallest Occluder** from 5 to 1. The **Smallest Hole** value can be left as is by default. Finally, select the **Bake** button. This process may take a while, so be patient.

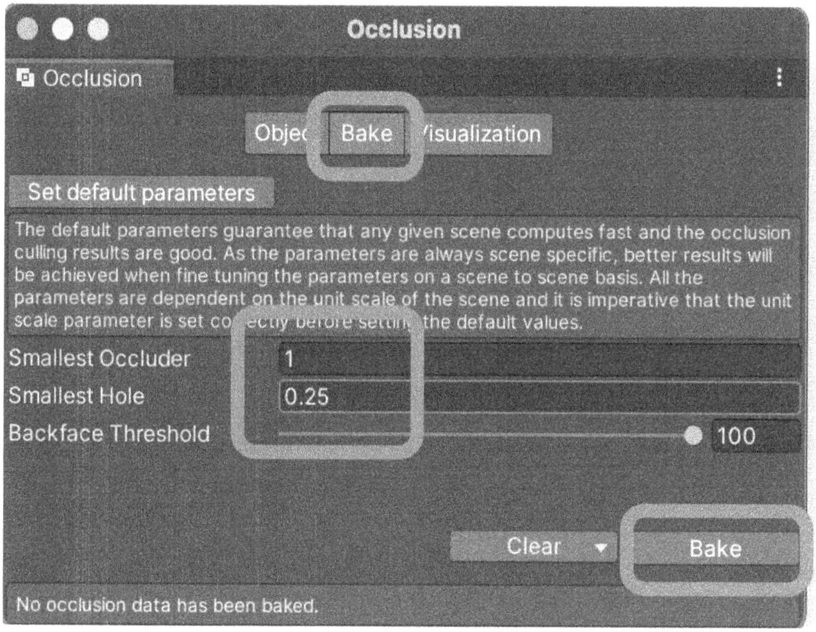

Figure 4.8 – Bake tab in Occlusion panel

3. As you can see, a blue mesh has been generated in the scene, something similar to the following figure.

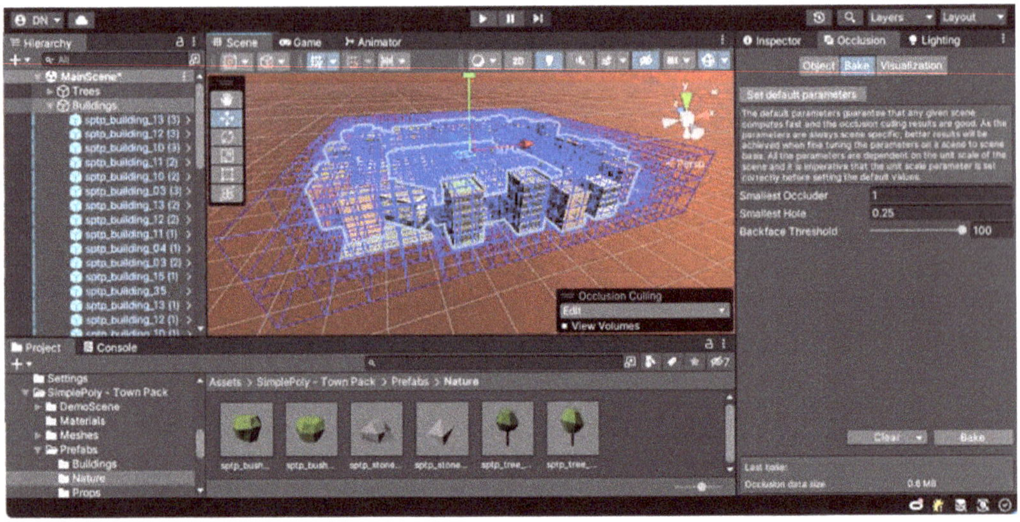

Figure 4.9 – Occlusion area previsualization

4. Now, click on the **Visualization** tab. You will notice that only objects that are within the camera's viewing angle will render in the scene.

Figure 4.10 – Occlusion area visualization

5. Finally, we need to apply the changes to the camera of the scene. For this, select the camera called **Main Camera** in the **Hardware** panel. Click the **Bake** button to finish the process.

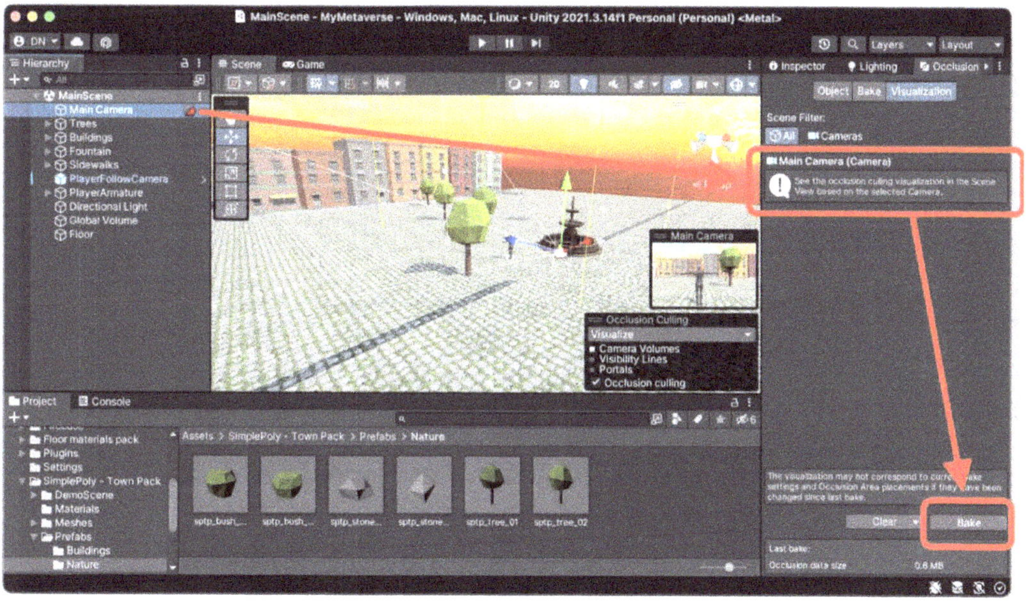

Figure 4.11 – Occlusion Bake button

Fantastic! You have managed to correctly apply the camera occlusion optimization; let's take a look at the real-time statistics to see whether we notice any changes. Click the **Play** button and open the **Statistics** window.

> **Tip: Common mistake**
>
> It is a common mistake to have problems with Unity culling and baking when a GameObject is not marked as static because the Unity engine assumes that objects can change their position, rotation, or scale at runtime. As a result, the engine cannot perform effective culling and baking calculations, which can lead to over-representation and unnecessary calculations. This can negatively impact game performance, especially in complex scenes or on resource-constrained devices. Marking relevant objects as static enables the engine to perform significant optimizations, such as culling and lighting baking, which helps avoid these problems and improves the overall game efficiency.

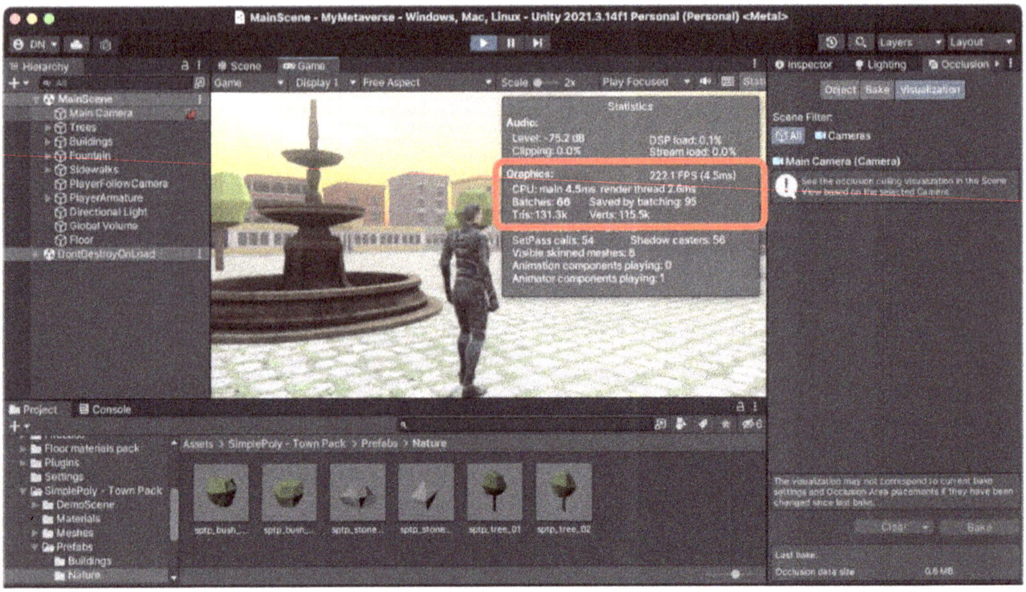

Figure 4.12 – Performance stats

At first glance, we can see that we now have 222.1 FPS, and before applying optimization, we had approximately 140 FPS. It can also be seen that the **CPU: main** values have dropped from `6.9ms` to `4.5ms`. We have certainly done a good optimization. Good job.

> **Note – values in Statistics**
>
> The values that appear in the **Statistics** window are subject to the technical characteristics of the computer. That is, the values that appear in this book do not have to be the same as on your computer; we are only comparing values before and after optimization.

You have been able to see how easy it is to optimize a scene. Our end users will appreciate all the optimization tasks we do. The next thing we are going to look at is undoubtedly the most important optimization task; then, we will learn about content optimization.

Light baking

The lights are undoubtedly the most dangerous thing in terms of the performance of a game. Lights have a big impact on rendering speed, so the lighting quality must be balanced by the frame rate.

Lighting in Unity can be so complex that we could dedicate an entire book to it. Let's summarize the concepts to understand the essentials.

In Unity, there are three ways to render a light:

- **Real-time**: Dynamic lights are, without a doubt, the most expensive lights (in terms of performance) in a video game. If they are real-time, this means that they can change at any time, be it their color, their angle, their intensity of projection, or the shadows they generate. This type of light is recalculated in each frame, so they are heavier in computation time.
- **Baked**: With a process called light baking, the shadows generated by the lights are transferred to the texture. Let's say the shadow is drawn in the texture, in a fixed way. This prevents the object from having to constantly calculate the shadow when receiving light.
- **Mixed**: Mixed lights, a mix that offers the best of each type, are capable of generating dynamic shadows on baked textures. Unity has a special method to optimize this type of lights, so they are not as expensive as real-time lights, nor as cheap as honed lights, in terms of performance.

> **Tip: Advanced information about Unity lighting**
>
> If you want to know more about the three ways to render lights, I recommend you navigate to the official Unity documentation at `https://docs.unity3d.com/Manual/LightModes.html`.

Next, we will put into practice the optimization of the lights; what we will do is apply the light baking technique to the directional light that exists in our scene and that simulates the Sun.

The way to apply maximum optimization in the lights is by configuring the lights with the **Baked** type. However, this leads to us losing visual quality. It is true that the scene will be super-fast, but on the other hand, we will sacrifice details such as the real-time shadow of the player.

The following screenshot shows the **Statistics** window after applying light baking with the lights taking the **Baked** type.

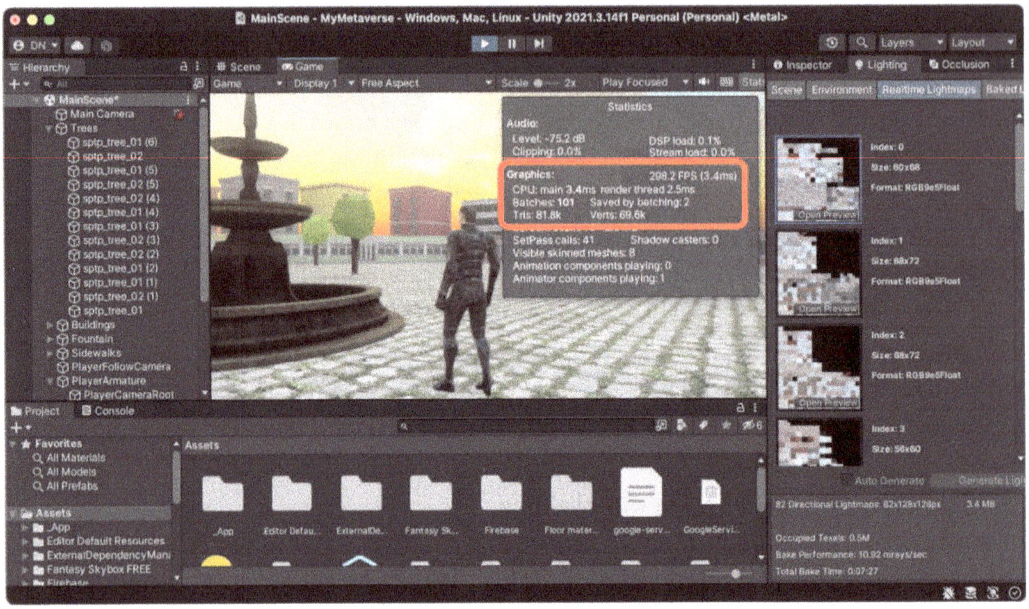

Figure 4.13 – Performance stats

As you can see, the player has lost their shadow in real time and the scene has become a little less vivid in terms of colors. However, the FPS has been increased by approximately 76 FPS, and the computation time in **CPU: main** has been reduced from `4.5ms` to `3.4ms`.

It is a great optimization, but we must sacrifice visual quality and details. Now, let's compare by applying the **Mixed** type setting to the light of the scene.

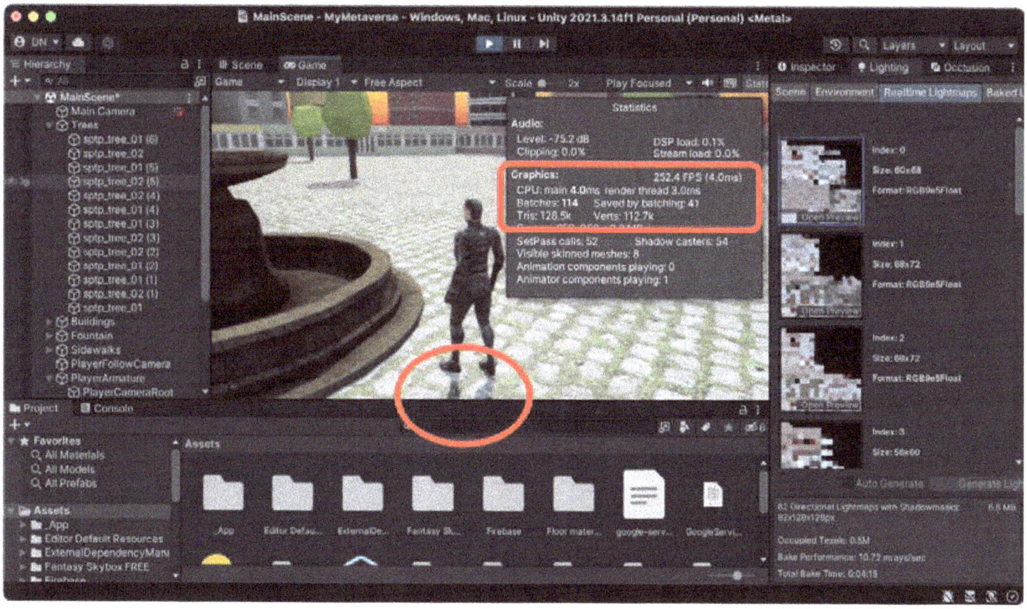

Figure 4.14 – Performance stats

By applying the optimization of **Light Baking** with the light taking the **Mixed** type, we see that the result in **Statistics** is better than not having optimization and lower than having a total optimization with the light taking the **Baked** type.

We have gained 30 FPS compared to the data when we were not optimized, and we have improved the **CPU: main** values from `4.5ms` to `4.0ms`. On the other hand, the visual quality is very close to what we had before applying **Light Baking**, and as you can see, we retain the shadow of the player.

In my opinion, the configuration depends on the level of quality and requirement we are looking for in a project, but for our metaverse, we can sacrifice some FPS in order to enjoy better graphic quality.

Next, we will see how to optimize lights; I will leave it in your hands to determine the quality you want to achieve for your project. This will not affect the process followed for the rest of the book at all.

To do this, follow these steps:

1. Locate the light of the scene called **Directional Light**, select it, and in the **Inspector** panel, change the value of the **Mode** property to **Mixed** (more visual quality but somewhat less optimization) or **Baked** (less visual quality but more optimization). In my case, I selected **Mixed**.

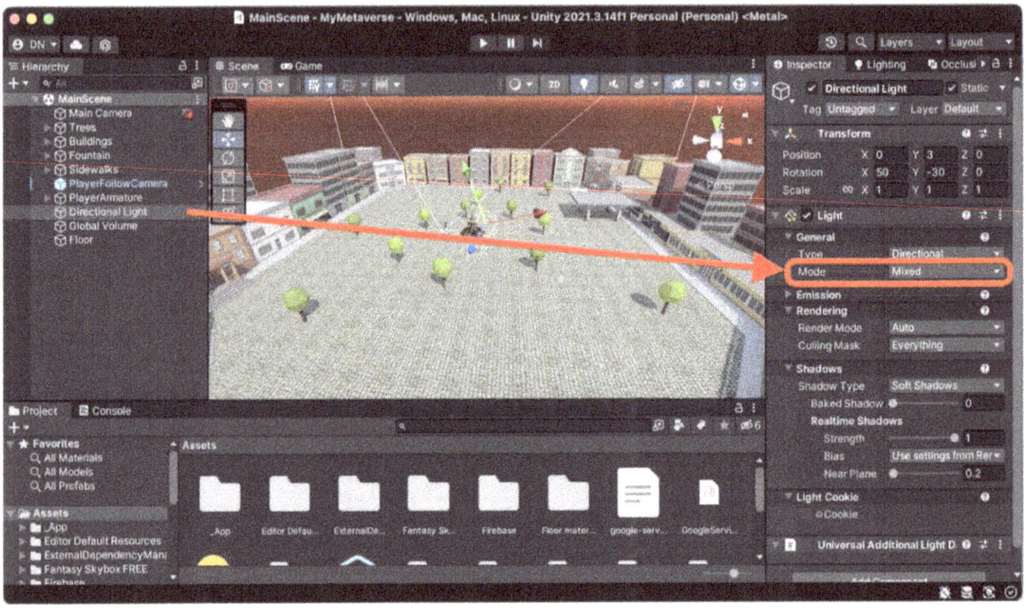

Figure 4.15 – Light mode option

2. From the main menu bar, select **Window** | **Rendering** | **Lighting** to open the **Lighting** panel.

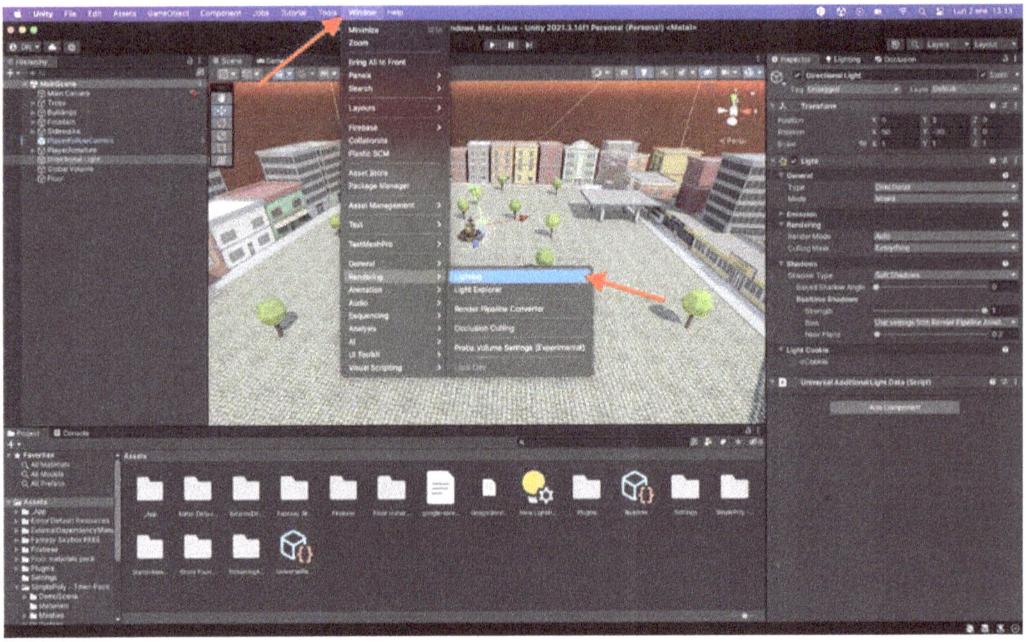

Figure 4.16 – Lighting menu

3. In the **Lighting** panel, click on the **New Lighting Settings** button to add a new light configuration arch.

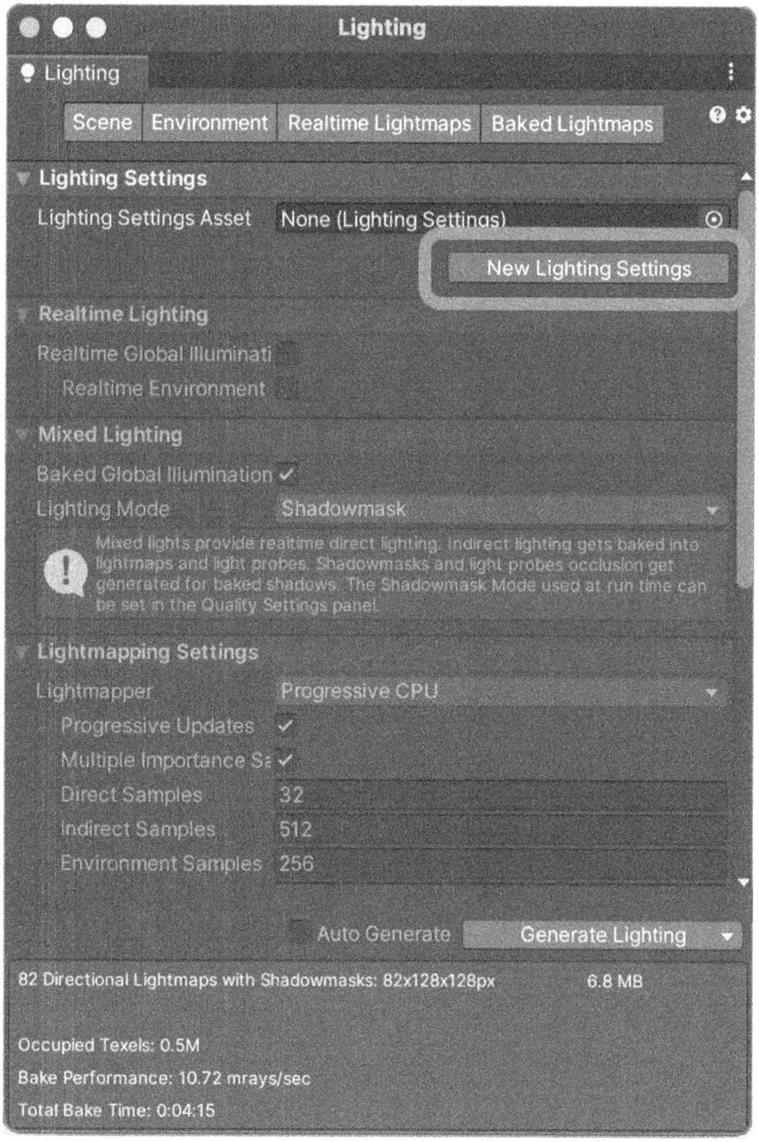

Figure 4.17 – New Lighting Settings button

4. After the configuration file is created, select the **Realtime Global Illumination**, **Realtime Environment Light**, and **Baked Global Illumination** checkboxes.

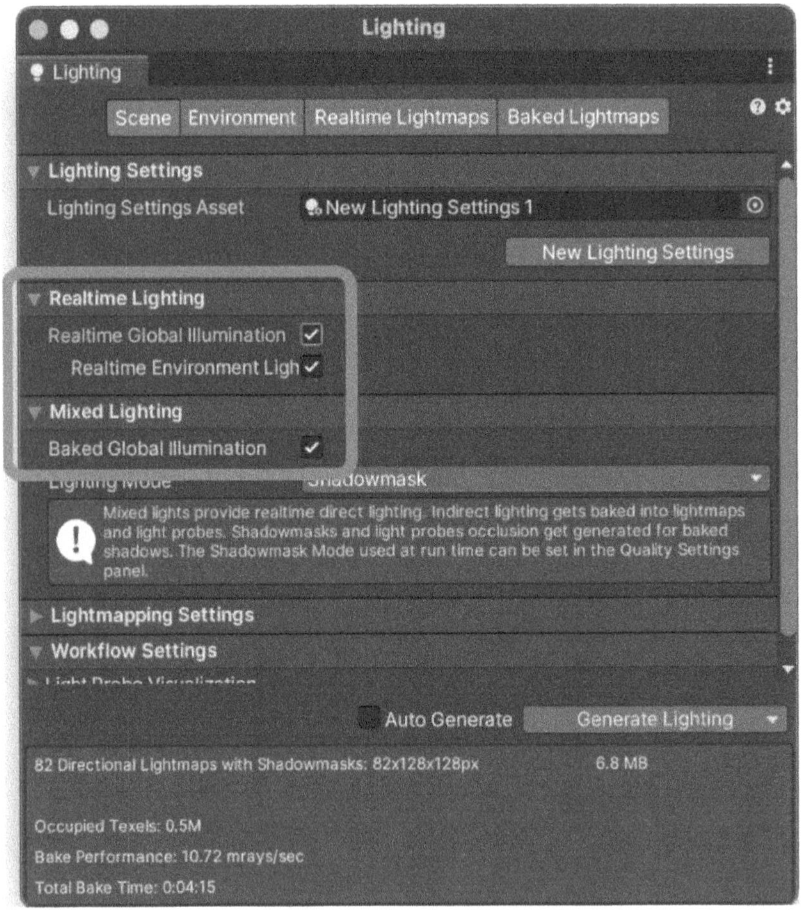

Figure 4.18 – Lighting configuration

5. In the **Lightmapping Settings** section, change the value of **Max Lightmap Size** to `512` and **Lightmap Compression** to **Low Quality**. Finally, activate the **Ambient Occlusion** checkbox.

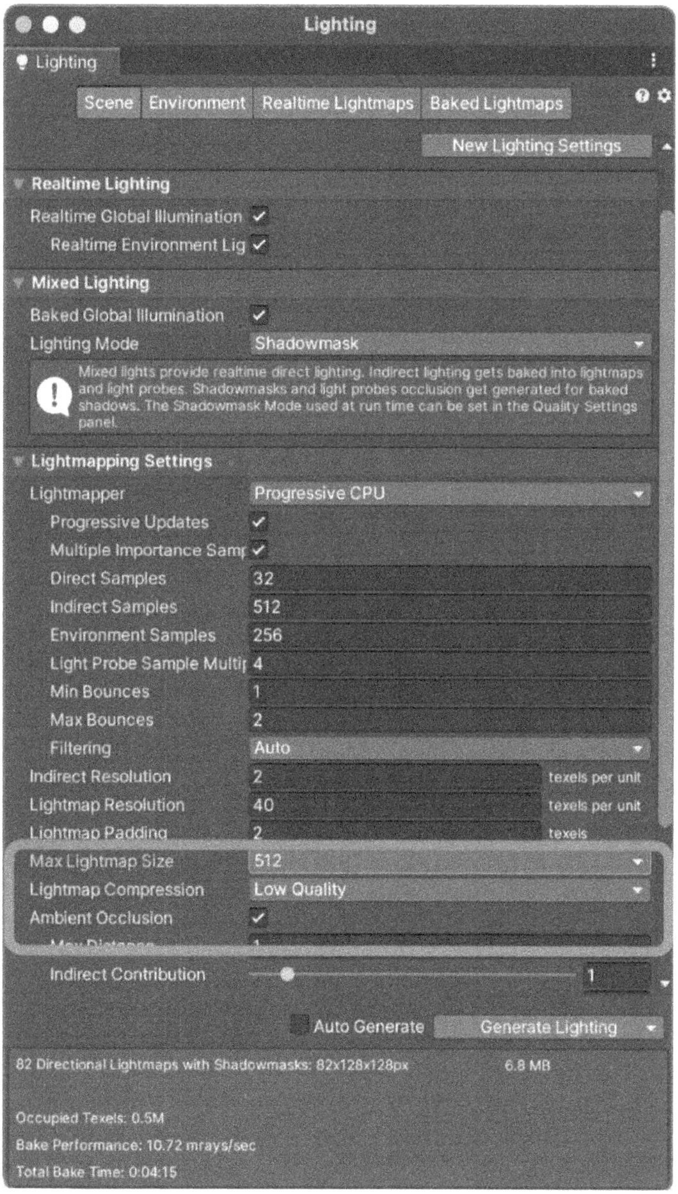

Figure 4.19 – Lighting configuration

These values that we have changed are related to the quality of the result. With higher values, your computer will need more time to process the **Light Mapping** task. The values we have selected offer optimal quality and the completion time of the task is acceptable.

I invite you to try higher configurations and check different qualities for yourself. Finally, I want to explain the operation of the **Ambient Occlusion** value that we have activated.

This gives a more realistic appearance to parts of objects where ambient light is blocked out or occluded. Activating it always gives a special touch to the scene, making it more realistic, and offers more visual volume:

1. In the **Lighting** panel, click on the **Environment** tab and drag **Directional Light** into the **Sun Source** box.

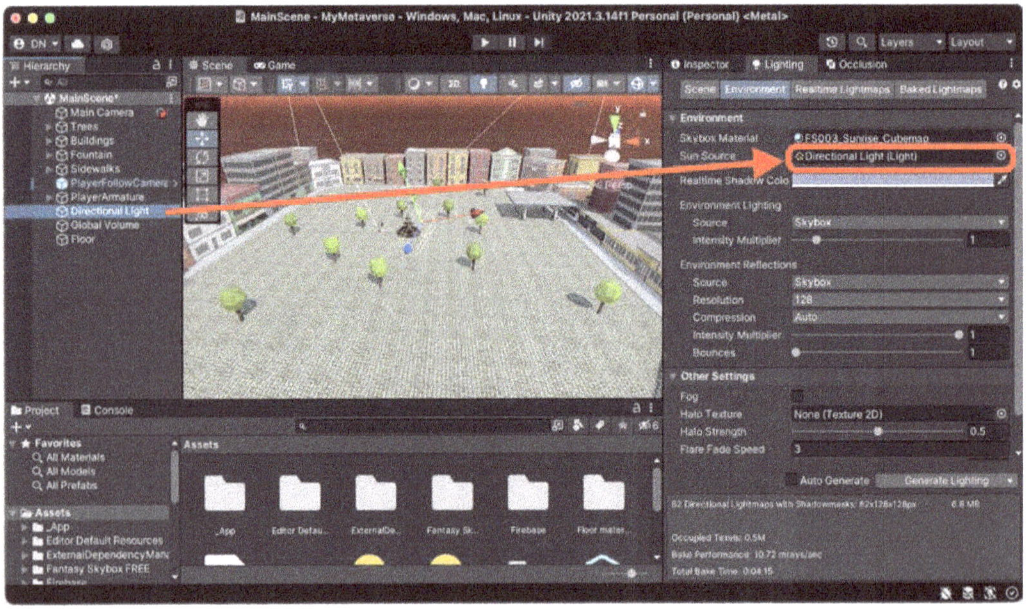

Figure 4.20 – Sun Source configuration

2. Finally, click the **Generate Lighting** button. This process, depending on the power of your computer, can take between 12 and 30 minutes (a progress bar will appear indicating the remaining time and the status of the task). Now is a good time to make a coffee.

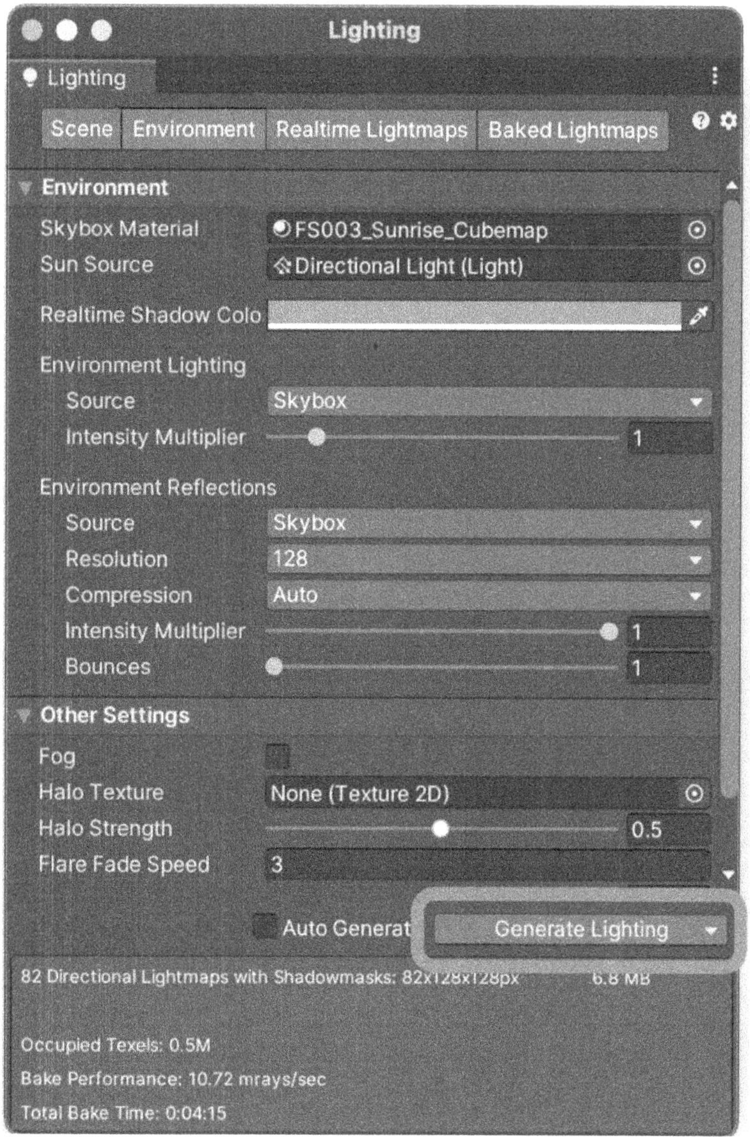

Figure 4.21 – Generate Lighting button

Congratulations, you've finished one of the most important optimization tasks in creating games with Unity. As we have learned, the lights now affect the performance of the scene less and we have managed to maintain the highest possible quality.

Remember that this task must be applied to all the scenes you create in the future.

Tip: Warning about the light optimization task

The task of optimizing lights will be carried out on the current objects of the scene. If these objects undergo any modification, such as a change of scale, position, or rotation, or if you add a new object, this optimization will be outdated, and you must launch it again.

I recommend that you leave this task to the end in your future projects. When the scene is totally to your liking and you think it will not undergo new changes, this is when you should execute the light baking optimization.

We can't conclude the section on optimizations without talking about colliders. As we have seen before, our buildings and objects used a Mesh Collider to detect collisions in all their geometry.

Colliders

As far as possible, we should avoid using Mesh Colliders in our GameObjects. This is because they are more expensive in terms of performance. For this, in most cases, Mesh Colliders can be replaced with Primitive Colliders. The most common are boxes, spheres, and capsules.

Let's look at a practical example by looking at one of our trees:

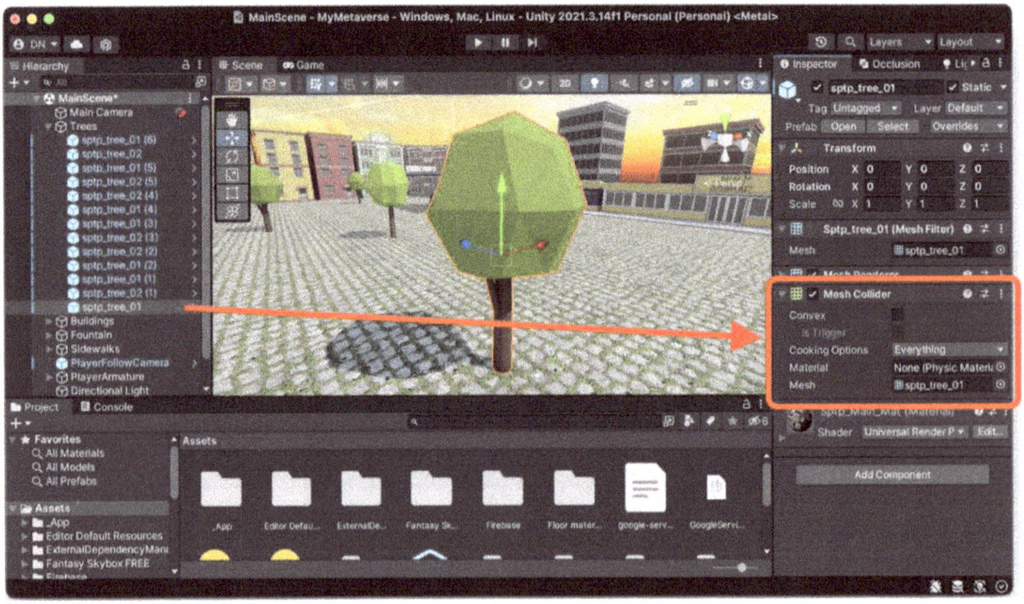

Figure 4.22 – Mesh Collider component

Now, let's reflect; our trees have a Mesh Collider. That means when the player comes into contact with it, they won't be able to collide with or pass through it. The question we must ask ourselves with each object in our scene is, *will there ever be a situation where the player can potentially pass through the leaves on the tree?* Can we simplify the Mesh Collider for a collider to only apply to the trunk?

Of course, we don't want our player to pass through the tree, but a Primitive Collider on the trunk will be sufficient since it is impossible for the player to reach the top of the tree.

To simplify this collider on the tree, let's follow these steps:

1. In the **Mesh Collider** component, click the button with the three dots and select the **Remove Component** option.

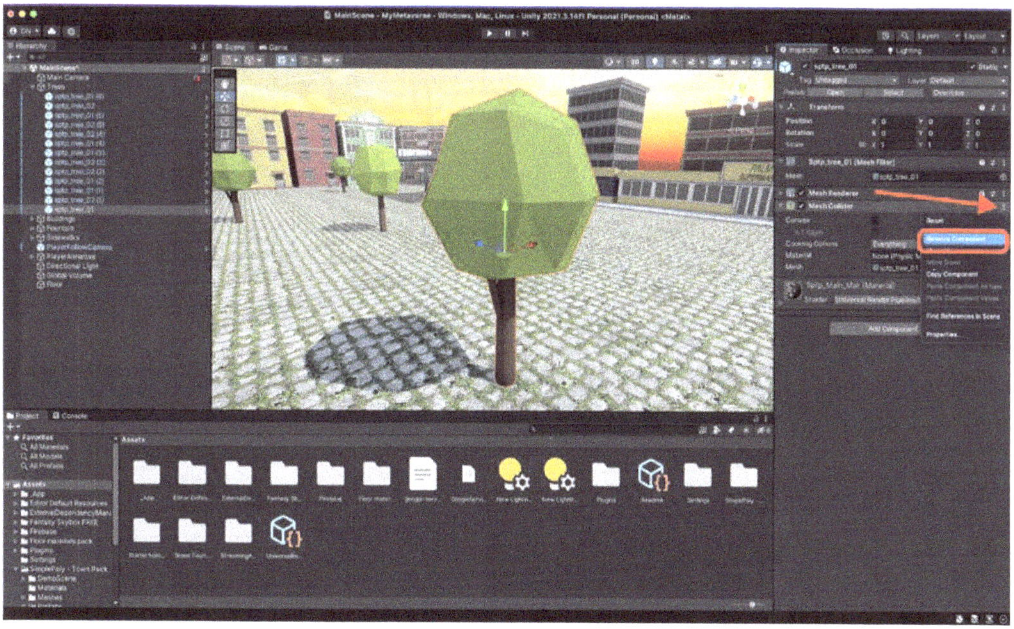

Figure 4.23 – Remove Component option

2. In the **Inspector** panel, click the **Add Component** button, type `collider` into the textbox, and select **Box Collider**.

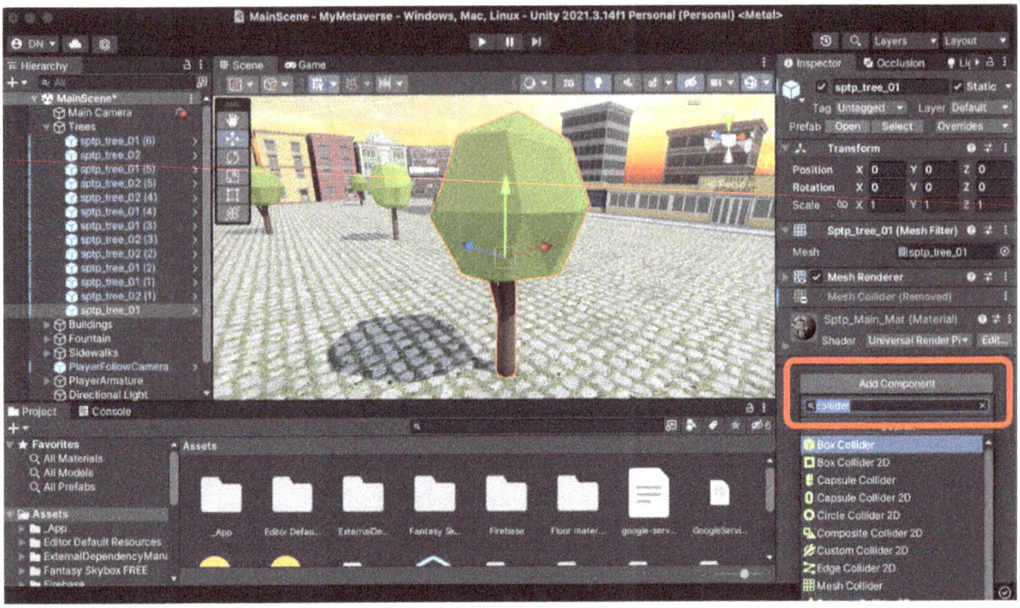

Figure 4.24 – Add Component option

3. You have added a Box Collider. You can see its graphical representation by wrapping the tree. The green lines give an indication of its size.

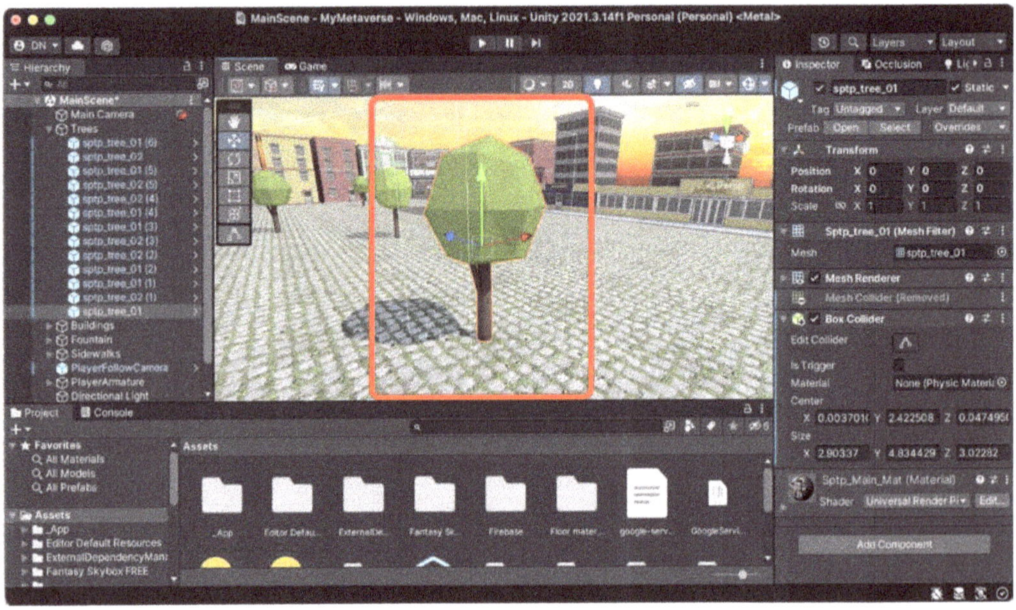

Figure 4.25 – Box Collider previsualization

4. Now, if we click on the **Play** button and direct our player to this tree, we will see that our player's body collides with the trunk of the tree. This is because the **Box Collider** component acts as a physical barrier.
5. To adjust the collider and make it more realistic, we want the player to be able to approach the trunk and be unable to break through it. In the **Inspector** panel, the **Box Collider** component has values that define its size and position.
6. We change the values of the **Size** property to **X**: `0.6`, **Y**: `2`, and **Z**: `0.6` and the values of the **Center** property to **X**: `0`, **Y**: `1`, **Z**: `0`. As you change the values, you will see how the collider is redrawn in the **Scene** panel.

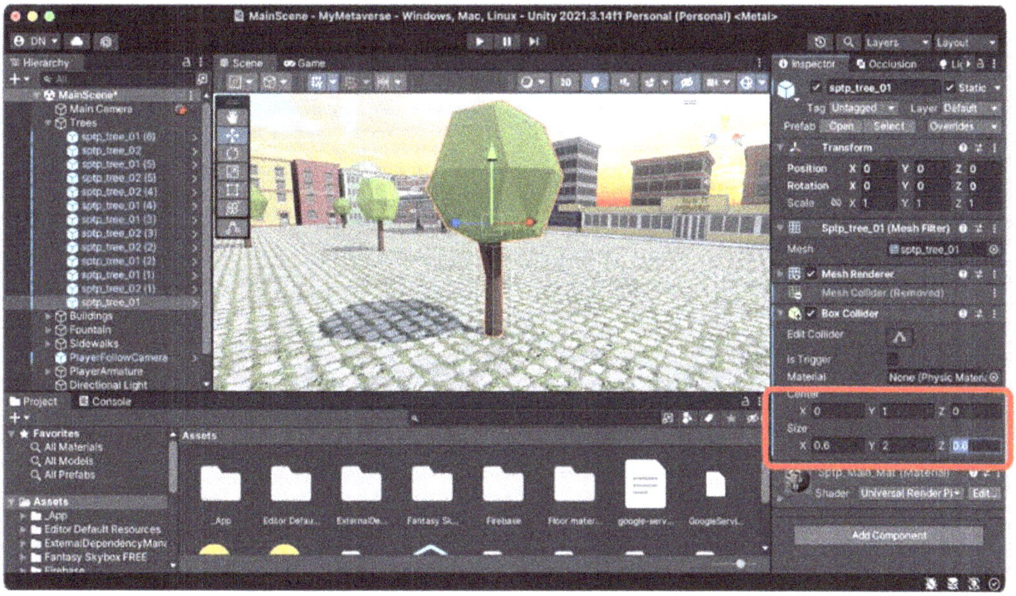

Figure 4.26 – Changing the Box Collider properties

7. Finally, one of the great advantages of using Prefabs is that if a change is made to one of them, you can easily propagate it to the others. This avoids us having to remove the Mesh Collider from each tree.
8. Add the Box Collider and change its values again. To do this, in the **Inspector** panel, there is a button that says **Overrides**. This button gives us information about the changes that have been made compared to the original version. Click on the **Overrides** button.
9. Then, a small modal window opens informing us that we have removed the **Mesh Collider** component and added a new **Box Collider** component. Click on the **Apply All** button.

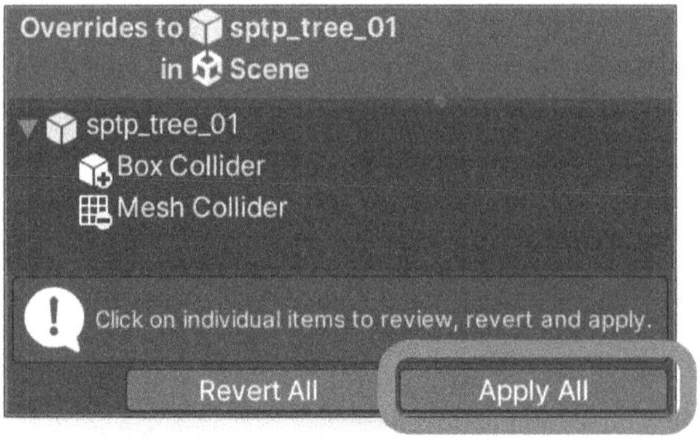

Figure 4.27 – Apply Prefab changes confirmation window

> **Tip: Information about the tree**
>
> You have just modified the collider of a tree in our scene. You have applied the changes so that they are reflected in all the trees that use the same Prefab. Specifically, my example has been on the tree called `sptp_tree_01`. Remember that we have used two types of trees in the scene. The changes have been applied to one of the two models (`sptp_tree_01` in my case). You must perform the same process for the other tree model, the tree called `sptp_tree_02`.

Impressive, you have just learned how to optimize colliders. Now that you know how to do it, you can do it on other objects, such as buildings and sidewalks, too. Remember, if any building or object is complex and you find it tedious to optimize the collider, just use the **Mesh Collider**. It will not lead to a dramatic change in the performance of our metaverse if we have some Mesh Colliders in the scene. Just avoid them whenever you can.

> **Tip: Advanced information about Colliders**
>
> The world of colliders in Unity has enough material and information to write an entire book. We have reviewed the most basic and important aspects. If you want to explore more technical details, I recommend you visit the following page: `https://docs.unity3d.com/560/Documentation/Manual/CollidersOverview.html`.

We've done a great job together; we've come a long way so far. If you take a look back, you will see that you have learned in this chapter how to design a beautiful scene from scratch. You have learned about important performance optimization techniques. To close this chapter, we are going to add a reference to the scene in the Firebase database.

Adding the first world to the Firebase database

As a culmination of this wonderful chapter in which we have learned a lot of things, we need to register a reference to the main scene in Firebase. We will do this because in *Chapter 5, Preparing a New World for Travel*, and *Chapter 6, Adding a Registration and Login Form for Our Users*, we will need to consult the database to know what worlds we have active to offer our users a place to travel.

For now, it will be enough to register a record in the database. Follow these steps:

1. Go to the Firebase console: `https://console.firebase.google.com`.
2. In the **Build** section, click **Firestore Database** and then click the **Create Database** button.

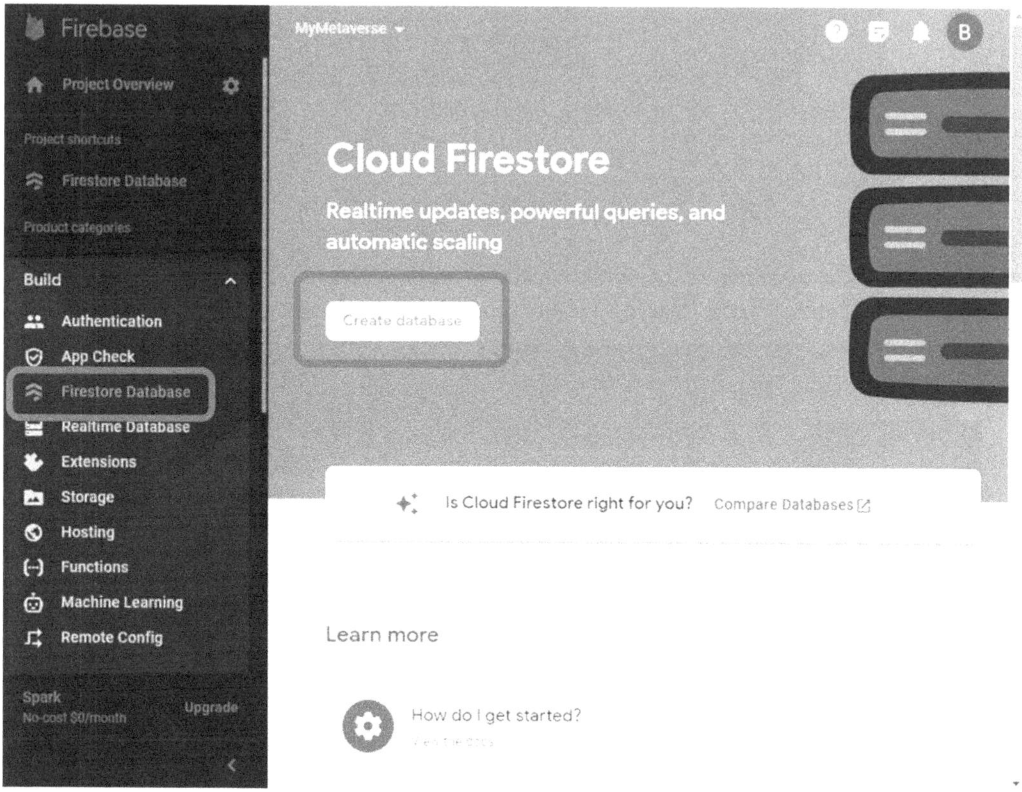

Figure 4.28 – Firebase Firestore page

3. A window will open with a configuration-type form for our first database. The first step that is shown is for write and read permissions for third-party applications. For now, we will leave it as the default and click the **Next** button.

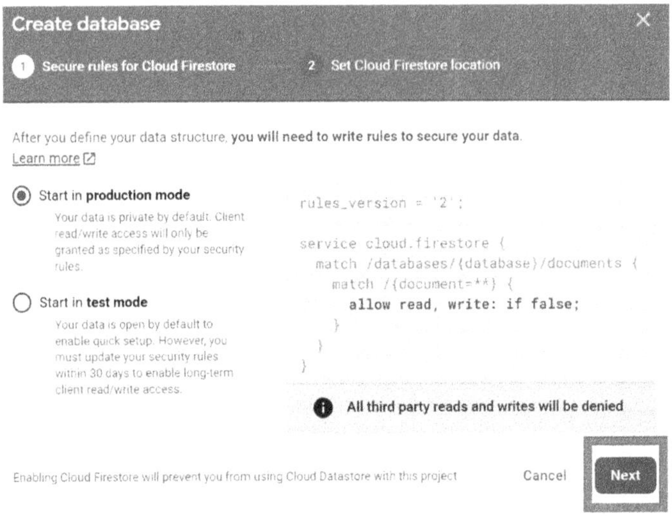

Figure 4.29 – Firebase Firestore activation window

4. The next step is to configure the physical location of the server of our database. We'll leave it as the default and click the **Enable** button to finish the configuration.

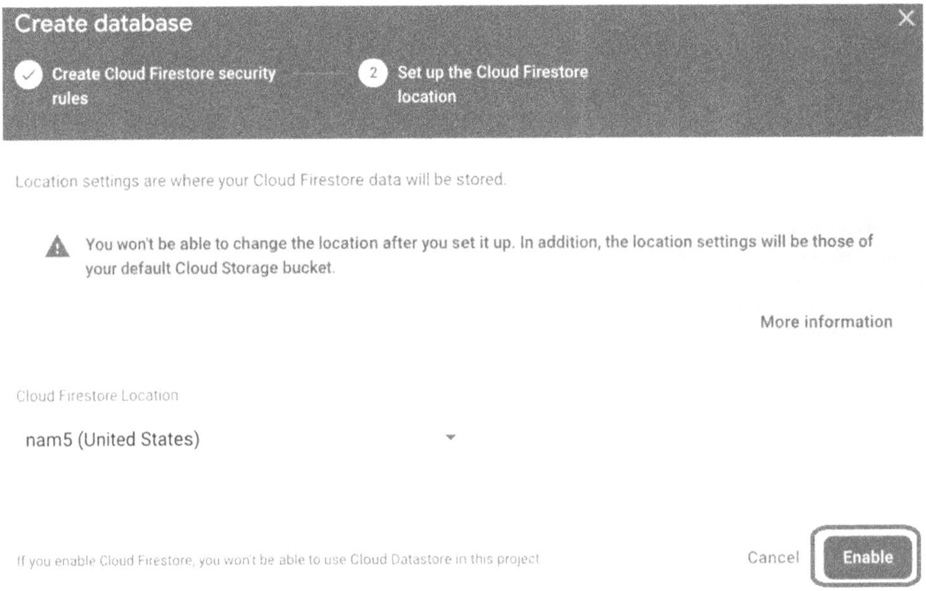

Figure 4.30 – Firebase Firestore creation window

5. The creation process may take a while. When it finishes, you will see a page similar to this:

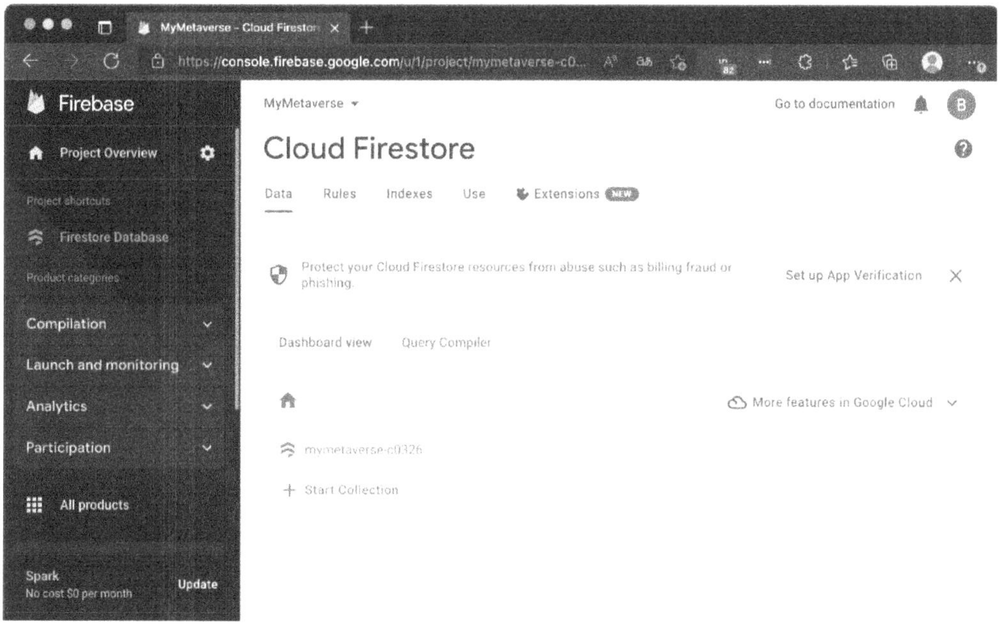

Figure 4.31 – Firebase Firestore dashboard page

6. In Firebase Firestore, collections are the equivalent of tables in standard databases. To create our first collection, click on the **Start Collection** button.

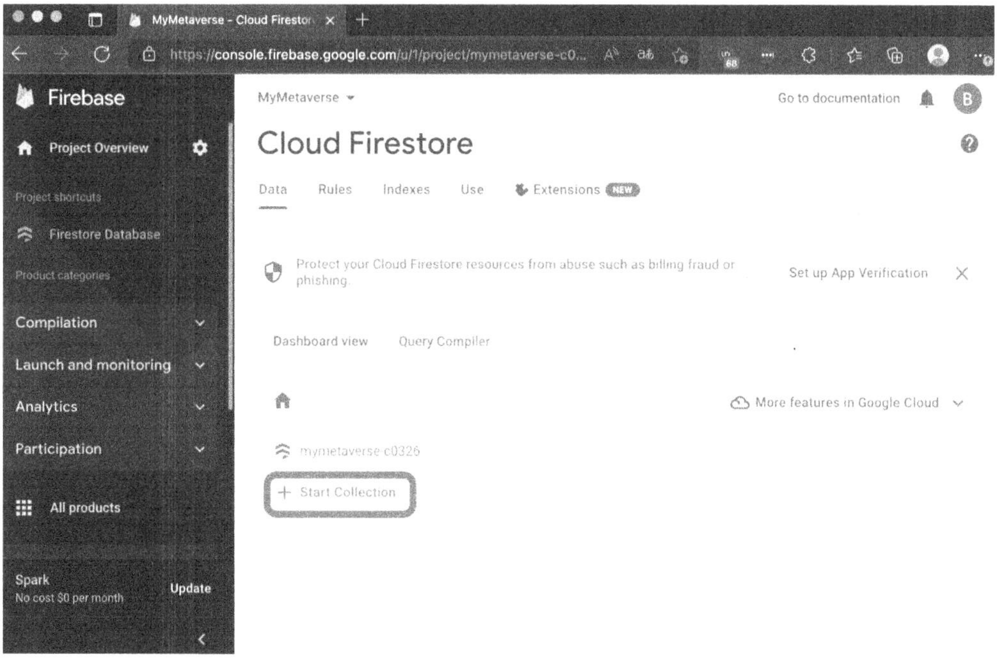

Figure 4.32 – Firebase Firestore Start Collection button

7. A creation form will open. We type the name of our collection. In this case, we will call it `Worlds` and click the **Next** button.

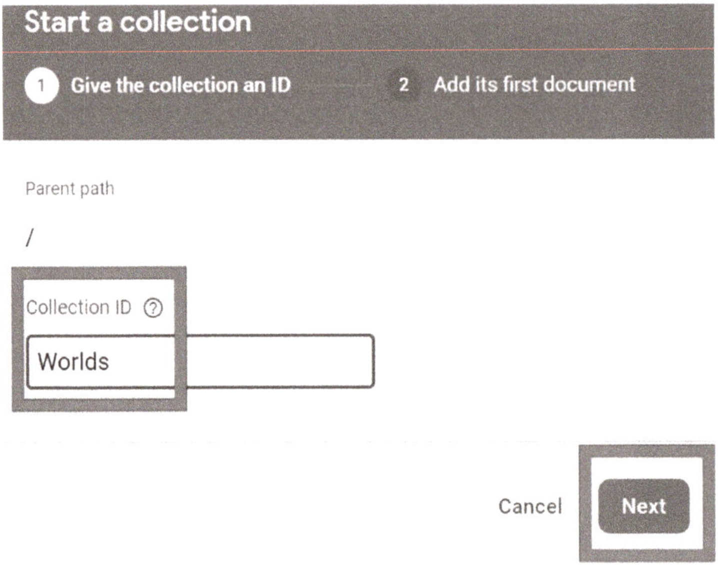

Figure 4.33 – Collection creation window

8. Well, the next step is to configure the first item of the collection. In our case, it is the first world of our **Worlds** collection. As you can see, the first field that it asks us to fill in is **Document ID**. This field is a unique ID for each record in the collection. We click the **Auto-ID** button so that Firebase generates one.

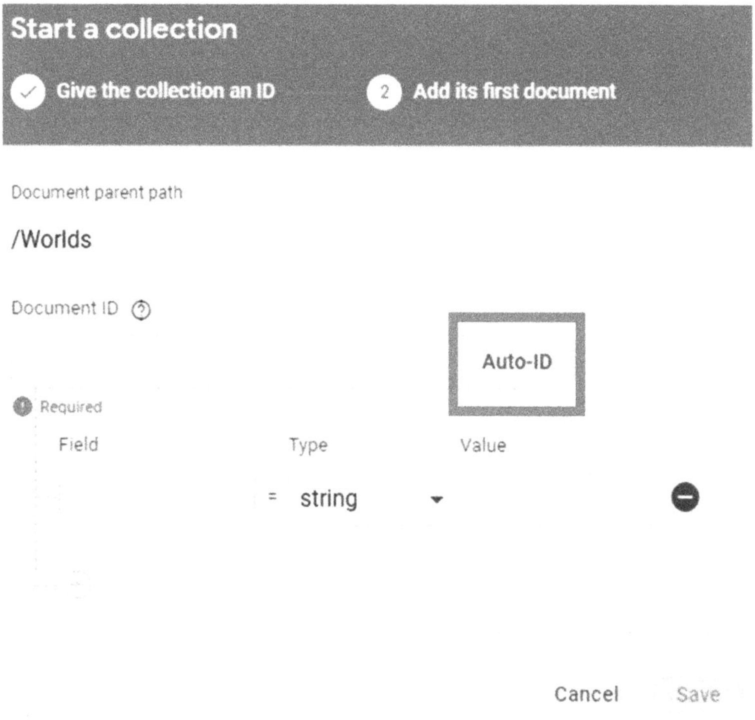

Figure 4.34 – Button to auto-generate a collection ID

9. Great, you may have noticed that Firebase has populated the field with a unique autogenerated ID. Now, we have to fill in the information of our world, that is, with information that we will need in the future. For now, we will need to provide the name, description, and name of the scene. Type `Name` in the **Field** textbox and `Meeting Point` in the **Value** textbox. You will have something similar to this:

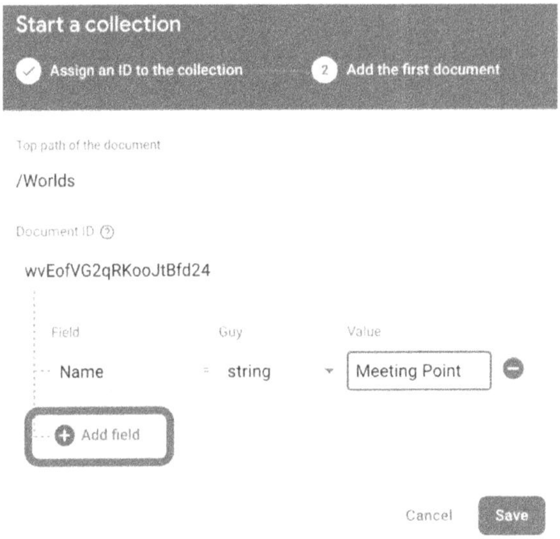

Figure 4.35 – Button for adding a new field to the collection

10. Click on the **Add field** button to add a new property to the record. In this case, we will fill in the **Field** textbox with `Description` and the **Value** property with `This is the main place of my Metaverse`.

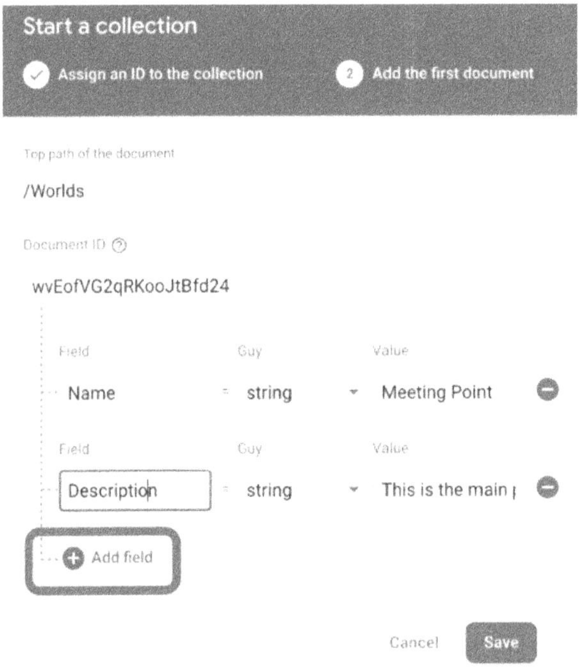

Figure 4.36 – Adding another property in the collection creation window

11. Click the **Add field** button again and add `Scene` to the **Field** textbox and `MainScene` to the **Value** textbox.

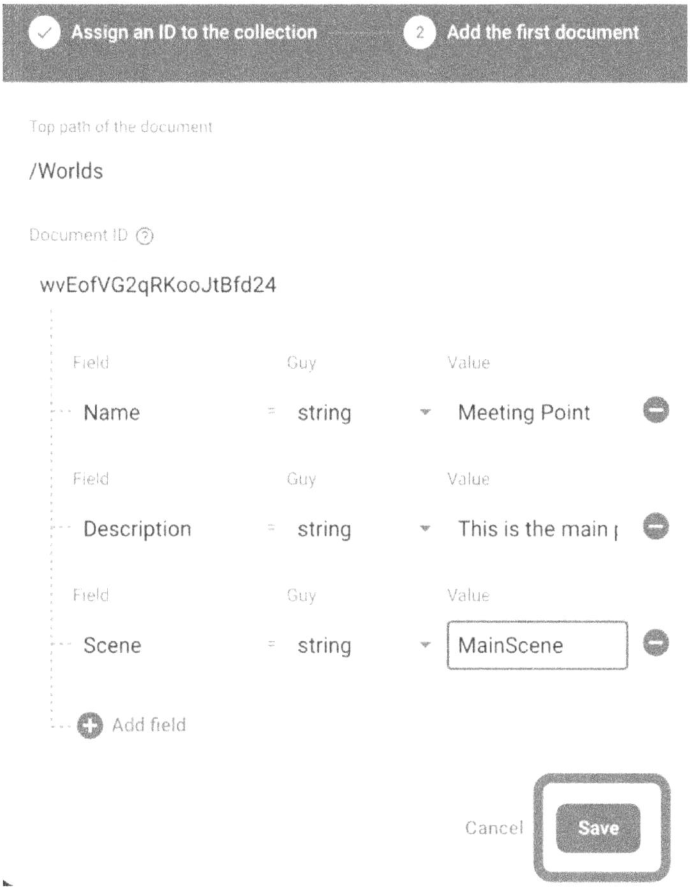

Figure 4.37 – New collection Save button

12. Finally, click the **Save** button to finish. You will then see the page with the Firestore dashboard and the registry you just created.

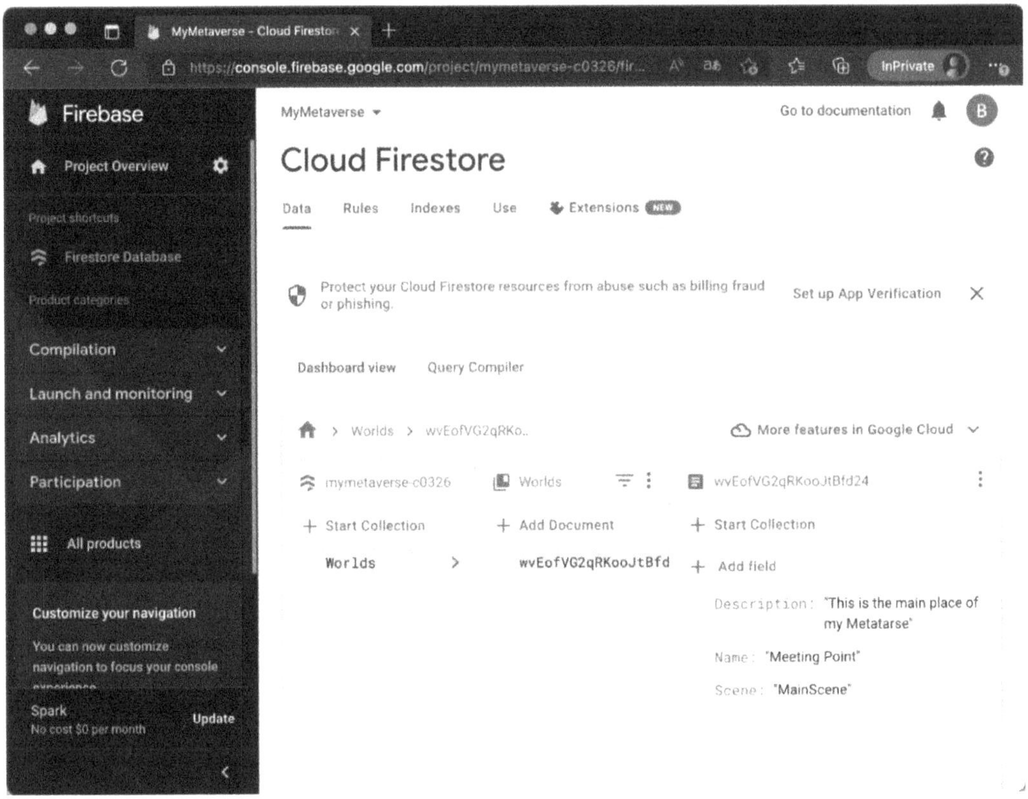

Figure 4.38 – Firebase Firestore dashboard page

With this last step, you have just created your first record in your first collection in Firebase Firestore. Impressive work!

There is a detail that you have surely discovered. In the **Scene** property of the record that we have just added, we have put exactly the same name that our scene has. This is not trivial. The name must be the same because in future chapters, we will use this value to dynamically navigate between scenes... Sounds interesting, right? I think we did a pretty good job in this chapter.

Summary

We have reached the end of this interesting chapter, in which we learned the most basic and important optimization techniques. Now you know how to improve performance with static objects, how to improve graphics performance with the camera occlusion culling technique, and how to optimize lights with the Bake technique. Finally, we learned more about the different types of Colliders and how they can be improved. All these techniques help us to obtain a higher FPS rate. Your future users will appreciate it.

Now that we have our initial world, in which every player will appear at the start of the experience, we need to offer another place to explore. In the next chapter, we will build another world for our users, in which we will include houses that can be purchased by our players.

5

Preparing a New World for Travel

After having designed an impressive meeting point in *Chapter 3*, *Preparing Our Home Sweet Home: Part 1*, we will need another scene for our users to travel to. In this second world we are going to design, we will focus on creating a neighborhood-type scene, with houses, sidewalks, lampposts, and some more details. The future intention of this scene is that the players will be able to purchase these houses. We will see this in *Chapter 8*, *Acquiring a House*.

The big difference in this chapter, compared to *Chapters 3* and *4*, is that the buildings will be loaded dynamically in the scene; that is to say, we will create records in the Firebase database with references to Prefabs in the project, and when executing the scene, they will be created.

This offers us great versatility to have a dynamic world. Loading one building Prefab or another will be customizable from the database.

We will cover the following topics:

- Designing the scene
- Adding the world to the Firebase database
- Adding houses dynamically
- Optimizing a dynamic scene

Technical requirements

This chapter does not have any special technical requirements, but as we will start programming scripts in C#, it would be advisable to have a basic knowledge of this programming language. We need an internet connection to browse and download an asset from the Unity Asset Store.

We will continue with the project we created in *Chapter 1, Getting Started with Unity and Firebase.* Remember that we have a GitHub repository, `https://github.com/PacktPublishing/Build-Your-Own-Metaverse-with-Unity/tree/main/UnityProject`, which contains the complete project that we will work on here.

You can also find the complete code for this chapter on GitHub at: `https://github.com/PacktPublishing/Build-Your-Own-Metaverse-with-Unity/tree/main/Chapter05`

Designing the Scene

We have finally arrived at the chapter where we will start programming. In addition to designing a nice scene that will serve to attach houses to our world, we will add functionalities through C# scripts. These functionalities will give the scene the power to obtain buildings dynamically from our Firestore database. But what does this mean?

We will create functionality that will help us to manage the slots in the scene that can contain a building. This way, we will not have to perform the tedious task of creating dozens of records in the database.

We, as creators of worlds, will design a scene resembling a neighborhood. In the holes where the houses will go, we will create a special object with a script in C# that will register and update these slots, linking them to Firebase.

Whether we press **Play** in the scene or when the metaverse is compiled and published, when the user enters the game, these scripts will connect to the database, consult the building, and if it exists, it will automatically be placed in the scene, in the position where we previously placed the special object that we created.

Think about the multiple advantages of the system that we are going to build. What if, in the future, we build an administration web console where we can manage these buildings? From there, we could modify the building registry and, for example, change the Prefab that should automatically load the player when they re-enter the Metaverse, will see the building with the new look, and all this without having to recompile and publish. Sounds amazing and magical, doesn't it? This is what we will do together in this chapter.

Creating the scene

Are you ready? Let's start by creating a scene and designing it, with what we previously learned in *Chapter 3, Preparing Our Home Sweet Home: Part 1*, and *Chapter 4, Preparing Our Home Sweet Home: Part 2*. We will be able to do it without any problem.

Follow these steps to create a new scene in our project:

1. In the main menu bar of the Unity Editor, click on **File** | **New Scene**.
2. A window will open with the scene types you can create. Select **Standard (URP)** and click on the **Create** button.

3. Next, Unity will ask you for the name of the new scene and where to create it. We will call it `World1`. Follow the **Assets** | **_App** | **Scenes** path, then click on the **Save** button to finish.

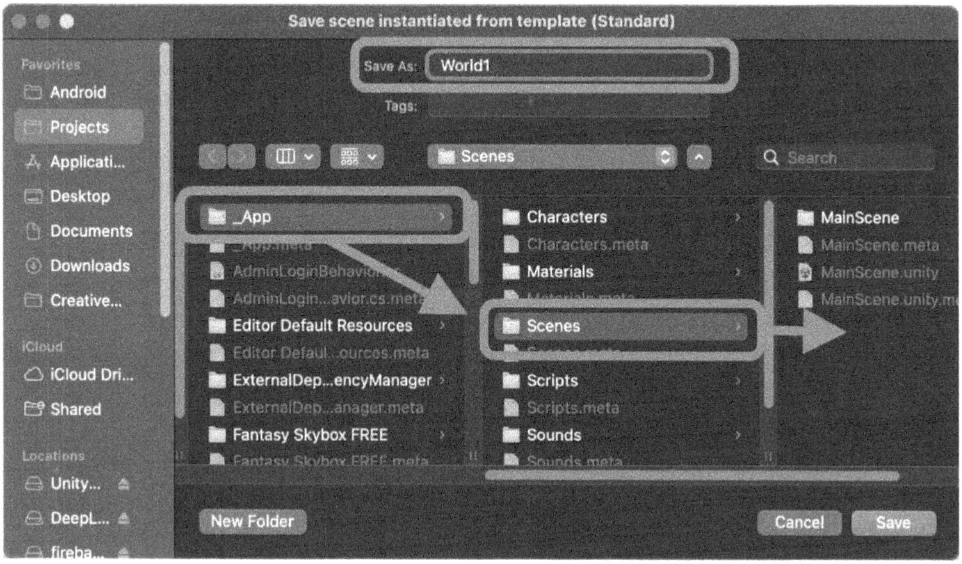

Figure 5.1 – Destination folder for a new scene

4. As you will see, a new scene called **World1** has been created and a folder with the same name contains the configuration.

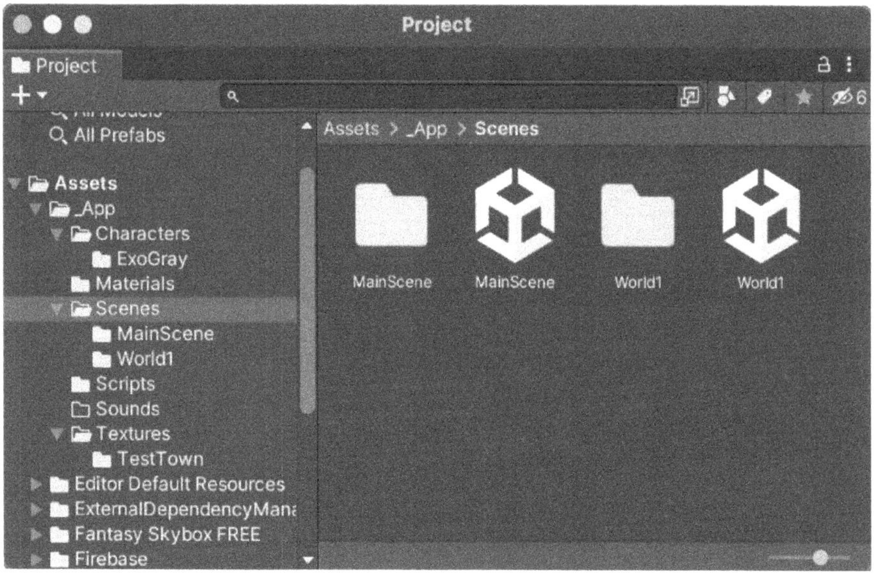

Figure 5.2 – Newly created scene

Fantastic! Once we have finished creating the scene, we can proceed with constructing the floor.

Creating the floor

After finishing the scene creation process, as we saw previously in *Chapter 1*, *Getting Started with Unity and Firebase*, Unity will open this new scene to start working on it. The first thing we will do is place a floor:

1. In the main menu bar, click on **GameObject** | **3D Object** | **Plane**.
2. A new GameObject will be created in the **Hierarchy** panel. Rename it `Floor`.

   With the **Floor** GameObject selected, change its **Scale** property to `10`, `10`, `10` so that your scene looks similar to as in the following screenshot.

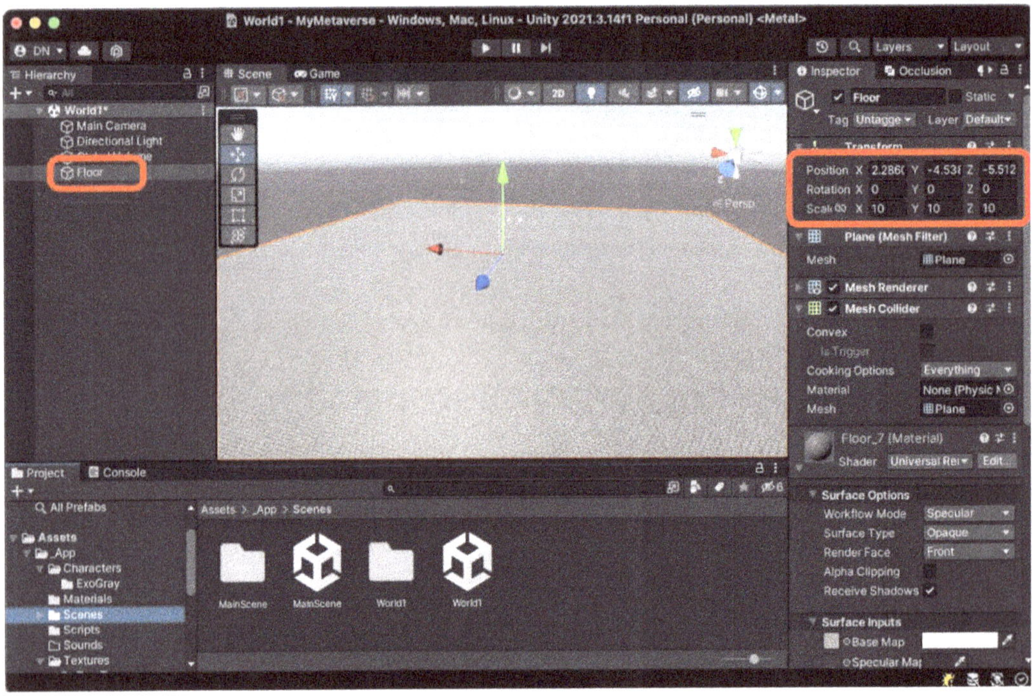

Figure 5.3 – Resized Floor GameObject

To give it a nicer look, let's assign a material to it.

3. In the **Project** panel, navigate to the **Assets** | **Floor materials pack** | **Material** path and drag the material named **Floor_7** to drop it onto the **Floor** GameObject:

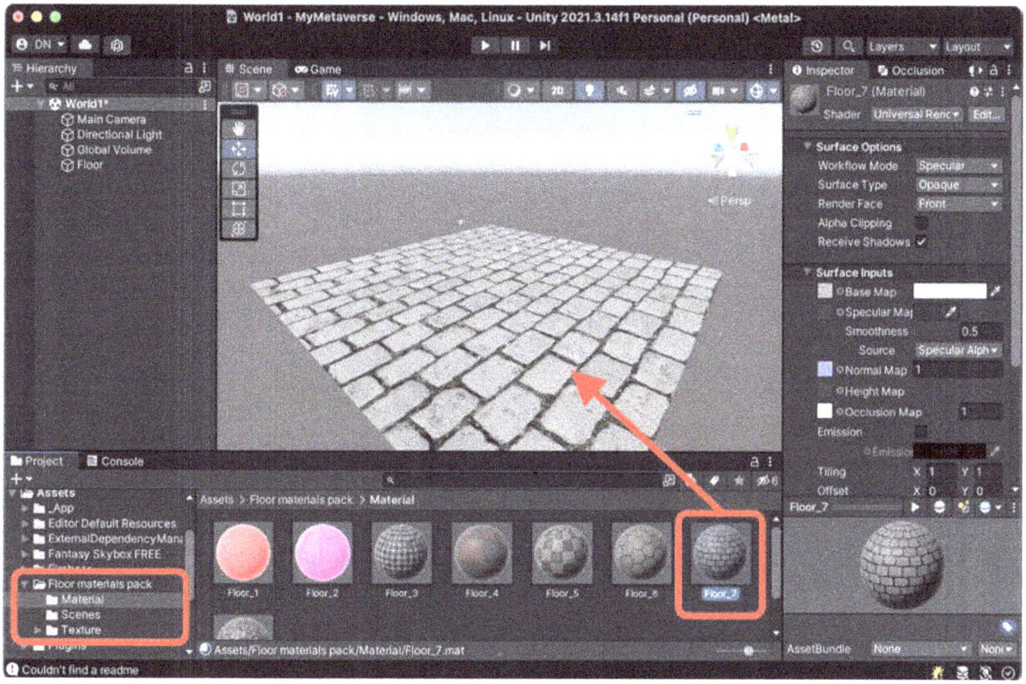

Figure 5.4 – Assigning a material to the floor

4. As you see, the aspect has changed, but the tiles are very big. We must customize the **Tiling** values so that the texture repeats more times. With the **Floor** GameObject selected, scroll down in the **Inspector** panel. Inside the **Material** component, change the **Tiling** values to **X**: `25`, **Y**: `25`, and the texture will adjust to the size of the floor.

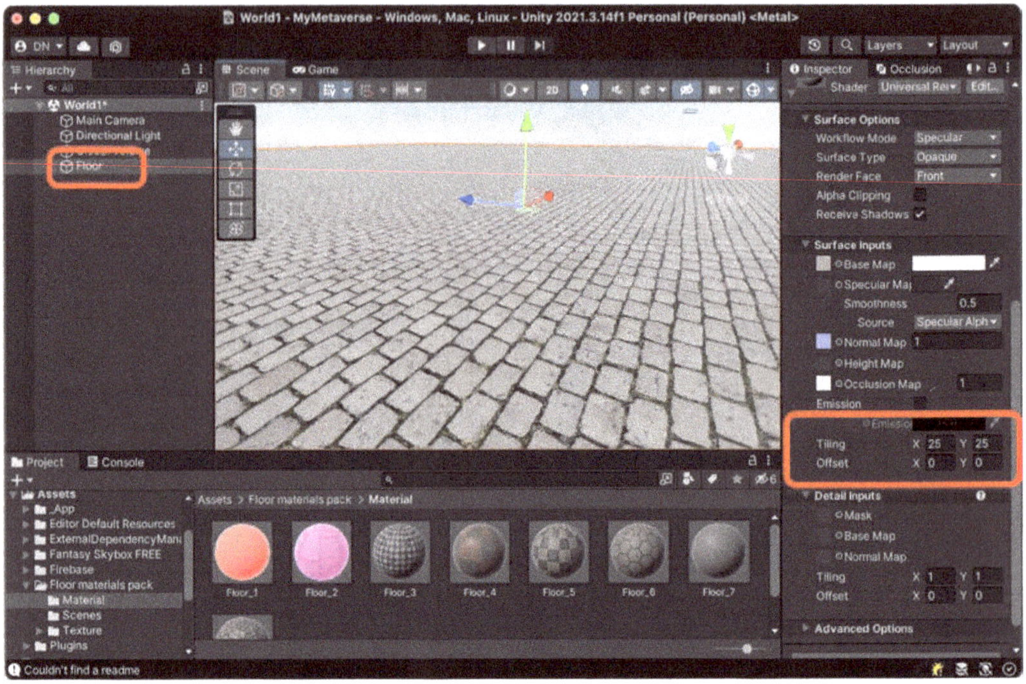

Figure 5.5 – Changing the floor's Tiling values

Easy, isn't it? With the ground built and perfectly configured, we can move on. Next, we will explore a free asset with beautiful houses that we will use.

Creating houses

Well, we now have a decent floor, but before we continue decorating the scene, we need to know the approximate size of the future houses we will place. The reason to know the size of the house is to know how big the holes we leave should be. The idea is to decorate the scene with sidewalks, but we don't want to place the sidewalks in a way that they disturb the houses.

Execute the following steps to create the houses:

1. I have selected a wonderful free asset in the Unity Asset Store. It is a low-poly house kit. To obtain this asset, go to `https://assetstore.unity.com/packages/3d/environments/low-poly-buildings-lite-98836` or search for `Low Poly Buildings Lite` in the Unity Asset Store browser text box, as we have previously done with other assets.

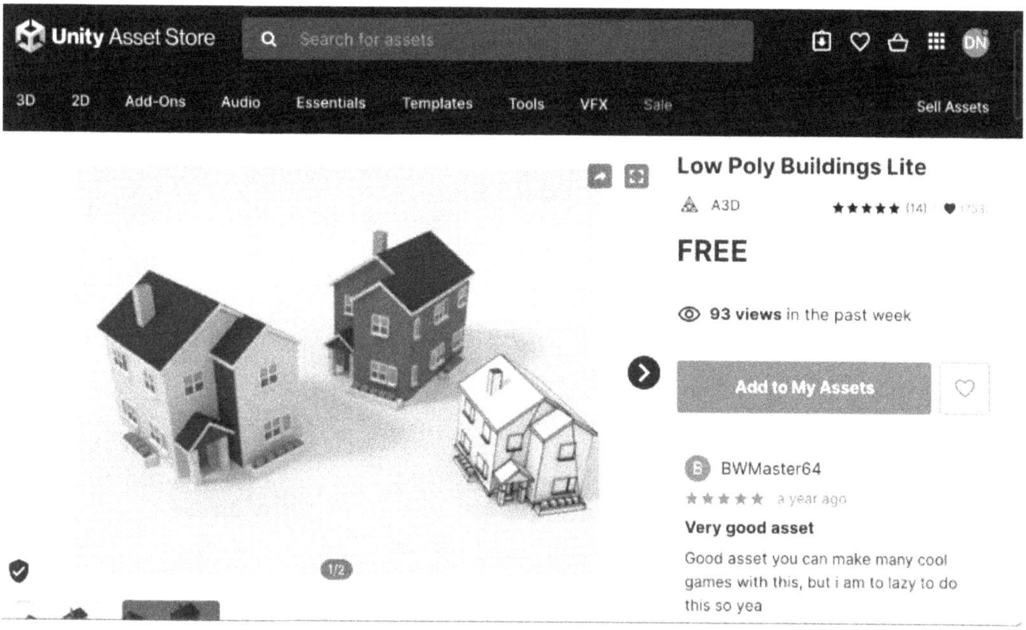

Figure 5.6 – Low Poly Buildings Lite pack asset download page

2. Click on the **Add to My Assets** button.
3. Click on the **Open in Unity** button. The **Package Manager** window will open.

4. Finally, click on the **Download** button and then click on the **Import** button to finish.

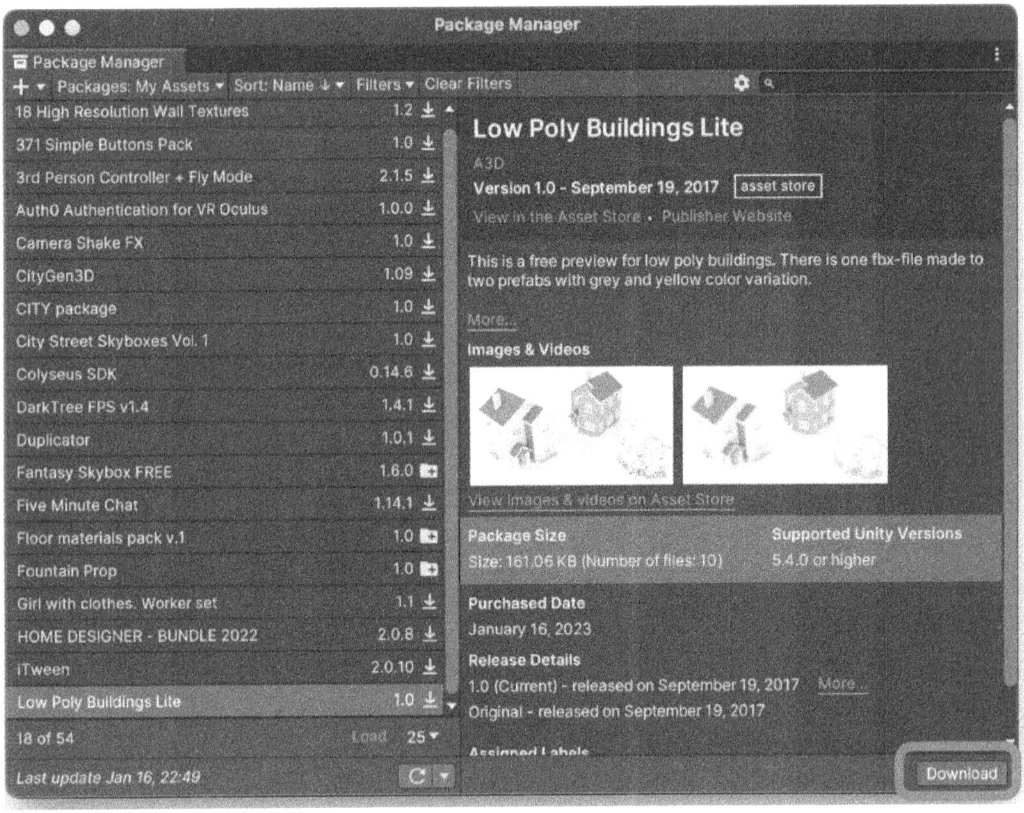

Figure 5.7 – Low Poly Buildings Lite pack in the Package Manager

5. Great, now we have this asset in our project. You can find it by following the **Assets** | **Low Poly Buildings Lite** path. In the `Prefabs` folder, you can find the two available models; as you can see, the Prefabs appear in a fuchsia color. You need to convert the materials withURP.

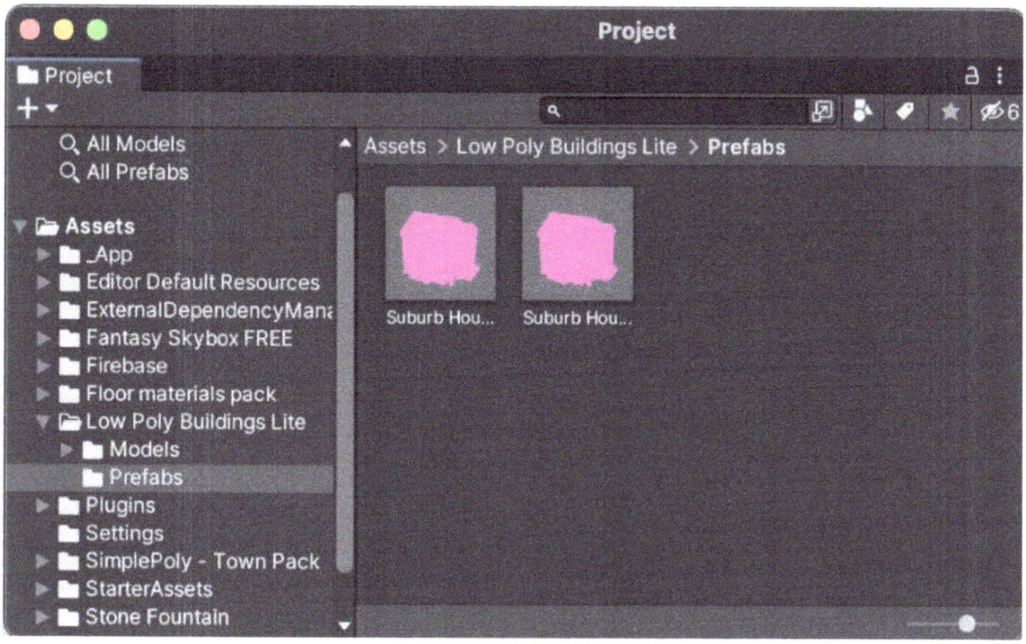

Figure 5.8 – Preview of Prefabs with URP compatibility problems

6. The materials are located in the **Assets** | **Low Poly Buildings Lite** | **Models** | **Materials** path. Simply select the materials that appear in fuchsia, then click in the main menu bar on the **Edit** | **Rendering** | **Materials** | **Convert Selected Built-in Materials to URP** option, and finally, click on the **Proceed** button to complete the operation.
7. If we go back to the `Prefabs` folder where the two houses are, they should be displayed correctly. If not, save the scene by pressing *Ctrl* + *S* on Windows or *Cmd* + *S* on Mac. It will refresh immediately.

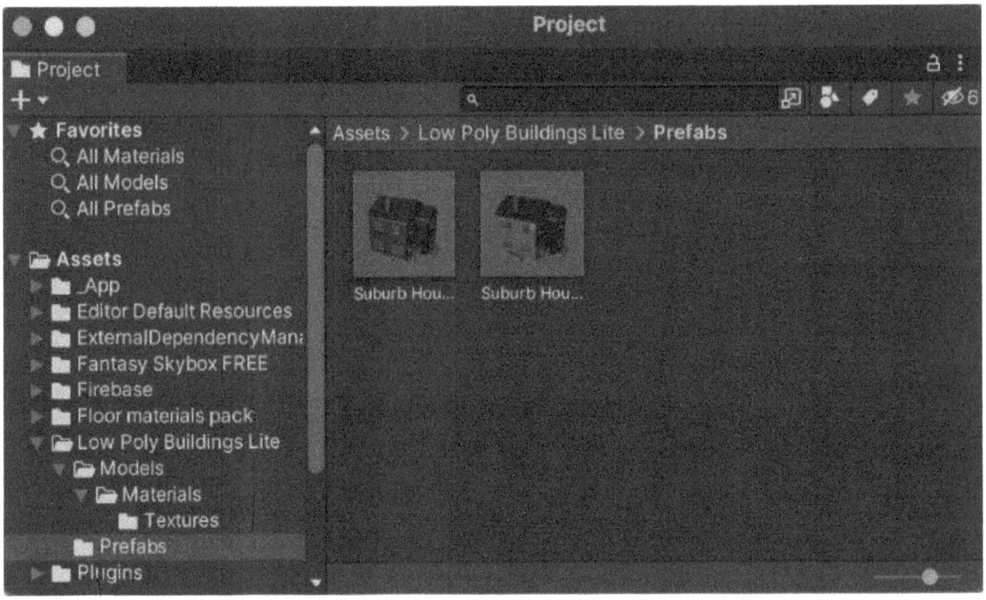

Figure 5.9 – Preview of Prefabs with corrected URP compatibility issues

Now that we have the model materials corrected, we can freely drag one of the Prefabs into the scene to see how big it is relative to the ground.

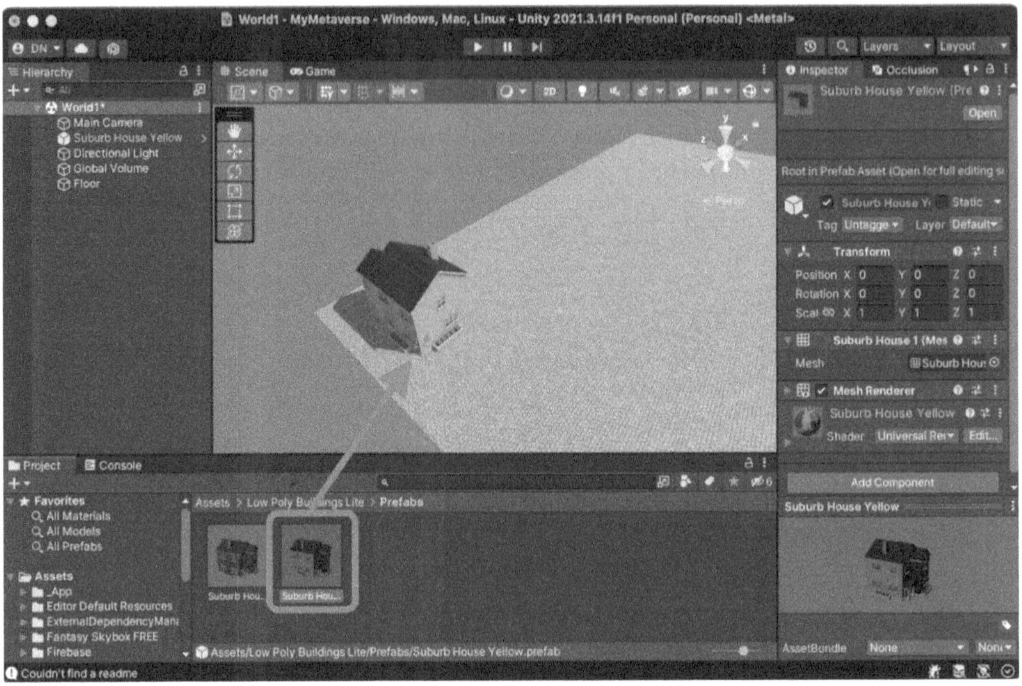

Figure 5.10 – Dragging Prefab to Scene

It seems that the house is of a reasonable size, but we must compare it with the size of the player.

8. If we follow the **Project** panel | **Assets** | **StarterAssets** | **ThirdPersonController** | **Prefabs** path and drag the Prefab named **PlayerArmature** to the scene near the house, we can observe the proportions.

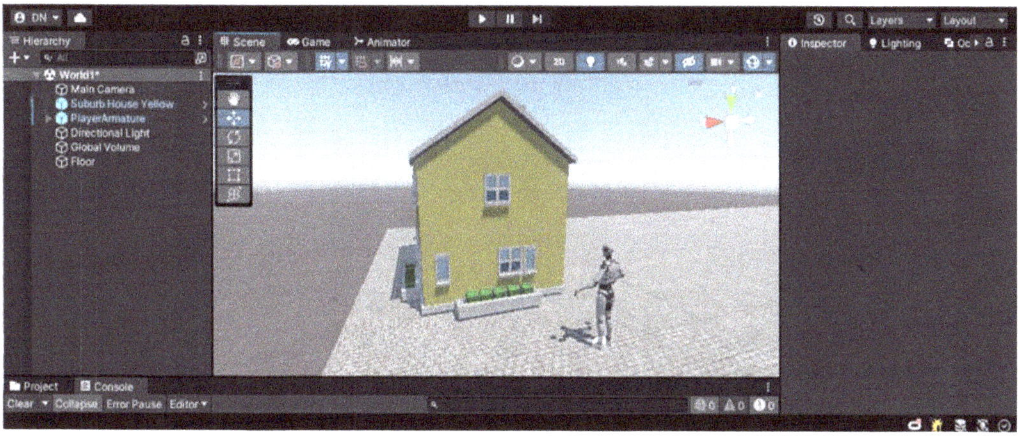

Figure 5.11 – Comparing the size of the player with the size of the house

9. Indeed, the house is too small. To correct this, select the Prefab of the house, and in the **Inspector** panel, change **Scale** to `2.5`, `2.5`, `2.5`. You can place the player Prefab next to the door of the house to check that it is now correctly scaled.

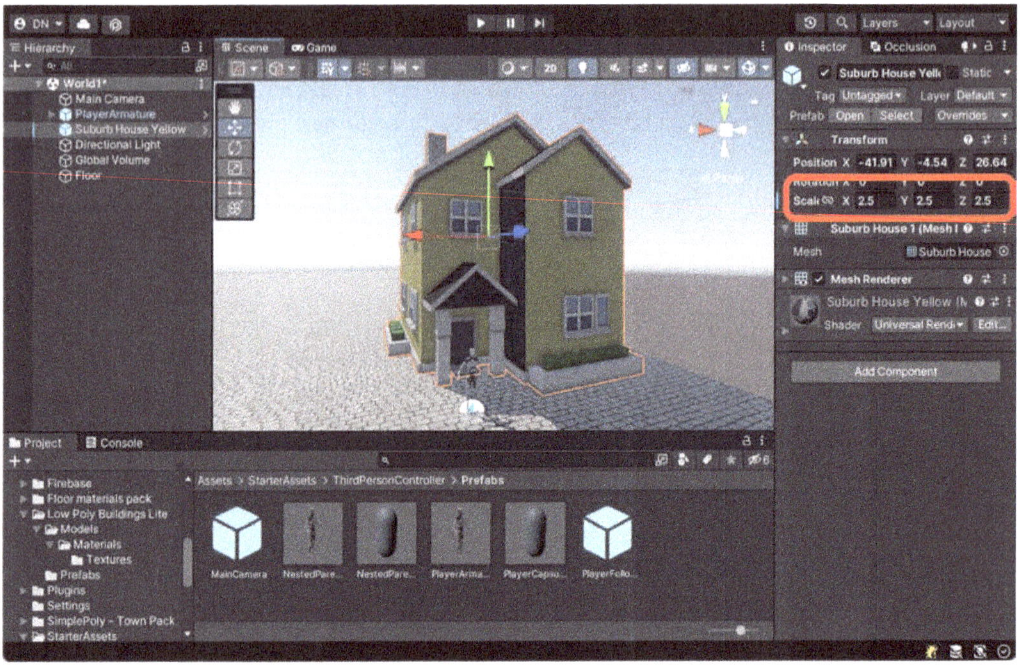

Figure 5.12 – Scaling the size of the home

10. As you can see, the Prefab of this house does not have a collider. To fix this, click the **Add component** button in the **Inspector** panel and add a Mesh Collider. It is also important that you mark the object as static.
11. To save the changes we have made to the Prefab, click on the **Overrides** button and then click on the **Apply All** button to finish the save action. You can do the same process for the second house model that is included in the **Low Poly Buildings Lite** pack.

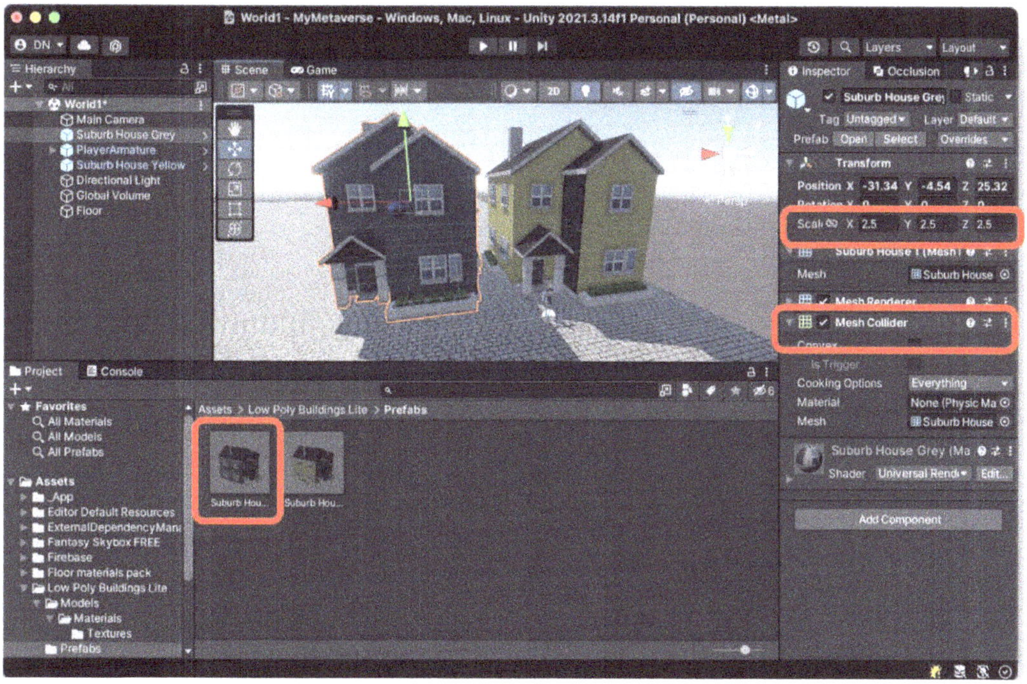

Figure 5.13 – Changing the Prefab scale

Perfect. Before we continue decorating the scene, we must make a small configuration in our project. We must lay the foundations that we will use to have order in the assets that we will use throughout the chapter.

Applying order

As we described at the beginning of the chapter, our scene will have the ability to load objects dynamically when it is executed. For this to be possible, Unity needs the Prefabs that will be loaded at runtime to be in a specific folder. This folder is called `Resources`. It is a special folder; we can only load objects from this folder when running the project.

We are going to create a series of folders to organize the Prefabs that we'll load later. To do so, follow these steps:

1. Create the `Resources` folder in the **Assets** | **Resources** path.
2. Inside the newly created `Resources` folder, create another folder called `Prefabs`.
3. Inside Prefabs create a folder called `Buildings`.

4. Finally, drag the two Prefabs of the houses from the **Assets** | **Low Poly Buildings Lite** | **Prefabs** path to the **Assets** | **Resources** | **Prefabs** | **Buildings** path and it will look like this:

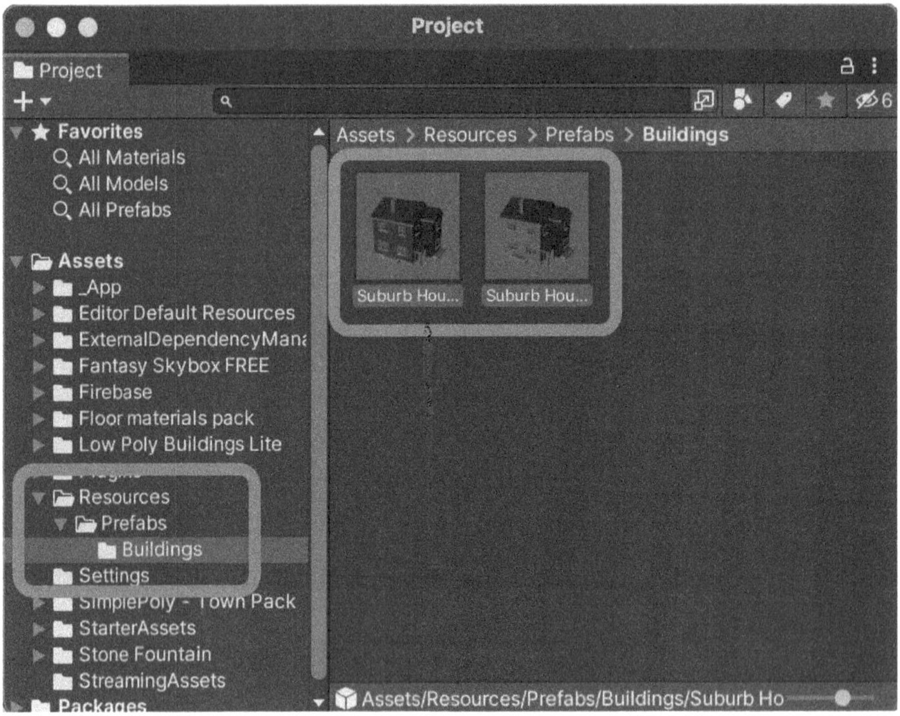

Figure 5.14 – Moving prefabs to another folder

Perfect. We have everything ready to continue with the design of the scene. To begin, we will do a little cleaning of the tests we have done. Remove the two houses from the scene and the player prefab so that only the floor remains.

Limiting the Scene

The first thing we are going to do is create a physical limitation on the perimeter, with the intention that players cannot jump into the void. It occurs to me that a good (and elegant) way to do this is with a **hedge**.

We are going to design a hedge that covers the length and width of the ground. To do so, follow these steps:

1. In the main menu bar, click on **GameObject** | **3D Object** | **Cube** to add a **Cube** GameObject to the stage.

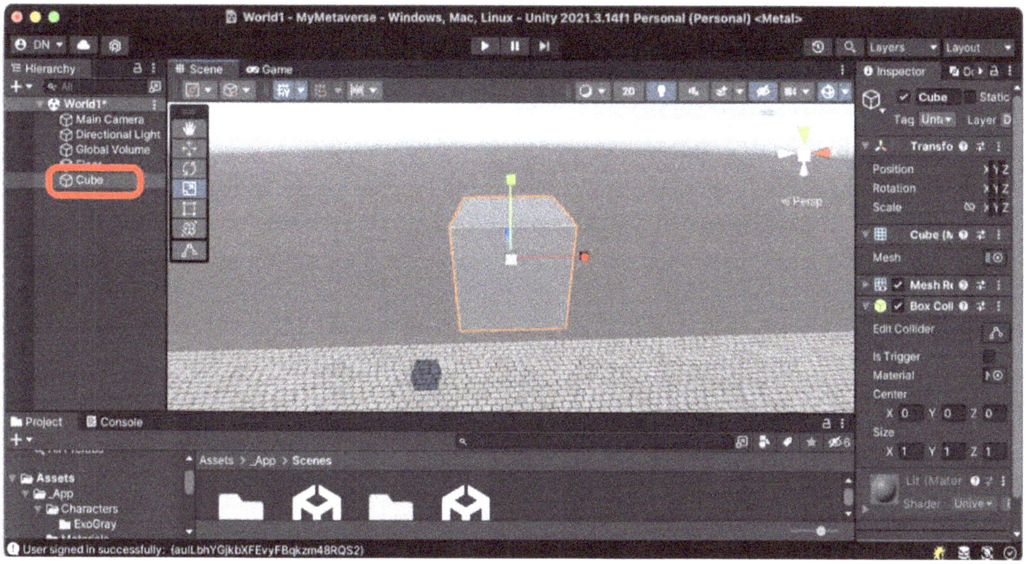

Figure 5.15 – Adding a Cube to the scene

2. We must resize the cube to turn it into a kind of wall; for example, we can apply a scale of `27`, `3`, `1` and place the cube so that it is at the beginning and the end of the floor.

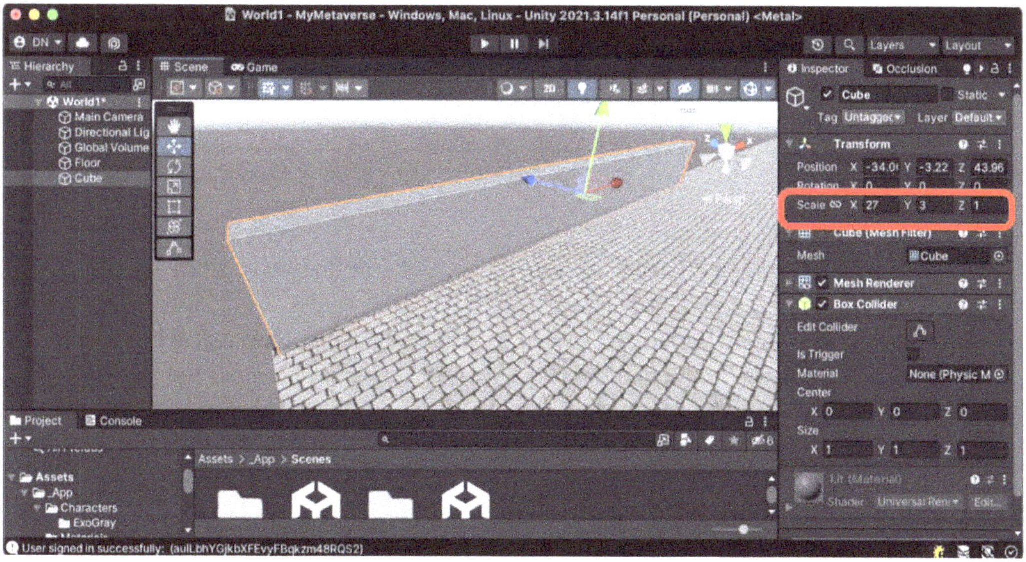

Figure 5.16 – Resizing the Cube

3. Now, we must find a nice texture with a hedge appearance. We will go to `Unity Asset Store https://assetstore.unity.com`.
4. In the search engine, type `Grass` and, as we have done on other occasions, order by **Price**, from lowest to highest, to see the free ones first.
5. After scrolling a little, I chose the asset called **Hand Painted Grass Texture**. Click on it, and then we'll proceed to download it and open it in Unity as we have seen on other occasions.

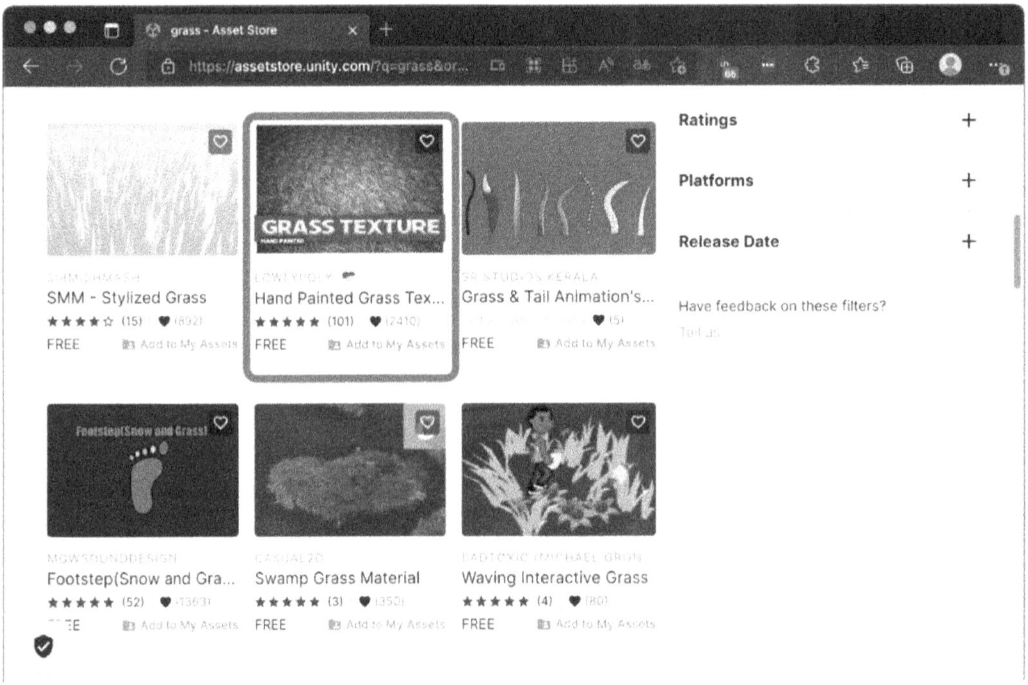

Figure 5.17 – Hand Painted Grass Texture, which is free in Unity Asset Store

6. Once downloaded and imported to the project, we go to the new folder created in the **Project** panel located in **Assets** | **Hand Painted Grass Texture** | **Material** and apply the conversion to URP of the materials that appear in fuchsia.

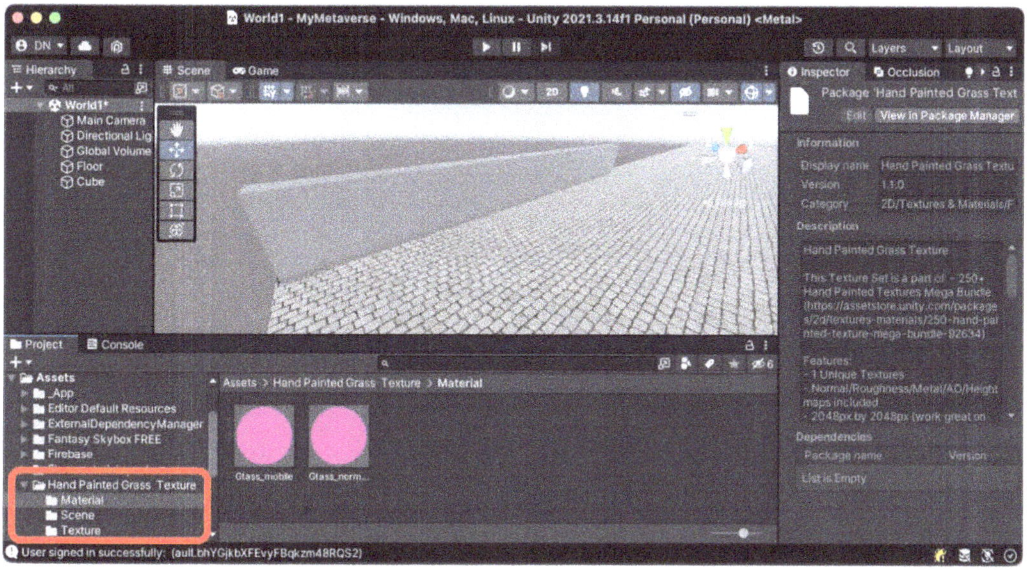

Figure 5.18 – Materials with URP compatibility issues

7. Drag the material named **Grass_mobile** to the **Cube** GameObject to apply the texture.

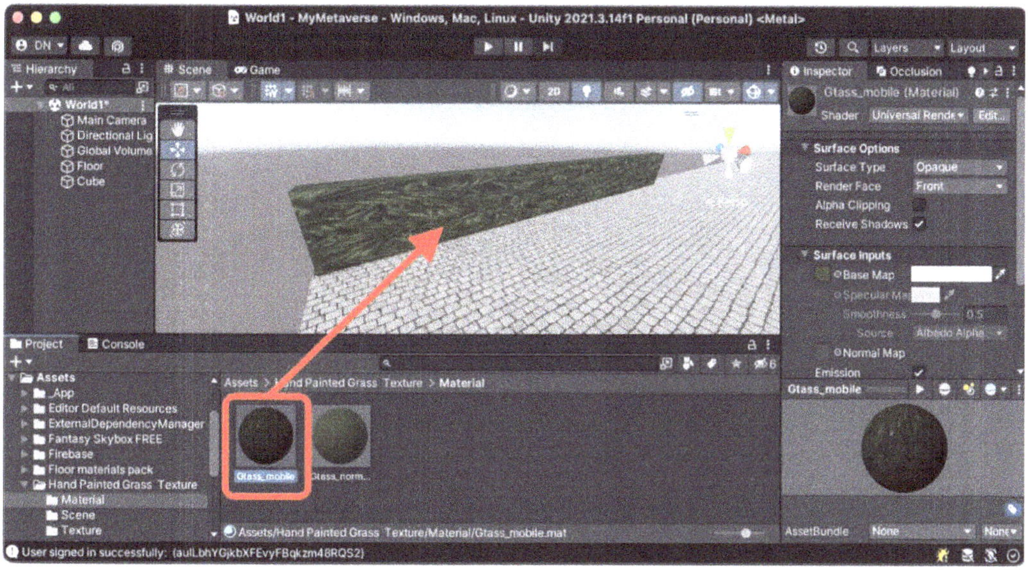

Figure 5.19 – Assigning new material to the Cube GameObject

8. As we can see, the texture is distorted. That is because our cube is longer. To correct this, we change the **Tiling** property in the **Material** component of the cube. For example, an optimal value would be **X**: `60`, **Y**: `30`. These values can be changed according to your taste; it will not affect the future performance of the project at all.

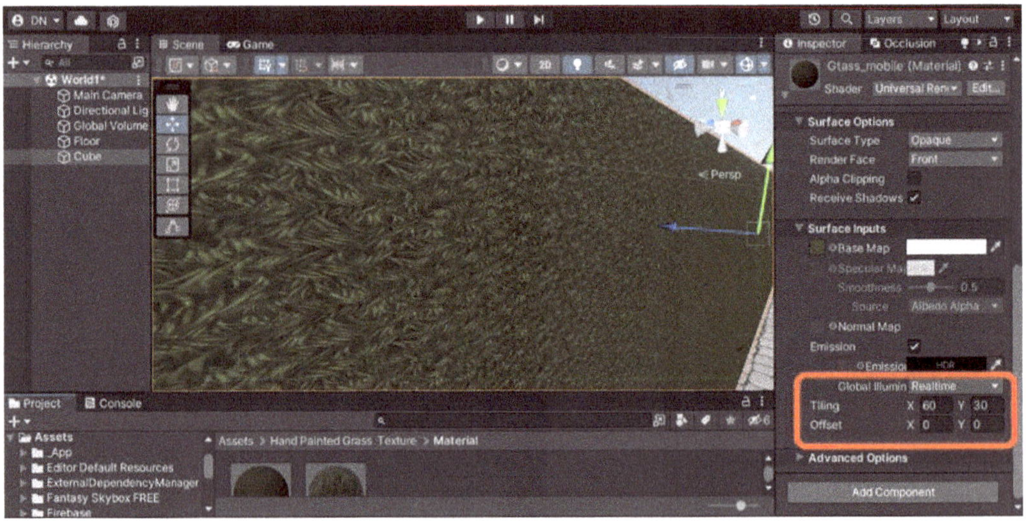

Figure 5.20 – Changing the Tiling property of the material

9. Perfect. We already have a hedge (or something similar). Now, before continuing with the final step, we will change the name of the GameObject. We will change **Cube** to `Hedge`, for example.
10. Finally, we duplicate the **Hedge** GameObject until it covers all four edges of the floor.

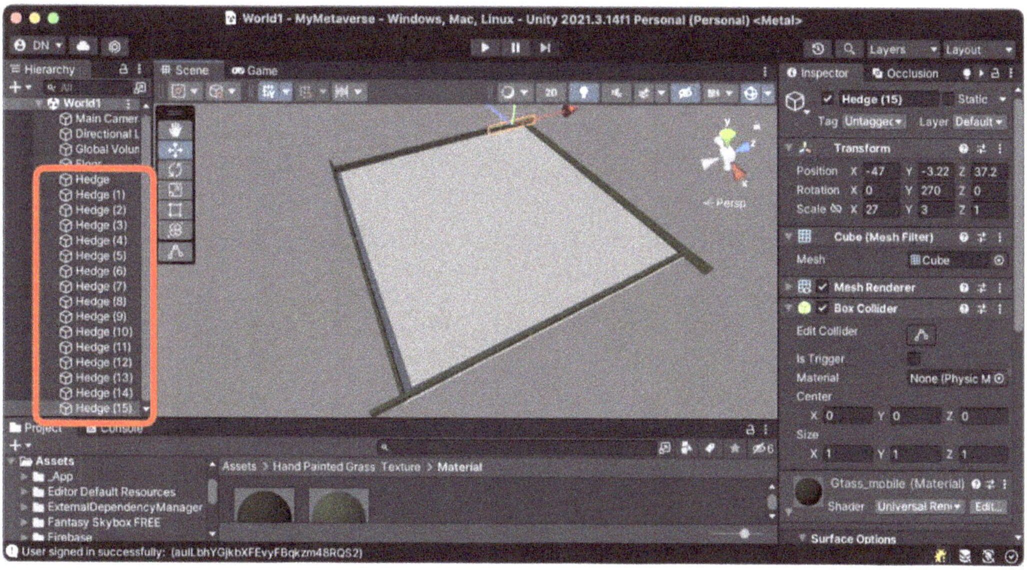

Figure 5.21 – Duplicating and placing the objects around the floor

Tip: Why did we make the hedge in sections?

You may have wondered why we have made the hedge in small pieces instead of one large one for each flank. The reason is optimization. If you remember the Camera Occlusion Culling optimization we learned about in *Chapter 3, Preparing Our Home Sweet Home: Part 1*, objects that are not in the camera's viewing area are not rendered. If we had put a large piece of hedge on each flank, they would always be seen by the camera. Instead, by dividing them into small pieces, we would only show the fragments within the viewing area; the others will be excluded.

11. Before we continue, let's not lose the good practices. Let's order all the GameObjects of the hedges in a parent called **Hedges**. Don't forget to mark as static all the GameObjects of Hedges and save the scene.

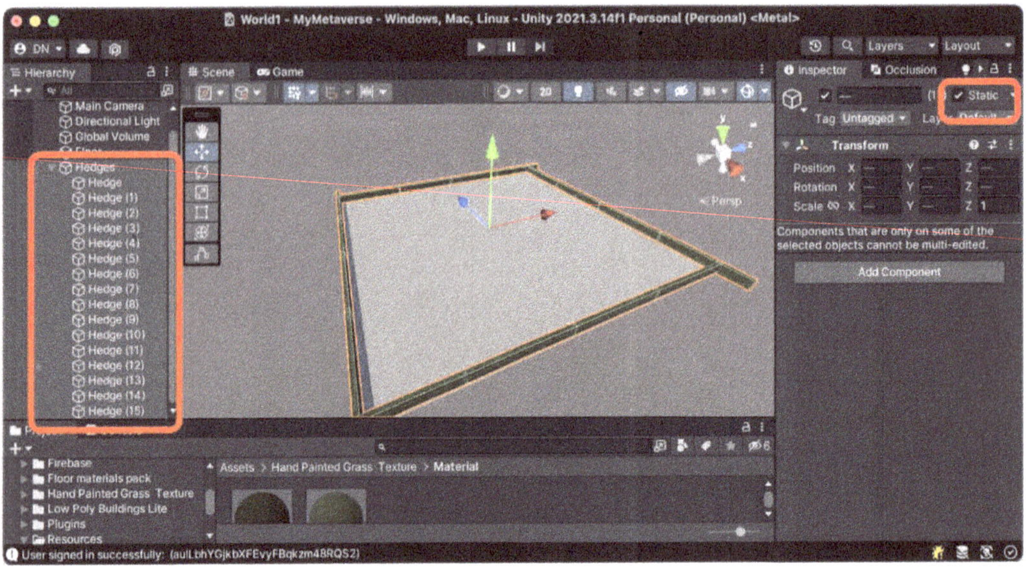

Figure 5.22 – Regrouping similar objects into a parent object

Great, the scene is getting nice. With the hedges along the edges of the scene we will prevent the player from going out of bounds. Continuing with our construction, we will now add some more assets to the scene to emulate a street.

Adding sidewalks

Now let's continue adding some sidewalks and asphalt. We will take advantage of the assets we downloaded in *Chapter 3, Preparing Our Home Sweet Home: Part 1*, found in the **Assets** | **Simple Poly – Town Pack** | **Prefabs** | **RoadSegments** path:

1. To place some nice sidewalks and asphalt, drag the Prefab named **sptp_road_segment_03** into the scene and place it in a position similar to as in the following screenshot.

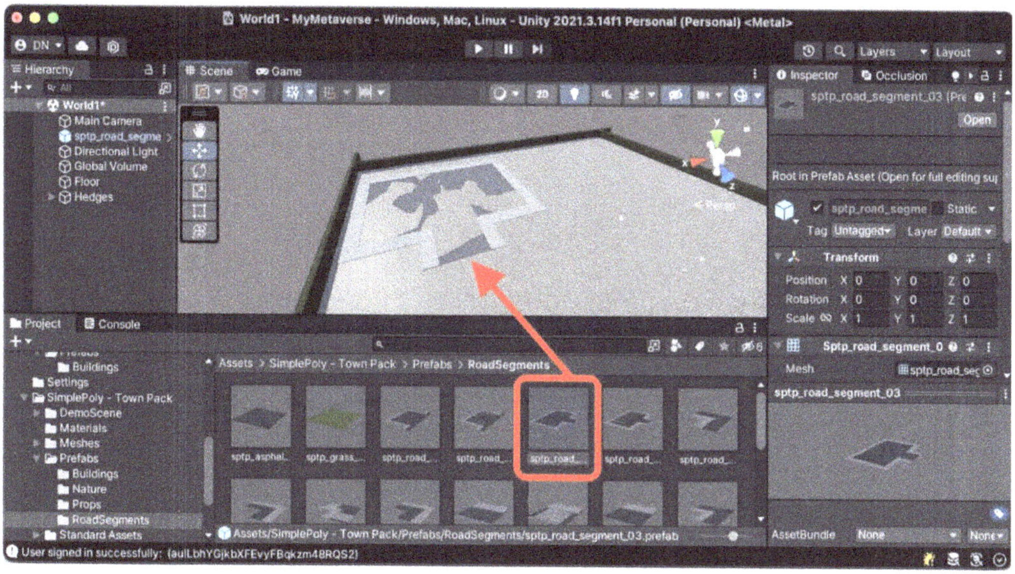

Figure 5.23 – Placing a sidewalk Prefab in the scene

As you can see, the asphalt texture is mixed with the ground texture. This is because both are at the same level.

2. To fix it, select the **sptp_road_segment_03** GameObject from the **Hierarchy** panel, and with the **Move** tool, raise it a little, very little, just enough until they stop intertwining.
3. Duplicate the GameObject to create an object similar to the right of the first one. Your scene will look something like the following screenshot.

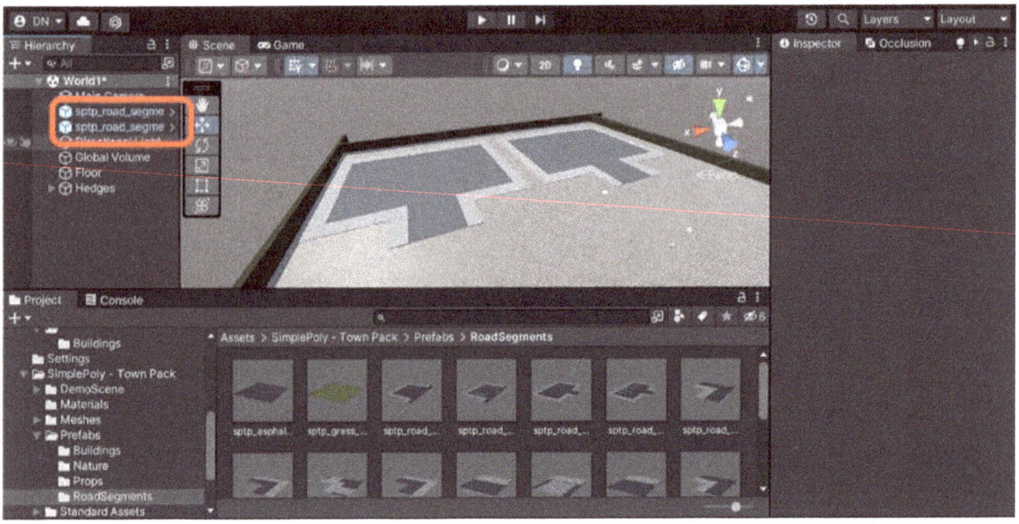

Figure 5.24 – Duplicating pavement objects in the scene

4. Duplicate the two GameObjects again, and with the **Rotate** tool, rotate them 180 degrees. Then, with the **Move** tool, place them in front of each other, as follows.

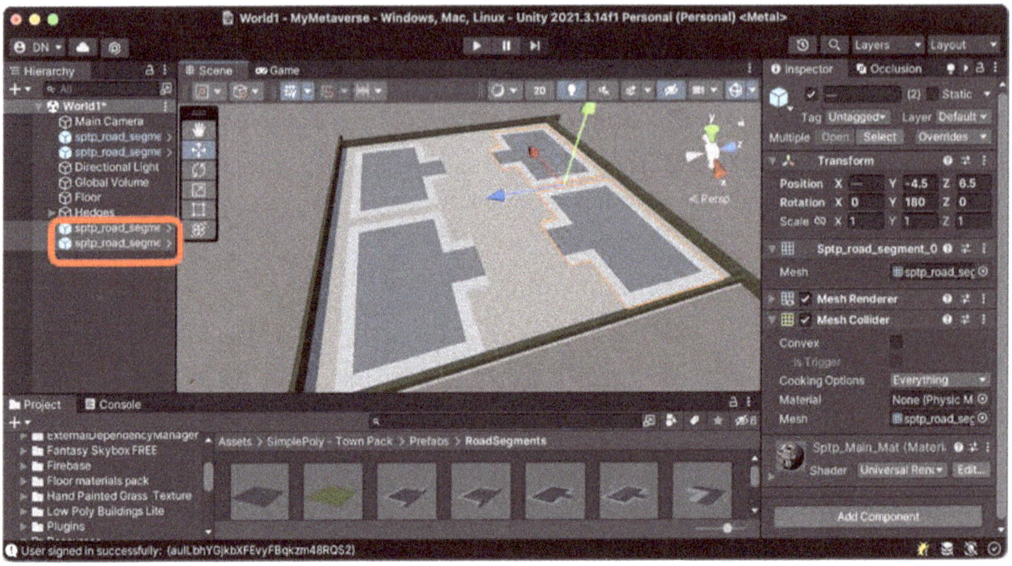

Figure 5.25 – Duplicating pavement objects in the scene

Perfect, we already have the spaces where the houses will go. In this scene, there will be space for four buildings.

5. Finally, to finish the design of the sidewalks, we will close the gaps that have been left between the four pieces of sidewalks that we have placed. We will use the **sptp_road_segment_11** Prefab.

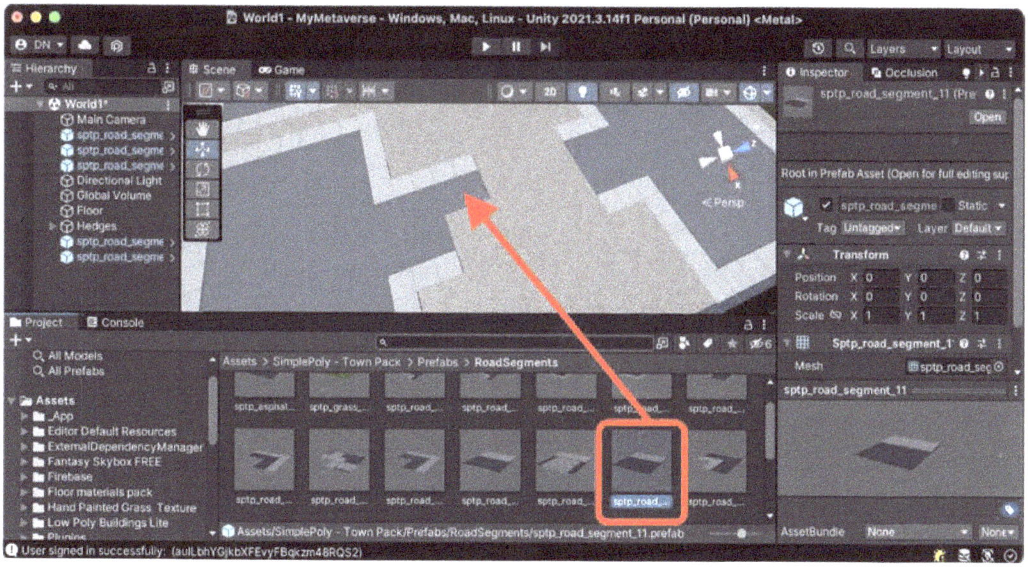

Figure 5.26 – Placing sidewalks on the scene

6. We will duplicate this GameObject to cover the length. Then, with the **Rotate** tool, we will fill the other side. We will do the same with the other area. Your scene will look like this.

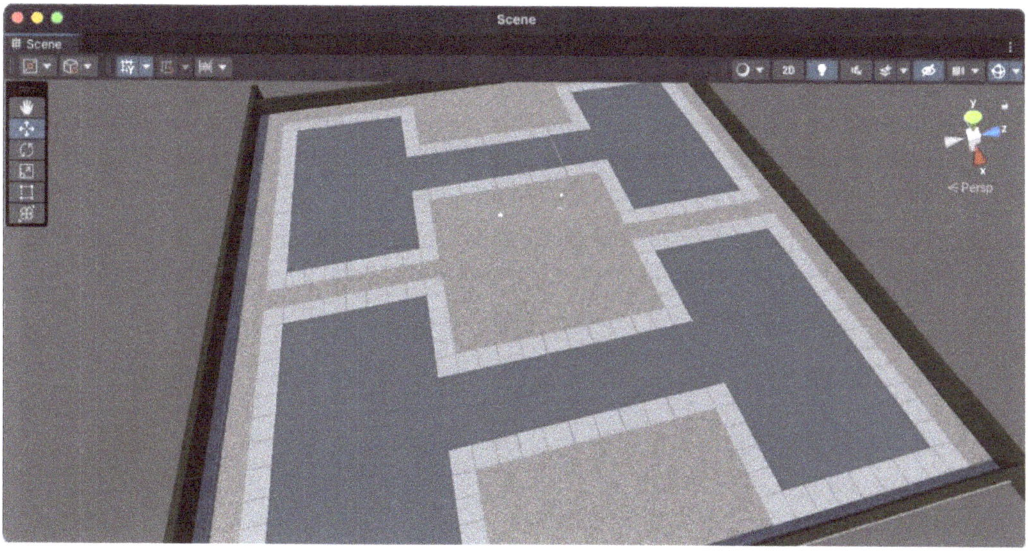

Figure 5.27 – Preview of the finished sidewalks

7. Make sure to organize all sidewalk pieces in a parent, called, for example, `Sidewalks`, and mark all GameObjects as static.

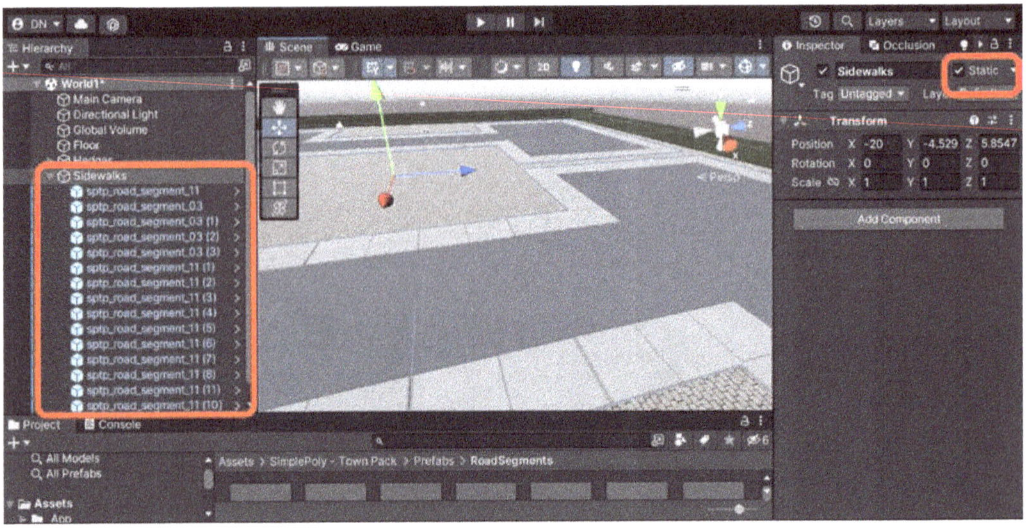

Figure 5.28 – Grouping sidewalk prefabs

Awesome, we are getting to the end of the scene design. Before we start programming, we'll add some trees and objects to make the scene more fun.

Adding trees

As we have done previously with the pavements, we will follow these steps to add some trees to the scene:

1. For the trees, we will choose a Prefab from the **Assets** | **SimplePoly – Town Pack** | **Prefabs** | **Nature** path and drag Prefabs where we like. Always remember to mark as static all the decorative elements that you incorporate into the scene.
2. I have decided to place trees with the Prefab called **sptp_tree_02**, as shown in the following screenshot. But you don't have to do exactly what I did. Unleash the artist within you!

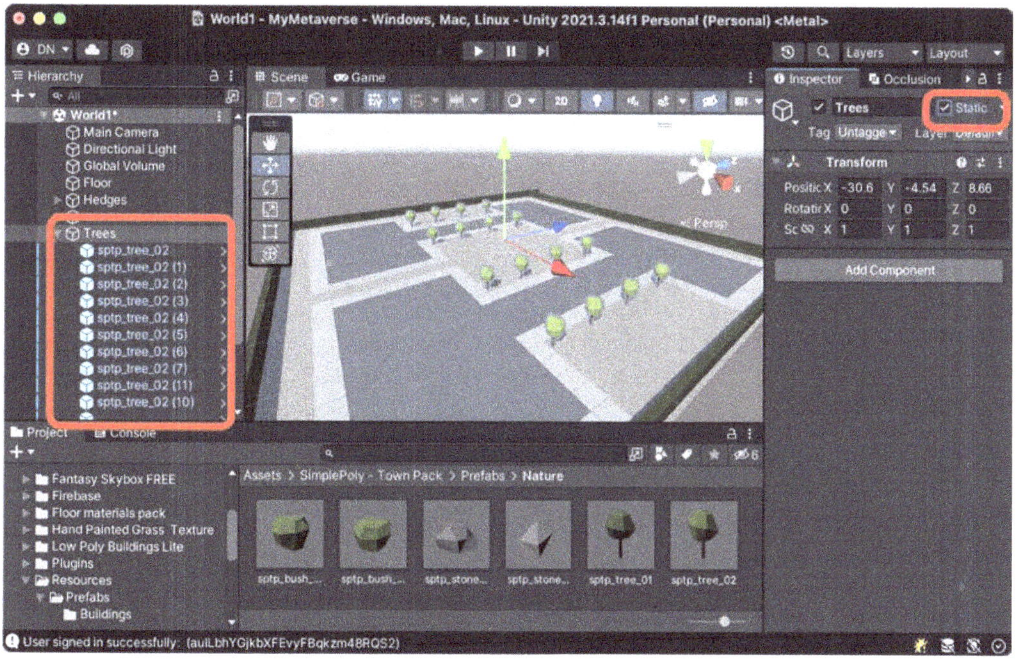

Figure 5.29 – Grouping tree prefabs

3. Don't forget to group all the tree GameObjects in a parent called **Trees** and mark them all as static.

The last modification that we will do in this scene will be to change the sky, and with this, we will finish this section.

Customizing the sky

Remember that in *Chapter 4, Preparing Our Home Sweet Home: Part 2*, we downloaded an asset, whose skybox can be found by following the **Assets** | **Fantasy Skybox FREE** | **Cubemaps** path.

I am going to use the Skybox called **FS002_Sunset_Cubemap**, which can be found by following the **Assets** | **Fantasy Skybox FREE** | **Cubemaps** | **FS002** path. To do this, drag the material to the sky. The appearance of the Skybox will change immediately.

Figure 5.30 – Replacing material for the skybox

With this last step, we finish the task of designing the scene. We have done a great job here. Now, our new world has physical barriers with a hedge-like appearance and a well-decorated ground with sidewalks and asphalt, all this under a beautiful sunset sky.

The next thing we will do in this chapter will be to add a reference to this new world in the Firestore database. So far, all the steps we have followed were learned in the previous chapter, but soon we will start programming. Your vision of the Metaverse you are building will expand with the new techniques we will learn from here.

Adding the world to the Firebase database

I know you are eager to start programming; I promise to be quick in this section. We will simply do the same as in *Chapter 4*, *Preparing Our Home Sweet Home: Part 2*, we will add a new world to our **Worlds** collection.

To do this, follow these steps:

1. Go to Firebase Console at `https://console.firebase.google.com`.
2. Click on the **Firestore Database** link.

3. In **Firestore Console**, select the **Worlds** collection and click on the **Add Document** button.

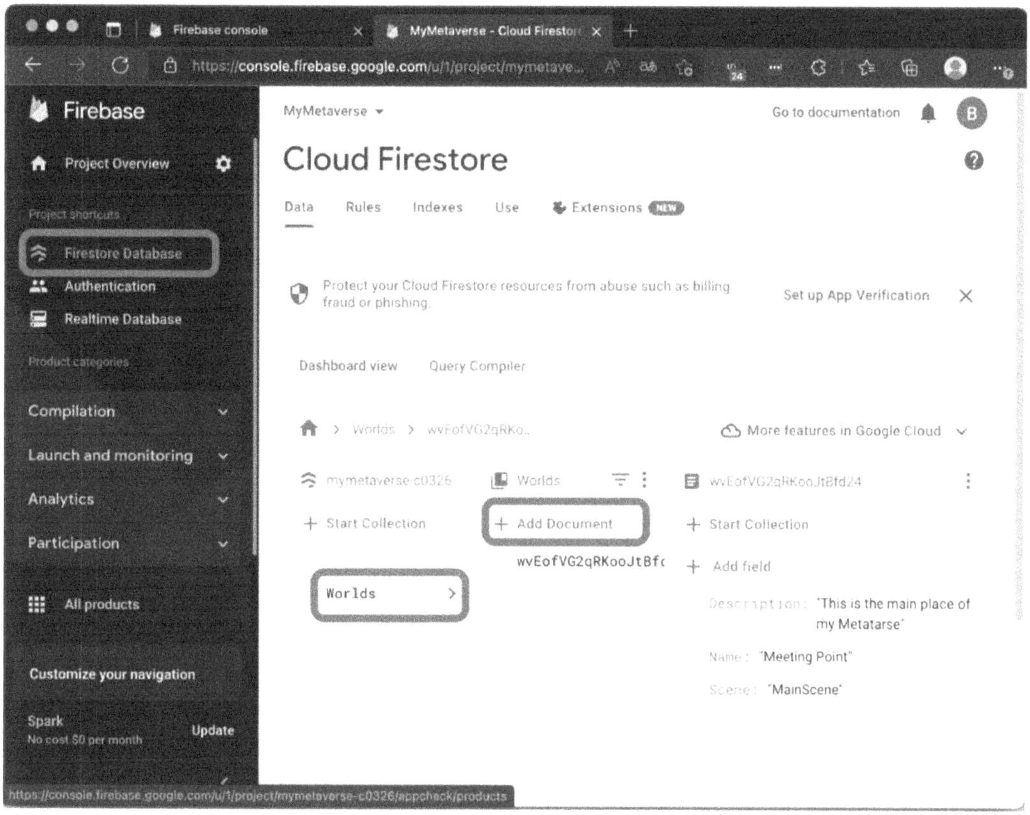

Figure 5.31 – Adding a new world document to the Worlds collection

4. In the document creation window, click on the **Auto-ID** button to obtain a unique ID and add the `Description` property with the value `The first neighborhood of my metaverse`, the `Name` property with the value `World1`, and the `Scene` property with the value `World1`, as shown in the following screenshot.

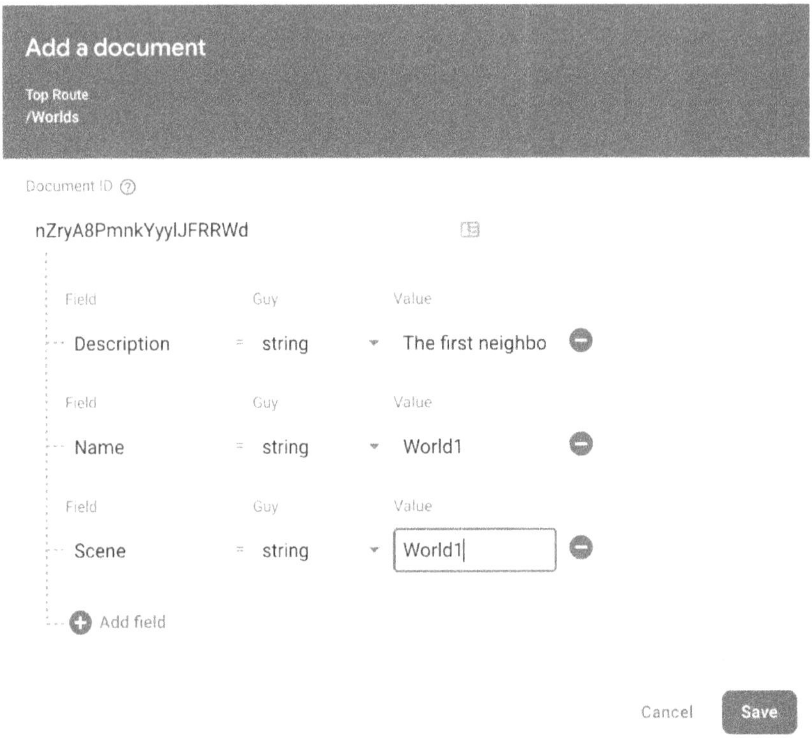

Figure 5.32 – Filling in the new document form

5. Finally, we click on the **Save** button to finish.

Perfect. We have created a reference to the new world in Firestore. With this step finished, we have the foundation ready to start programming. Next, we will see how to manage the creation and update of buildings in the Firestore database from the Unity Editor.

Adding objects dynamically

It is time to get into programming. In this section, we will create C# scripts that will allow us to connect to the database and create or update records. The intention of these scripts is to make our scene dynamic. How can it benefit us to have a scene with dynamic access to a database?

Consider that, in the future, when your Metaverse is in production, with hundreds of users, you may want to implement a system that allows your users to improve their houses. To achieve this, the scripts we are going to create in the next pages will follow this flow:

1. They will connect to the database and query the building we want instances of in the scene.
2. If our user improves their building, we will update the property where the Prefab is stored, for the new building type.
3. In this way, we can dynamically load one Prefab or another in the scene depending on the modifiable value that is outside the project in Firebase.

Impressive, isn't it? Surely you can think of a multitude of improvements and applications with this that we will learn next. Well, let's get to work. The roadmap of the scripts that we will use is as follows:

- A script that identifies us as administrators to be able to read and write in Firestore
- A script that allows us to read and write in Firestore, with all the necessary logic to manage a building
- A script that adds to the latter some visual buttons to manage the creation and editing

Identifying ourselves in the system

Read and write operations in Firebase require the user performing the action to be logged in. It is possible to open up security so that operations can be performed by anonymous users without identifying themselves, but this is not recommended at all.

We must make a couple of configurations in Firebase to activate the identification system and configure the database to confirm whether our identity has been verified.

To do this, follow these steps:

1. Go to Firebase Console at `https://console.firebase.google.com`.
2. Click on the **Authentication** section.
3. Click on the **Sign-in method** tab.

4. Click on the **Email/password** option.

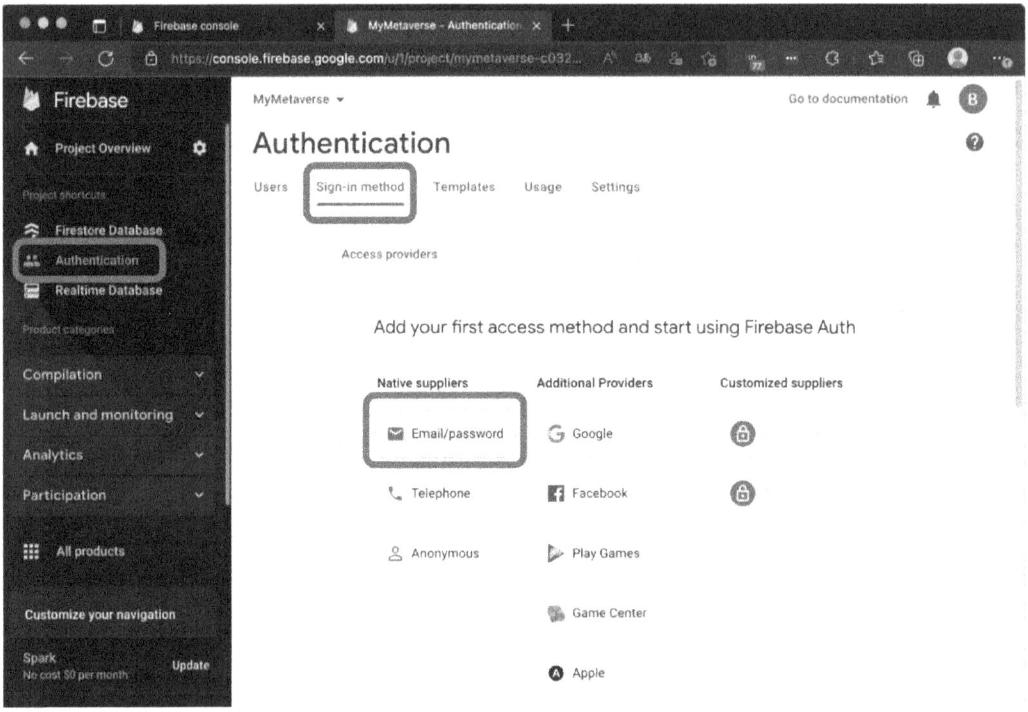

Figure 5.33 – Auth suppliers in Firebase Authentication

5. Activate the **Enable** option and click on the **Save** button. What we have just done is activated the system so that users can create new accounts and identify themselves in the future. We can also create accounts manually.

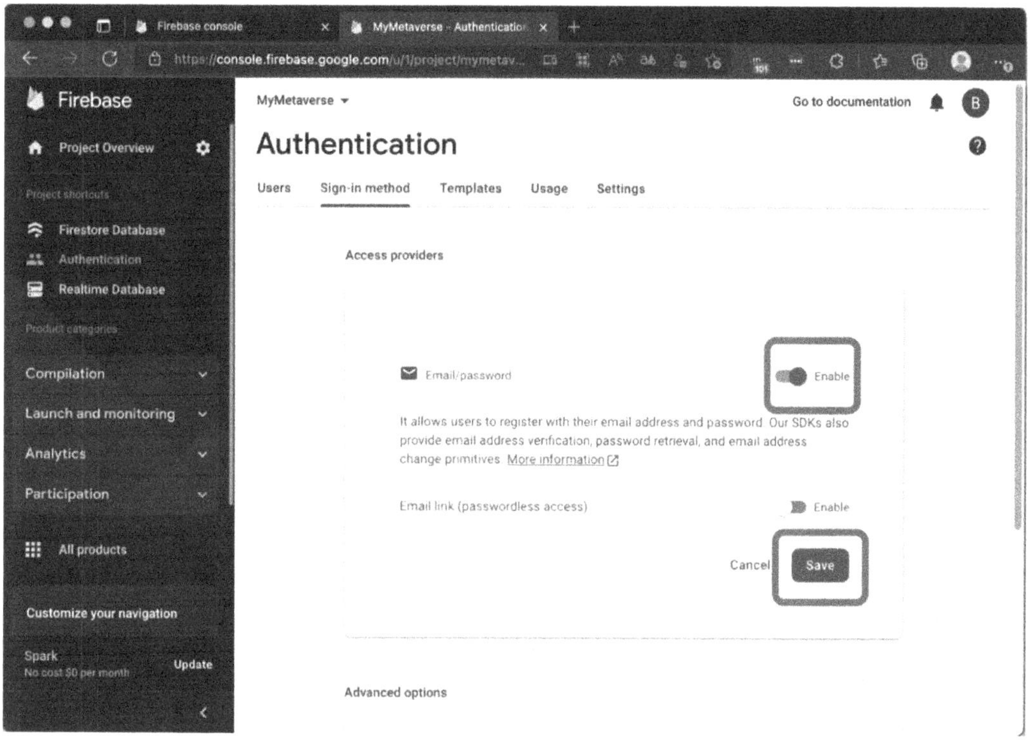

Figure 5.34 – Enabling the Email/password auth supplier

6. Next, we are going to create a username and password that identifies us as an administrator. For that, return to the **Users** tab.
7. Click on the **Add User** button.
8. Fill in the username field with `admin@mymetaverse.com` (or one that you prefer, even if the email does not really exist) and the password `123456` (or one that you prefer), and click on the **Add User** button to finish.

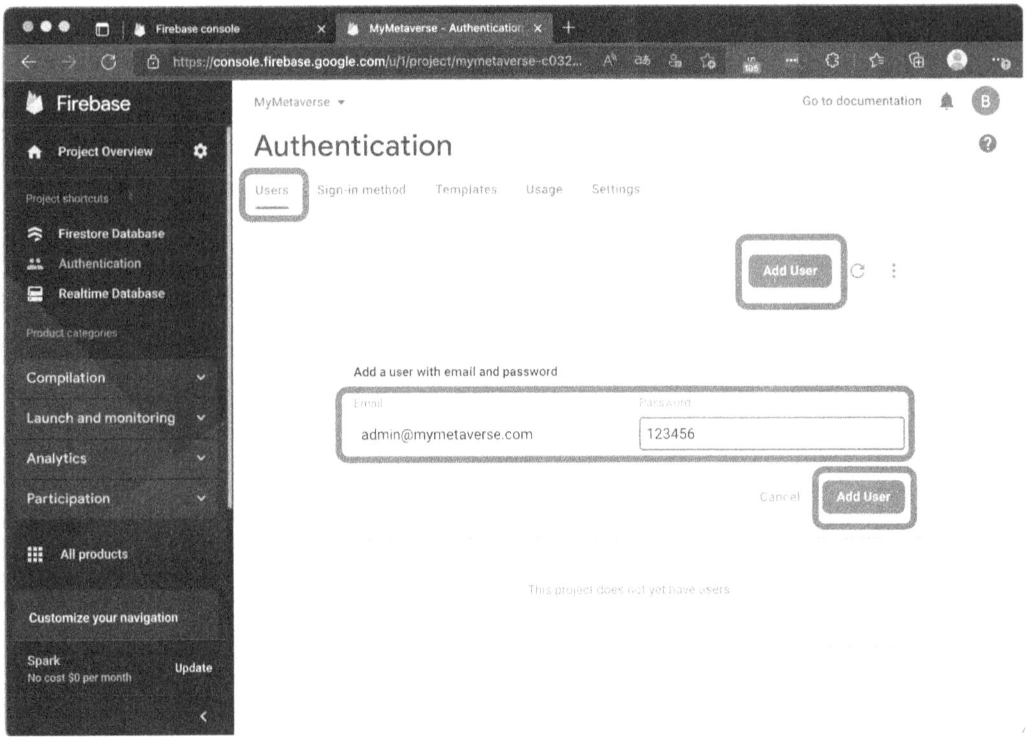

Figure 5.35 – Creating a new user in Firebase Authentication

Great. As you can see, after clicking the **Add User** button, a user has been created with the information you have entered, and now appears in the list of available users. When your users start registering in your Metaverse, their usersnames will appear in this section. Now, we must configure the security rules of the Firestore database to avoid being given access if we are not identified.

9. Click on the link to **Firestore Database**.
10. Click on the **Rules** tab.
11. Replace the code that appears with the following code:

```
rules_version = '2';
service cloud.firestore {
  match /databases/{database}/documents {
    match /{document=**} {
      allow read;
      allow write: if request.auth != null;
    }
  }
}
```

12. Click on the **Publish** button to save the changes.

If we look at the code we have just pasted, we can see the line `allow read` which means that any unidentified user can query the database, but in the `allow write` line, we add a condition that requires the incoming request to have the **auth** property, which means that it forces us to be identified.

In this book, we will assume that you have a basic knowledge of C# programming in Unity; otherwise, I recommend you read the following links:

- **Introduction to C# programming in Unity**: `https://unity3d.com/learning-csharp-in-unity-for-beginners`
- **Working with scripts (tutorial)**: `https://learn.unity.com/tutorial/working-with-scripts`
- **Introduction to scripting (course)**: `https://learn.unity.com/tutorial/working-with-scripts`

Admin Login script

Perfect. Everything is ready. The first thing we will do is create the script that will identify us in the Unity Editor.

Follow these steps to create a script:

1. With the **World1** scene open, click, in the main menu bar, on the **GameObject** | **Create Empty** option, and rename this new GameObject `AdminAuthManager`.

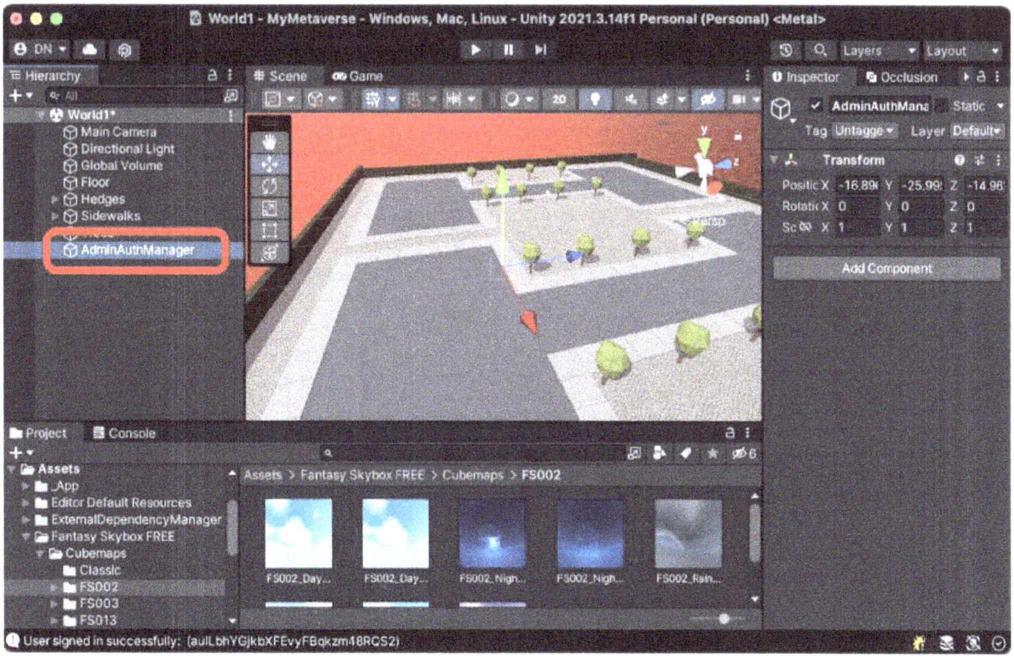

Figure 5.36 – Creating an empty GameObject

2. Well, this object that we have just created will be the one that will contain the script. To add a new script, from the **Inspector** panel, click on the **Add component** button.
3. In the search engine of the window that appears, type `AdminAuthManager`. As there is no component or script with this name, the **New script** option will appear. Click and confirm to finish the creation.

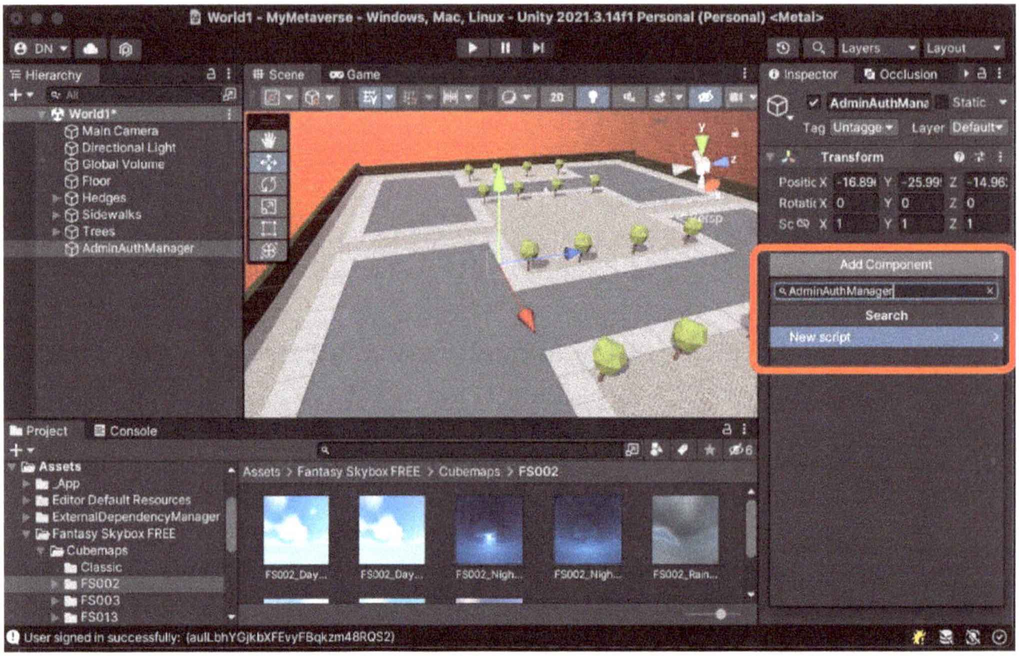

Figure 5.37 – Creating a new script in an empty GameObject

4. After the creation is complete, you can confirm that it has been created successfully by viewing the script in the **Inspector** panel.

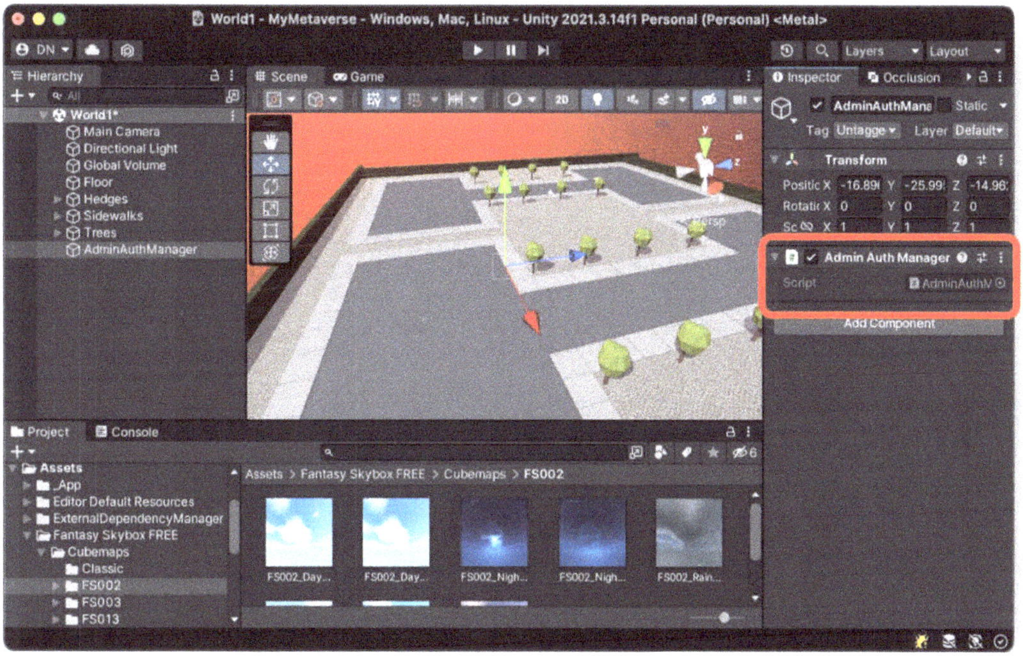

Figure 5.38 – Preview the script created as a component

5. To edit the contents of the script, click on the button with three dots and select the **Edit Script** option.

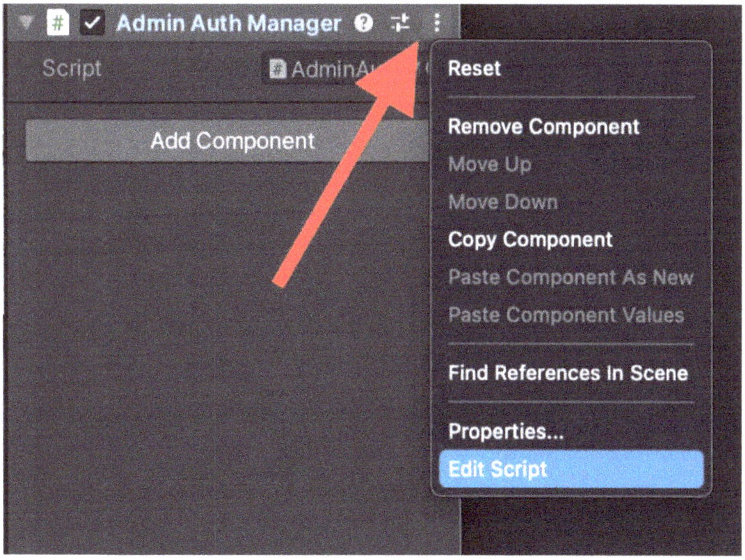

Figure 5.39 – Choosing the Edit Script option

The script will then open in Visual Studio and look something like this:

```
using System.Collections;
using System.Collections.Generic;
using UnityEngine;

public class AdminAuthManager : MonoBehaviour
{
    // Start is called before the first frame update
    void Start()
    {
    }

    // Update is called once per frame
    void Update()
    {
    }
}
```

This is the basic initial structure of any Unity script. As you can see, at the beginning of the code, there are the `import` statements to include other libraries that may be needed. These imports may vary throughout the development of the script.

Then, we find the declaration of the `AdminAuthManager` class. The class must have the same name as the file; otherwise, Unity will show a compilation error. Every script that you want to use in a GameObject of the project must inherit from `MonoBehaviour`.

The `Start` function will be executed automatically when the script is started when the scene is executed. The `Update` function, on the other hand, will be executed once per frame.

> **Tip: More information about MonoBehaviour life cycles**
>
> `MonoBehaviour` offers a multitude of system functions and each one is executed in a certain order. If you want to find out more about what this class can offer, I recommend you visit `https://docs.unity3d.com/ScriptReference/MonoBehaviour.html` and `https://docs.unity3d.com/Manual/ExecutionOrder.html`.

Now that we've explained the basics, we will continue extending our script so that it fulfills its functionality.

In order to access the Firebase Authentication methods, we need to add a new import in the script. To do this, add the following line to the `import` area of the script:

```
using Firebase.Auth;
```

Great, the next thing is to add an attribute to the script that will allow the script to run only in the Unity Editor while editing. It will not be able to run in Play mode or in production when the Metaverse is published. This is exactly what we want. It is a tool for us as administrators. The login system that we will later implement for the rest of the users will use another system, which we will see in *Chapter 6, Adding a Registration and Login Form for Our Users*.

To do this, add `[ExecuteInEditMode]` on the line before the class declaration. Your code should look something like this:

```
using System.Collections;
using System.Collections.Generic;
using UnityEngine;
using Firebase.Auth;

[ExecuteInEditMode]
public class AdminAuthManager : MonoBehaviour
{
    // Start is called before the first frame update
    void Start()
    {
    }

    // Update is called once per frame
    void Update()
    {
    }
}
```

Continuing, to be able to log in, we are going to need two variables, which we will have to configure in the **Inspector** panel. These are the username and password. We will also need a variable of the `FirebaseAuth` type, in which we will store the instance to be able to use the `FirebaseAuth` SDK methods.

> **Tip: Something important to know about ExecuteInEditMode**
>
> Adding this attribute to the script completely changes the way some methods of the `MonoBehaviour` class behave. For example, the `Update` method stops executing once per frame. Now, it will only execute every time there is a change in the scene. For more information about what has changed by adding this attribute, please visit `https://docs.unity3d.com/ScriptReference/ExecuteInEditMode.html`.

Add the following variables to the script:

```
using System.Collections;
using System.Collections.Generic;
using UnityEngine;
using Firebase.Auth;

[ExecuteInEditMode]
public class AdminAuthManager : MonoBehaviour
{
    public string username;
    public string password;
    FirebaseAuth auth;

    // Start is called before the first frame update
    void Start()
    {
    }

    // Update is called once per frame
    void Update()
    {
    }
}
```

Now, we need to add the key function of our script: the function that will execute the call for the ID. Add the following code fragment:

```
    void Login()
    {
        // We call the SignInWithEmailAndPasswordAsync
           method of the SDK to identify ourselves with the
           user and password that we will store in the
           previously declared variables.
        auth.SignInWithEmailAndPasswordAsync(username,
            password).ContinueWith(task =>
        {
            if (task.IsCanceled)
            {
                Debug.LogError("SignInWithEmailAndPasswordA
                    sync was canceled.");
                return;
            }
            if (task.IsFaulted)
```

```
            {
                Debug.LogError("SignInWithEmailAndPasswordA
                    sync encountered an error: " +
                        task.Exception);
                return;
            }

            // At this point, the user is correctly
               identified.
            FirebaseUser newUser = task.Result;
            Debug.LogFormat("User signed in successfully:
                {0} ({1})", newUser.DisplayName,
                    newUser.UserId);
        });
    }
```

As you will have seen from the comments inside the code, the `SignInWithEmailAndPasswordAsync` function will launch the identification against Firebase Authentication with the username and password that we will store in the declared variables.

The next method that we must add is the one in charge of instantiating the `auth` variable and calling the `Login` method. Add the following code fragment to the script:

```
    void InitializeFirebase()
    {
        // If the auth variable is not instantiated or the
           user's session has expired, we re-identify
           ourselves.
        if (auth == null || auth?.CurrentUser == null)
        {
            auth = FirebaseAuth.DefaultInstance;
            Login();
        }
    }
```

Finally, we will call the `InitializeFirebase` method inside the `Update` function. The completed class is shown here:

```
using System.Collections;
using System.Collections.Generic;
using UnityEngine;
using Firebase.Auth;

[ExecuteInEditMode]
public class AdminAuthManager : MonoBehaviour
```

```
{
    public string username;
    public string password;
    FirebaseAuth auth;

    void Update()
    {
        InitializeFirebase();
    }

    void InitializeFirebase()
    {
        if (auth == null || auth?.CurrentUser == null)
        {
            auth = FirebaseAuth.DefaultInstance;
            Login();
        }
    }

    void Login()
    {
        auth.SignInWithEmailAndPasswordAsync(username,
            password).ContinueWith(task =>
        {
            if (task.IsCanceled || task.IsFaulted)
                return;
            FirebaseUser newUser = task.Result;
        });
    }
}
```

Fantastic, you have made your first script in C#. It is time to see it in operation. Before continuing, save the changes to the script. Then, return to the Unity Editor and observe that the text inputs have appeared in the component that we have created with our script:

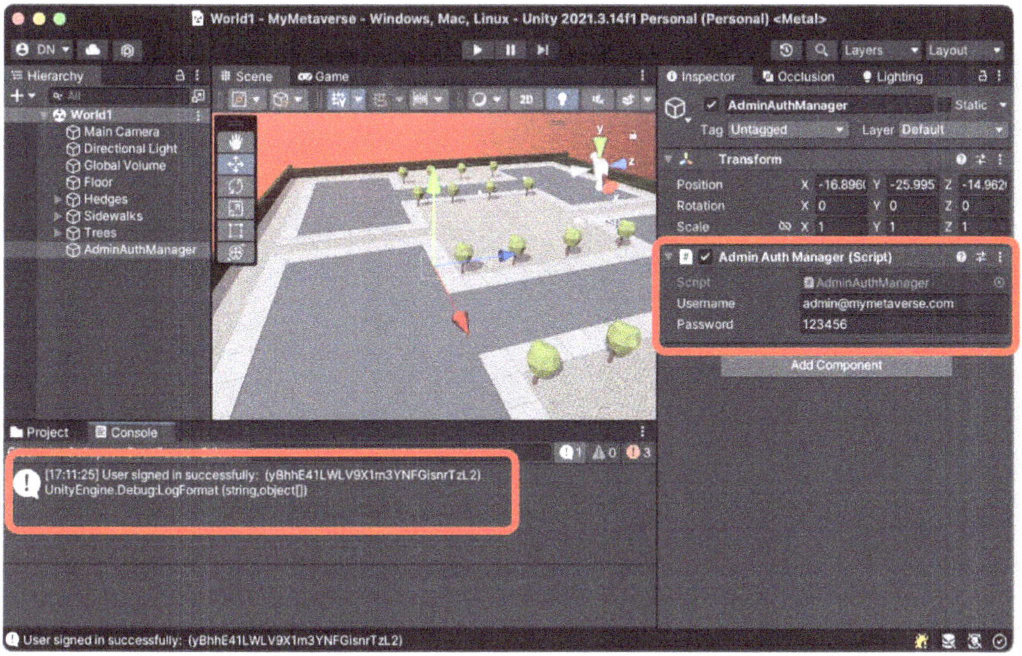

Figure 5.40 – Admin Auth Manager script working

Do you remember that pages ago, in Firebase, we created some administrator credentials? It is here where we must introduce them. Type `admin@mymetaverse.com` (or whatever you used) in **Username** and `123456` (or whatever you used) in **Password**.

If you save the scene, Unity will detect that a change has occurred in the scene and launch the `Update` function of our script, so the identification function will be executed, and in the **Console** panel, we will see a message telling us that everything went well.

It is important to remember that all the scripts that we make ourselves must be organized (for our future selves' sake). By default, in Unity, when you create a script from the **Inspector** panel, it is created in the **Assets** root. We will locate the script in the root and move it to our **Assets** | **_App** | **Scripts** path.

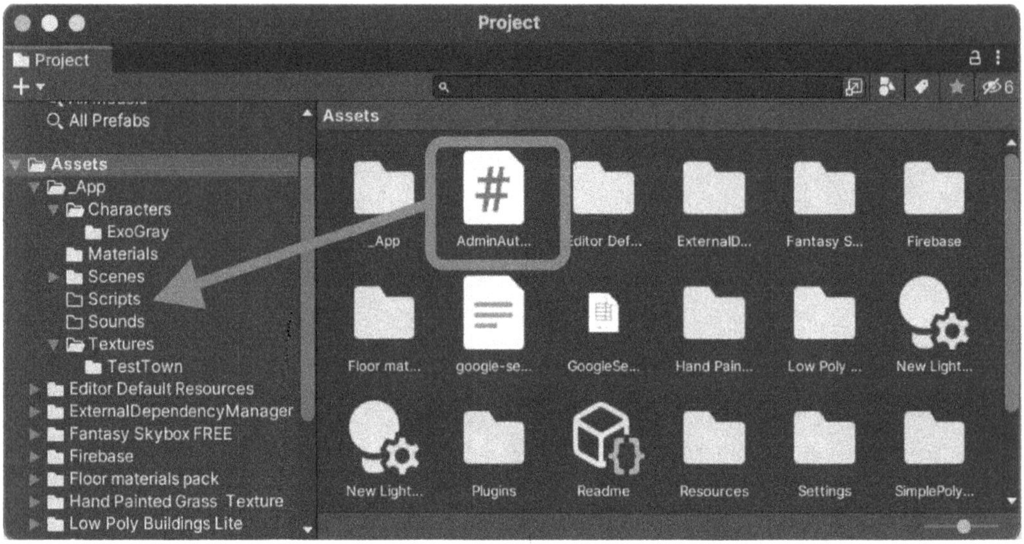

Figure 5.41 – Moving a script to another folder

Good job! We have successfully created the `AdminAuthManager` script, which will manage our login so that the following connections to Firestore will be successful.

BuildingInstanceManager script

The `BuildingInstanceManager` script that we are going to program in the following pages will be responsible for instantiating an existing building in the scene, creating a new document in the **Buildings** collection or updating it if it already exists in Firestore. But for that, we first need to create other classes that are necessary.

Creating the Building class

But first of all, we need to create a class that defines what properties we want to store for our buildings. This class, called `Building`, will represent a building in our Firebase database. All the properties we define in it will be persisted in a Firestore document. We will now follow these steps to create it:

1. Let's take this opportunity to learn another way to create scripts. Go to our `Scripts` folder located in **Assets** | **_App** | **Scripts** and right-click on the empty space of the content, as shown in the following screenshot.

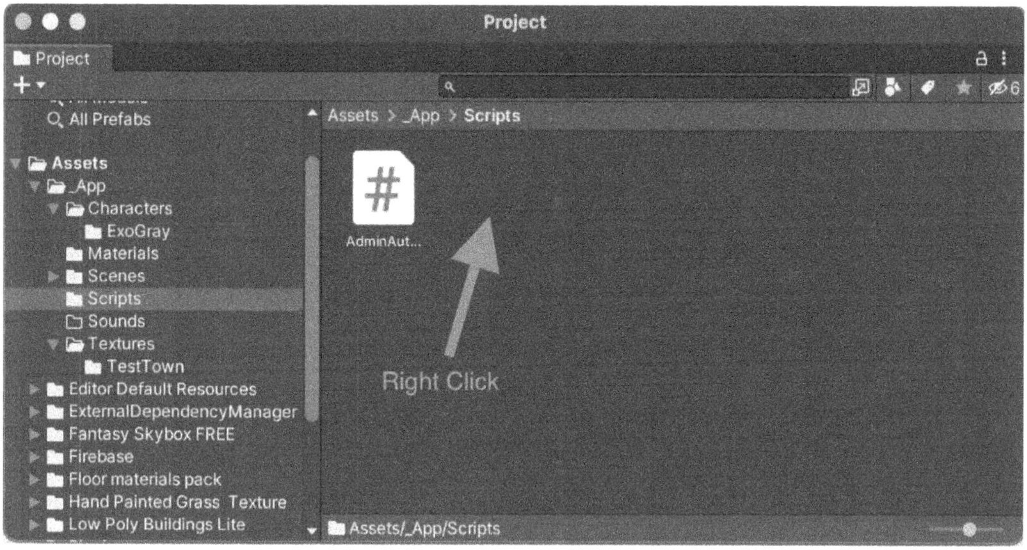

Figure 5.42 – Another way to create new scripts

2. Select the **Create** | **C# Script** option. A new object will be created in the folder. Rename it to `Building`.

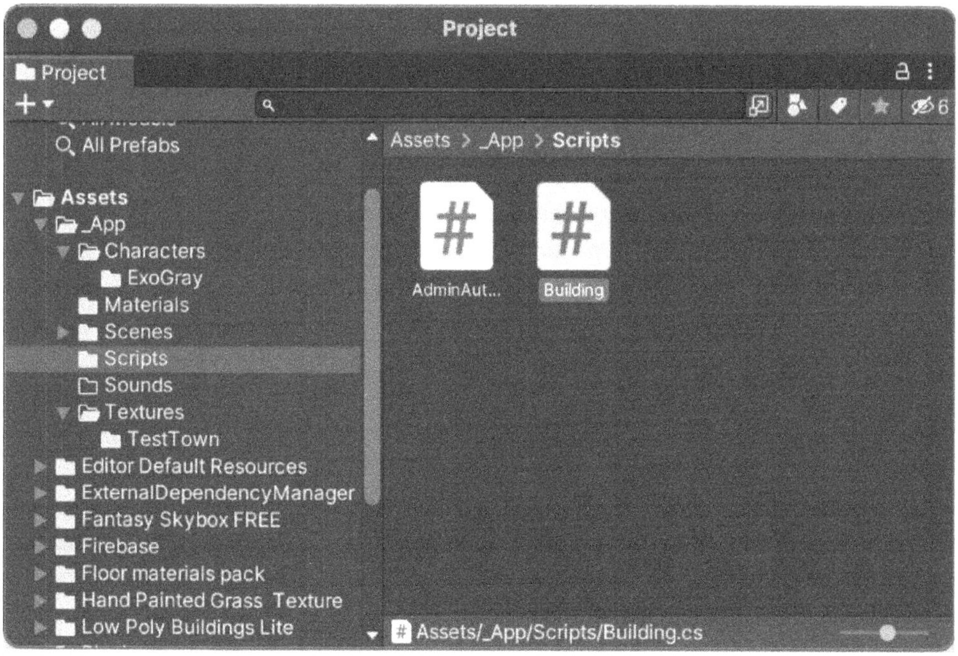

Figure 5.43 – Recently created Building class

3. Double-click on the newly created `Building` file to edit it in Visual Studio. This will not really be a script, but a pure class. The default code that appears is not useful. We'll replace it completely with the following code:

```
using Firebase.Firestore;

[FirestoreData]
public class Building
{
    [FirestoreProperty]
    public string Id { get; set; }

    [FirestoreProperty]
    public string Prefab { get; set; }

    [FirestoreProperty]
    public string WorldId { get; set; }

    [FirestoreProperty]
    public float PosX { get; set; }

    [FirestoreProperty]
    public float PosY { get; set; }

    [FirestoreProperty]
    public float PosZ { get; set; }

    [FirestoreProperty]
    public string OwnerUserId { get; set; }

    public Building()
    {
    }
}
```

Well, as you may have noticed, this script is a simple class, but some things may not ring a bell, such as the `[FirestoreData]` and `[FirebaseProperty]` attributes.

These attributes make an ordinary class intelligible, serializable, and deserializable for Firebase. With the `[FirestoreData]` property, we indicate that the class can be transformed into a format that Firestore understands, and with the `[FirestoreProperty]` attribute, we indicate that it is a property of the collection.

Now, let's review the properties that we have declared and what we intend with each of them:

- `Id`: This will be a unique ID for each building, in string format, that we will generate in the next script
- `Prefab`: This is the name of the Prefab object that will represent our building
- `WorldId`: This is the unique ID of the world where the building is instantiated
- `PosX`, `PosY`, and `PosZ`: These are the coordinates where the building is positioned
- `OwnerUserId`: We will store the ID of the user who owns this building

We have a class on which the following script depends, so we can continue. Now, we will create the code in charge of performing the actions on the buildings in the database.

Creating the BuildingInstanceManager script

With this script being a big one, we are going to create it step by step, explaining in detail all the changes we are going to make in each step. Let's start:

1. Right-click again on the content of the `Scripts` folder and select the **Create | C# Script** option. Rename the new file `BuildingInstanceManager` and finally double-click on it to edit it.

   We will see the typical initial default code:

   ```
   using System.Collections;
   using System.Collections.Generic;
   using UnityEngine;

   public class BuildingInstanceManager : MonoBehaviour
   {
       void Start()
       {
       }
       void Update()
       {
       }
   }
   ```

   This script has the peculiarity that it will be able to be executed in the Unity Editor without **Play** mode, in **Play** mode, and in production. This is because we need its functionality when we are editing the scene and as well as when it is executed in **Play** mode, so that it instantiates the building in the scene correctly according to the specifications.

2. To make it work in the three modes, we add the `[ExecuteAlways]` attribute just above the class declaration:

```
[ExecuteAlways]
public class BuildingInstanceManager : MonoBehaviour
{
  ...
```

3. Next, we will declare the variables that we will need during the course of development. Add the highlighted code to the script:

```
using System.Collections;
using System.Collections.Generic;
using Firebase.Firestore;
using Firebase.Extensions;
using System;
using UnityEngine;

[ExecuteAlways]
public class BuildingInstanceManager : MonoBehaviour
{
    FirebaseFirestore db;
    public string buildingId;
    public string prefabName;
    public string worldId;
    public string ownerUserId;
    void Start()
    {
    }
    void Update()
    {
    }
}
```

As you will have observed, the variables contain the same information as the `Building` class. These properties of the script will be transferred to a `Building` class to send it to Firestore.

Creating auxiliary methods

It is time to create three auxiliary methods that will be used by the main functions: `PlacePrefab`, `CleanPrefab`, and, finally, `GetBuilding`.

The first one will be in charge of searching in our `Resources` folder for the Prefab with the name indicated by the parameter and instantiating it in the scene. The second one, as its name indicates, will be used to clean the Prefab of the scene if it already exists. Finally, the `GetBuilding` method will be in charge of transforming the information of the script variables into a `Building` class:

1. Add the `PlacePrefab` method to the script:

```
private void PlacePrefab(string prefabName)
{
    try
    {
        GameObject buildingInstance =
            Instantiate(Resources.Load(
                "Prefabs/Buildings/" +
                    prefabName), transform) as
                        GameObject;

        buildingInstance.transform.parent =
            transform;
        buildingInstance.transform.localPosition =
            Vector3.down;
        gameObject.Getcomponent<MeshRenderer>()
            .enabled = false;
        gameObject.Getcomponent<Collider>()
            .enabled = false;
    }
    catch(Exception e)
    {
        Debug.LogError(e.Message);
    }
}
```

   The operation is simple: we pass a string with the name of the prefab by parameter, then we look in the `Prefabs/Buildings` path for a Prefab with the same name as the `prefabName` variable. It is necessary to say here that the route is obviating the parent folder called `Resources`. This is because Unity 3D, by default, uses this folder to search for the assets that, by script, we want to instantiate in the scene.

2. After this, what we do is instantiate the Prefab as a child of the GameObject in which we have attached this script, which will be a **Cube** GameObject. We will see it graphically later.
3. We also copy the position of the cube to the new Prefab so that it loads in the same place, precisely where we placed the cube with the script.
4. Finally, we disable the visibility of the cube and its collider so that it does not conflict with the new Prefab.

5. It is time to create another auxiliary method, the `CleanPrefab` function. It is very simple; just add this code fragment to your script:

```
private void CleanPrefab()
{
    int childCount = transform.childCount;

    for (int i = childCount - 1; i >= 0; i--)
    {
        DestroyImmediate(transform.GetChild(i)
            .gameObject);
    }

    Debug.Log("Building prefab cleaned");
}
```

   The operation is very simple: we obtain the number of children attached to the **Cube** GameObject and destroy them.

6. The last auxiliary method, `GetBuilding`, will get the value of the script variables and return a new object of the `Building` type, ready to send to Firebase. Add the following code snippet to the script:

```
// We create a Building class with the local data
   of the object
private Building GetBuilding()
{
    return new Building()
    {
        Id = buildingId,
        PosX = transform.position.x,
        PosY = transform.position.y,
        PosZ = transform.position.z,
        Prefab = prefabName,
        WorldId = worldId,
        OwnerUserId = ownerUserId
    };
}
```

Well done, we are almost there. Now we are going to create a method that initializes the data of a building.

Creating the InitializeBuilding method in the script

This method is fundamental since we need to obtain a unique ID before inserting it in Firestore:

1. To automatically obtain a unique ID from the `InitializeBuilding` function. Copy and paste the following code snippet into your script:

```
public void InitializeBuilding()
{
    // We get a new unique ID for this building
       and assign it to the object in the scene
    string uniqueId = Guid.NewGuid().ToString();
    gameObject.name = uniqueId;
    gameObject.isStatic = true;

    buildingId = uniqueId;
}
```

   This method must be called before performing any write operation on a building. As you can see, we obtain a unique ID through the `Guid.NewGuid().ToString()` function and we programmatically change the name of the GameObject with the obtained ID. Thus, we're guaranteed to have a unique ID that we can obtain later to consult in Firestore.

   We mark the instantiated object as static and assign the unique ID to the `buildingId` variable.

2. It is now time to create the function in charge of creating or updating a building in the database. It will be called `CreateOrUpdate` and this is its code; add it to the script:

```
public void CreateOrUpdate()
{
    // We check that we have instantiated the
       variable "db" that allows access to the
       Firestore functions, otherwise we
       instantiate it
    if (db == null)
        db = FirebaseFirestore.DefaultInstance;
    if (string.IsNullOrEmpty(buildingId))
        return;
    if (string.IsNullOrEmpty(worldId) ||
        string.IsNullOrEmpty(prefabName))
        return;
    DocumentReference docRef =
        db.Collection("Buildings")
        .Document(buildingId);
    docRef.SetAsync(GetBuilding(),
        SetOptions.MergeAll)
```

```
                .ContinueWithOnMainThread(task =>
                {
                    if (task.IsCanceled || task.IsFaulted)
                        return;
            });
        }
```

As you can see, this script is a bit more complex, but totally intelligible. The flow is actually very clear:

I. First, we make sure that we have instantiated the Firestore SDK in the variable called `db`.

II. Next, we check that the `buildingId` variable has content; otherwise, it will cause an error, as it means that we have not initialized the building first.

III. We also check that a value has been added to the `worldId` and `prefabName` variables; otherwise, it causes an error, as we cannot add a building to the database with inconsistent data.

IV. If all the preceding goes well, we point to the **Buildings** collection and a particular document with the ID we have generated for the building. You may think: I have not created the **Buildings** collection in Firestore, let alone a document inside the **Buildings** collection. What are we doing? The answer is that the Firestore SDK will create both the collection and the document for you if it does not exist previously. Wonderful, isn't it?

V. Finally, when we call the `SetAsync` method, we pass as a parameter to the function an object of the `Building` type that we obtain thanks to our auxiliary `GetBuilding` method. If everything goes well, **Firestore** will create a new document, with the `ID` that we have autogenerated, and within the **Buildings** collection.

Creating the GetExistingBuilding method in the script

The other great method of this script is the one in charge to consult in data base a concrete building and then to instantiate it. It is called `GetExistingBuilding`. Its code is as follows; add it to your script:

```
public void GetExistingBuilding()
{
    if (db == null)
        db = FirebaseFirestore.DefaultInstance;
    if (transform.childCount > 0)
        CleanPrefab();

    string buildingId = gameObject.name;

    DocumentReference docRef =
        db.Collection("Buildings")
        .Document(buildingId);
```

```
        docRef.GetSnapshotAsync().ContinueWithOnMainThread(
            task =>
            {
                if (task.IsCanceled || task.IsFaulted)
                    return;

                DocumentSnapshot buildingRef = task.Result;

                if(!buildingRef.Exists)
                    return;

                Building building =
                    buildingRef.ConvertTo<Building>();

                buildingId = building.Id;
                prefabName = building.Prefab;
                worldId = building.WorldId;
                ownerUserId = building.OwnerUserId;
                PlacePrefab(prefabName);
            }
        );
    }
```

The operation of this function is similar to the previous one. Its flow can be described as follows:

1. We make sure that we have an instance of the Firestore SDK in the `db` variable.
2. If an instantiated building already exists as a child in this GameObject, we destroy it.
3. We get the unique ID of the building, which we have stored in the name of the GameObject containing this script.
4. We create a reference to the **Buildings** collection and to a document inside it whose ID is the same as the one we have obtained.
5. Using the `GetSnapshotAsync` method, we obtain the results of the query.
6. If the query has failed, or if for any reason it has been canceled or it has not found any building in the indicated path, we will cause an error and exit the function.
7. If all the preceding has gone well, we obtain the query data in the `buildingRef` variable, by means of the `buildingRef.ConvertTo<Building>()` line statement, converting the query information into a class of the `Building` type.
8. Finally, we copy the values obtained from the database to the script variables and call the `PlacePrefab` auxiliary method to instantiate the object in the scene.

9. To finalize the flow, we call the `GetExistingBuilding` method in the `Start` function:

```
void Start()
{
    GetExistingBuilding();
}
```

The following shows what the complete script code looks like:

```
...
[ExecuteAlways]
public class BuildingInstanceManager : MonoBehaviour
{
    FirebaseFirestore db;
    public string buildingId;
    public string prefabName;
    public string worldId;
    public string ownerUserId;

    void Start()
    {
        GetExistingBuilding();
    }
    public void CreateOrUpdate()
    {
        if (db == null)
            db = FirebaseFirestore.DefaultInstance;

        if (string.IsNullOrEmpty(buildingId))
            return;

        if (string.IsNullOrEmpty(worldId) ||
            string.IsNullOrEmpty(prefabName))
            return;
        DocumentReference docRef =
            db.Collection("Buildings")
                .Document(buildingId);
        docRef.SetAsync(GetBuilding(),
            SetOptions.MergeAll)
            .ContinueWithOnMainThread(task =>
            {
            if (task.IsCanceled || task.IsFaulted)
                return;
            });
```

```
}
public void InitializeBuilding()
{
    string uniqueId = Guid.NewGuid().ToString();
    gameObject.name = uniqueId;
    gameObject.isStatic = true;
    buildingId = uniqueId;
}
public void GetExistingBuilding()
{
    if (db == null)
        db = FirebaseFirestore.DefaultInstance;
    if (transform.childCount > 0) CleanPrefab();

    string buildingId = gameObject.name;

    DocumentReference docRef =
        db.Collection("Buildings")
            .Document(buildingId);
    docRef.GetSnapshotAsync()
        .ContinueWithOnMainThread(task =>
        {
            if (task.IsCanceled || task.IsFaulted)
                return;
        DocumentSnapshot buildingRef =
            task.Result;

        if (!buildingRef.Exists)
            return;

        Building building =
            buildingRef.ConvertTo<Building>();
        buildingId = building.Id;
        prefabName = building.Prefab;
        worldId = building.WorldId;
        ownerUserId = building.OwnerUserId;
        PlacePrefab(prefabName);
        });
}
private void CleanPrefab()
{
    int childCount = transform.childCount;
```

```
            for (int i = childCount - 1; i >= 0; i--)
            {
                DestroyImmediate(transform.GetChild(i)
                    .gameObject);
            }
        }
        private void PlacePrefab(string prefabName)
        {
                GameObject buildingInstance =
                    Instantiate(Resources.Load(
                        "Prefabs/Buildings/" +
                            prefabName), transform) as
                                GameObject;

                buildingInstance.transform.parent =
                    transform;
                buildingInstance.transform.localPosition =
                    Vector3.down;
                gameObject.Getcomponent<MeshRenderer>()
                    .enabled = false;
        }

        private Building GetBuilding()
        {
            return new Building()
            {
                Id = buildingId,
                PosX = transform.position.x,
                PosY = transform.position.y,
                PosZ = transform.position.z,
                Prefab = prefabName,
                WorldId = worldId,
                OwnerUserId = ownerUserId
            };
        }
    }
```

Good job! We have completed the script, but something important is missing. These functions that we have added must be invoked from somewhere, some type of button that appears in our component and facilitates this management.

Creating the BuildingInstanceEditor script

To get some buttons to help us invoke these methods from the Unity interface, Unity has a library called **CustomEditor**, which allows us to add elements such as buttons, sliders, checks, and texts to our components.

> **Tip: Learn more about Unity's CustomEditor**
>
> Unity's **CustomEditor** can help us to create amazing scripts that display your own graphical interface in the **Inspector** panel. This can make our life easier when programming executable scripts from the Editor. To learn more about this, I recommend you visit `https://docs.unity3d.com/Manual/editor-CustomEditors.html`.

We create a new script in the same location, rename it `BuildingInstanceEditor`, and double-click to edit it. Replace all the default code with the following:

```
#if UNITY_EDITOR
using UnityEngine;
using System.Collections;
using UnityEditor;

[CustomEditor(typeof(BuildingInstanceManager))]
public class BuildingInstanceEditor : Editor
{
    public override void OnInspectorGUI()
    {
        DrawDefaultInspector();

        BuildingInstanceManager myScript =
            (BuildingInstanceManager)target;

        if (GUILayout.Button("Initialize Building"))
        {
            myScript.InitializeBuilding();
        }

        if (GUILayout.Button("Create Or Update in
            Firestore"))
            {
                myScript.CreateOrUpdate();
            }
    }
}
#endif
```

If you look at the brief code, in the `OnInspectorGUI` method, we inject two buttons, **Initialize Building** and **Create Or Update in Firestore**. These buttons also call the functions that we have implemented in the previous script. Save the script changes and let's see it in action.

Creating the BuildingManager Prefab

Great, we have everything ready to see it working. Now we follow a few steps to enjoy these functionalities that we have programmed:

1. We are going to create our own Prefab to be able to enjoy this functionality in other future scenes. For that, create a new folder called `Prefabs` in the **Assets** | **_App** | **Prefabs** path.
2. Inside the `Prefabs` folder we just created, right-click on the content and select the **Create** | **Prefab** option. Rename the new element `BuildingManager`.

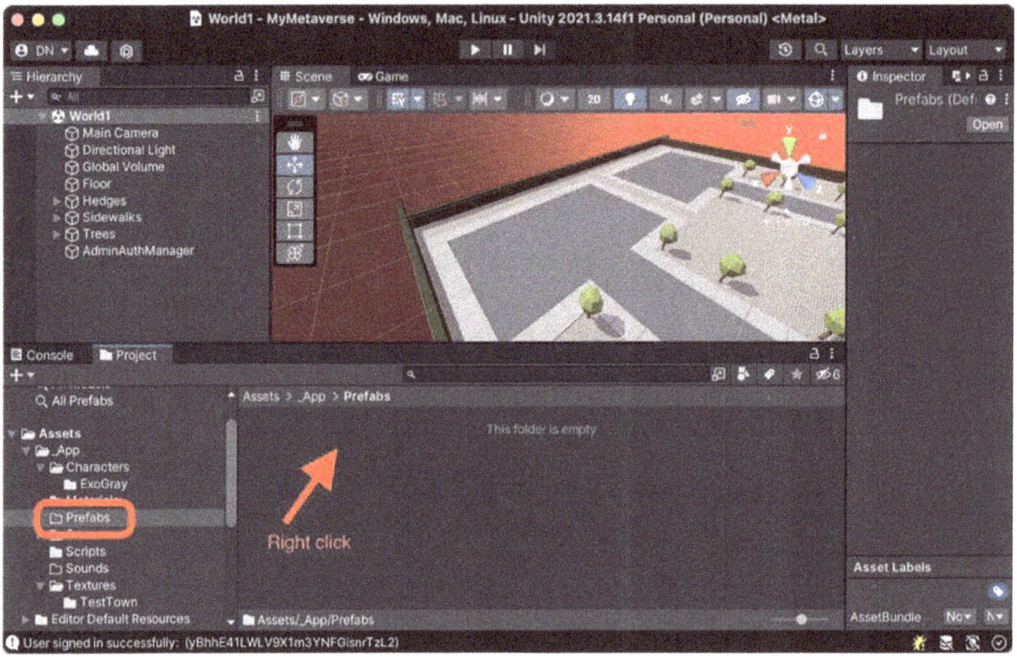

Figure 5.44 – Creating a new prefab

3. Select the **BuildingManager** Prefab and click on the **Add component** button in the **Inspector** panel.

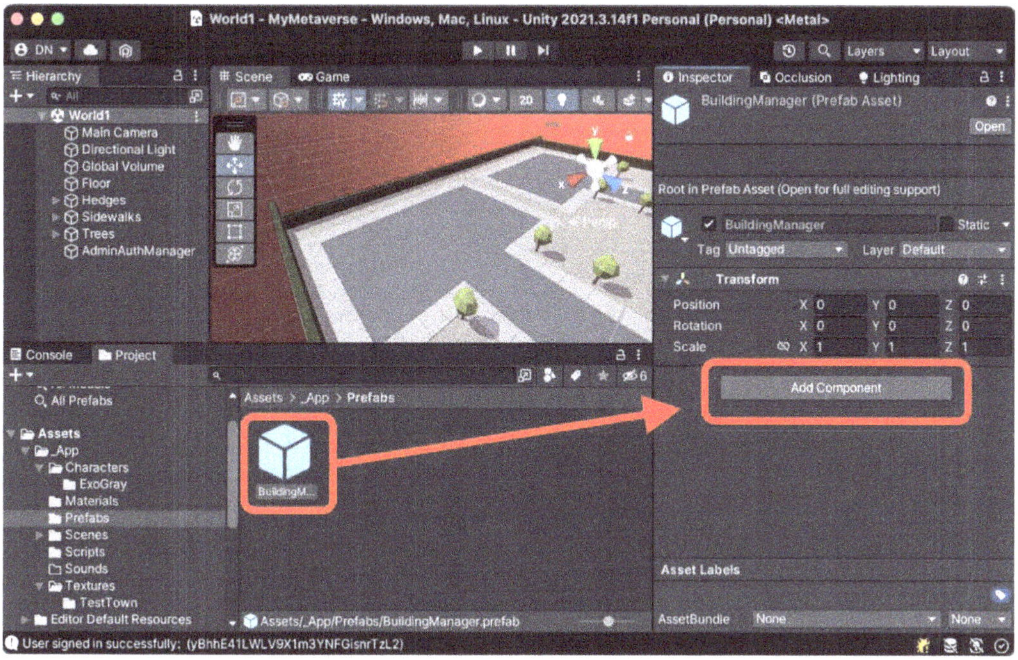

Figure 5.45 – Adding a new component to a prefab

4. Type `BuildingInstanceManager` and double-click on the result to add it.

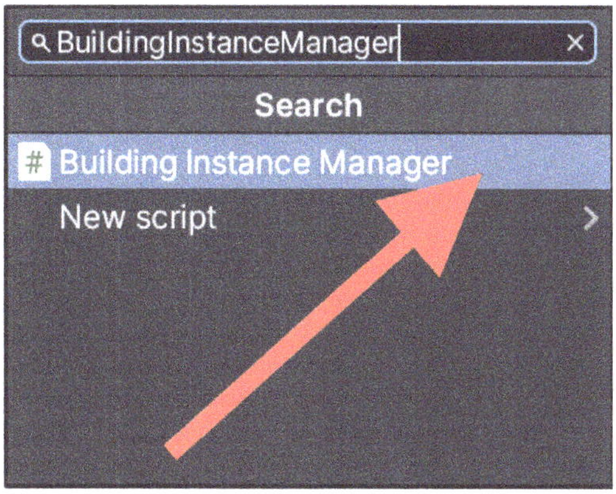

Figure 5.46 – Searching for our BuildingInstanceManager script

5. To have a graphical preview of the Prefab in the scene, we will add a **Mesh Renderer** component.
6. Finally, we will add a **Mesh Filter** component. Click on the button to search for a shape, type `Cube` in the search engine, and select the Cube that appears in the list.

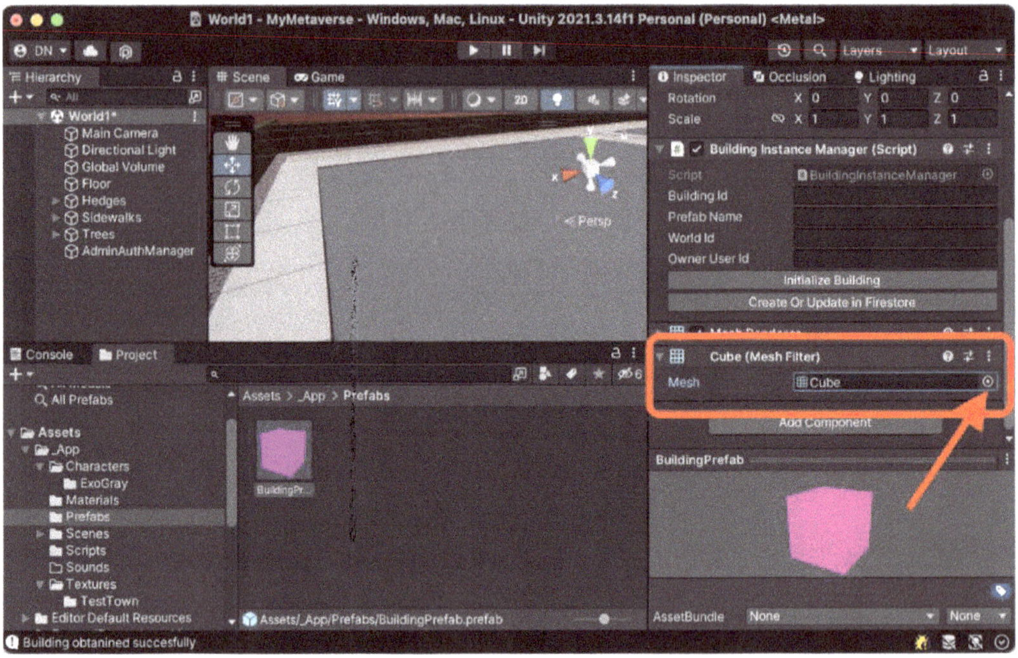

Figure 5.47 – Choosing a Cube as the prefab form

Great! You have created your first Prefab and connected it to the `BuildingInstanceManager` script we created earlier. Now, we can drag our `BuildingManager` Prefab to the scene and configure it.

7. Drag the `BuildingManager` Prefab to the scene, approximately in the same place as in the following screenshot:

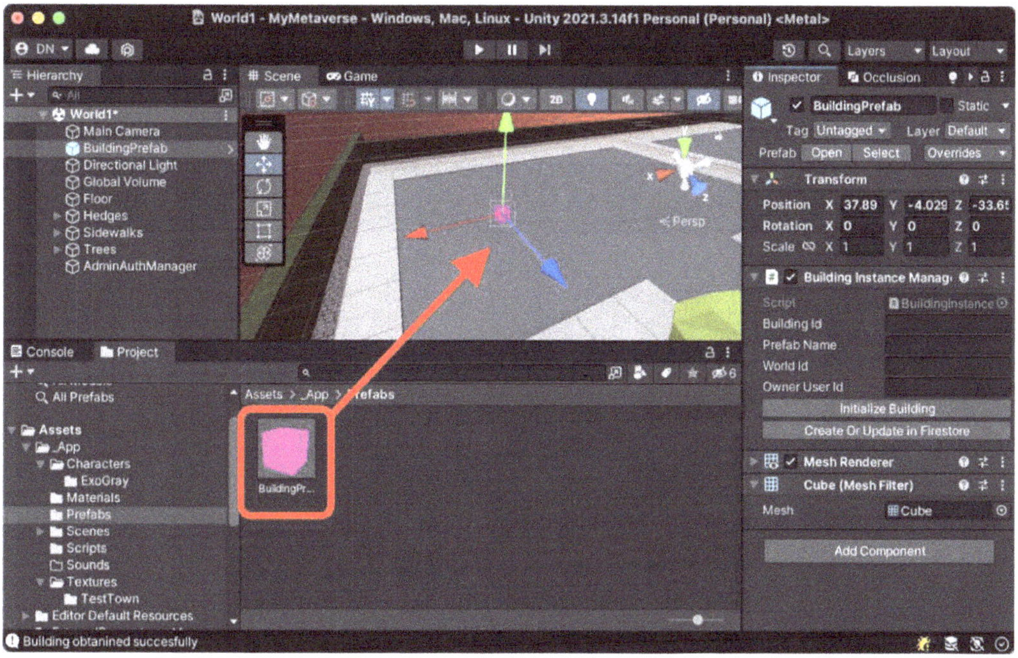

Figure 5.48 – Placing our Prefab in the scene

8. We selected the **BuildingPrefab** GameObject in the **Hierarchy** panel and in our **Building Instance Manager** component. We can observe that we have text boxes to fill in the necessary information and, above all, the two buttons that we have created to invoke the functions. Click on the **Initialize Building** button.
9. As we had programmed, now the `Building Id` variable has been loaded with a unique value automatically, and also, the GameObject has changed its name to that of the ID. This works!
10. If we remember, the `Prefab Name` and `World Id` variables are mandatory. In the `Prefab Name` variable, we will type `Suburb House Yellow`, which was one of the two Prefabs we placed in the **Assets** | **Resources** | **Prefabs** | **Buildings** path.

11. To get the value to write in the **World Id** property, we must go to Firebase Console, and then get the document ID of the last world we have added. You can find this value in the **Firestore Database** section, inside the **Worlds** collection. Select the **World1** document. In my case, the value I have to copy and paste into the **World Id** variable is `nZryA8PmnkYyylJFRRWd`.

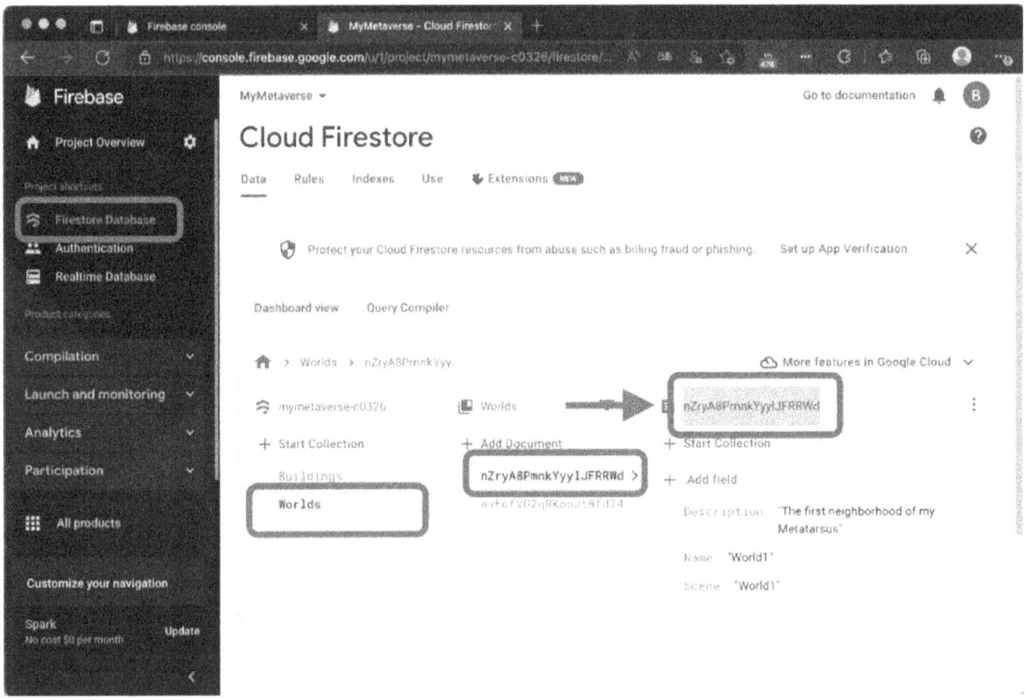

Figure 5.49 – Coping the document ID from Firebase Firestore

12. With all the variables initialized, we can click on the **Create Or Update in Firestore** button.

If everything went well, you will see a message with the text **Building created or updated successfully**. You can also again visit Firestore Console and see that a new collection called **Buildings** has been created. Inside it, we have our building.

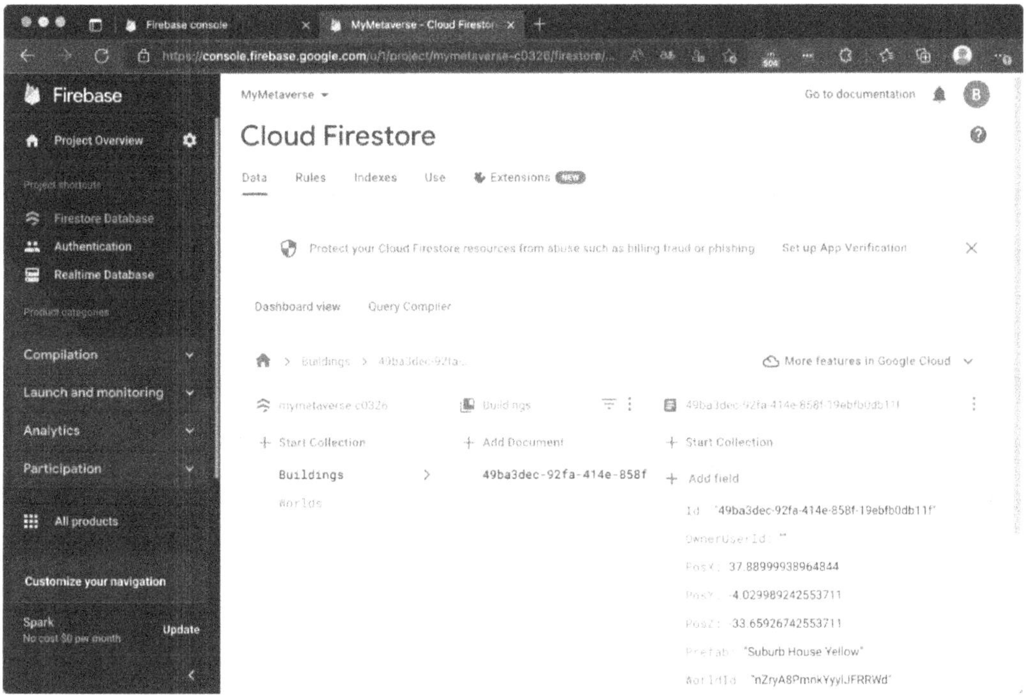

Figure 5.50 – Created building in Firebase Firestore

Now, you can press the **Play** button and see the result in the **Scene** panel tab. The building we wanted has been loaded, in the place we wanted. Congratulations!

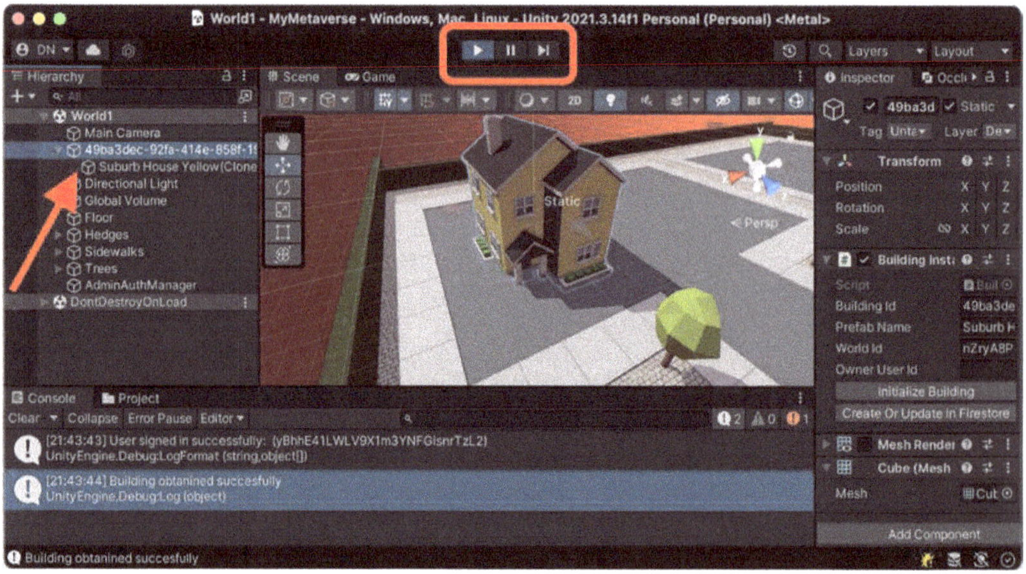

Figure 5.51 – Preview of the script in operation

Once you have finished editing this scene, don't forget to delete the **MainCamera** GameObject, to avoid conflicts with multiple cameras in future chapters.

Now you can perform the same steps to add more **BuildingManager** Prefabs elsewhere in the scene, wherever you'd like them.

You have just learned a powerful functionality in this chapter; you have learned how to create from scratch a new component to be able to read and write on a Firestore database. I'm sure that right now you have in your mind ideas about functionalities that you can implement in the future, taking advantage of the innumerable possibilities that the world of Firebase with Unity offers us.

Optimizing a dynamic Scene

We are coming to the end of the chapter, but we can't end without recalling the benefits of optimizing the scene. This scene is not very different from the one we designed in *Chapter 4*, *Preparing Our Home Sweet Home: Part 2*. It has static elements too. The only difference is that the Prefabs are added at runtime.

It is important that you apply the same optimization techniques that we saw in *Chapter 4, Preparing Our Home Sweet Home: Part 2*. We have programmed these dynamic elements to be instantiated statically, so we are already applying optimization in terms of lights.

Although these buildings are added later, when the scene is run, the **Camera Occlusion Culling** optimization will remove and display them depending on whether or not they are in the viewing area.

This type of scene may result in less FPS due to not having baked shadows on the dynamic buildings, but it is not a concern. I recommend that you analyze the FPS as your scene grows and evaluate how many buildings you can keep, or maybe you want to make smaller scenes, rather than a single large one.

Summary

In my humble opinion, I think this has been a very fruitful chapter. We have learned a lot in these pages. After designing a new scene from scratch and registering it as a new record in the Firestore database, we (finally) learned how to program several scripts.

In these scripts, and in a very didactic way, we learned how to use the Firebase Authentication SDK to identify ourselves as administrators and later create another script that, through the Firestore SDK, we have been able to read and write from the Unity Editor.

The lesson learned here is that there are infinite possibilities in front of us, having learned how to program scripts that communicate with Firebase services, which will have awakened in your mind a storm of ideas and new possibilities.

In the next chapter, we will learn how to integrate another great feature of Firebase, Firebase Authentication. We will include a system that will allow users to log in or create a new account. For this, we will design some screens with the Unity GUI and make some scripts that are able to connect to this Firebase user identification tool.

Part 2: And Now, Make It Metaverse!

In this second part, we will go deeper into programming. We will integrate Firebase Authentication to add security to our project, forcing users to log in or create a new account, all with screens created with the Unity GUI. We will also create our first **Non-Player Character** (**NPC**), a fun 3D animated character, who will offer an indispensable service to us: provide the ability to travel between worlds.

We will also develop functionality that allows the user to acquire a building, link it to the database, and then create an NFT with ChainSafe technology. Finally, we will learn how to integrate Photon SDK, to turn our Metaverse into a multiplayer world, with a voice and text chat that allows users to communicate in real time.

This part has the following chapters:

6
Adding a Registration and Login Form for Our Users

Access control to a system through credentials is a fundamental feature in any user-based project. In our metaverse, we will provide the user with a welcome screen where they can register or log in. To carry out this task, we will rely on the Firebase Authentication service SDK, which we previously learned about in *Chapter 5, Preparing a New World for Travel*.

When a user registers for the first time, we will store basic information, collected from the registration form. This data will be stored in a collection of users in Firebase Firestore for later use.

Throughout this chapter, we will learn about a new set of tools in Unity, called **Unity UI**. These tools will allow us to create graphical interface features with which the user can interact, such as buttons, inputs, and labels.

We will cover the following topics:

- Creating a sign-up screen
- Creating a sign-in screen
- Placing the player

Technical requirements

This chapter does not have any special technical requirements, but we will be programming with scripts in C#. It would be advisable to have basic knowledge of this programming language. We need an internet connection to browse and download an asset from the Unity Asset Store.

We will continue with the project we created in *Chapter 1*. Remember that we have a GitHub repository, `https://github.com/PacktPublishing/Build-Your-Own-Metaverse-with-Unity/tree/main/UnityProject`, which contains the complete project that we will work on here.

You can also find the complete code for this chapter on GitHub at: `https://github.com/PacktPublishing/Build-Your-Own-Metaverse-with-Unity/tree/main/Chapter06`

Creating a sign-up screen

Unity UI is a new concept that you will read about often throughout these pages, but before we continue, we must establish new knowledge about this tool.

Unity UI is a set of tools and resources that facilitates the creation of UIs in the game. The most common elements we will work with will be the following:

- **Canvas**: The canvas is the working area and the root that contains a set of UI elements. It can be configured to automatically fit and scale to any screen.
- **Button**: A button can contain only text or include an image. With a button, we can activate functions of our scripts.
- **Text**: This allows us to display text of any size, with multiple configurations that allow high customization.
- **Image**: This is simply an image that can serve as a background or decorative element anywhere in our interface.
- **Panel**: This serves as a parent element for grouping and organizing multiple UI elements.
- **Input Field**: This allows the user to write text in the interface.

> **Tip: More documentation on Unity UI**
>
> In the previous list, we described the most used and most basic elements of the Unity UI tool. There are more elements. In this chapter, we will see, step by step, how to create and use the elements that we will need. But if you want to know more or you feel that you do not have the necessary knowledge about Unity UI, I recommend you navigate to the documentation page at `https://docs.unity3d.com/Manual/UIToolkits.html` and follow the tutorial that you will find at the following web address: `https://learn.unity.com/tutorial/using-the-unity-interface`.

In the following pages, we will create the entry point for all users. Every time they access our metaverse, they will have to go through this so that they can identify themselves. These identification functions will be accessible from a welcome screen, where we will show different buttons and textboxes that will allow the user to write their credentials or create a new account in order to continue.

The first thing we will do is create a new scene, this time a simple scene since we will not need lights or sky. To do this, go to **Assets** | **_App** | **Scenes** and right-click on the content to click on the **Create** | **Scene** option.

Next, a new scene file named **New Scene** will appear. Rename it `Welcome`.

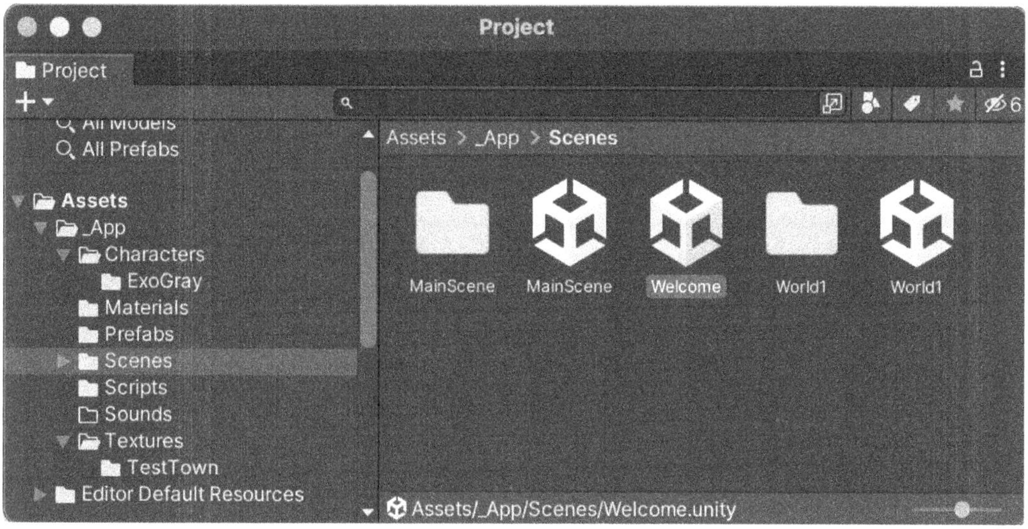

Figure 6.1 – New scene created

Finally, double-click on our new **Welcome** scene to open it.

Before we continue, we need to stop and think about what features and functionality this initial scene will have:

- A button that displays textboxes to identify yourself, with a button to confirm
- A button that displays a registration form, with a **Confirm** button
- A button to go back and select another option

> **Tip: A better way to work in 2D scenes**
>
> The **Welcome** scene we are working with is a 2D projection scene; that is, all the elements we will show on the screen do not need a 3D perspective – the camera will focus on them only from the front. To work more easily with scenes of this type, it is advisable to activate 2D mode, which you will find in the toolbar of the **Scene** panel.

This option will block the option to change the perspective of the scene and will be limited to focusing on all Unity UI elements from the front. In the following screenshot, we can see how from the view options bar, we can easily switch between 2D and 3D views.

Figure 6.2 – Scene view options

As we mentioned before, every UI needs to be created under a **Canvas** element. To create ours, in the Unity main menu bar, click on the **GameObject** | **UI** | **Canvas** option.

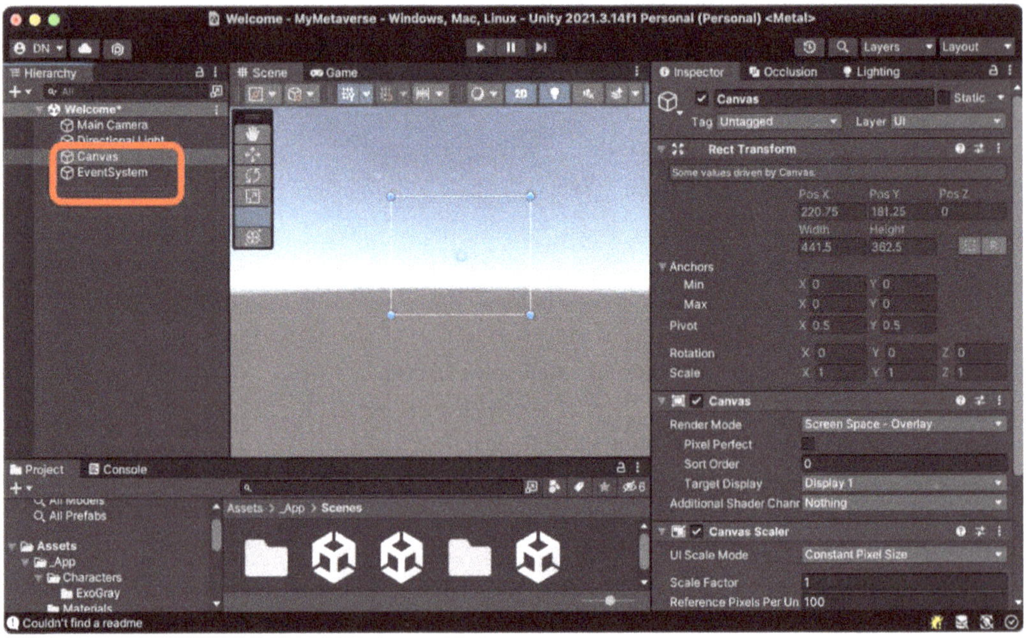

Figure 6.3 – Canvas element in the scene

As you may have noticed, the **Canvas** element has been created in the scene, but also, a new GameObject called **EventSystem** has appeared, which will be in charge of capturing the user's input events, for example, button clicks or text input.

The **EventSystem** GameObject is created only the first time if one did not existed previously, as well as the Canvas element.

If we select **EventSystem** in the **Hierarchy** panel, we will see a warning in the **Inspector** panel that we need to configure the scene to capture the input events with the new system, which we installed in *Chapter 1, Getting Started with Unity and Firebase.*

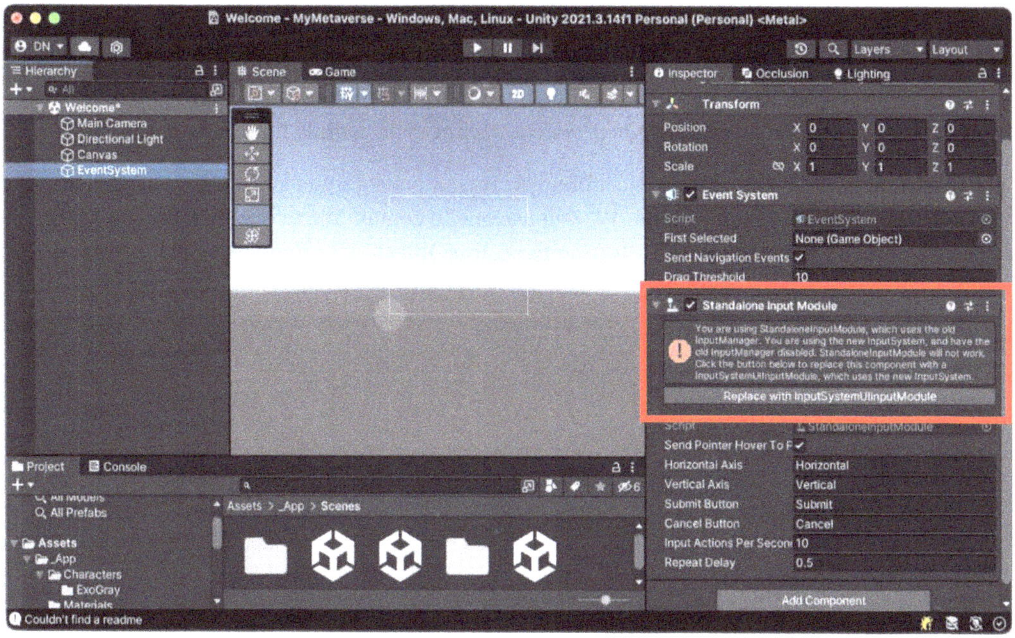

Figure 6.4 – Warning about input module

Simply click on the button with the text **Replace with InputSystemUIInputModule** to apply the configuration.

To be able to continue with the design of our interface, we need to apply one last configuration, this time to the canvas; this configuration will help the Canvas to automatically adjust to the screen when we click the **Play** button. For this to happen, we select the GameObject **Canvas** in the **Hierarchy** panel, and we apply the following configuration to the **Canvas Scaler** component.

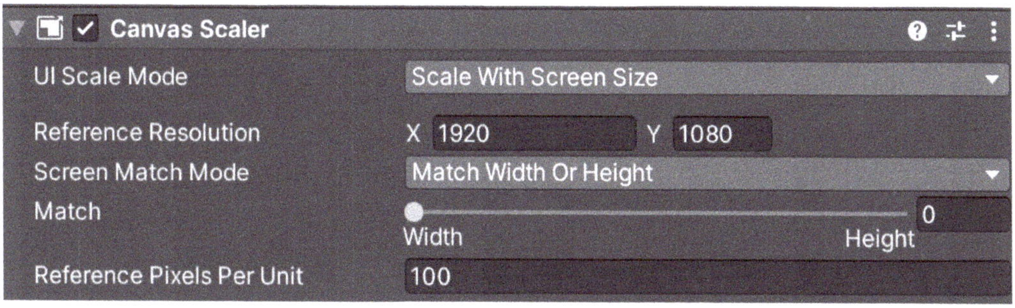

Figure 6.5 – Canvas Scaler component

With this configuration, we are indicating that it scales up or down, depending on the size of the screen. We also indicate that as a reference, our design will be for **Full HD** screens (1,920x1,080 pixels) and that it also takes as a reference the width of the screen to be scaled; that is, we want it to scale up to the width of the screen, so it will keep the proportion without deforming the content.

Well, now we have everything we need to continue. As we saw previously, in Unity UI, we have elements of the **Panel** type, which group content. We will use one to contain the two main buttons. For this, we need to create a **Panel** element by selecting the **GameObject** | **UI** | **Panel** option. Once created, we will rename it `MenuButtons`.

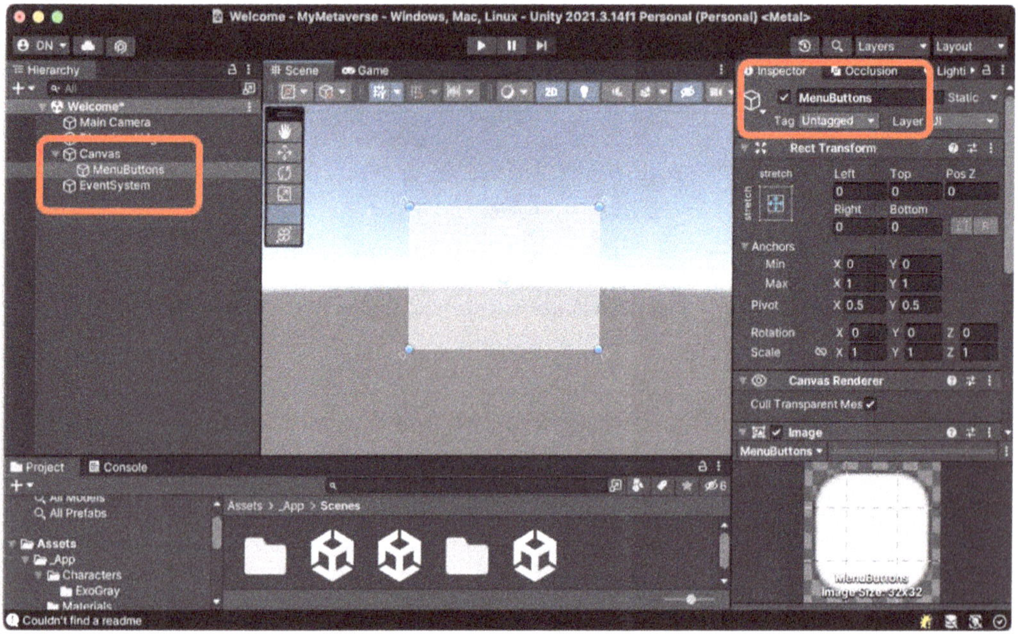

Figure 6.6 – MainButtons panel

By default, the panel created takes the same size as **Canvas**, expanding completely. Our intention is to make it smaller and always centered on **Canvas**. For this, we need to perform two operations on the panel:

1. With the **MenuButtons** GameObject selected, in the **Inspector** panel, select the alignment button found in the **Rect Transform** component and select the **center**.

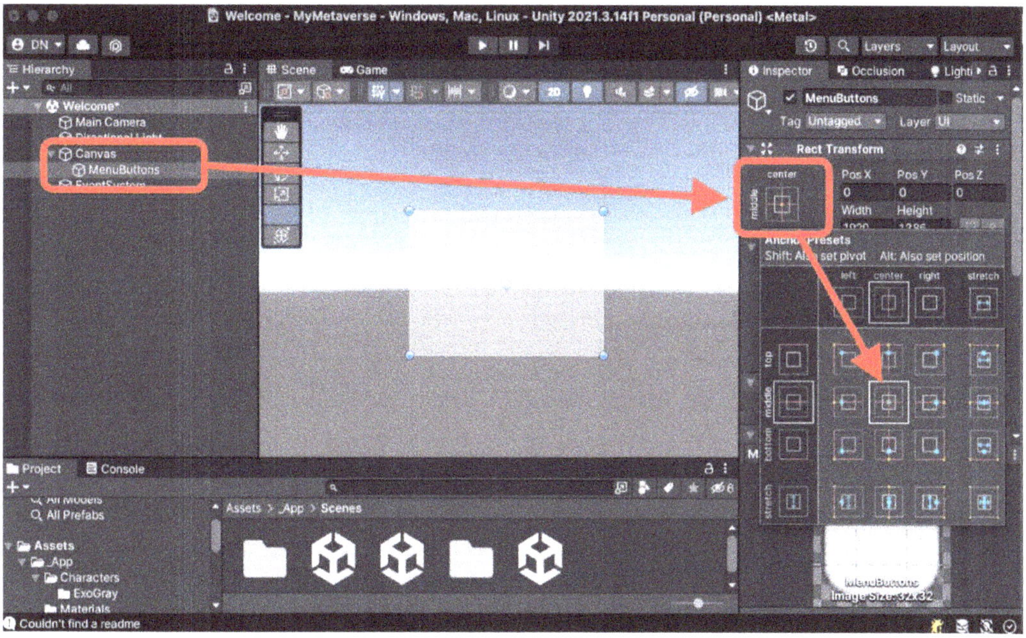

Figure 6.7 – MainButtons alignment option

2. To make the panel smaller, with the **Rect Transform** tool and holding down the *Alt* key, select the bottom edge of the panel and make it smaller. Do the same operation on the side edge. It will look something like this:

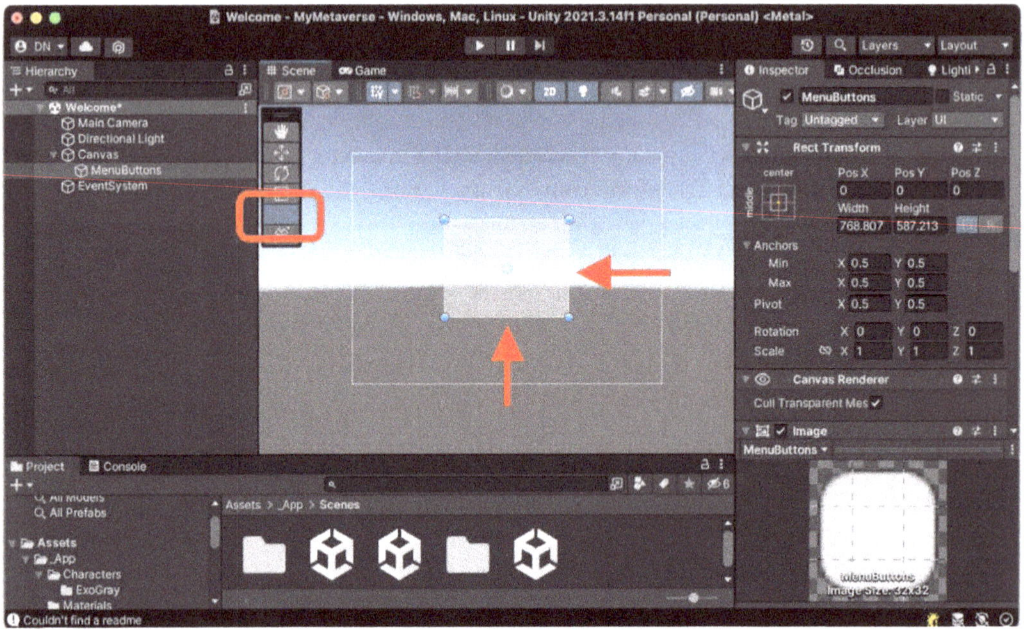

Figure 6.8 – Resized MainButtons panel

Great, we have our first panel that will host the **Sign In** and **Sign Up** buttons. First, we will create the **Sign Up** button. For this, with the **MenuButtons** GameObject selected in the **Hierarchy** panel, right-click on it and select **UI** | **Button TextMesh Pro**. This will create a new item inside **MenuButtons**.

Oops, a warning message has appeared.

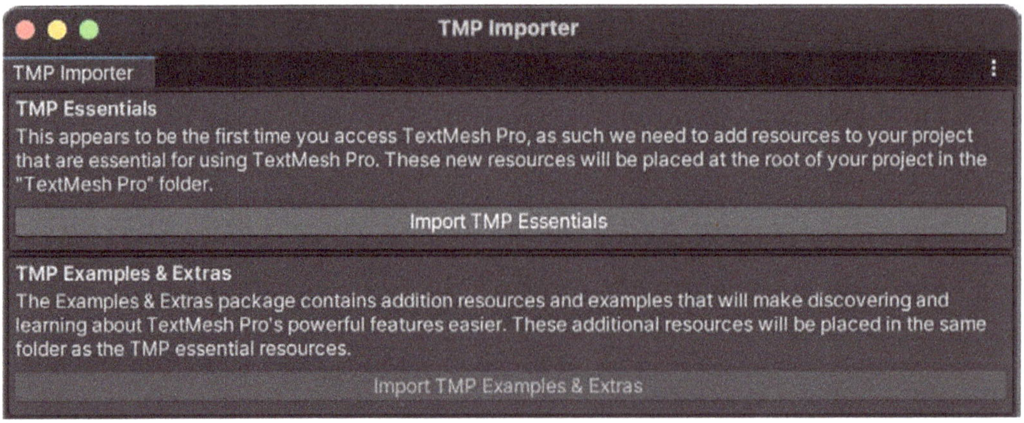

Figure 6.9 – TMP Importer dialog

Don't worry; for some Unity UI elements, such as Button and Text, we use a library called **TextMesh Pro** (**TMP**), which allows us to apply advanced customization to the text, such as borders, fonts, and gradients.

Simply click on the **Import TMP Essentials** button to continue. Once the progress bar indicating the import status has finished, you can close the pop-up window.

The button has been created as planned. Rename it `Sign Up button`.

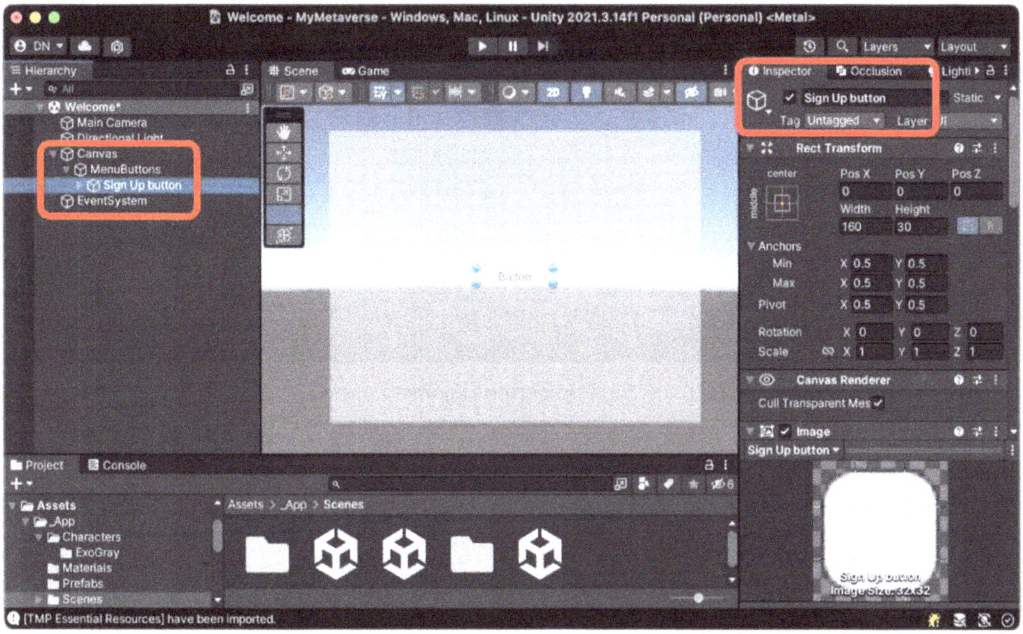

Figure 6.10 – Sign Up button element

Again, let's apply a little configuration to the button to adjust its appearance. To do this, perform the following steps:

1. In the **Rect Transform** component, select the alignment button, but this time we will select **top center**.

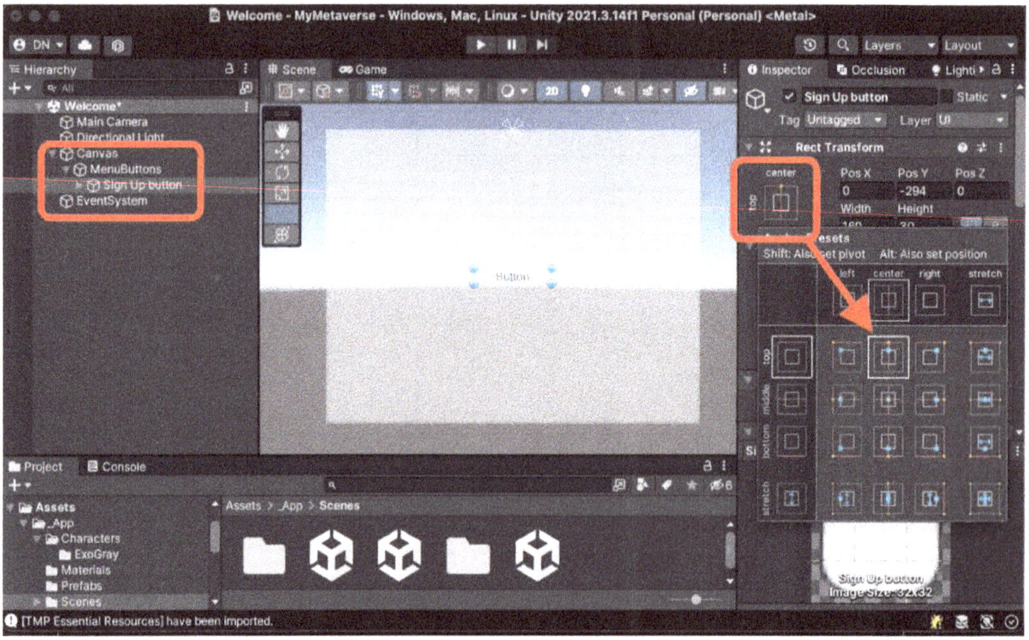

Figure 6.11 – Sign Up button alignment option

2. With the **Rect Transform** tool and the *Alt* key pressed, expand the bottom edge and then the side edge to make the button larger.

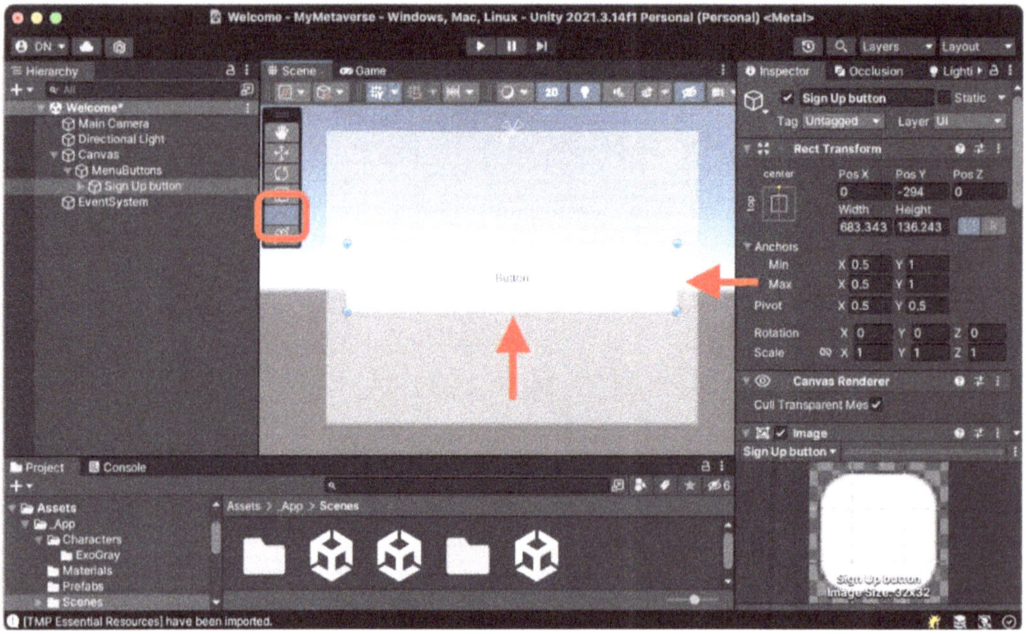

Figure 6.12 – Resized Sign Up button

3. With the **Move** tool, we move the button's positioning up a little. The intention is to leave enough space below to place the **Sign In** button in the next section.
4. Finally, we deploy the **Sign Up button** GameObject and we will see that within it, there is a **Text (TMP)** element, which is in charge of showing text inside the button. Select the **Text (TMP)** GameObject, and in the **TextMeshPro – Text (UI)** component, change the text to `Sign Up` and check the **Autosize** checkbox. It should look similar to the following screenshot.

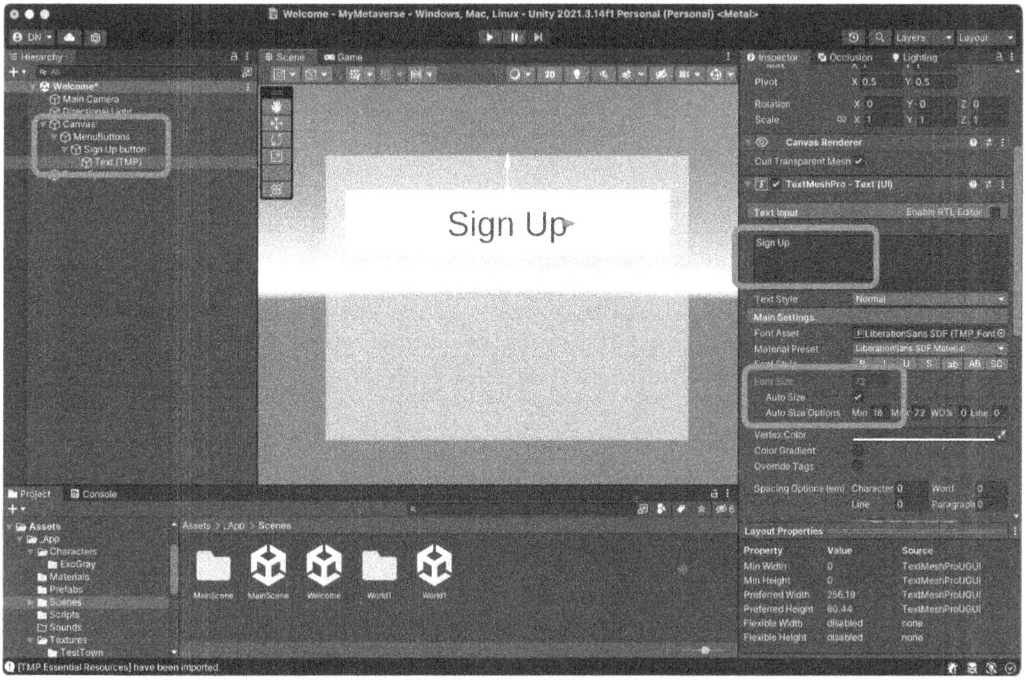

Figure 6.13 – Renamed button label

5. Repeat the previous steps to create another similar button but with the text **Sign In** and move it with the **Move** tool below the **Sign Up** button, similar to as in the following screenshot:

Figure 6.14 – Duplicating the button

> **Tip: Duplicating a button**
>
> Once you have created a button and applied a style, such as font size, button size, and colors, you can duplicate the GameObject and then change its name and label text, so you don't have to do the whole process from scratch.

6. Finally, you can deactivate this panel so that when we create the next one, it will not overlap and create confusion. Don't worry; at the end of the chapter, we will create a script that will automatically activate and deactivate the multiple panels that we will create.

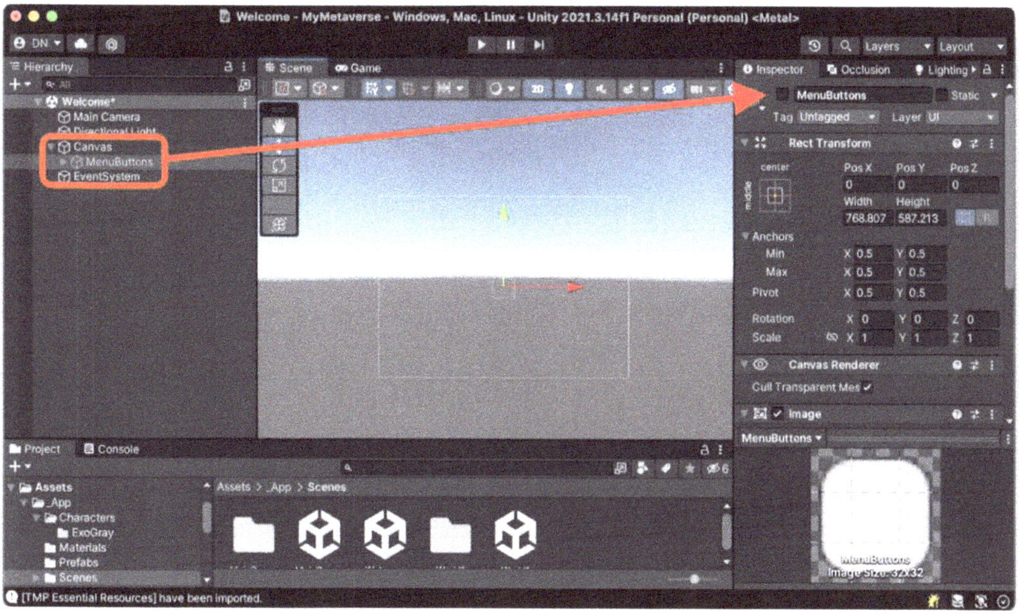

Figure 6.15 – Deactivating the MenuButtons panel

Fantastic, we have already created and configured the first button. Now, we will create another panel. This one will be in charge of grouping the registration form.

7. For it, in the main menu bar, select the **GameObject** | **UI** | **Panel** option and rename it `Sign Up Form`.

   We will apply the same configuration that we did to the **MenuButtons** panel.

8. In the **Inspector** panel, select the **center** alignment.

9. With the **Rect Transform** tool, we will resize the panel until it looks like the following screenshot:

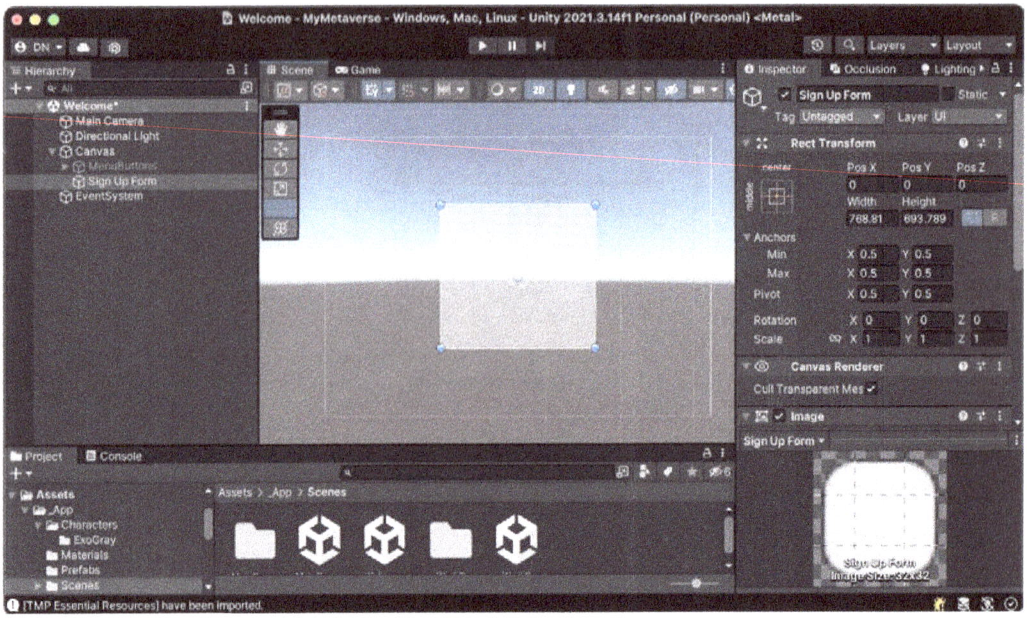

Figure 6.16 – Resizing the Sign Up Form panel

Very good, it looks great. Now that we have mastered the creation of Unity UI elements, let's take the opportunity to learn something new.

Understanding Vertical Layout Group

In Unity UI, in addition to the elements we've seen previously, there are components that make it easier for us to organize the design we want to implement and how it looks. For example, for the registration form that we are going to create next, we could use *something* to help us organize the textboxes and the **Submit** button. It would be a tedious job to manually arrange each element with the **Move** and **Scale** tools. Besides, it would be much more difficult to maintain in different screen sizes.

That's why Unity UI offers us a component called **Vertical Layout Group**. Vertical Layout Group will order, place, and scale the child elements that are hanging inside the parent that carries this component; fantastic, isn't it?

Let's see it in action:

1. To create a vertical layout group, select the GameObject of the newly created **Sign Up Form** panel, and in the **Inspector** panel, click on the **Add Component** button and type `Vertical Layout Group`. Select it from the list of results.

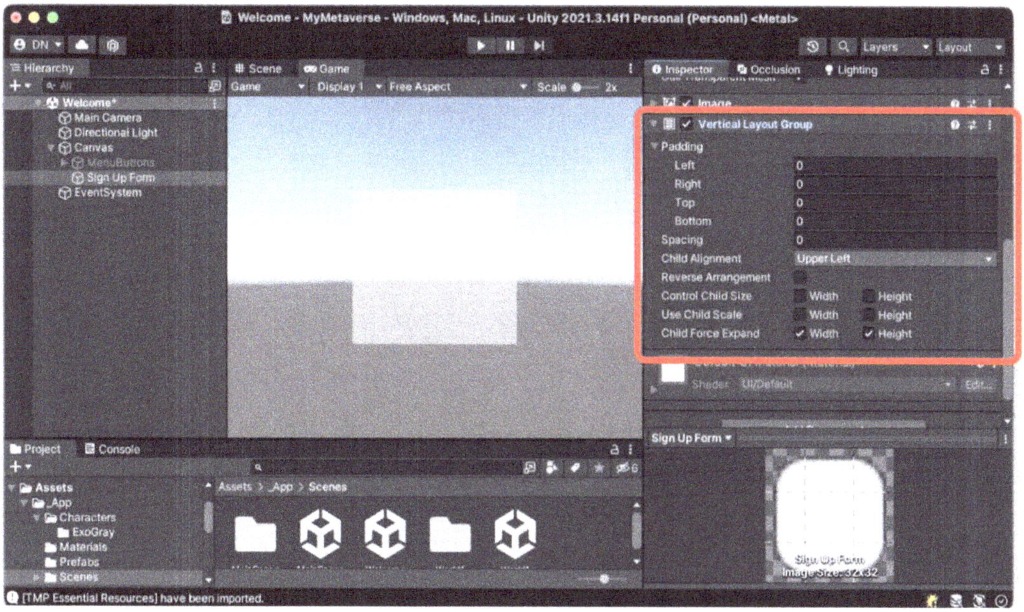

Figure 6.17 – Vertical Layout Group properties

As you can see, the component offers a configuration to organize the items automatically. To see how it works practically, let's add the necessary elements for our form.

2. Right-click on the **Sign Up Form** panel and select the **UI** | **Input Field** – **TextMeshPro** option. Rename it `NicknameInput`.
3. Repeat the same process to add input fields called `UsernameInput` and `PasswordInput`.

It should look something similar to the following screenshot:

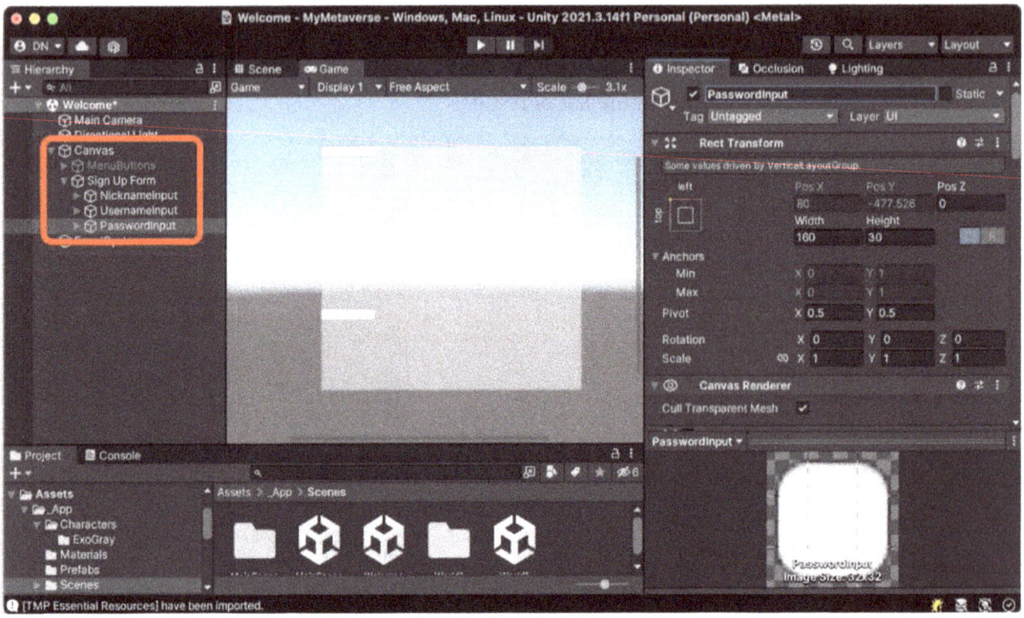

Figure 6.18 – Adding inputs to the Sign Up Form panel

As you can see, the inputs are very small and separated. Don't worry about this for the moment.

4. The last element that we need in our registration form is a button. Right-click on the **Sign Up Form** panel, select the **UI** | **Button - TextMeshPro** option, and rename it `Sign Up Submit`.
5. Well, now comes the magic of the **Vertical Layout Group** component. We select the **Sign Up Form** panel in the **Hierarchy** panel and change the configuration of the component to this one:

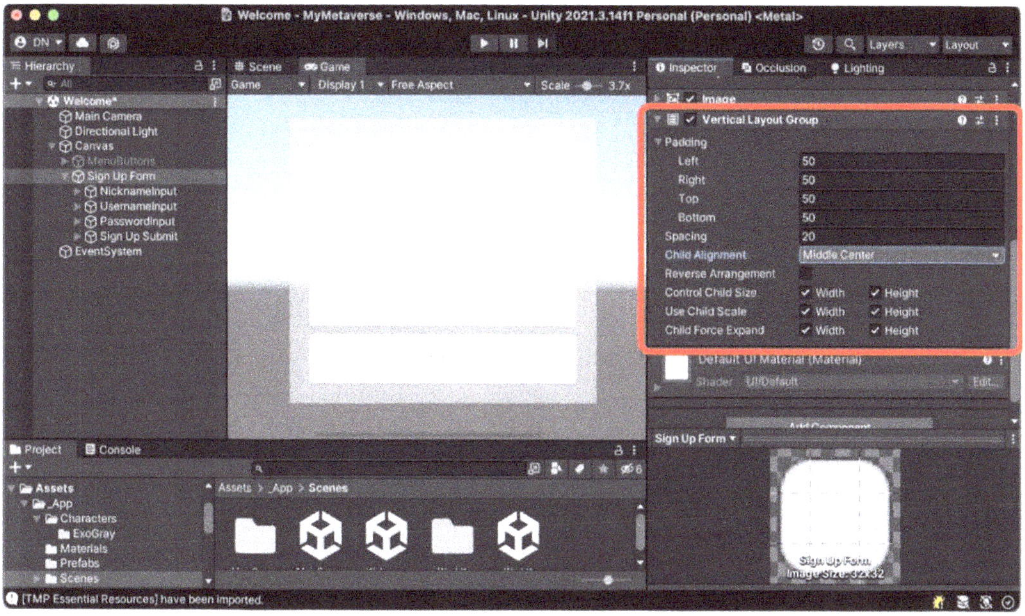

Figure 6.19 – Setting Up Vertical Layout Group

We can see the following in the preceding screenshot:

- In the **Padding** section, apply the value `50` to all sides. This makes it apply space to each side.
- In **Spacing**, we apply the value `20`. This adds a separation between each element.
- We activate all the checkboxes so that the component automatically controls the size of the child elements.

Impressive, isn't it? This component opens an impressive range of possibilities for a faster and more optimal design.

There is one pending issue to finish designing our form. As you may have noticed, we need to make some changes to the appearance of the form.

6. To enlarge the font of the placeholder (the text that indicates to the user what to write in the field) and the text that the user writes when selecting it in the **Hierarchy** panel, we have a property called **Point Size**, inside the **TextMeshPro – Input Field** component. Change it to `50`, for example. You will immediately see a change in the size of the text.
7. If you expand the GameObject of the input, you will see that there is a child called **Placeholder**. Here, type some text so that the user knows what needs to be written, for example, `Nickname`, `Username`, and `Password`, respectively.

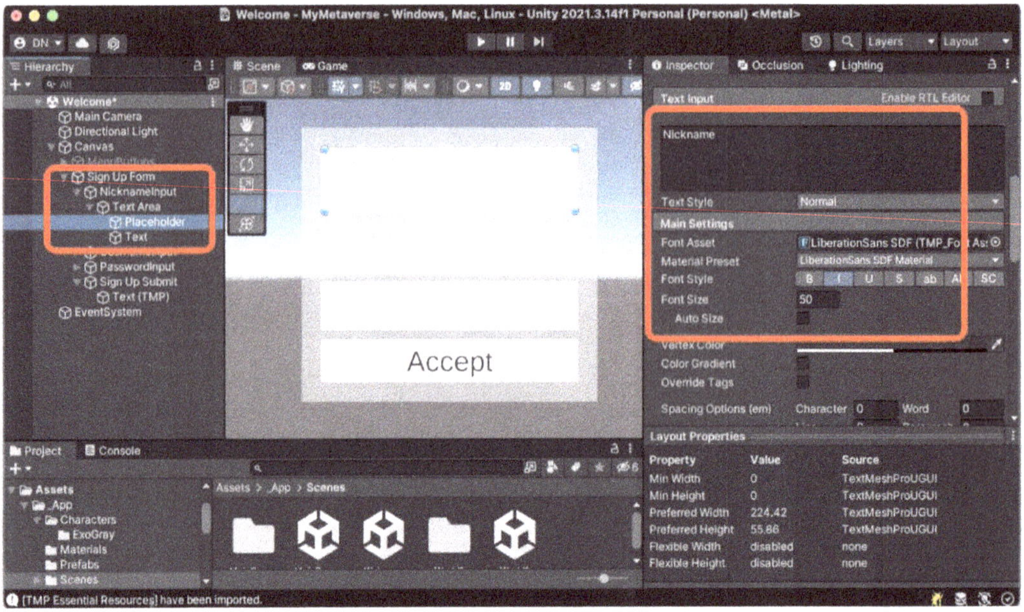

Figure 6.20 – Renaming the input placeholder

8. For the **Sign Up Submit** button, change the **Text Input** property in the GameObject named **Text (TMP)** to **Accept**, and finally, check the **Autosize** checkbox.

Good job, you just finished the design of the registration form. I invite you to explore all the properties of the UI elements. There are many features that can be customized to give a more artistic look to the interface, such as the colors of the text and the background colors of the buttons. Remember to save the scene often so as not to lose changes.

It is time to program again. This time, we will create a script that manages the user's registration through the data entered in the form.

Follow these steps to create the script:

1. We will create a small class that will store the necessary user data to save it later in Firestore. Following the **Assets** | **_App** | **Scripts** path, right-click and select the **Create** | **C# Script** option and call it `UserData`, then double-click on it to edit it.
2. Replace all the auto-generated code with the following:

```
using Firebase.Firestore;

[FirestoreData]
public class UserData
{
```

```
        [FirestoreProperty]
        public string Uid { get; set; }

        [FirestoreProperty]
        public string Nickname { get; set; }

        [FirestoreProperty]
        public string Username { get; set; }

        public UserData()
        {

        }
    }
```

3. Now, we will create a new script that will be in charge of managing the registration form. Again, following the **Assets | _App | Scripts** path, right-click and select the **Create | C# Script** option. Rename it `SignUpManager`.
4. Replace all the auto-generated code with the following:

```
    ...
    public class SignUpManager : MonoBehaviour
    {
        public TMP_InputField nicknameInput;
        public TMP_InputField usernameInput;
        public TMP_InputField passwordInput;
        FirebaseAuth auth;
        FirebaseFirestore db;


        void Start()
        {
            InitializeFirebase();
        }


        void InitializeFirebase()
        {
            if (auth == null)
                auth = FirebaseAuth.DefaultInstance;
            if (db == null)
                db = FirebaseFirestore.DefaultInstance;
        }
```

```
public void CreateUser()
{
    if (string.IsNullOrEmpty(nicknameInput.text)
       || string.IsNullOrEmpty(usernameInput.text)
       ||string.IsNullOrEmpty(passwordInput.text)
       )
       return;


    auth.CreateUserWithEmailAndPasswordAsync(
        usernameInput.text, passwordInput.text)
            .ContinueWithOnMainThread(
                async task =>
    {
        if (task.IsCanceled || task.IsFaulted)
            return;
        FirebaseUser newUser = task.Result;
        await UpdateUserProfile();
        await CreateUserDataDocument();
    });
}


private async Task UpdateUserProfile()
{
    var currentUser =
        FirebaseAuth.DefaultInstance.CurrentUser;
    UserProfile userProfile = new UserProfile()
        {DisplayName = nicknameInput.text  };
    Await currentUser
        .UpdateUserProfileAsync(userProfile);
}


private async Task CreateUserDataDocument()
{
    var currentUser =
        FirebaseAuth.DefaultInstance.CurrentUser;
    DocumentReference docRef =
        db.Collection("Users")
            .Document(currentUser.UserId);
    UserData userData = new UserData()
```

```
                    { Uid = currentUser.UserId,
                        Nickname = nicknameInput.text,
                            Username = usernameInput.text};


                await docRef.SetAsync(userData,
                    SetOptions.MergeAll)
                        .ContinueWithOnMainThread(task =>
                {
                    if (task.IsCanceled || task.IsFaulted)
                        return;
                    SceneManager.LoadScene("MainScene");
                });
            }
        }
```

Great, you have created the two scripts needed so far to register a user. Let's stop for a moment to review what the preceding code flow does:

1. There are some variables to which we will bind the form inputs.
2. When we start the scene, the `Start` method will check that we have the instances of the variables of the Firebase SDK.
3. We have a function called `CreateUser`, which we will invoke when we click the **Accept** button, which calls the `CreateUserWithEmailAndPasswordAsync` method of the Firebase Authentication SDK with the username and password entered into the form.
4. If the creation went well, we will update the nickname on the user's profile. This action is encapsulated in the `UpdateUserProfile` function.
5. We create a reference to the user in the database with the `CreateUserDataDocument` function. In the `Users` collection, a new document will be created with basic user information, such as `ID`, `nickname`, and `username`.
6. Finally, when the whole process has finished successfully, we will redirect the user to the main scene, `MainScene`.

Now, we will link the created script to the form, but before that, we must set up a small configuration in the project, and I will explain why.

In Unity, when we want to change a scene through code, that is to say, in our script, we indicate this with the `SceneManager.LoadScene("MainScene")` statement.

If we execute our script right now, it will fail. Why is that? The explanation is very simple. Unity has a window called **Build Settings**, which we saw in *Chapter 1*, *Getting Started with Unity and Firebase*. In this window, a list of the scenes *available* to the user is kept; that is to say, we can have 100 scenes in the project, but only those that are active in the **Build Settings** window will be *navigable*.

Let's see it. In the main menu bar, click on **File** | **Build Settings**.

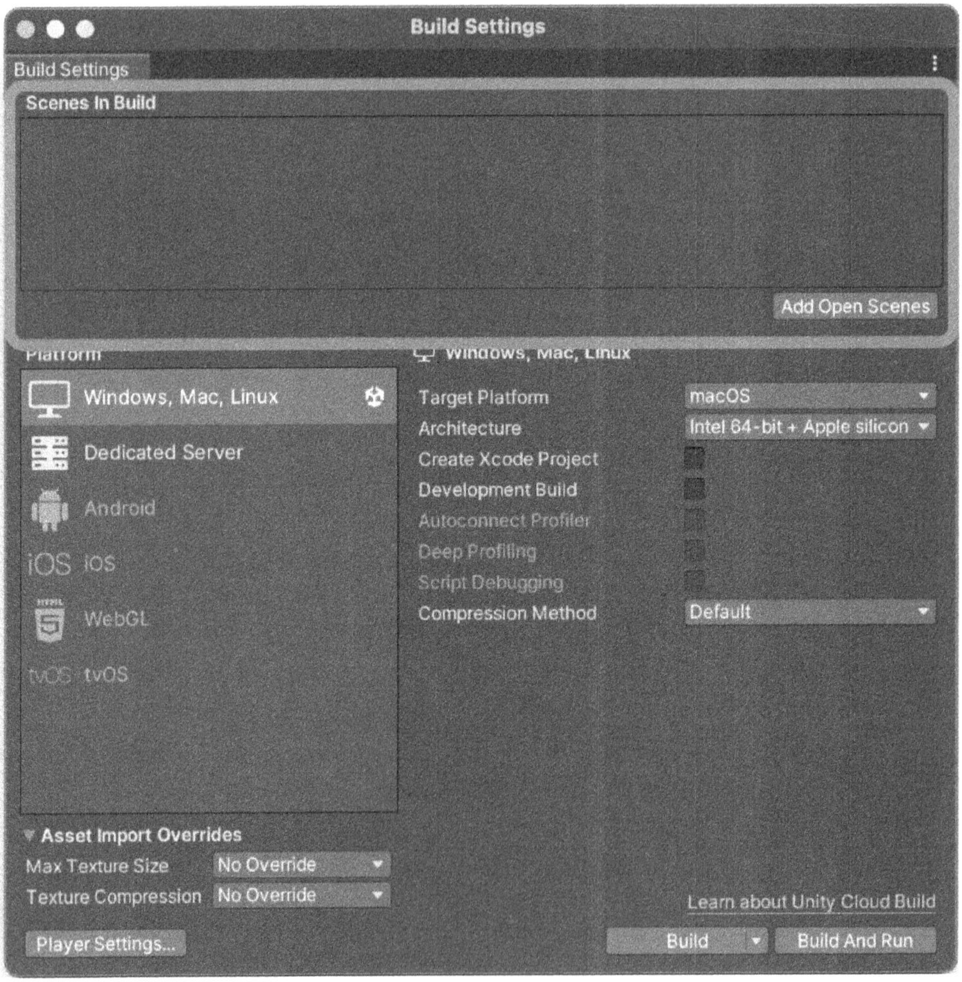

Figure 6.21 – Scene list in the Build Settings window

The list of available scenes is empty, so our script will not find a scene called **MainScene** in this list.

To correct it, it is very simple. We go to the **Assets** | **_App** | **Scenes** path and drag all the scene files onto this list. It will look like this:

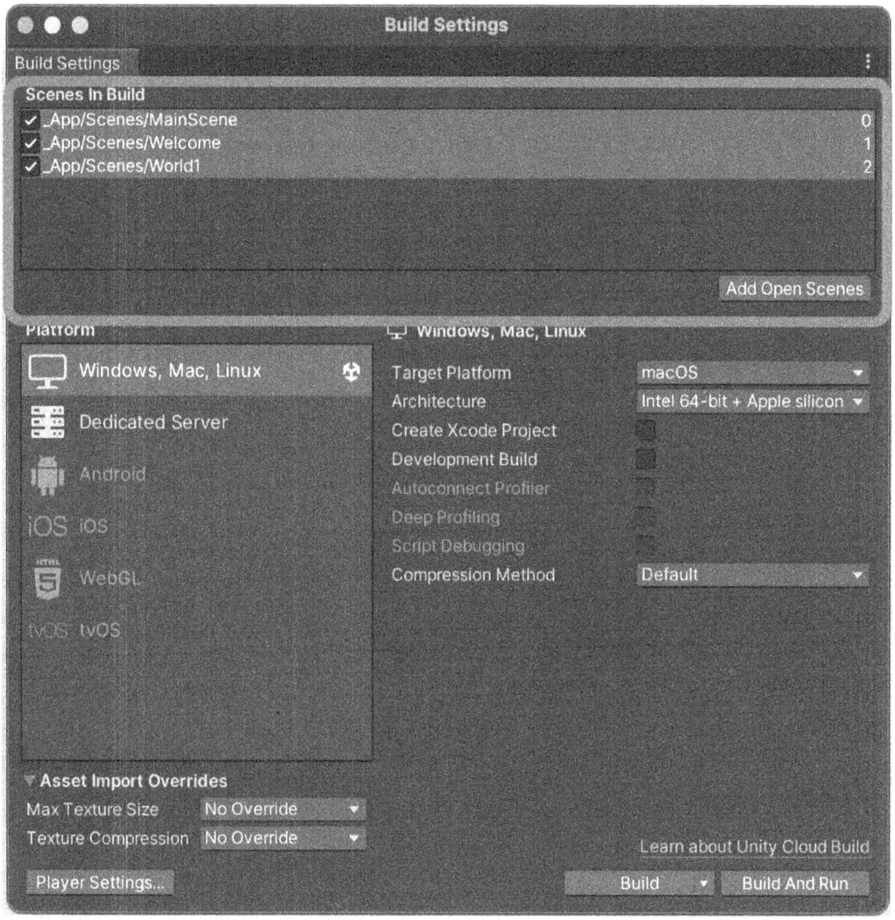

Figure 6.22 – Imported scenes in the Build Settings window

Easy, isn't it? Remember to add any future scenes you create here. If you have noticed, to the right of the list, there are some numerical values, **0**, **1**, and **2**. These values are indexes. They help us if we want to change the scene by index; for example, instead of calling the instruction `SceneManager.LoadScene("MainScene");`, you could call it with `SceneManager.LoadScene(0);`.

I don't particularly like to call scenes by their index. The reason is that in the future, the list can be altered, and the index can change. This can cause chaos in a big project.

Tip: Something important about the order of the scenes

When we compile our project to put it into production, Unity will take the first scene in the list as the starting point. As we want all our users to go through the **Welcome** scene first to allow them to identify themselves, we must change the order of the list. Simply select the **Welcome** scene, drag it, and place it in the first place.

In the following screenshot, we can see the list of imported and active scenes in the **Build Settings** window.

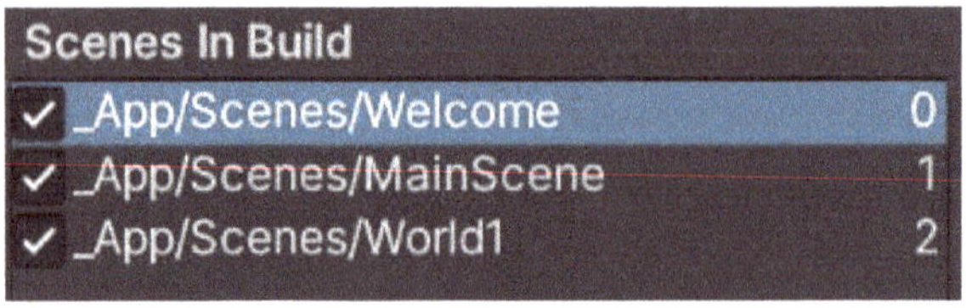

Figure 6.23 – Index number in the scenes list

Well, once we have configured the navigable scenes, we can continue with our script. In the **Hierarchy** panel, select the panel called **Sign Up Form** and follow these steps:

1. In the **Assets** | **_App** | **Scripts** path, drag the `SignUpManager` script into the **Sign Up Form** **Inspector** panel.

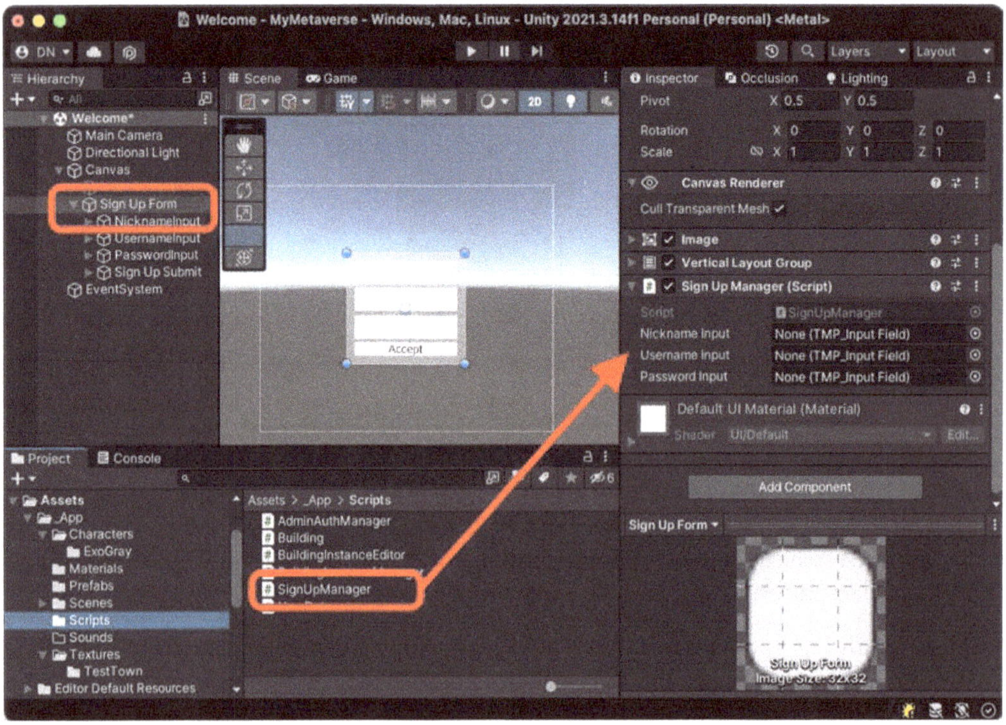

Figure 6.24 – Adding the SignUpManager script to Sign Up Form

2. Drag and drop **NicknameInput** into the **Nickname Input** slot, **UsernameInput** into the **Username Input** slot, and **PasswordInput** into the **Password Input** slot of the script.

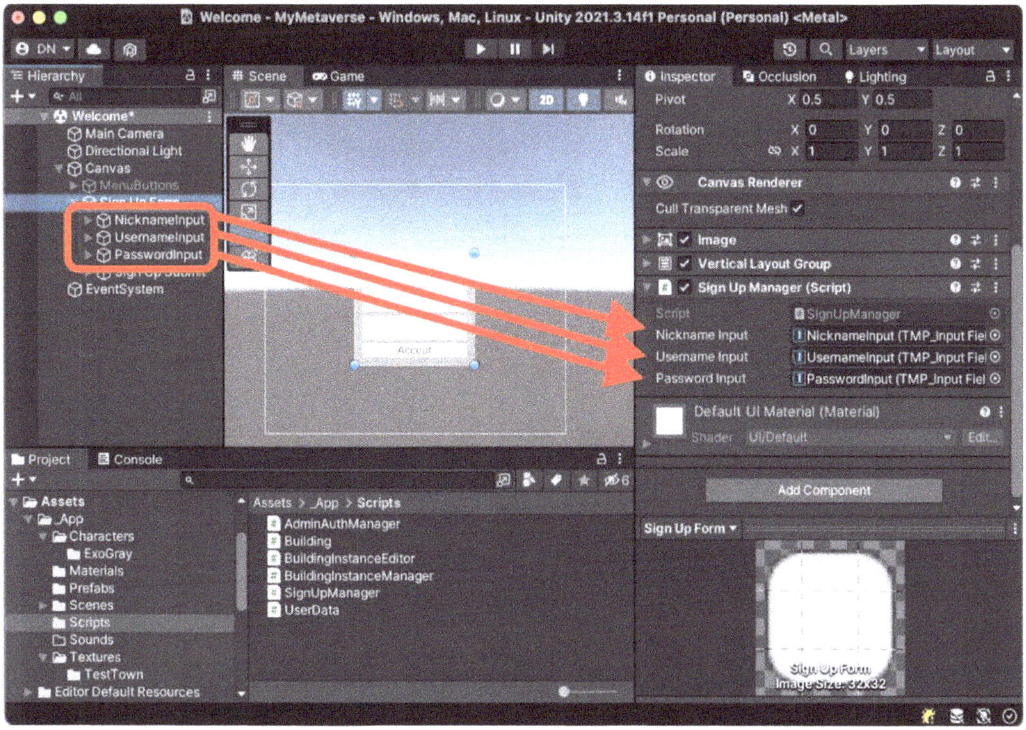

Figure 6.25 – Assigning inputs to Sign Up Manager slots

3. Now, select the **Sign Up Submit button** GameObject. In the **Inspector** panel, you will find a component called **Button**. You will find a section called **On Click**. Click on the + button on the right to make the empty slot appear. Drag the **Sign Up Form** GameObject into the empty slot.

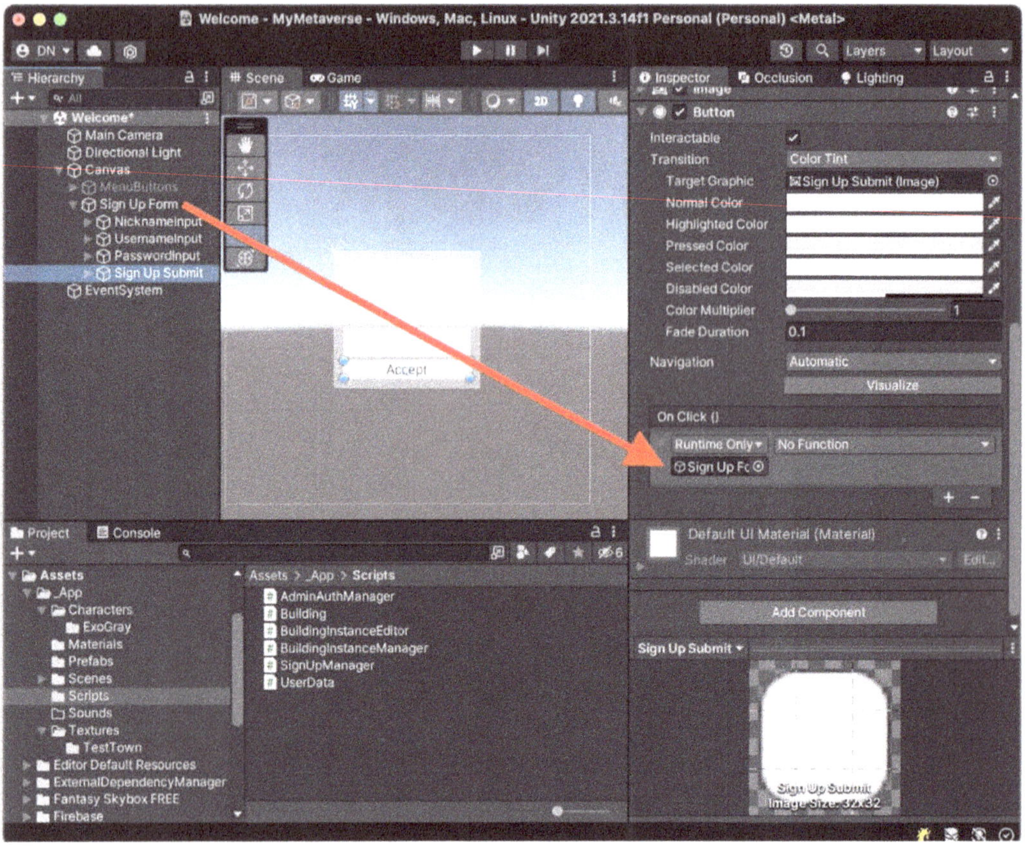

Figure 6.26 – Assigning the Sign Up Form GameObject to the Sign Up Submit button slot

4. Finally, in the same **On Click** section, there is a dropdown with the text **No Function**. Click on it and select **SignUpManager.CreateUser**.

Figure 6.27 – Final result of the On Click property

Congratulations, you have just finished your registration form. Let's try it! Click on the **Play** button and fill in the form. Remember that the username must be in email address form. It doesn't matter if the one you enter doesn't exist. The information in the following screenshot corresponds to the user I have created; you should enter your own user details and password.

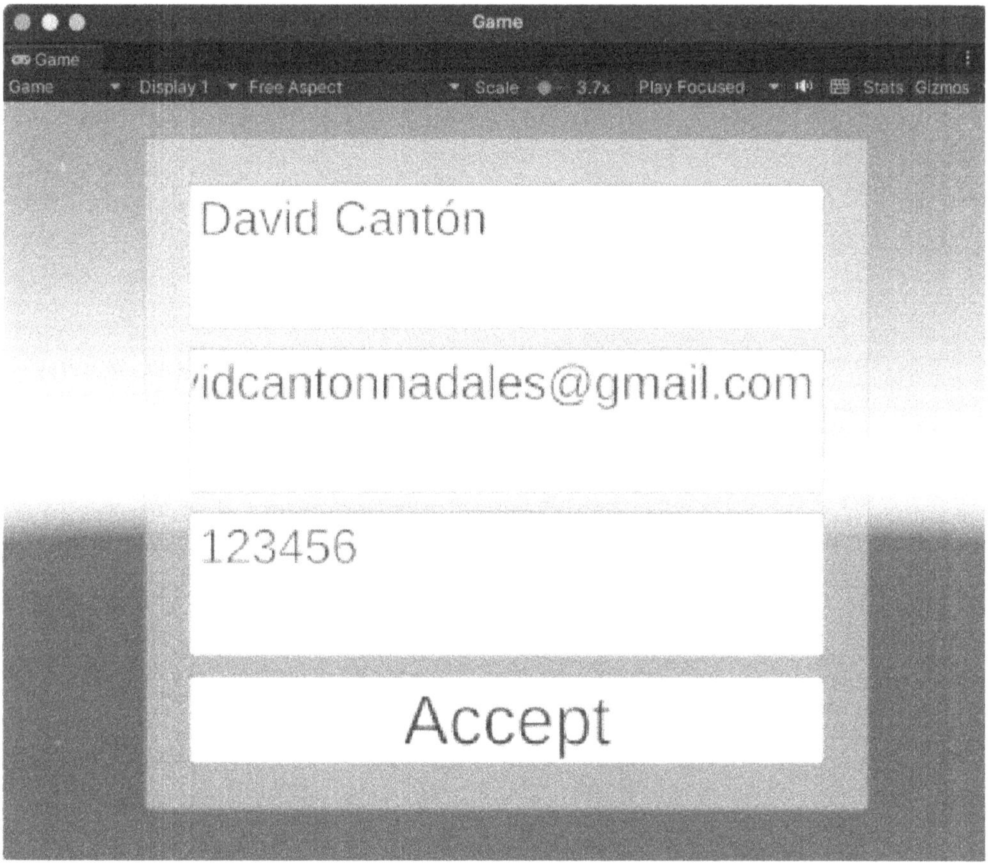

Figure 6.28 – Filled-in sign-up form

Finally, we click the **Accept** button on our form, and voilà!

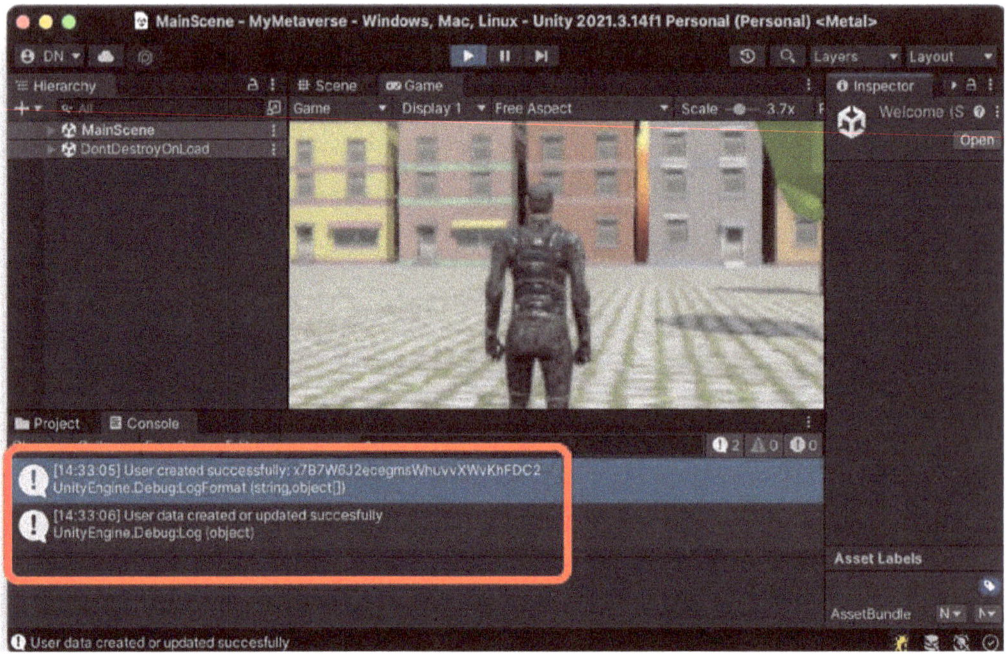

Figure 6.29 – MainScene after successful signup

A new user has been created. We see the evidence in the logs in the Unity console. As we wanted, we have appeared in **MainScene**. Excellent work!

If you look in Firebase Console now, you'll see a new collection called **Users** in Firestore with the registered user data.

We've successfully solved the task of allowing our users to register. Next, we'll see how to allow them to log in if they already have an account.

Creating a sign-in screen

We've learned a lot about creating UIs throughout this chapter. Now that you've successfully created a fully functional registration form, you'll find it much easier to create a login form:

1. Before we continue, as we did before, we'll select the **Sign Up Form** panel and disable it so that it won't get in the way when we create the new panel.

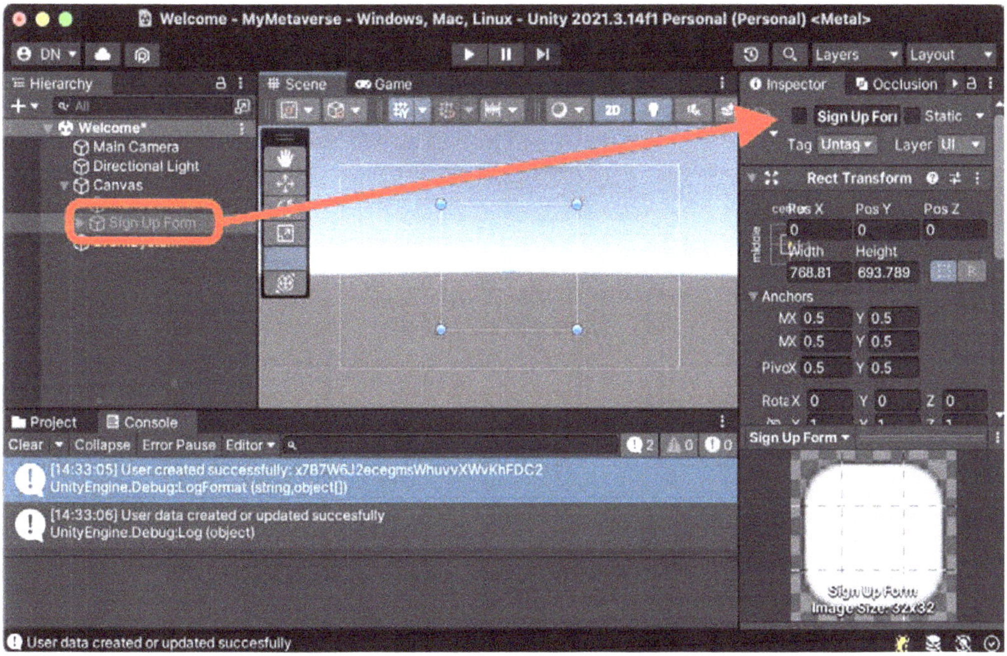

Figure 6.30 – Deactivating Sign Up Form

2. OK, we are ready. Let's start with the creation of our **Sign In** form. To do so, right-click on **Canvas** and select the **UI** | **Panel** option. Rename the new GameObject `Sign in Form`.

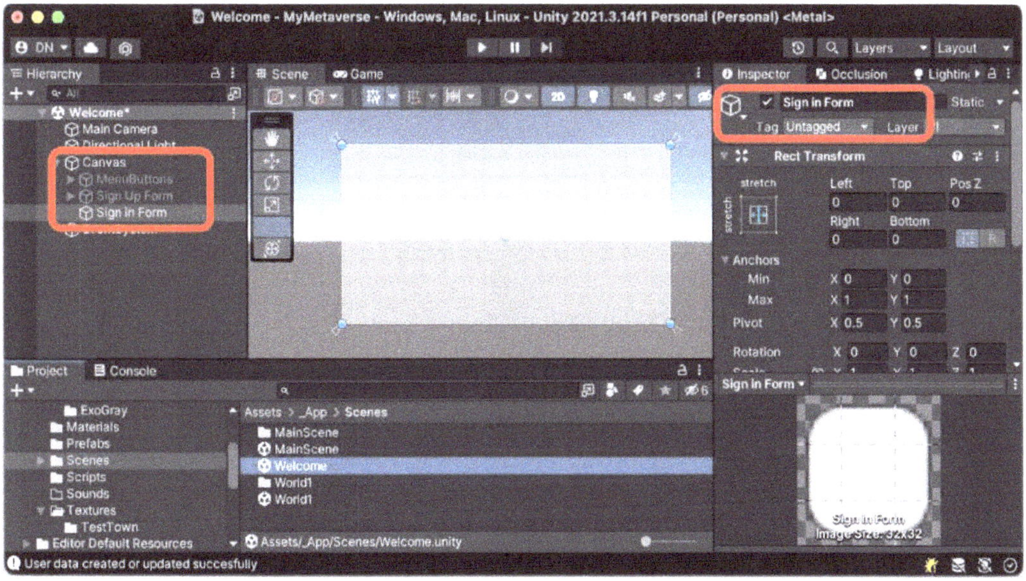

Figure 6.31 – Renaming panel

3. We will apply the same settings as we did with the **Sign Up Form** panel to modify its appearance:
   - In the **Inspector** panel, select the **center** alignment.
   - With the **Rect Transform** tool, we will resize the panel until it looks as in the following screenshot:

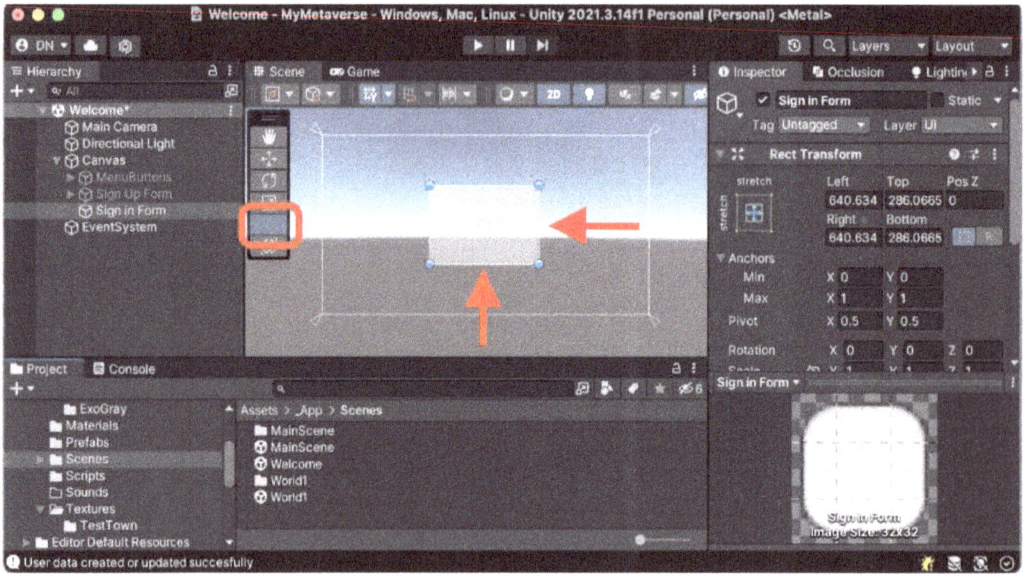

Figure 6.32 – Resizing Sign In Form

4. Perfect! Now, we will use **Vertical Layout Group** and add two inputs and a button, just like we did with the **Sign Up Form** panel. The result should look as in the following screenshot:

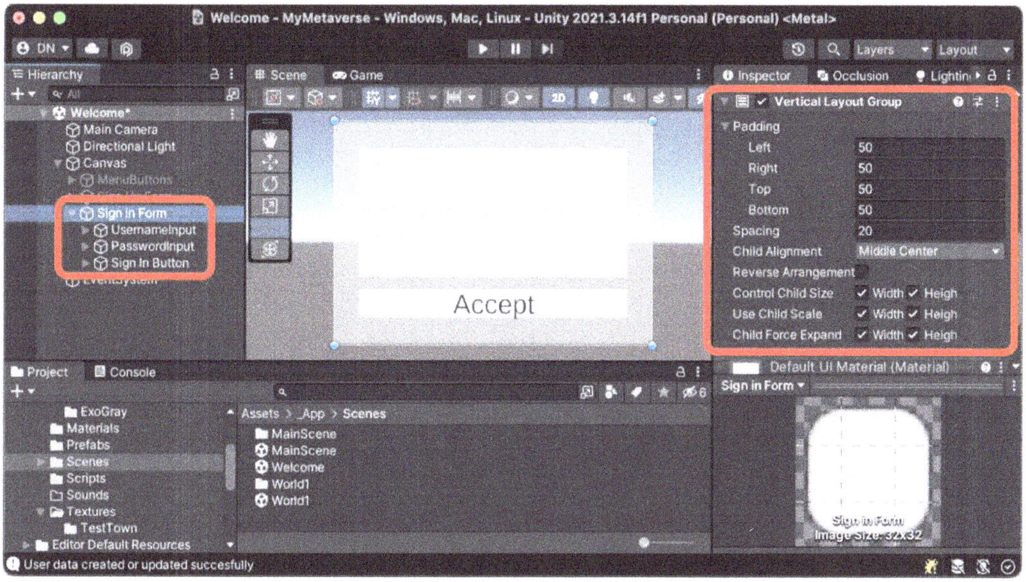

Figure 6.33 – Setting up Vertical Layout Group properties

5. Well, now that we have the login form designed, we will proceed to create the script that will manage it. To do this, follow the **Assets** | **_App** | **Scripts** path, right-click, and select the **Create** | **C# Script** option. Rename it `SignInManager`. Finally, double-click on it to edit it.

   Replace all the auto-generated code with the following and save the changes:

```
...
public class SignInManager : MonoBehaviour
{
    public TMP_InputField usernameInput;
    public TMP_InputField passwordInput;
    FirebaseAuth auth;


    void Start()
    {
        InitializeFirebase();
    }


    void InitializeFirebase()
    {
        if (auth == null)
            auth = FirebaseAuth.DefaultInstance;
```

```
        }


        public void LoginUser()
        {
            if (string.IsNullOrEmpty(usernameInput.text)
                || string.IsNullOrEmpty(passwordInput.text)
                )
                return;


            auth.SignInWithEmailAndPasswordAsync(
                usernameInput.text,
                    passwordInput.text)
                        .ContinueWithOnMainThread(
                            async task =>
            {
                if (task.IsCanceled || task.IsFaulted)
                    return;
                FirebaseUser existingUser = task.Result;
                SceneManager.LoadScene("MainScene");
            });
        }
    }
```

Tip: The importance of initialization

As you can see, in the **Start** function of our script, we have created a call to the `InitializeFirebase` method. This ensures that the Firebase SDK is initialized before performing any operation on its SDK. It is vital that we initialize any reference to the Firebase SDK before anything else.

6. Now, from the **Assets** | **_App** | **Scripts** path, drag the newly created `SignInManager` script into the **Inspector** panel of the **Sign In Form** GameObject.

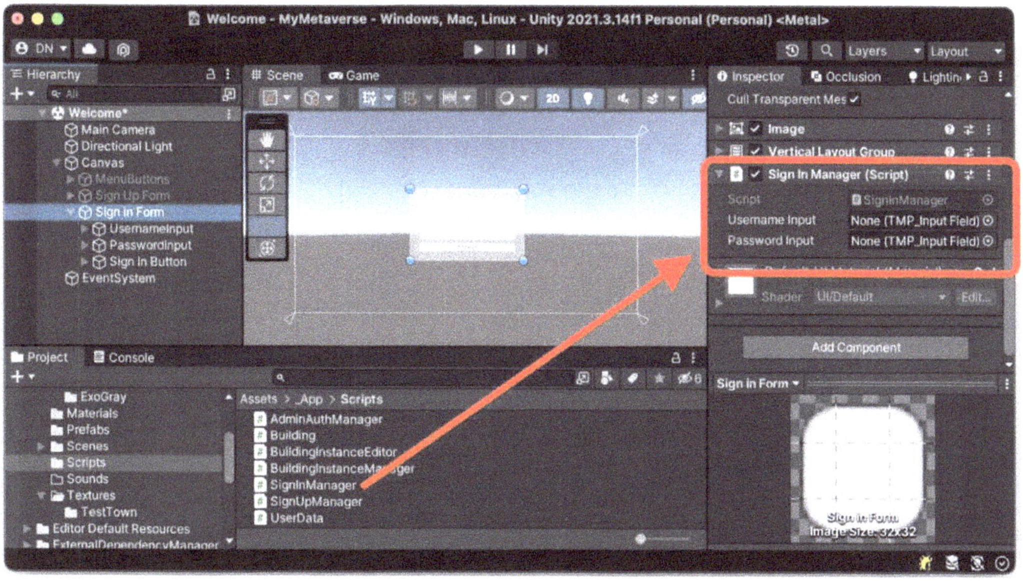

Figure 6.34 – Dragging SignUpManager to the Sign In Form panel

7. Now, drag the `UsernameInput` and `PasswordInput` inputs to the slots of the script. It will look like the following screenshot:

Figure 6.35 – Final result of the Sign In Manager component

Finally, for the script to work, we need to link the **Sign In** button to the `LoginUser` function of the script we created earlier.

8. Select the **Sign in button** GameObject, then in the **Inspector** panel, drag the **Sign in Form** GameObject to the slot of the **On Click** property.

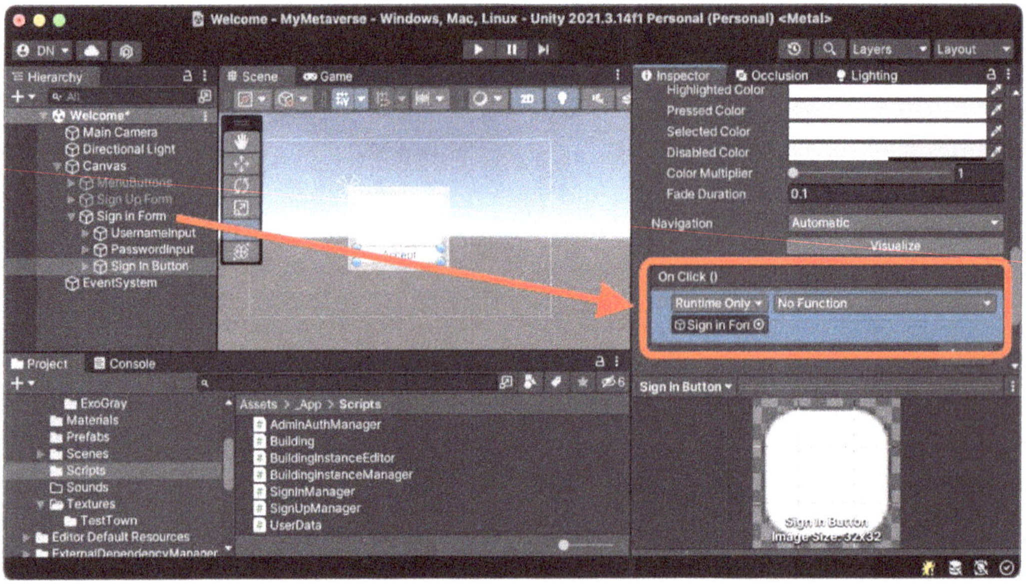

Figure 6.36 – Dragging Sign In Form to the Sign In button On Click slot

9. Select the dropdown with the text **No Function** and select **SignInManager.LoginUser**.

Figure 6.37 – Final result of the On Click property

10. Great, it should work now, so let's try it. Click on the **Play** button and enter the credentials of the user you registered earlier. In my case, it looks as follows:

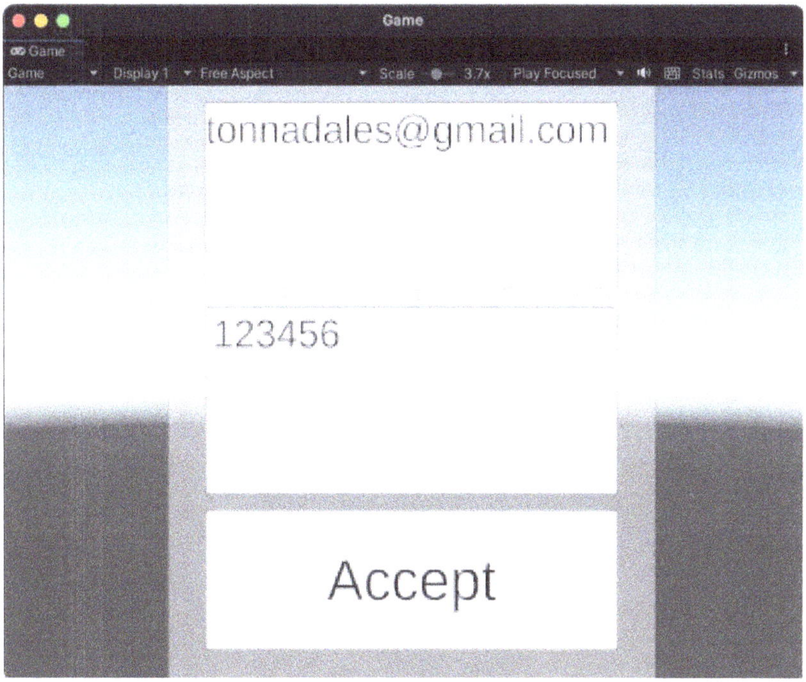

Figure 6.38 – Filled-in sign-in form

We click the **Accept** button and... magic!

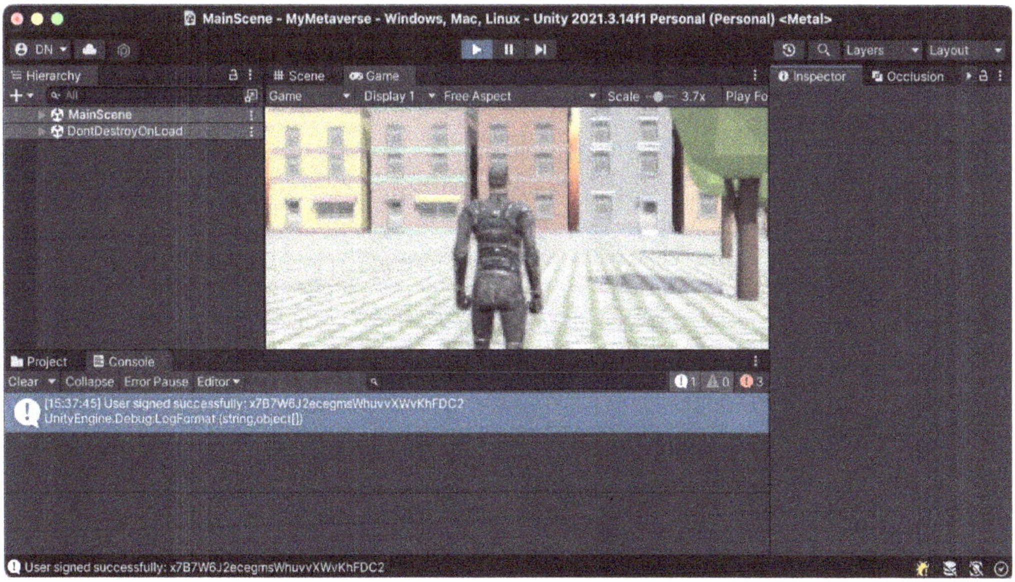

Figure 6.39 – MainScene after a successful login

Our user has been correctly identified; we can check the log in the Console, which confirms it. Now, our user can access the **MainScene** scene after Firebase has checked that the credentials are correct.

To close the flow of the whole interface we have designed, we must create a script that allows the user to choose between the **Sign In** and **Sign Up** buttons, as well as go backward.

To do this, follow these steps:

1. Deactivate the **Sign In Form** panel, so that all three are disabled.
2. Right-click on **Canvas** and select **UI** | **Button – TextMeshPro**. Rename it `Back Button`.
3. Change the button text to `Back`. At this point, your scene should look similar to in the following screenshot:

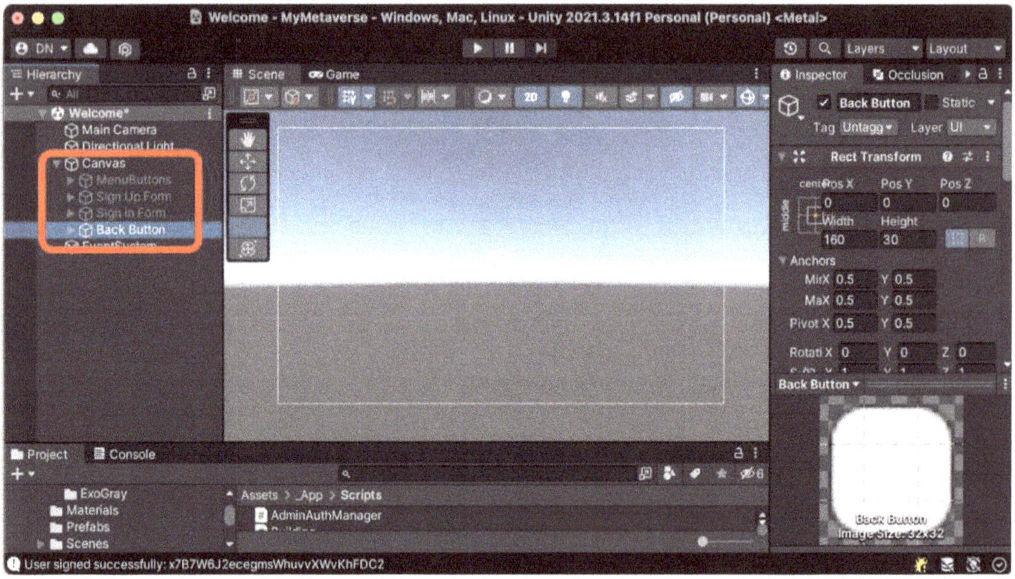

Figure 6.40 – Back button in scene

4. Select the alignment button and choose **bottom center**.

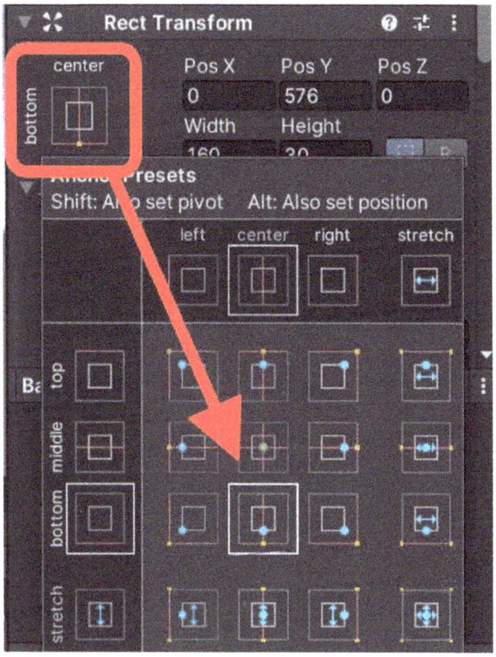

Figure 6.41 – Alignment option

5. Using the **Move** tool, move the button down until it looks like this:

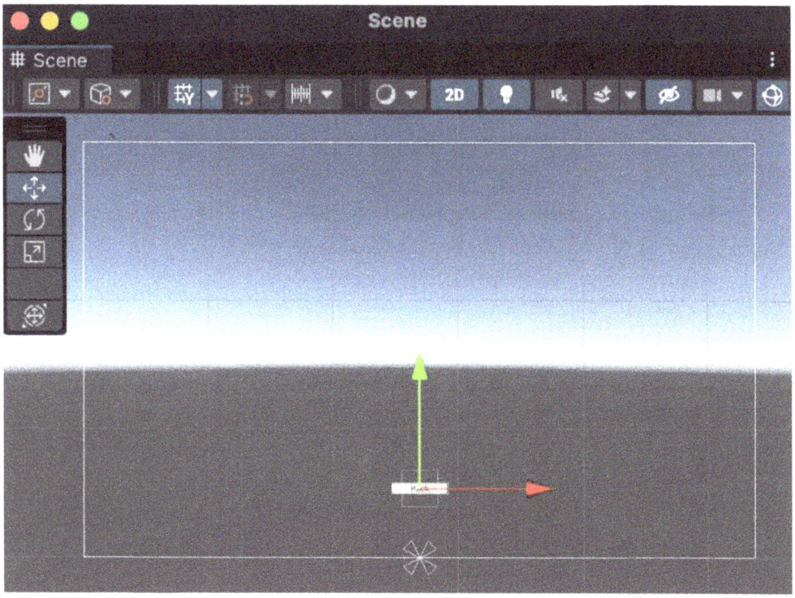

Figure 6.42 – Moving Back button

6. Now, following the **Assets** | **_App** | **Scripts** path, double-click and select the **Create** | **C# Script** option. Rename it `IdentificationUIManager` and double-click on it to edit it.
7. Replace all the auto-generated code with the following:

```
using UnityEngine;


public class IdentificationUIManager : MonoBehaviour
{
    public GameObject MenuButtonspanel;
    public GameObject SignUppanel;
    public GameObject SignInpanel;


    void Start()
    {
        ShowInitialScreen();
    }


    public void ShowInitialScreen()
    {
        MenuButtonspanel.SetActive(true);
        SignUppanel.SetActive(false);
        SignInpanel.SetActive(false);
    }


    public void ShowSignUppanel()
    {
        MenuButtonspanel.SetActive(false);
        SignUppanel.SetActive(true);
        SignInpanel.SetActive(false);
    }


    public void ShowSignInpanel()
    {
        MenuButtonspanel.SetActive(false);
        SignUppanel.SetActive(false);
        SignInpanel.SetActive(true);
    }
}
```

8. Drag the newly created script and drop it into the **Inspector** panel of the **Canvas** GameObject.

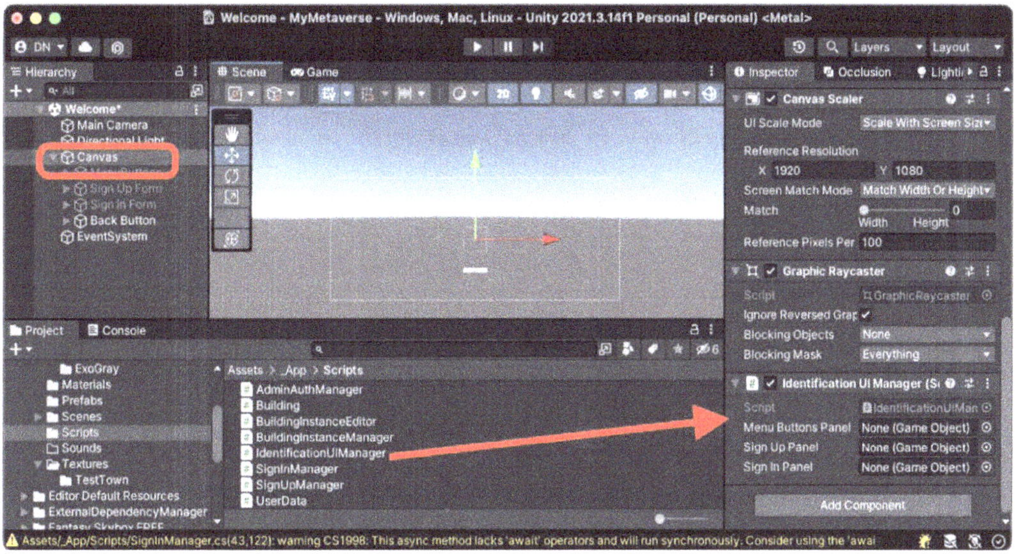

Figure 6.43 – Dragging the IdentificationUIManager script to the Canvas element

9. Drag the **Menu Buttons**, **Sign Up Form**, and **Sign In Form** panels to their respective slots in the **Identification UI Manager** component. It will look similar to in the following screenshot:

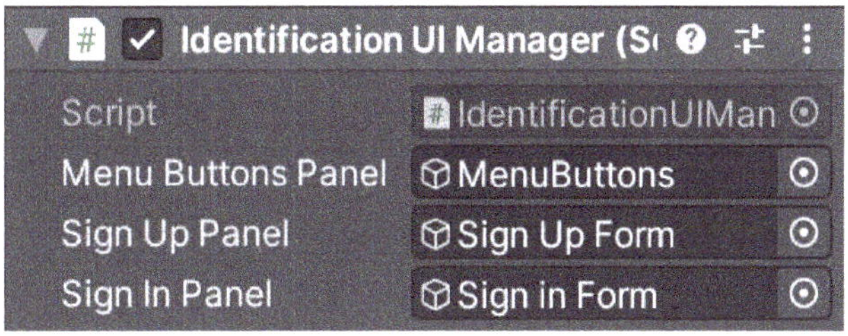

Figure 6.44 – Final result of the Identification UI Manager component

10. Now, select **Back Button**, and in its **On Click** property, drag the **Canvas** GameObject to the empty slot. In the dropdown with the text **No Function**, select **IdentificationUIManager.ShowInitialScreen**.

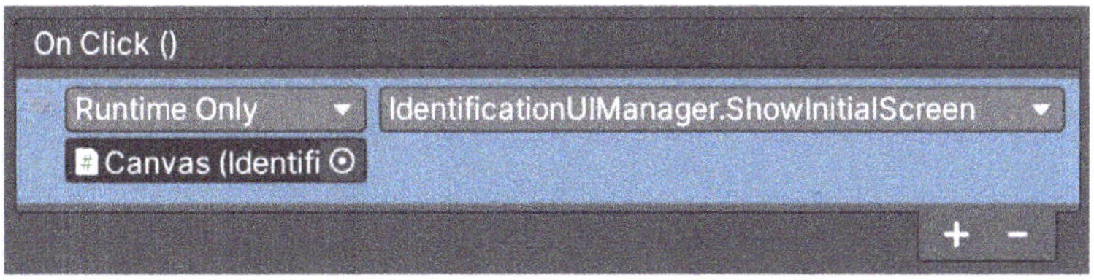

Figure 6.45 – Final result of Back Button On Click property

11. Finally, we deploy the **MenuButtons** GameObject, and for each button, we drag and drop the **Canvas** GameObject into the empty slot of the **On Click** property. For the **Sign Up** button, we will bind it with the `ShowSignUppanel` function, and for the **Sign In** button, we will bind it with the `ShowSignInpanel` function.

If everything went well, click the **Play** button and you should see the panel with the initial buttons. If you navigate through them, you will see that you can go back using the **Back** button. The navigation flow is now closed, and we allow the user to navigate freely through the options.

Good job! You have learned how to create a registration and login system from scratch using Unity UI elements and the Firebase SDK.

Keep in mind that the design we have applied to the forms is somewhat basic. We have focused on the functionalities for didactic purposes. You are totally free to customize the look and feel of the design to your liking. The appearance will not affect the proper functioning of the rest of the projects in the chapter.

Next, we will move on to the last point of this chapter, the dynamic instantiation of the player in the scene.

Placing the player

If you remember when we designed the initial scene in *Chapter 4*, *Preparing Our Home Sweet Home: Part 2*, we placed the Prefab with the player in the scene, and when we pressed **Play**, we took the perspective of the player so we could move around.

Our project will become multiplayer from *Chapter 9*, *Turning Our World into a Multiplayer Room*, onward, so we will progressively make adjustments to the current configuration to reach that goal. The way in which we currently instantiate the player would not be useful in the future in a multiplayer project, as there is only one instance of the Prefab player in the scene; it is fixed, and everyone who connects will use the same player. Crazy, right?

That's why in this final section, we will create a dynamic instancing system. Every time a player logs in and appears in the main scene, they will do so with a new Prefab player clone:

1. First of all, save the current changes. Then, we need to go back to **MainScene**. Double-click on the scene file called **MainScene** located in **Assets** | **_App** | **Scenes**.
2. Locate where we currently have the player Prefab.

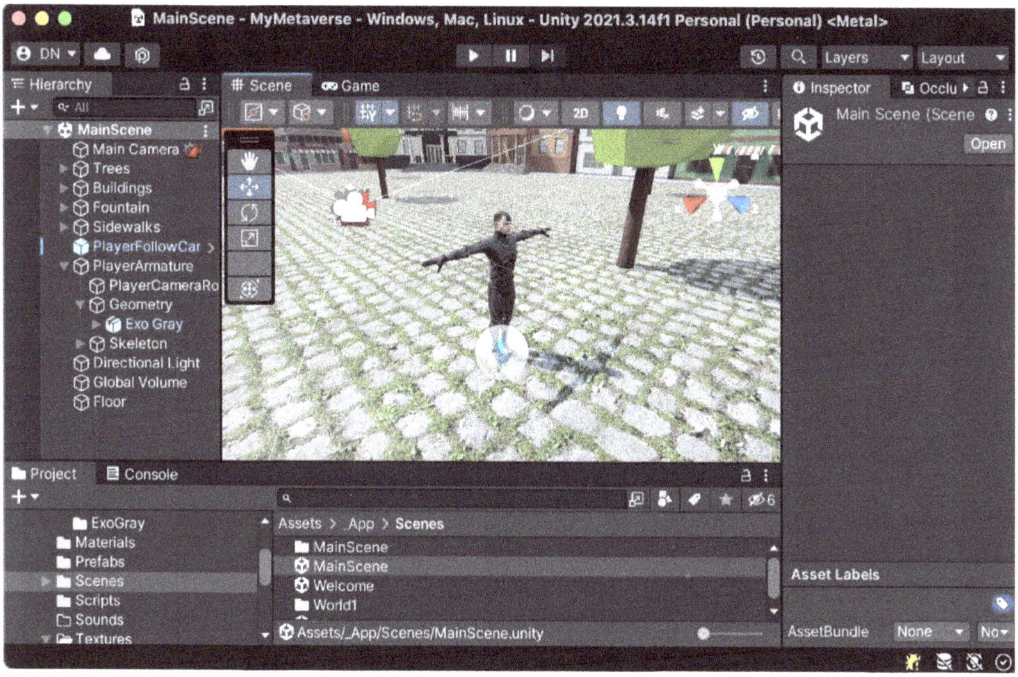

Figure 6.46 – Player in MainScene

If you remember, in the scene, there are three GameObjects that are necessary for the correct functioning of the player. One is the player themself (**PlayerArmature**). Another is a system for the camera to follow the player (**PlayerFollowCamera**), and finally the camera itself (**MainCamera**).

To make it easier to instantiate the player in any scene, we are going to unify these three elements in a single Prefab. To do so, follow these steps:

1. In the **Assets** | **_App** | **Prefabs** path, right-click and select the **Create** | **Prefab** option. Rename it `PlayerInstance`.
2. Select the three GameObjects mentioned previously and right-click on one of them to select the **Create Empty Parent** option.

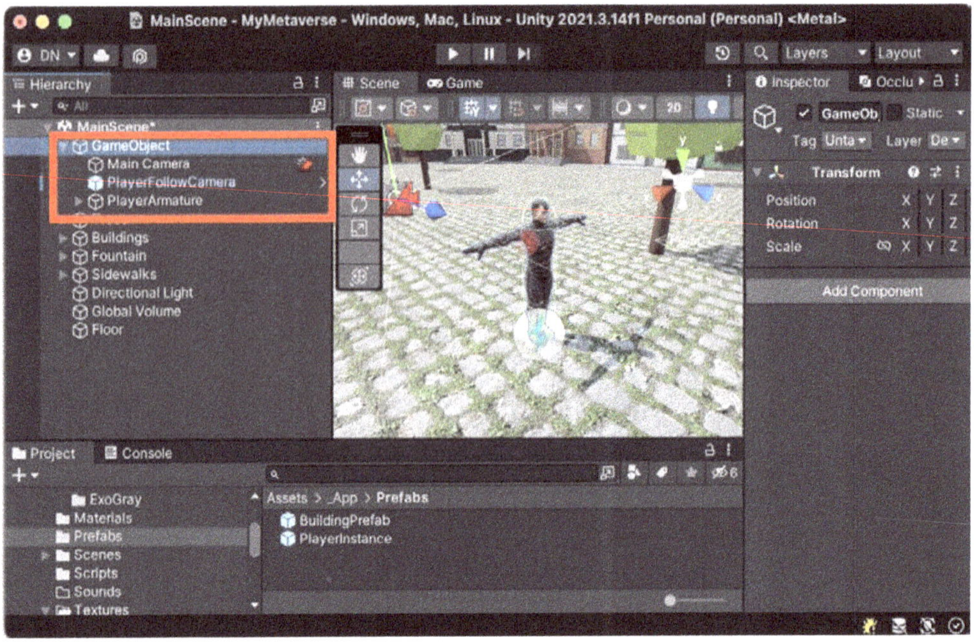

Figure 6.47 – Creating a parent GameObject to group other objects together

3. Select and drag the parent GameObject that has been created to the `PlayerInstance` Prefab; then a confirmation window will appear. Click on the **Replace Anyway** button.

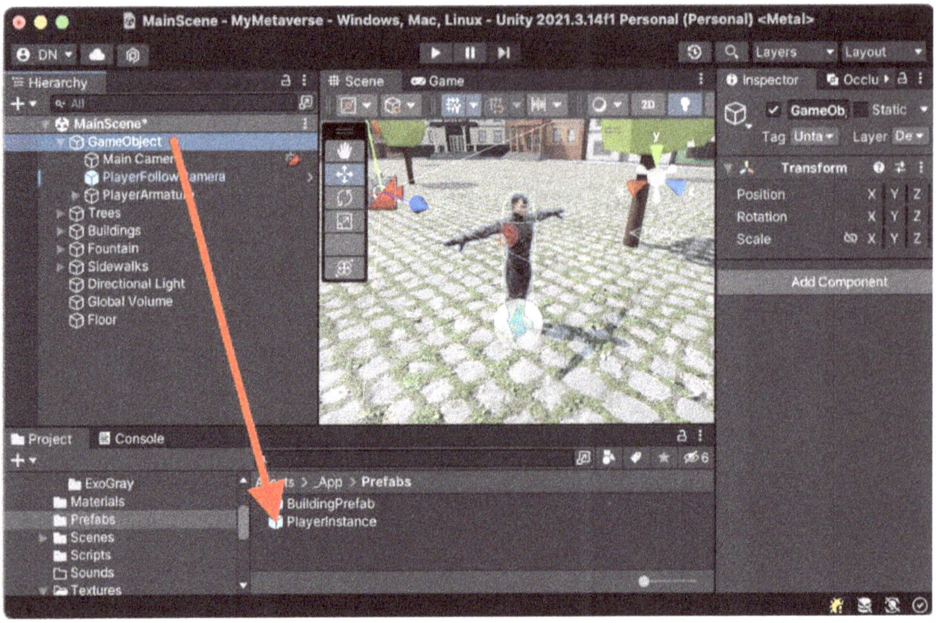

Figure 6.48 – Assigning a GameObject to a Prefab

4. Finally, remove the parent GameObject with all its elements from the scene. You will be left with a scene similar to as in the following screenshot:

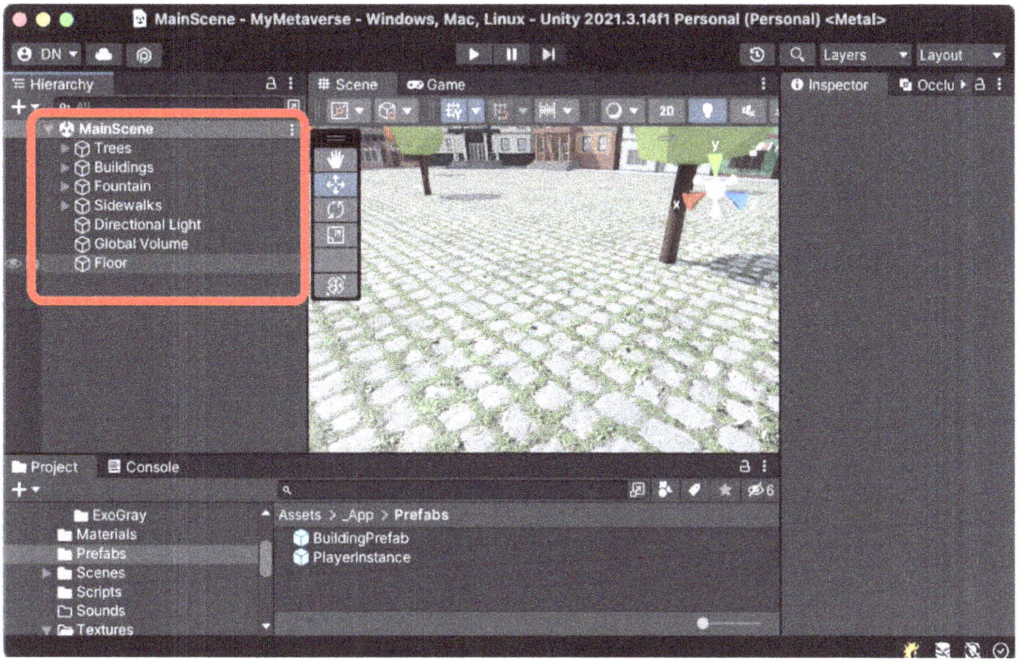

Figure 6.49 – Removing the player from scene

Great, we have optimized the three necessary elements of the player into a single Prefab. This will make it easier for us to create players dynamically through programming.

Next, we'll create a spawn point, which will serve as a starting point for all players. So, create a cube and place it wherever you want. Rename it `SpawnPoint`.

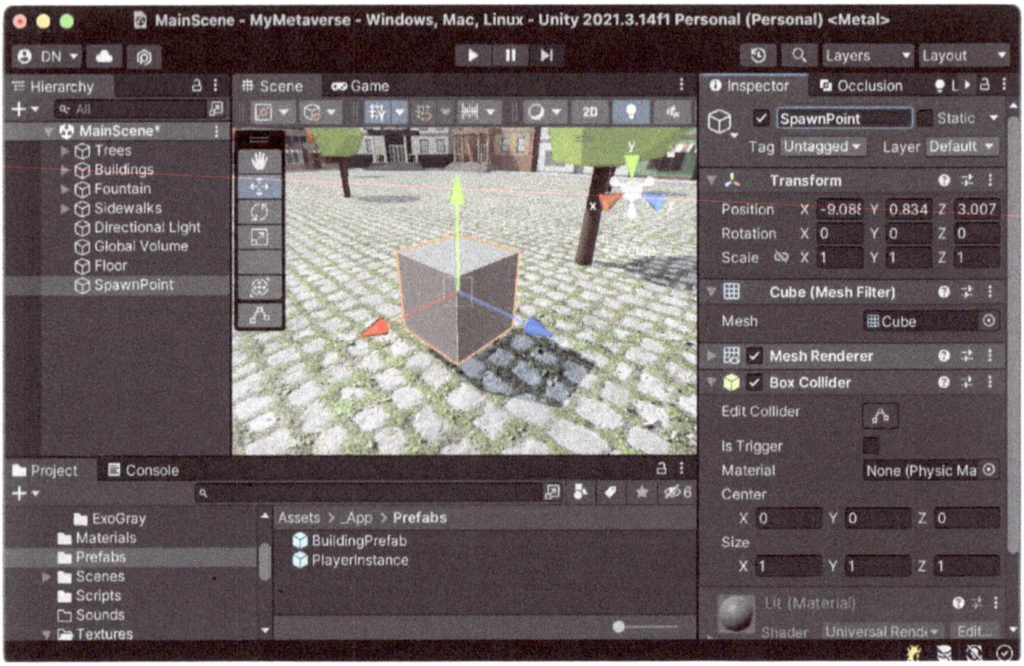

Figure 6.50 – Placing a cube as a player spawn point

In the GameObject cube, we are only interested in its position in the scene; we are not interested in its physical appearance. You can disable the **Mesh Renderer** and **Box Collider** components as we don't need them.

Great, now let's create a simple script that instantiates the Prefab we created for the player when the scene starts.

Following the **Assets** | **_App** | **Scripts** path, right-click and select **Create** | **C# Script**. Rename it `InstantiatePlayer`. Edit it by double-clicking on it.

Replace the auto-generated code with the following:

```
using UnityEngine;

public class InstantiatePlayer : MonoBehaviour
{
    public GameObject playerPrefab;

    // Start is called before the first frame update
    void Start()
    {
        Instantiate(playerPrefab, transform);
```

```
        }
    }
```

Now, we drag this new script to the **Inspector** panel of the **SpawnPoint** GameObject.

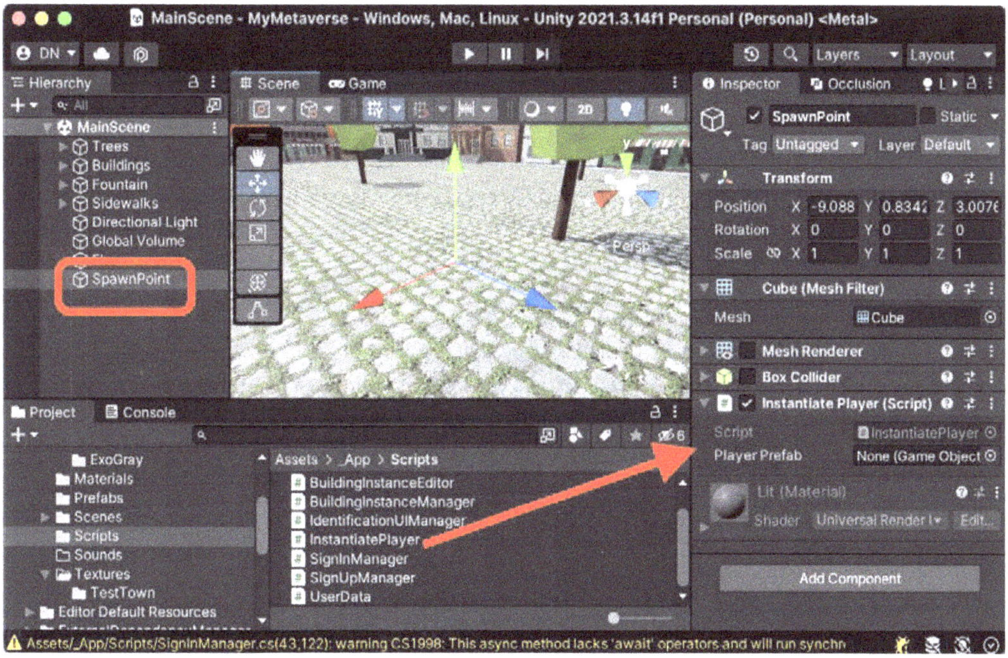

Figure 6.51 – Dragging the SignInManager script to the SpawnPoint GameObject

Finally, we drag the **PlayerInstance** Prefab located in **Assets** | **_App** | **Prefabs** into the empty slot of the **Instantiate Player** component.

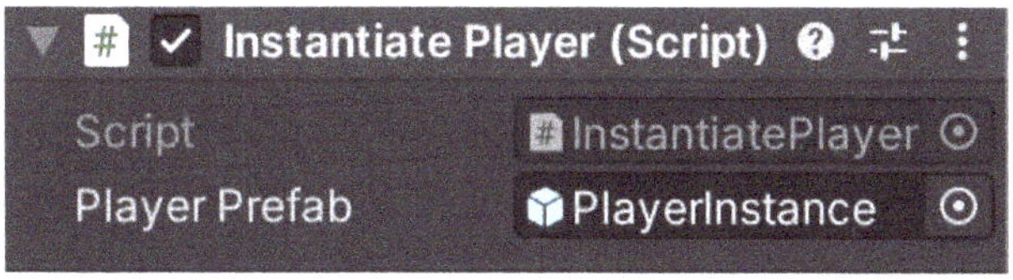

Figure 6.52 – Final result of the Instantiate Player component

Good work! You have successfully completed a new system for dynamically instantiating players in the scene. You can now click the **Play** button and see the player appear.

We have just laid an important foundation for converting the project into multiplayer mode later on.

In the next chapter, we will create a bot (**Non-Player Character** (**NPC**)) in our scenes. These NPCs will allow us to consult the available scenes in Firebase Firestore. Through some scripts that we will program in C#, we will show a pop-up window to the user so they can select any world from the list and travel to that scene. Sounds interesting, doesn't it? See you in the next chapter.

Summary

Throughout this chapter, we learned about the new tools offered by Unity UI, and we now know how to create UIs that allow the player to interact with our metaverse. In addition, we delved deeper into the Firebase SDK and our project now has a welcome screen, with scripts that allow the user to create a new account or log in. Finally, we took another step toward a multiplayer metaverse, creating a player Prefab that we can instantiate via code when the scene is started.

In the next chapter, we will learn how to create our first NPC. We will create a fun, animated character that will allow the player to travel to other worlds. For this, we will program a script that detects the presence of the player and shows a window to select the destination world.

7

Building an NPC That Allows Us to Travel

Non-Player Characters (**NPCs**) are characters that we usually find in video games. They have scripted behavior and serve mainly to offer the real player some kind of service. I am sure you have interacted with one of them in a video game.

In our metaverse, we will learn how to program our own NPC. For us, this character will serve as a gateway to other scenes. Her *script* will be to show us a list of available worlds. This will be achieved by launching a query to the Firestore Worlds collection.

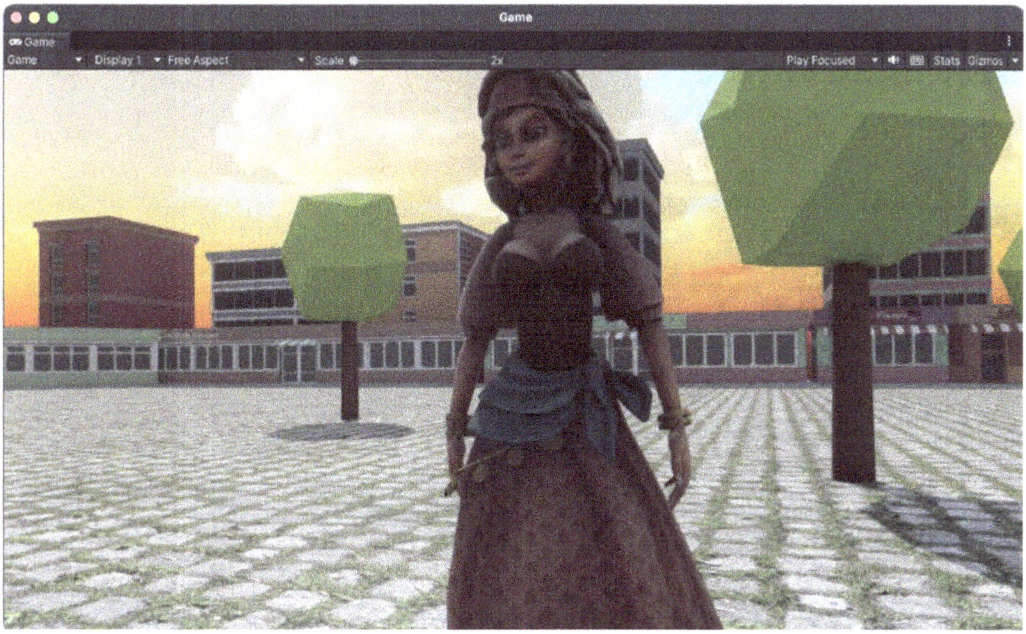

Figure 7.1 – Final result of the NPC created in this chapter

We will cover the following topics:

- Choosing an aspect
- Bringing the NPC to life
- Triggering when we are close
- Showing the available worlds in a window
- Traveling to the selected world

Technical requirements

This chapter does not have any special technical requirements, but we will be programming with scripts in C#. It would be advisable to have basic knowledge of this programming language. We need an internet connection to browse and download an asset from the Unity Asset Store. We will continue with the project we created in *Chapter 1*, *Getting Started with Unity and Firebase*.

Remember that we have a GitHub repository, `https://github.com/PacktPublishing/Build-Your-Own-Metaverse-with-Unity/tree/main/UnityProject`, which contains the complete project that we will work on here.

You can also find the complete code for this chapter on GitHub at: `https://github.com/PacktPublishing/Build-Your-Own-Metaverse-with-Unity/tree/main/Chapter07`

Choosing an aspect

To get a 3D character, we will go back to the Mixamo website, just like we did in *Chapter 2*, *Preparing Our Player*, for our main character. The steps we will take in this section are the same as required for any characters we will find on the Mixamo website. What I mean by this is you are totally free to choose whichever character you like the most.

Without further ado, let's get down to business.

The first thing we will do is choose a character in Mixamo; to do so, follow these steps:

1. Go to the website `https://www.mixamo.com`.
2. Sign in with your username or create a new account; it's free.
3. Click on the **Characters** tab and choose the one you like the most; in my case, I have chosen **Peasant Girl**. I think she has a magical touch, which is very appropriate for traveling between worlds.

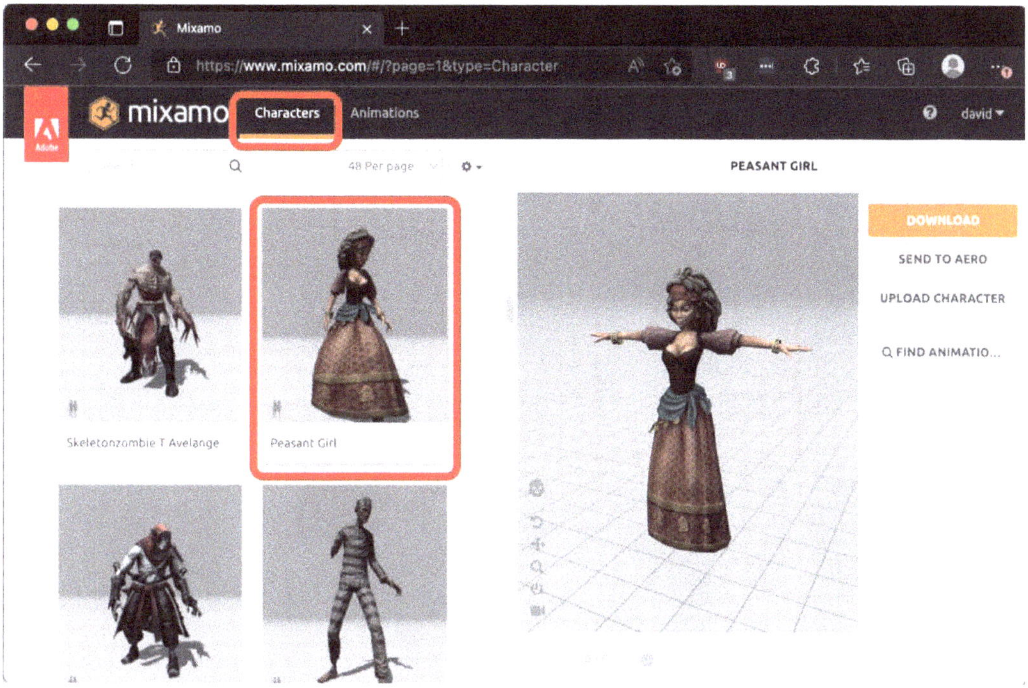

Figure 7.2 – Selecting an FBX model for our NPC in Mixamo

4. We will take the opportunity to implement a default animation to give the character some *life*. Click on the button on the right called **FIND ANIMATIONS**.
5. As you can see, there is a long list of animations available; the ideal thing is to implement an animation with a closed loop. The *idle*-type animations are perfect for this. Type the word `idle` into the search box and press the *Enter* key.
6. I have selected the animation called **Dwarf Idle**. Click on it in the list. You will see how it is applied to the character preview.

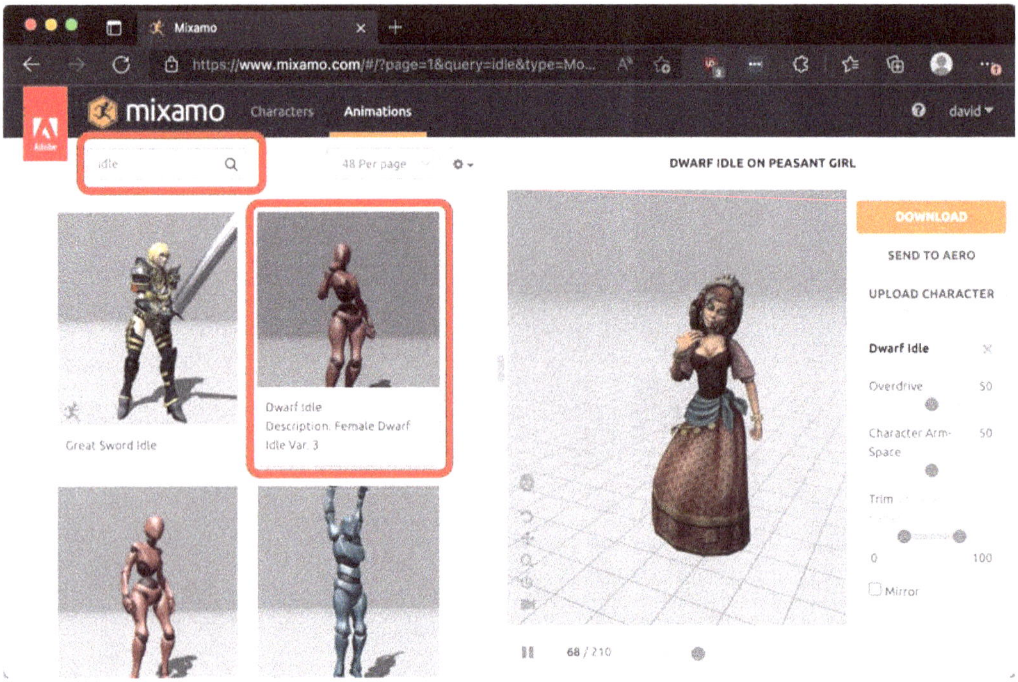

Figure 7.3 – Selecting an animation for our FBX model in Mixamo

7. Now, click on the **DOWNLOAD** button and a modal window will open with configuration options. Select **FBX for Unity(.fbx)** under **Format** and **With Skin** under **Skin**. Click again on the **DOWNLOAD** button.

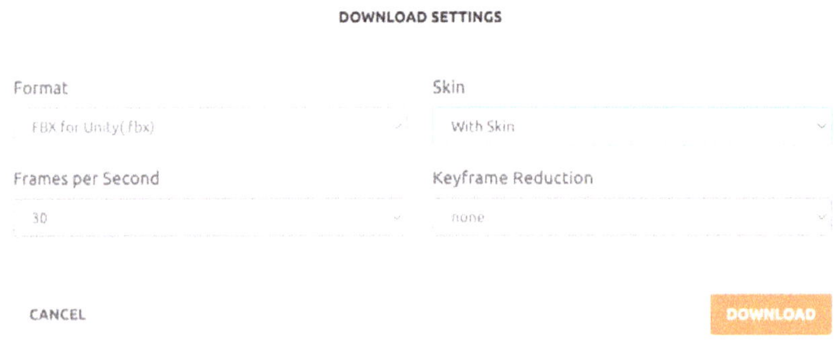

Figure 7.4 – Export options for the selected animation in Mixamo

Great. If all went well, a file called `Peasant Girl@Dwarf Idle.fbx` will have been downloaded to your `Downloads` folder. The filename may vary if you have chosen a different character or animation.

8. Now we are going to import this file into our project. To do so, following the **Assets** | **_App** | **Characters** path, create another folder, called `NPCs`, and inside it, create another folder with the name of the character you have chosen. Finally, drag the downloaded file into this folder:

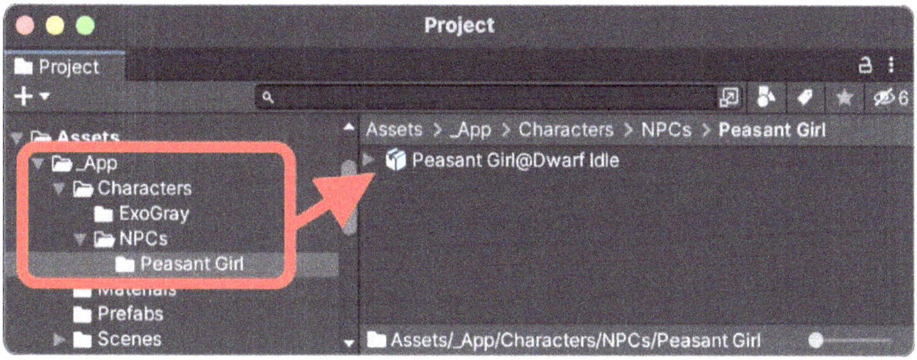

Figure 7.5 – FBX imported into our project

Good work; we have successfully imported the character file into the project. Now we need to apply a bit of configuration to make it work properly.

9. If you select the **Peasant Girl@Dwarf Idle** file in the **Project** panel, you will see the preview on the right, but the character appears completely white. To fix this, at the top of the **Inspector** panel, select the **Materials** tab and click on the **Extract Textures** button.

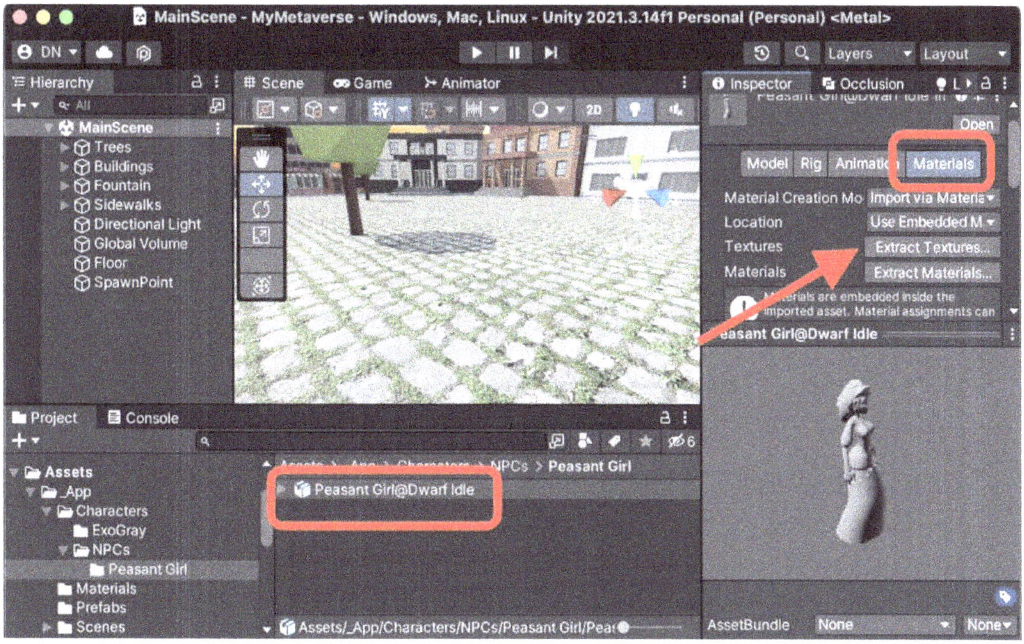

Figure 7.6 – Extracting textures from FBX

Unity will ask you for a destination folder to extract the textures from. By default, the same folder as the FBX file will be selected. Click **Accept**.

If everything goes well, you will see in the Peasant Girl folder the textures and the preview with character colors.

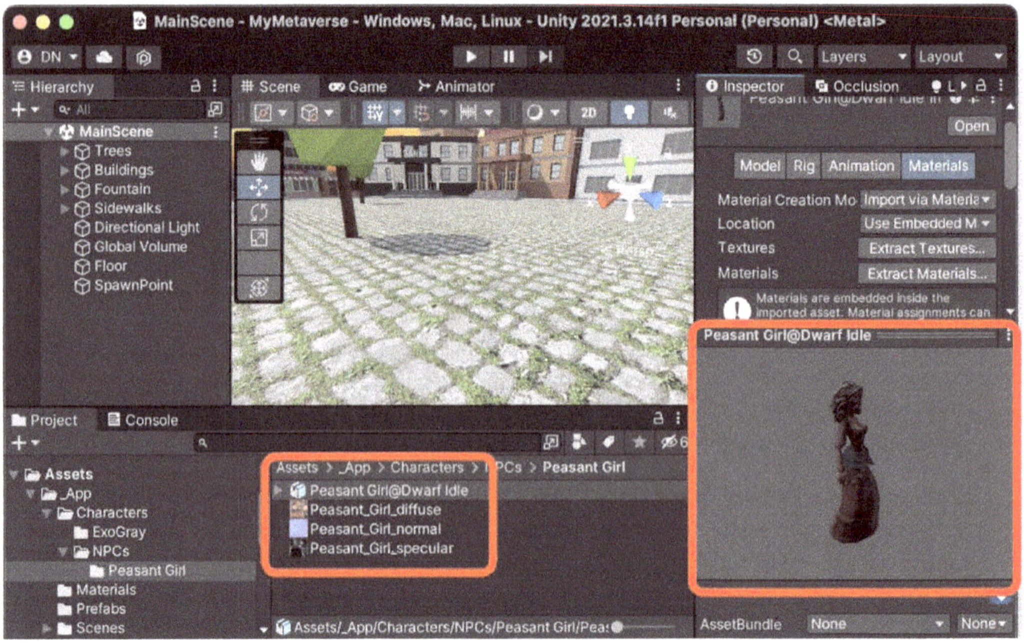

Figure 7.7 – Preview of the FBX with the applied textures

Our character is now set up, so we can continue working with it. Next, we will create a new Prefab to house this character and we will learn how to create an Animation Controller and an Animator component that will animate the character once the scene is executed.

Bringing the NPC to life

Our NPC will be present in all the scenes; otherwise, we will not be able to leave the world. To facilitate this task, we will create a Prefab that houses the character and its functionalities.

To achieve this, we will follow these steps:

1. Anywhere in **MainScene**, drag and drop the **Peasant Girl@Dwarf Idle** asset. If we click **Play**, we will see that the character is static; it has no life and it doesn't move with the animation we selected in Mixamo. This is because we need to add an **Animator** component and link the animation.

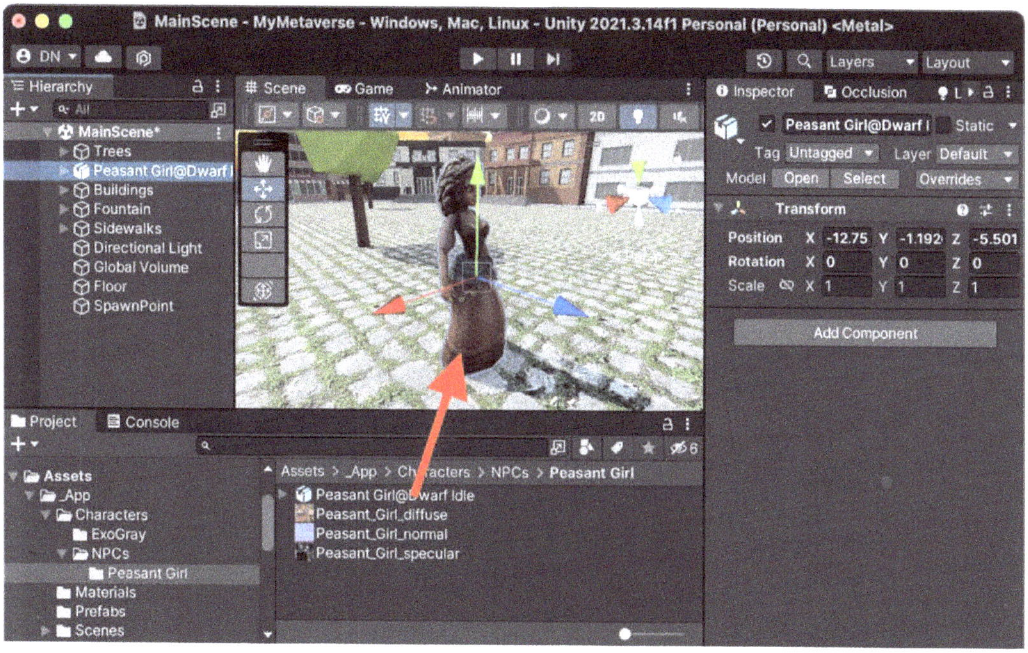

Figure 7.8 – Placing the FBX model in the scene

2. While in the **Peasant Girl@Dwarf Idle** GameObject, click on the **Add Component** button in the **Inspector** panel and add the **Animator** component.

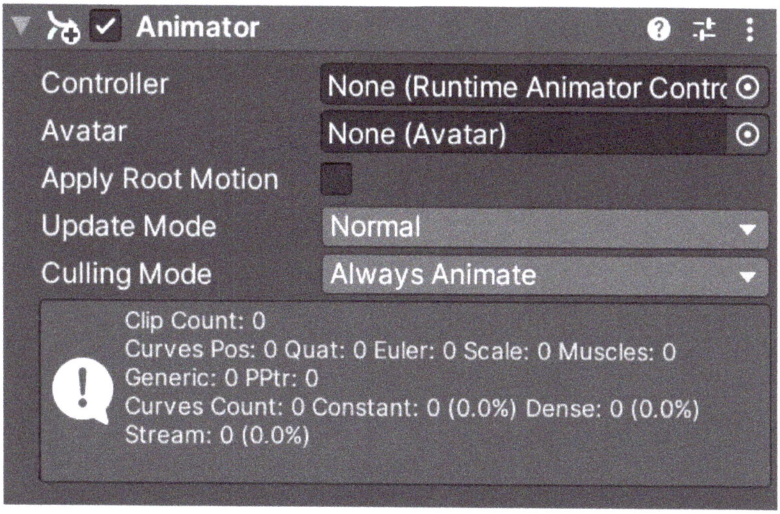

Figure 7.9 – Animator component

3. The **Animator** component needs an Animator Controller. To create an Animator Controller, in the **Assets** | **_App** | **Characters** | **Peasant Girl** folder, right-click and select the **Create** | **Animation Controller** option and rename it `Peasant Animation Controller`.
4. Double-click on the Animation Controller we just created. This will open the **Animator** window, which manages the character states and animations. Drag the **Peasant Girl@Dwarf Idle** file to any part of the graph.

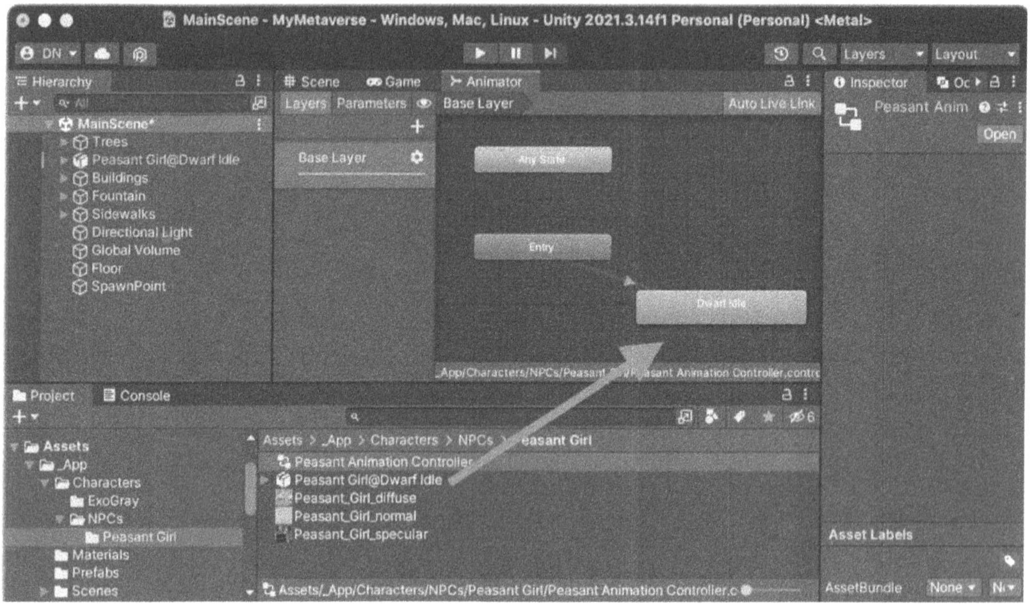

Figure 7.10 – State diagram of the Animator component

What we have just done is to tell the Animation Controller, when it runs, to launch the **Dwarf Idle** animation. By default, animations will run only once, but we want it to be an infinite loop:

1. To achieve an infinite loop in our animation, we select the **Peasant Girl@Dwarf Idle** asset in the **Project** panel, and in its **Inspector** panel, we select the **Animation** tab and activate the **Loop Time** checkbox.

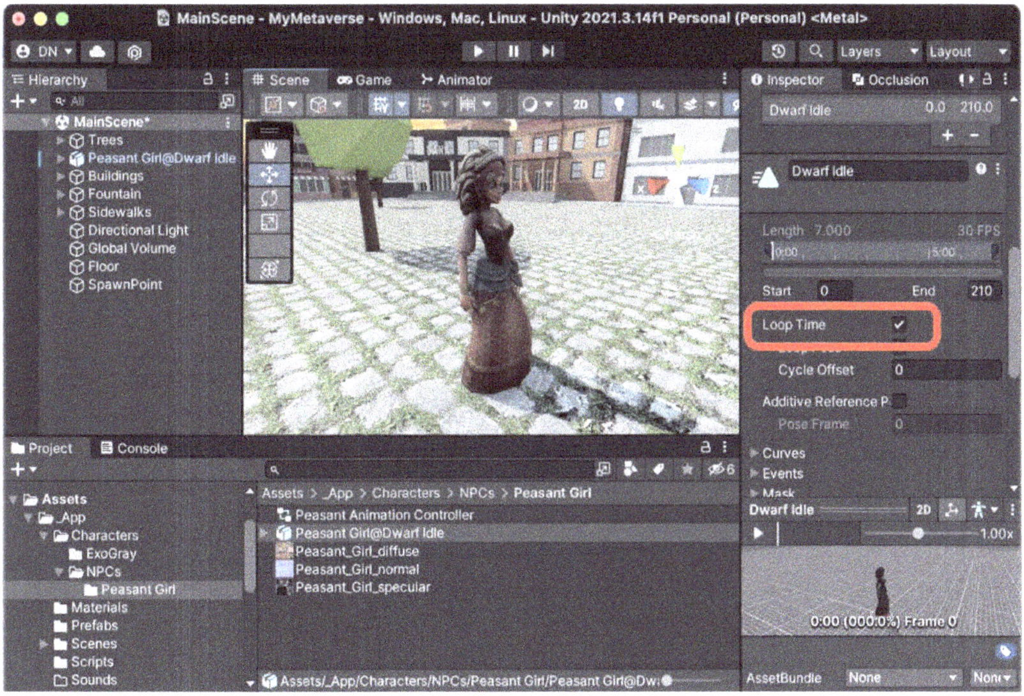

Figure 7.11 – Setting up the loop in the animation

2. Finally, to finish configuring the animation of our character, with the **Peasant Girl@Dwarf Idle** GameObject selected in the **Hierarchy** panel, we drag **Peasant Animation Controller** to the empty slot of the **Animator** component.

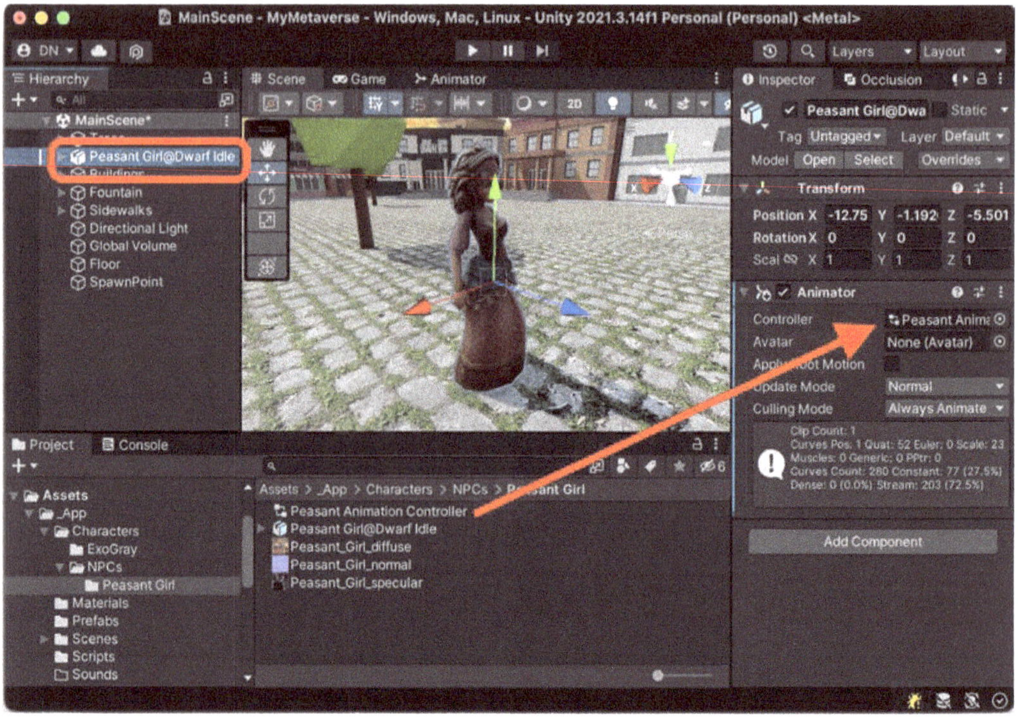

Figure 7.12 – Configuring the controller in the Animator component

Good job! Now the NPC is moving. Click the **Play** button and enjoy your creation.

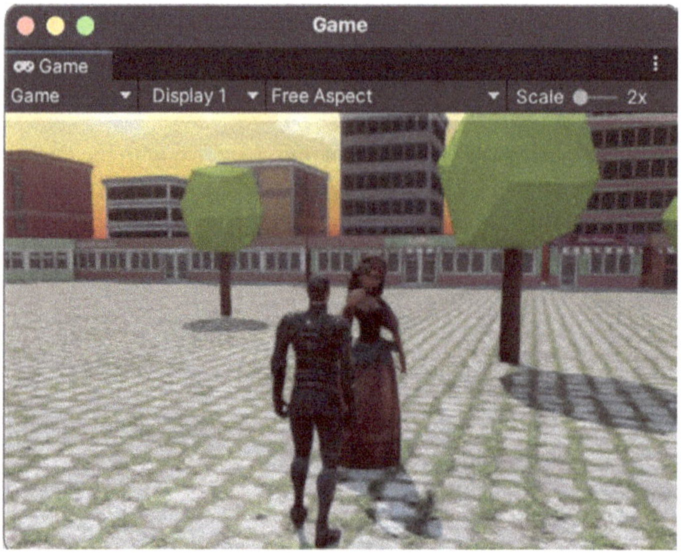

Figure 7.13 – Preview of animated model in the scene

Finally, we will encapsulate our NPC with the animation in a new Prefab.

1. Following the **Assets** | **_App** | **Prefabs** path, right-click and select the **Create Prefab** option. Rename it `Travel NPC`.
2. Select the **Peasant Girl@Dwarf Idle** GameObject from the **Hierarchy** panel and drag it to the **Travel NPC** Prefab.

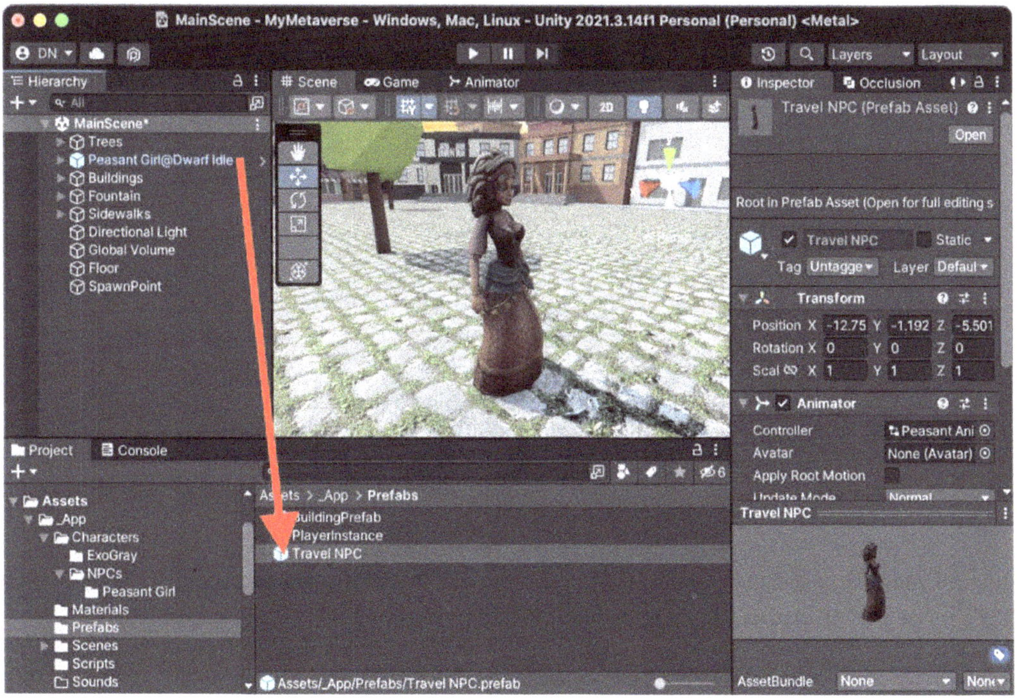

Figure 7.14 – Creating a Prefab from the GameObject

3. A confirmation modal will appear. Click on the **Replace Anyway** button and then another modal window will appear asking whether you want to create an original Prefab or a variant. Click on the **Prefab Original** button.
4. Finally, remove the **Peasant Girl@Dwarf Idle** GameObject from the scene and add the **Travel NPC** Prefab instead.

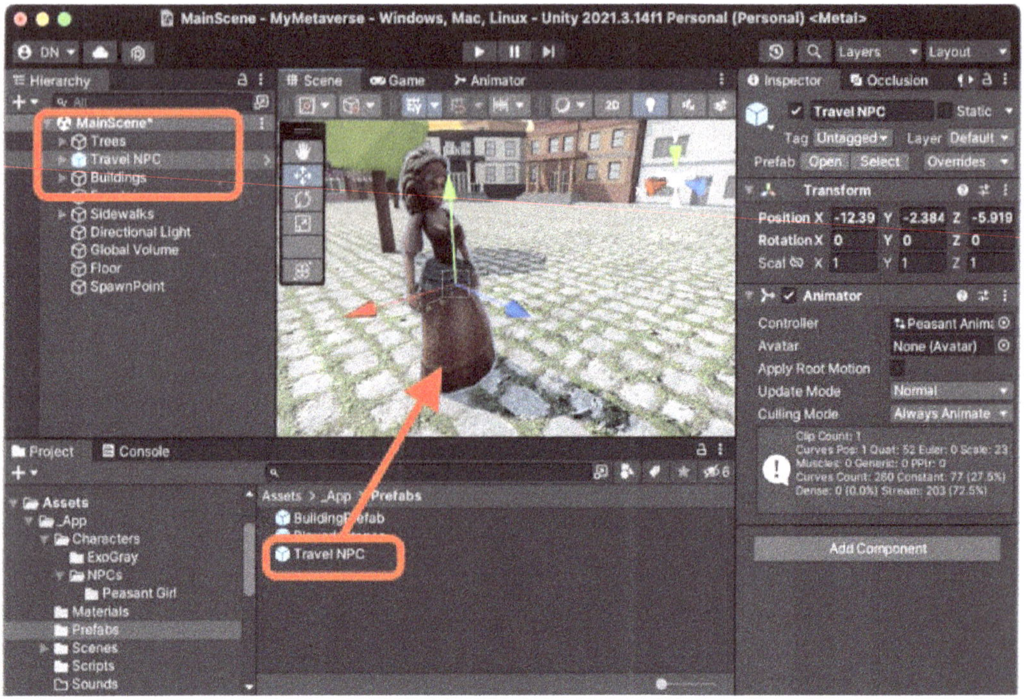

Figure 7.15 – Placing the created Prefab in the scene

Great job! You have successfully completed the creation of a Prefab that encapsulates the character in a single piece, with its **Animator** component.

Don't forget to save your changes. Next, we are going to work on programming its functionality. We will create a C# script that, when the player is close, will show text on the screen to invite us to interact with the NPC. Sounds interesting, doesn't it?

Triggering when we are close

Can you think of a game with an NPC where when you approach them, the typical text *Press the "E" key to interact* appears on the screen? That's exactly what we're going to program. We will create a radius of influence for the NPC. When the player enters this radius, we will show the message; when they leave it, we will hide the message.

The radius of influence will be a Collider. In Unity, Colliders emit a trigger when an object containing a **Rigidbody** component comes into contact with them. The same happens when the object leaves the Collider. We will use these events to create our dialogue flow:

1. To create our area of influence, we will first add a Collider. To do this, right-click on the GameObject of our NPC called **Peasant Girl@Dwarf Idle** and select **3D Object | Cube**. Rename it `Influence`.

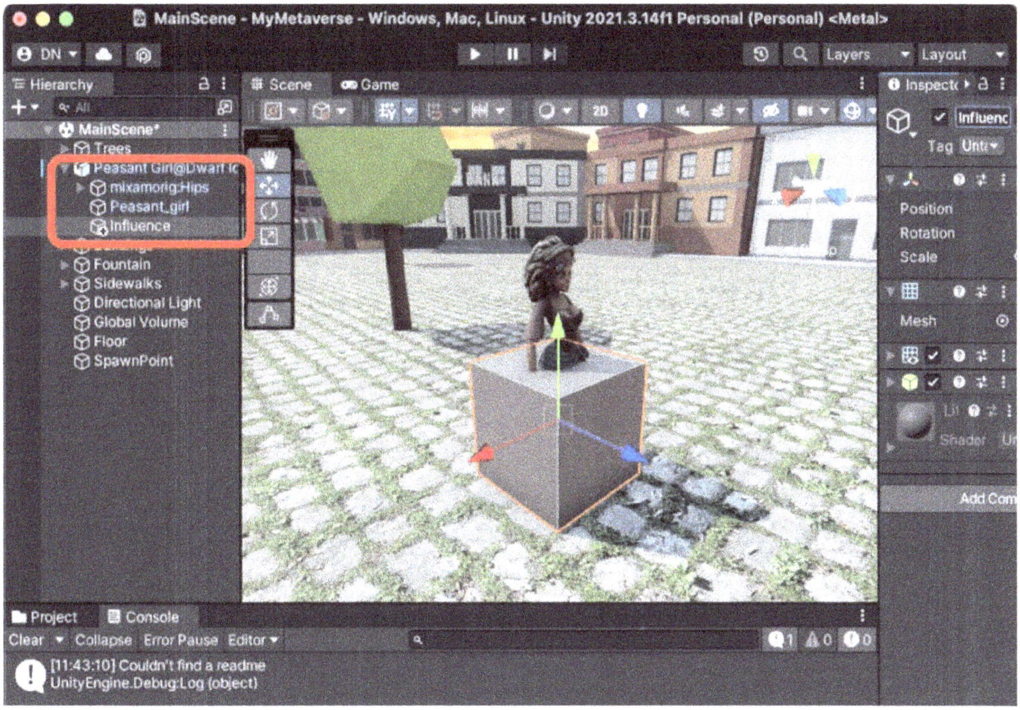

Figure 7.16 – Placing a cube on the scene

2. Use the **Scale** and **Move** tools to zoom and move the cube until it looks similar to as in the following screenshot:

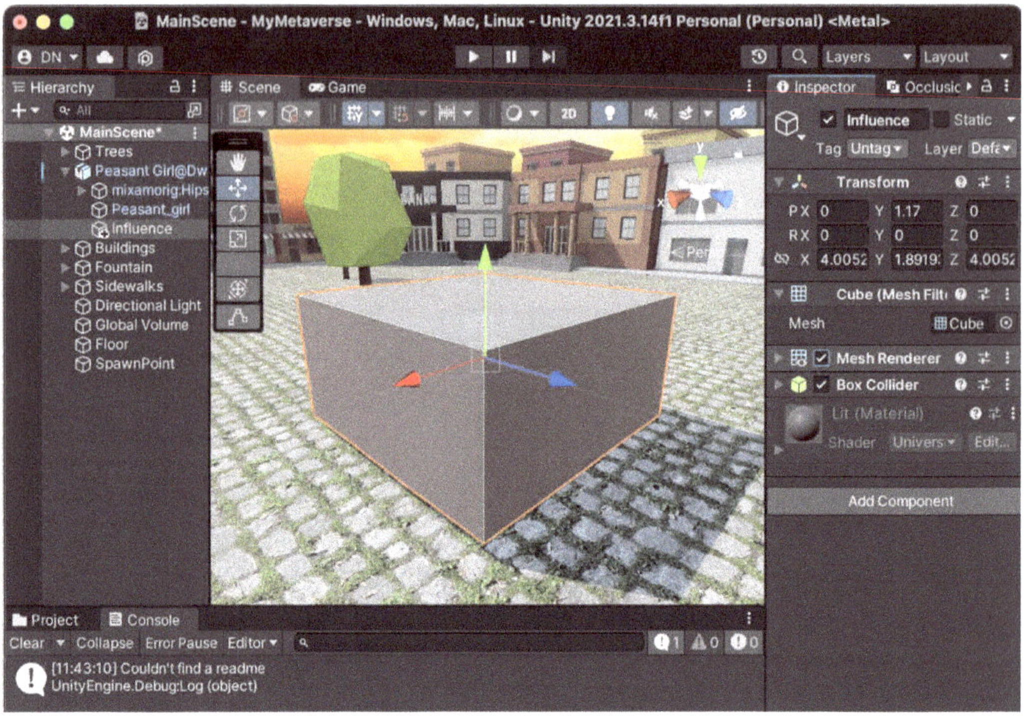

Figure 7.17 – Scaling and moving the cube

3. With the **Influence** GameObject selected, in the **Inspector** panel, deactivate the **Mesh Renderer** component. This will make the cube invisible, but we will keep its Collider, which is what we really need.

It's important to know that the size of the Cube doesn't have to be exactly the same as I've used here. You can use whatever you want; it won't affect the performance we're looking for in this chapter.

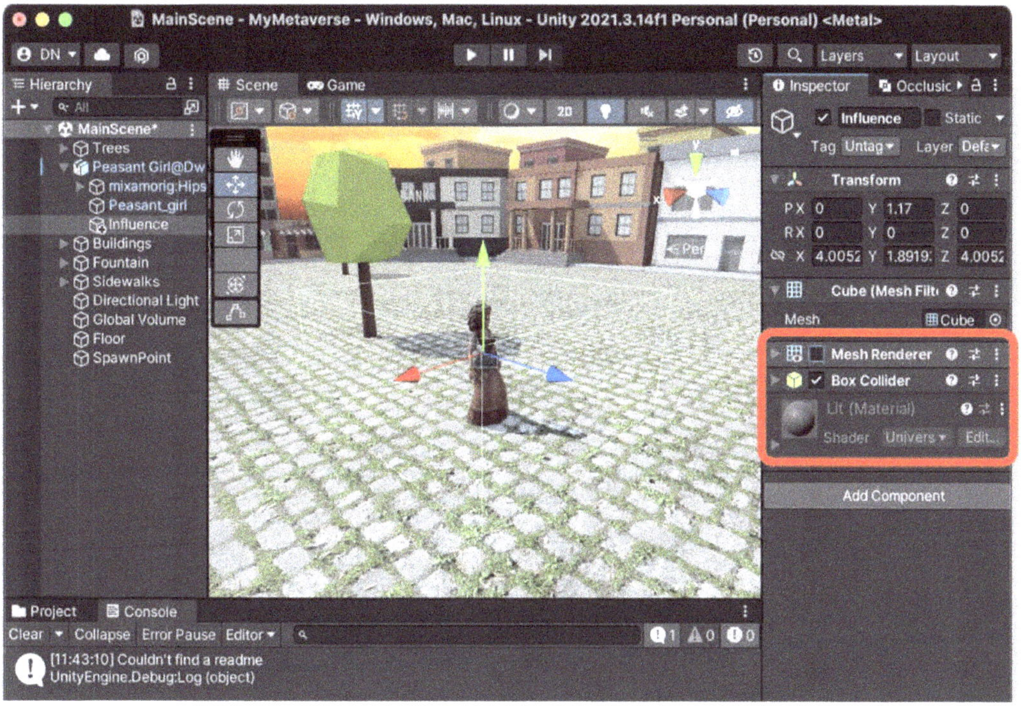

Figure 7.18 – Box Collider component added to the Cube

4. Finally, an important setting that we should always apply when we want to use a Collider as a *tip-off* and not as a physical barrier that prevents the player from crossing it is to activate the **Is Trigger** checkbox in the **Collider** component. This will completely change the behavior of the Collider and allow it to trigger events when an object enters or exits it.

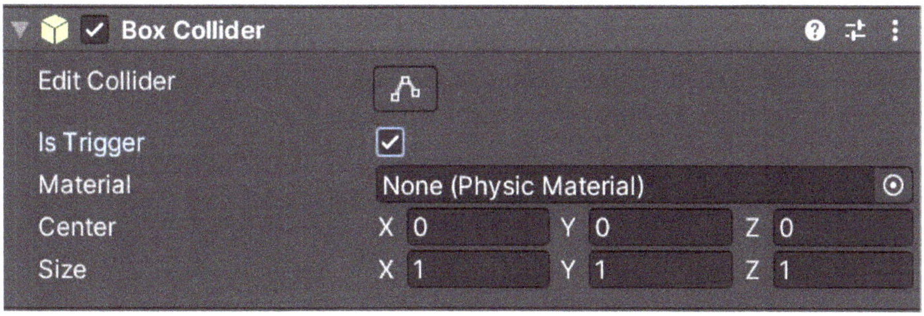

Figure 7.19 – Configuration of the Box Collider component

Great, we have created an area of influence. Before we start programming, we need to add a modification to our NPC. This is to place text over her head to show a warning when the player can interact.

5. We'll use a 3D text object; to add it, right-click on the NPC's GameObject and select **3D Object | Text – TextMeshPro**. Rename it `Message`.

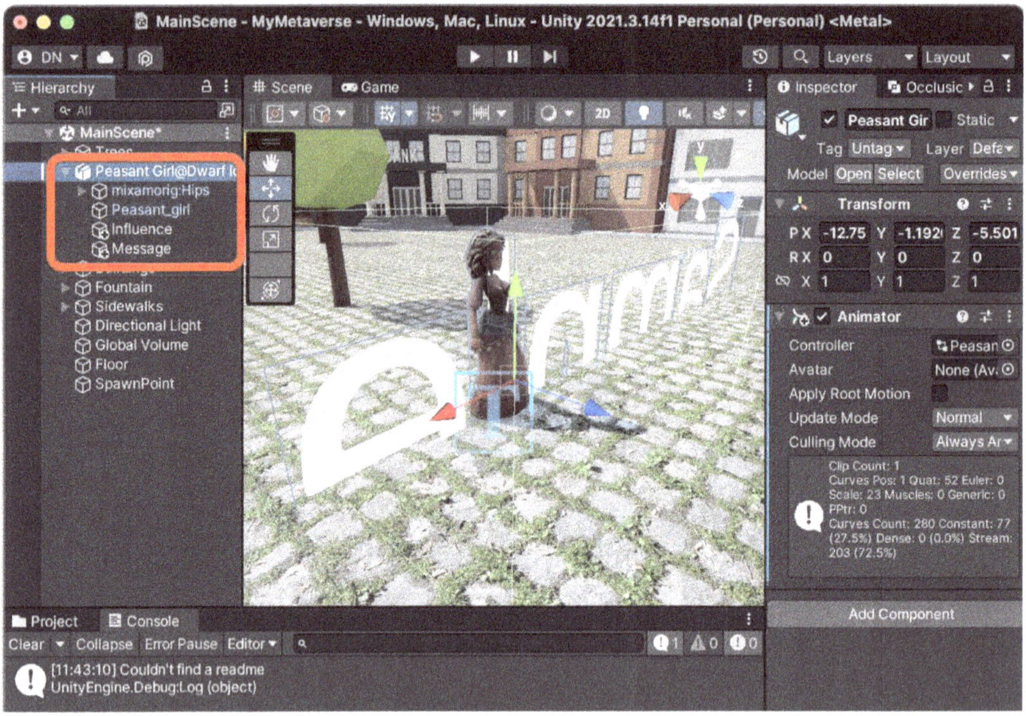

Figure 7.20 – Text – TextMeshPro GameObject added to the scene

We have just added some 3D text. This type of text, unlike the ones we have used before for UIs, behaves like an object that can be placed in the scene and its position, rotation, and scale can be changed.

This text object also includes a **TextMeshPro** component, so the text customization options will be familiar to you. Right, now that we've clarified the difference between this 3D text GameObject and the one we used previously in the UI, we can move on:

1. As you will have seen, our new **Message** GameObject has a scale and rotation that doesn't match the orientation of our NPC. Use the **Scale**, **Move**, and **Rotate** tools to make it look similar to as in the following screenshot.

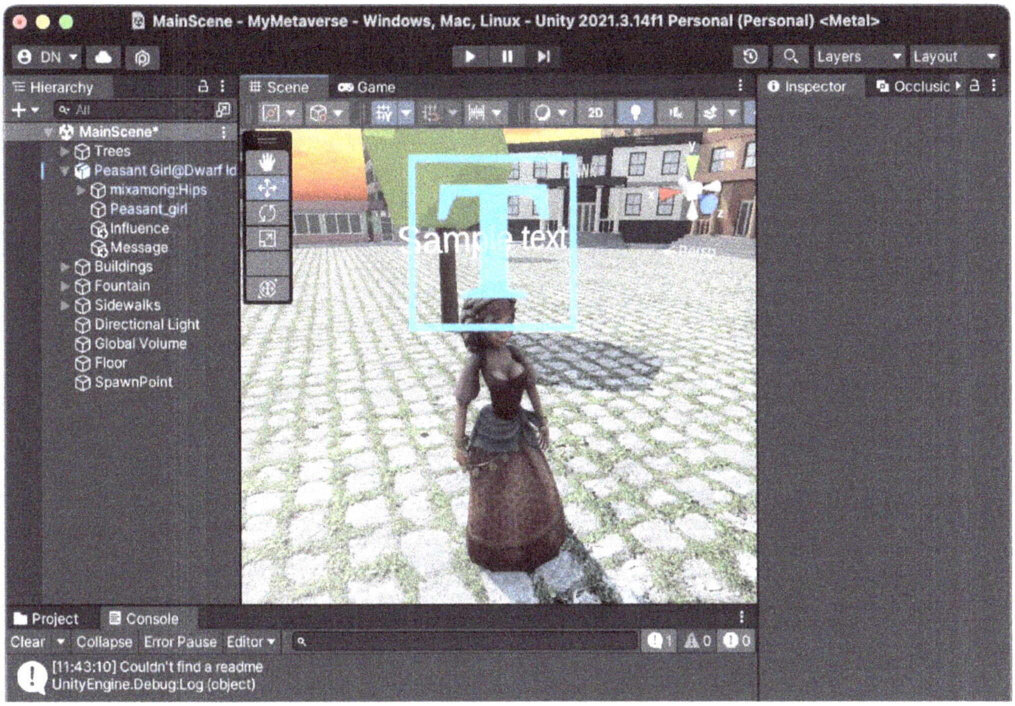

Figure 7.21 – Repositioning and scaling the GameObject message

Finally, the **Sample Text** text that comes by default in the component will be modified dynamically in the script that we create next. So, we must make sure that it looks good if the text is longer.

To do this, we will modify some properties in the **TextMeshPro** component of the **Message** GameObject:

1. We will apply a centered text justification. To do so, click on the button shown in the following screenshot.

Figure 7.22 – Text alignment options

2. Check the **Auto Size** box and change the minimum font size to `14` and the maximum to `24`.

Figure 7.23 – Text size options

3. Great, now we have all the ingredients to prepare our script. To create a new script, navigate to the **Assets** | **_App** | **Scripts** path and right-click to select **Create** | **C# Script**. Finally, rename it `TravelNPCTrigger` and double-click to edit it.

   Replace all the default code with the following:

```
using TMPro;
using UnityEngine;
public class TravelNPCTrigger : MonoBehaviour
{
    const string NPC_TEXT = "Do you want to
        travel?\nPress 'E' key to interact";
    public TextMeshPro messageGameObject;
    public GameObject travelWindow;
    private bool canInteract;
    private void Awake()
    {
        travelWindow.SetActive(false);
    }
    void Start()
    {
        messageGameObject.text = string.Empty;
    }
    private void OnTriggerEnter(Collider other)
    {
        if (other.gameObject.tag == "Player")
        {
            messageGameObject.text = NPC_TEXT;
            canInteract = true;
        }
    }
    private void OnTriggerExit(Collider other)
    {
        if (other.gameObject.tag == "Player")
        {
            messageGameObject.text = string.Empty;
            canInteract = false;
            travelWindow.SetActive(false);
        }
    }
    void Update()
    {
        if (Input.GetKeyDown(KeyCode.E) &&
            canInteract)
```

```
            {
                travelWindow.SetActive(true);
                Cursor.lockState = CursorLockMode.None;
            }
        }
    }
```

Now that we have the script programmed, we will assign it to the **Influence** GameObject and drag the **Message** GameObject to the empty **Message Game Object** slot.

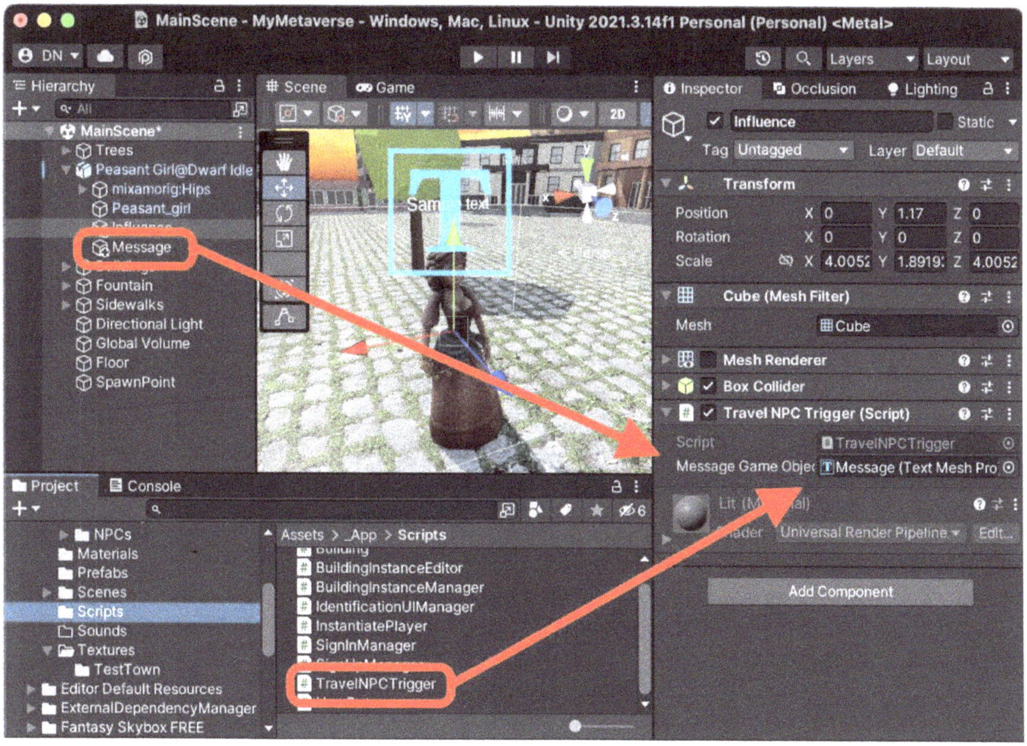

Figure 7.24 – Adding and configuring the TravelNPCTrigger script as a component

Good work! Before testing it, let's briefly explain how the script works:

I. By placing the script on a GameObject that has a Collider with the **Is Trigger** property enabled, it shall execute the `OnTriggerEnter` and `OnTriggerExit` functions automatically when a GameObject contacts and exits the GameObject.

II. When our player comes into contact, we will know it is them by comparing the `Player` tag that is assigned by default to the `PlayerArmature` Prefab.

III. We will show a message in the **Message** GameObject and delete it when the player exits the GameObject.

Now we can see it in action. Click the **Play** button and approach the NPC, then walk away.

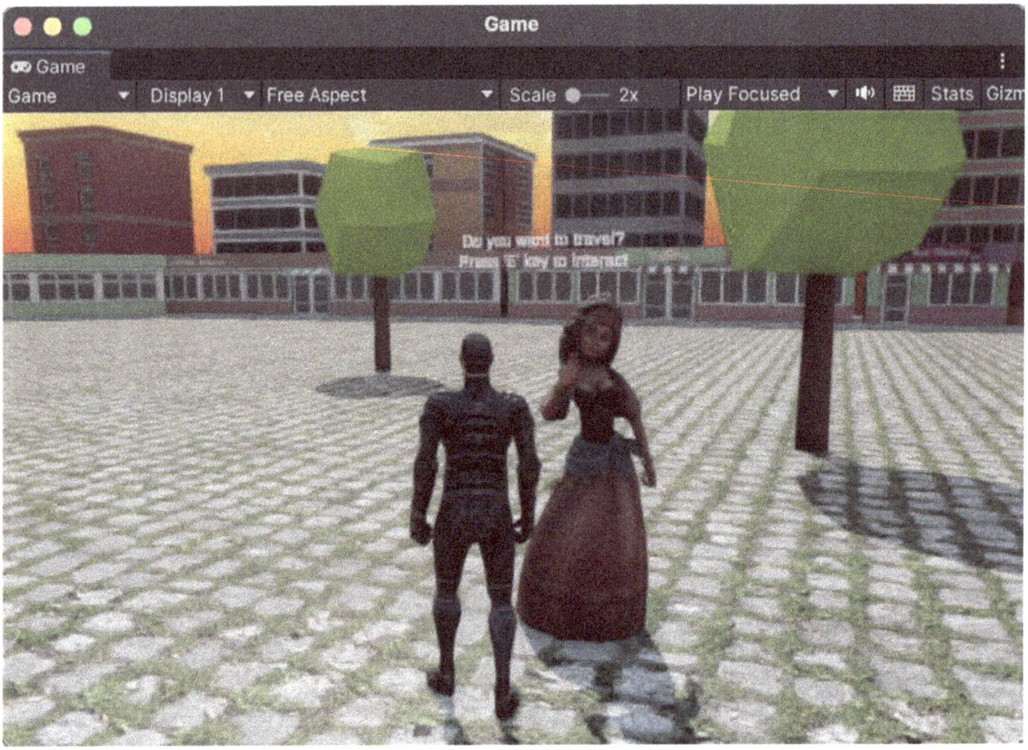

Figure 7.25 – Trigger operation preview

Excellent work! You have finished and implemented a system that allows our NPC to detect the player and display a message to invite them to interact. Knowing how a Trigger Collider works will open up an infinite range of possibilities for future improvements and functionalities in this and other projects.

In the next section, we will expand and improve upon our script to detect when the player presses the *E* key and display a window with the worlds we have available in the Firestore Worlds collection.

Showing the available worlds in a window

Throughout this section, we will learn about new Unity UI components that will allow us to build interfaces with new functionalities. In previous chapters, we learned how to implement buttons and text inputs. Now it's time to learn about the **Scroll Rect** component.

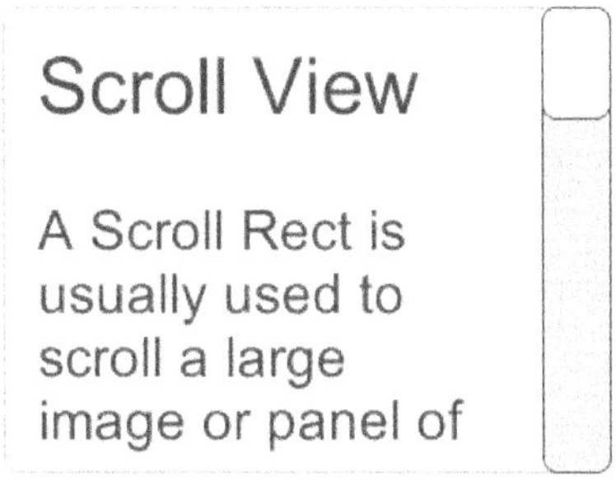

Figure 7.26 – Example of the Scroll View component

This component allows us to insert large content into a fixed-size container and make the excess content scrollable. It can be configured to scroll vertically, horizontally, or both.

We will use the **Scroll Rect** component to host the list of available worlds we want to display to the user.

> **Tip: Detailed information about Scroll Rect**
>
> If after getting to know this component you are curious about what it has to offer, I recommend that you visit `https://docs.unity3d.com/es/2018.4/Manual/script-ScrollRect.html`.

Now, we will start the creation of our popup with scrollable content. To do so, follow these steps:

1. First of all, as you know from previous chapters, to work on a graphical interface with Unity UI, we first need a canvas. In the Unity main menu bar, select **GameObject** | **UI** | **Canvas**. A GameObject **Canvas** and a GameObject **EventSystem** will appear in the **Hierarchy** panel.
2. In the **Canvas Scaler** component of the **Canvas** GameObject, set the **UI Scale Mode** property to **Scale With Screen Size** and **Reference Resolution** to `1920` and `1080`.

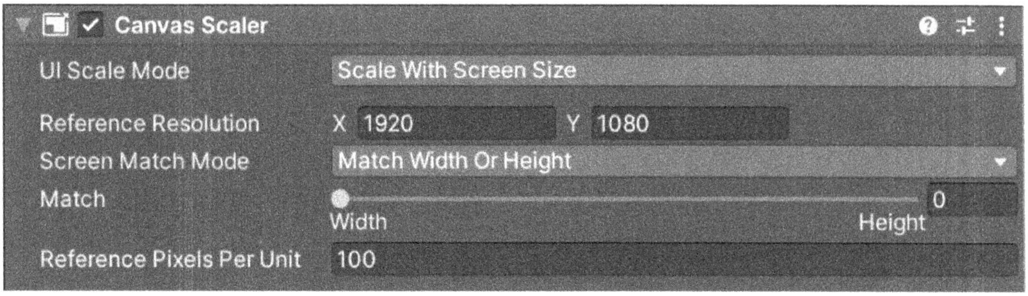

Figure 7.27 – Canvas Scaler component

3. We now need a panel to serve as a container. Right-click on the **Canvas** GameObject and select the **UI** | **Panel** menu.
4. Right-click on the **GameObject** panel and select **UI** | **Scroll View**. A new GameObject of type **Scroll View** will be added within the hierarchy of the **GameObject** panel. With the **Rect Transform** tool, we will expand its boundaries to make it approximately the same size as in the following screenshot. We will leave enough space at the bottom to place a **Close** button later.

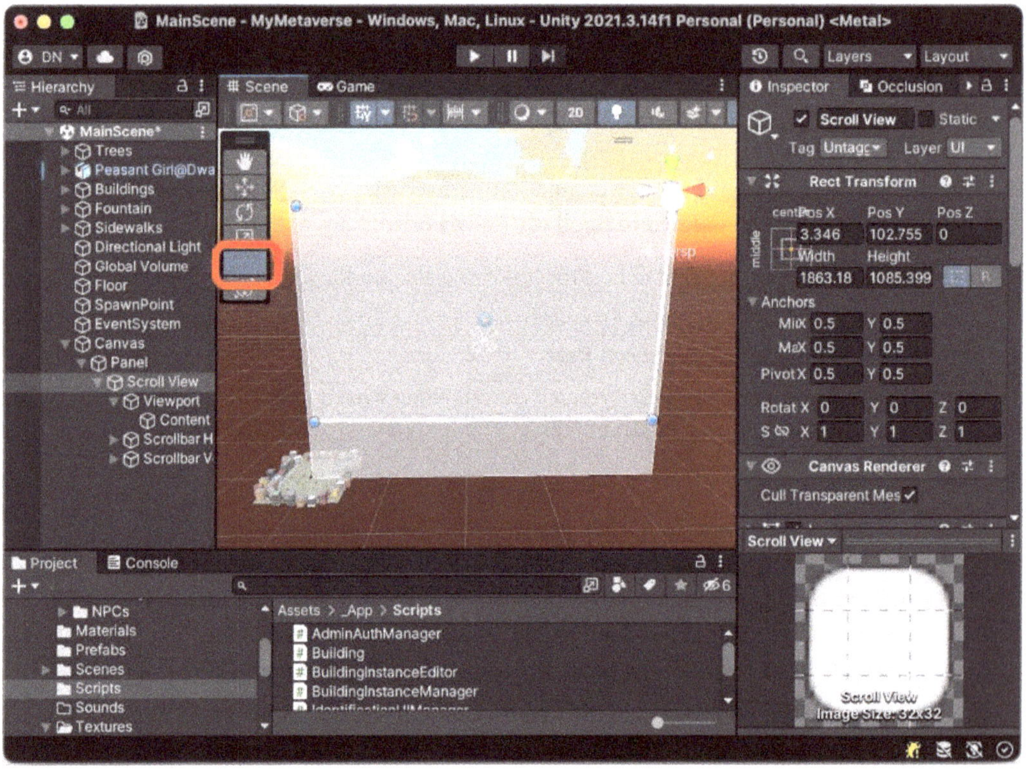

Figure 7.28 – Adjusting the size of the Scroll View component

5. Now, we will add a button called **Close** and change the text to `Close` as well. This button will allow the player to cancel the modal. Your scene will look similar to as in the following screenshot.

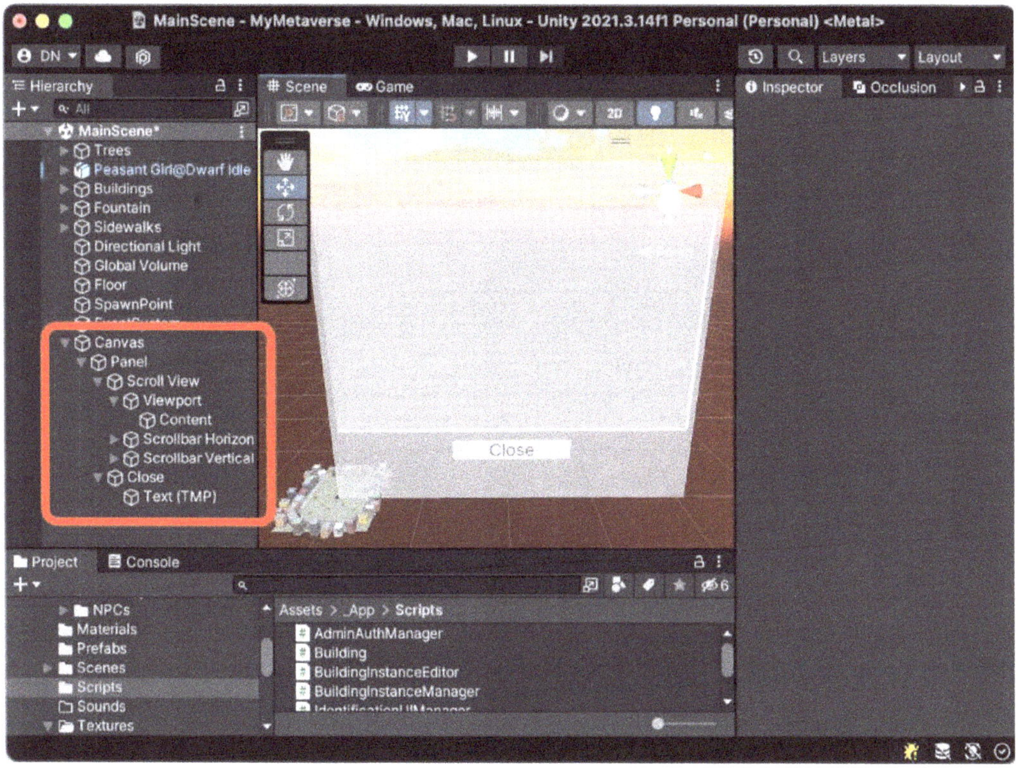

Figure 7.29 – Close button recently added

6. Finally, we will add a **Vertical Layout Group** component to the **Content** GameObject. This last one will be the parent of the elements that will be scrollable. We will need a **Vertical Layout Group** component that will be in charge of applying an automatic organization to all the elements that are added to the scroll.

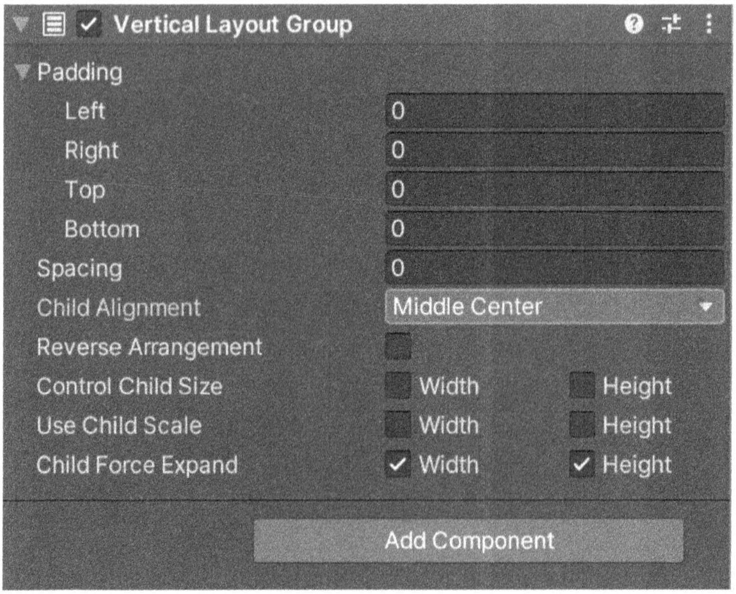

Figure 7.30 – Vertical Layout Group component

Good job, you have successfully created a **Modal** window with a Scroll View component. The next step is to create a Prefab that represents an element that we want to appear in the scroll and that contains information about the world we want to travel to, as well as a button to select it.

This Prefab will be the one that we will instantiate repeatedly from the script that we will make later, and as a whole, it will visually represent the Firestore Worlds collection.

To achieve this, we will follow these steps:

1. Right-click on the **Content** GameObject and select **UI** | **Panel**. Rename it `WorldItem`.
2. With the **WorldItem** GameObject selected, resize it in the **Inspector** panel and change the **Width** value to `1800` and **Height** to `300`. It will look similar to as in the following screenshot:

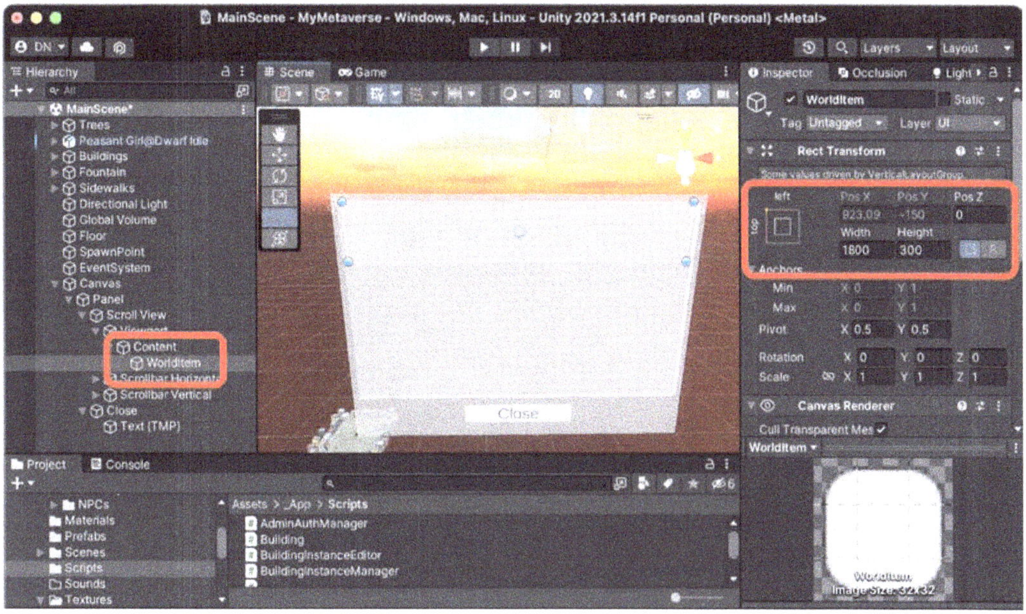

Figure 7.31 – Resizing the WorldItem GameObject

3. The **WorldItem** GameObject will contain visual information about the world and also a button to select it. That's why we will need to add a **Horizontal Layout Group** component to organize the content, add it, and change the alignment to **Middle Center**.

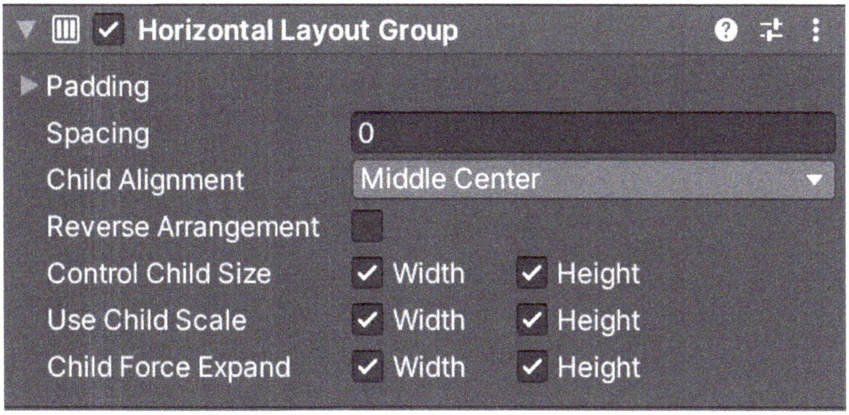

Figure 7.32 – Horizontal Layout Group component

4. The first thing we want to display is the name of the world. To do this, use a **Text** GameObject, right-click on the **WorldItem** GameObject, and select **UI** | **Text - TextMeshPro**. Rename it `Name`.
5. With the **Rect Transform** tool, enlarge the textbox, and in the **Inspector** panel, change the **Color** property to **Black**, activate **Auto Size**, and set **Alignment** to centered.

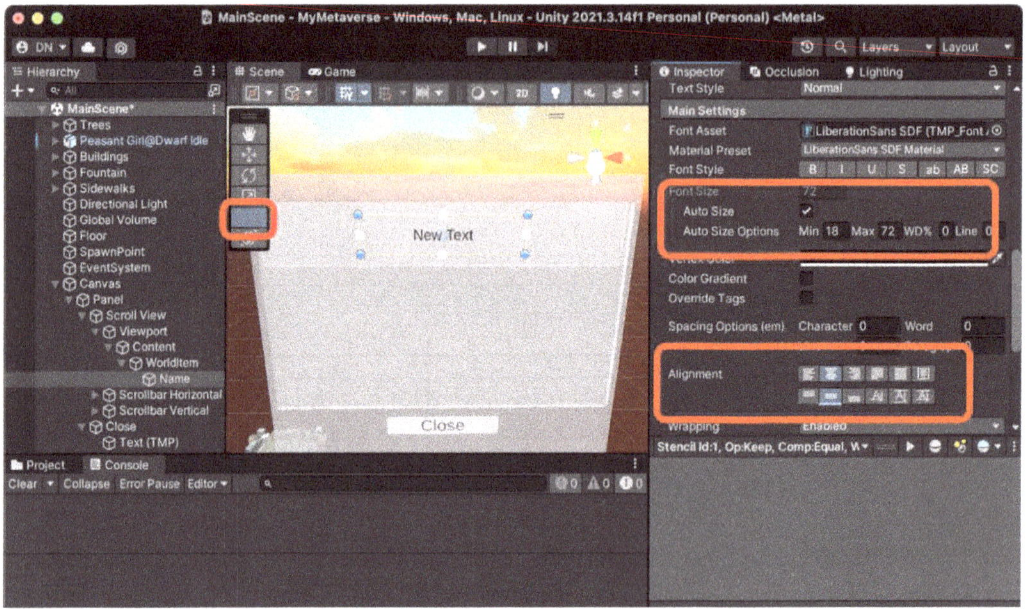

Figure 7.33 – Adjusting text size and alignment

6. Now, we will create the button that will allow us to select the world. Right-click on the **WorldItem** GameObject and select **UI** | **Button - TextMeshPro**. Rename it `Join`.
7. Change the text to `Join` and enable the **Auto Size** property on the **Text (TMP)** GameObject inside the **Join** button you just created. It will look something similar to as in the following screenshot.

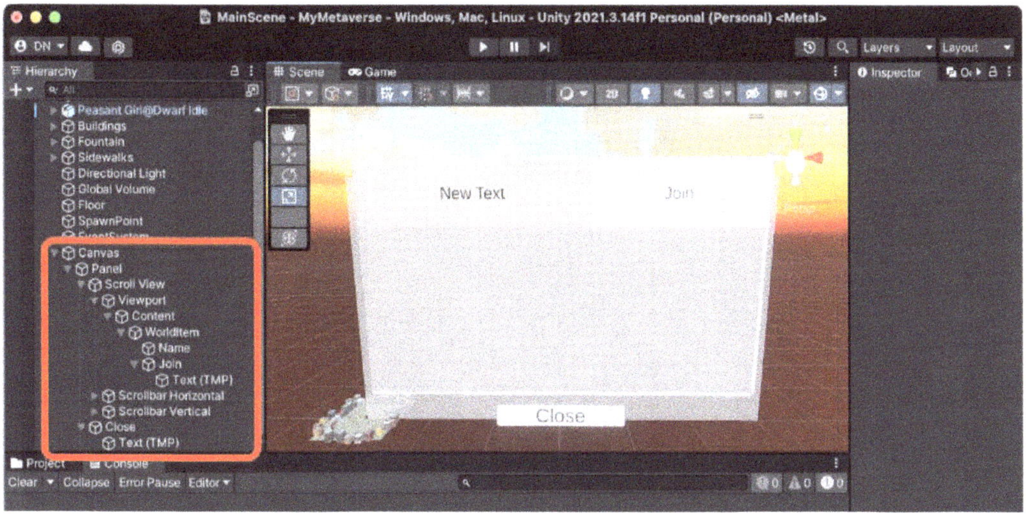

Figure 7.34 – Final appearance of the WorldItem GameObject

You're doing great! Now, to finalize the behavior of the **WorldItem** GameObject, we'll create a script to give it functionality before turning it into a Prefab.

8. To do this, go to **Assets** | **_App** | **Scripts** and right-click to select **Create** | **C# Script**. Rename it `WorldItemBehaviour`. Double-click on it to edit it.

Replace all the auto-generated code with the following code:

```
using UnityEngine;
using UnityEngine.SceneManagement;
using UnityEngine.UI;
public class WorldItemBehaviour : MonoBehaviour
{
    public Text worldNameGameOject;
    public string sceneName;
    public string worldName;
    void Start()
    {
        worldNameGameOject.text = worldName;
    }
    public void JoinWorld()
    {
        SceneManager.LoadScene(sceneName);
    }
}
```

9. Save the changes and add this script to the **WorldItem** GameObject, then drag and drop the **Name** GameObject into the empty **World Name Game Object** slot.

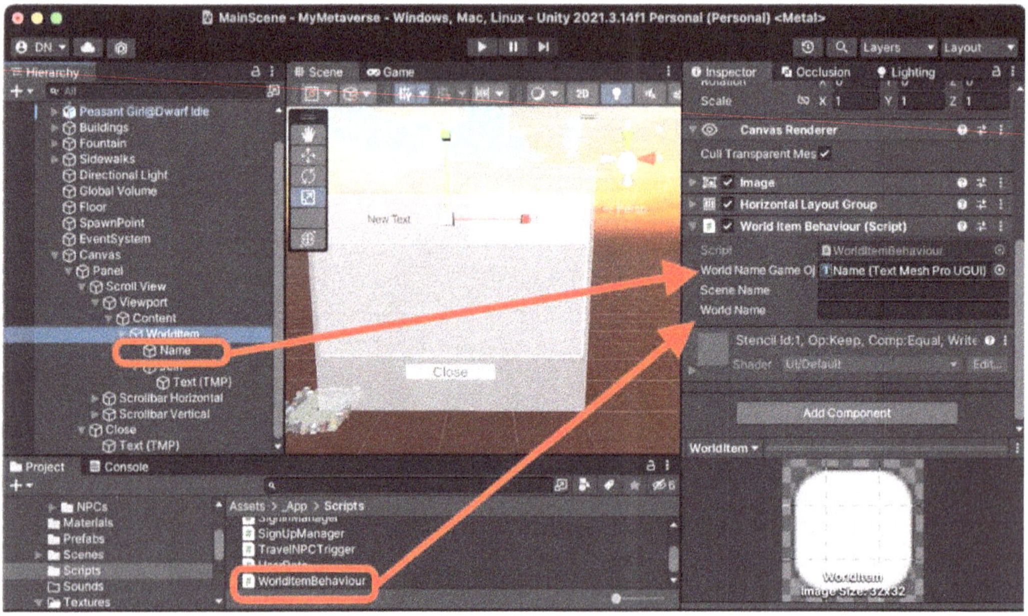

Figure 7.35 – Adding the WorldItemBehaviour script as a component in the WorldItem GameObject

10. Now, we must bind the `OnClick` event of the **Join** button to the `JoinWorld` function. To do so, select the **Join** button. In the **Inspector** panel, in the `OnClick` section, drag and drop the **WorldItem** GameObject into the empty slot and select from the **WorldItemBehaviour | Join World** dropdown.

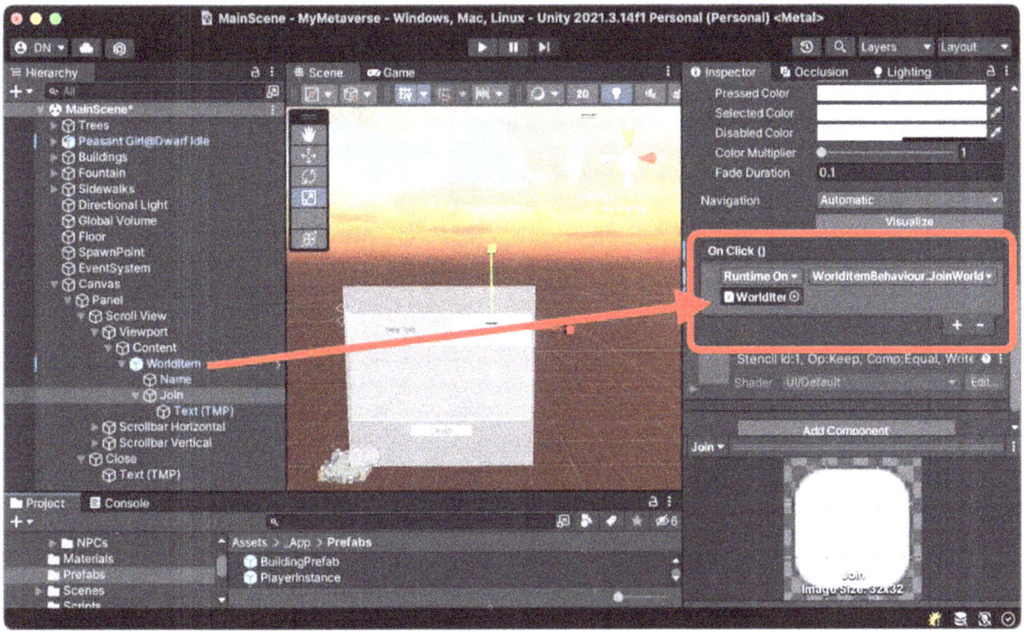

Figure 7.36 – Setting the button's OnClick event

Perfect. With this last step, we have now finished creatingthe **WorldItem** GameObject and we can convert it into a Prefab for further instantiation programmatically:

1. To turn it into a Prefab, go to **Assets** | **_App** | **Prefabs** and right-click to select **Create** | **Prefab**. Rename it `WorldItem`.

2. Drag the **WorldItem** GameObject onto the **WorldItem** Prefab, and in the confirmation popup, click the **Replace Anyway** button.

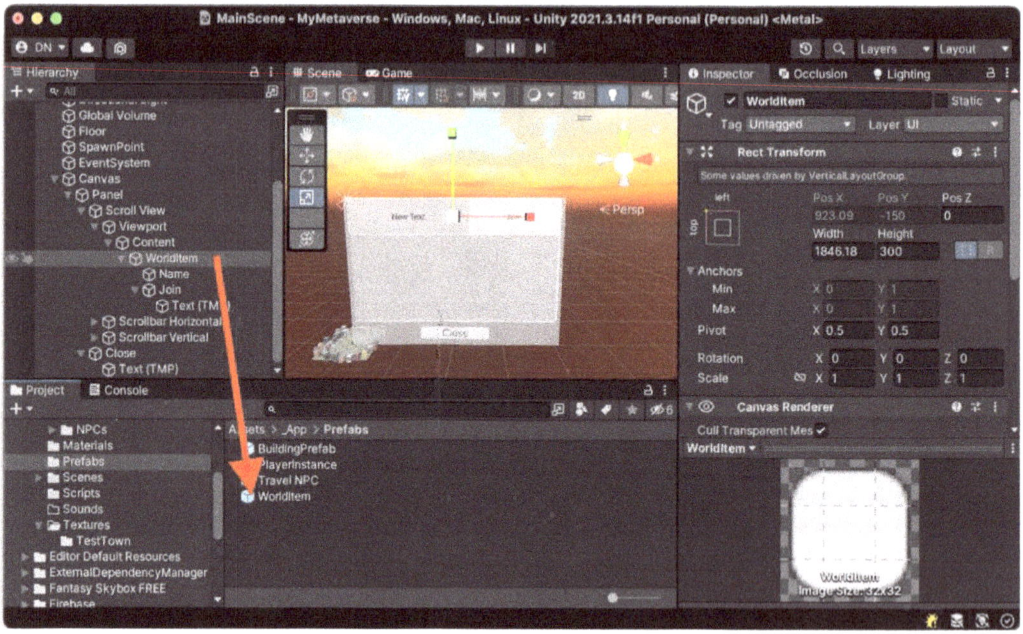

Figure 7.37 – Converting the WorldItem GameObject into a Prefab

Finally, we delete the **WorldItem** GameObject found in the **Content** GameObject. Don't forget to save the scene.

Now, we can create the final script, which will give the full functionality to the window:

1. First, we are going to need a class called `World`, which is used to convert a document from the Worlds collection into a C# class. To do this, right-click on the `Scripts` folder, located at the **Assets** | **_App** | **Scripts** path, and select **Create** | **C# Script**. Rename it to `World`, double-click on it, and replace all the auto-generated content with the following code:

```
using Firebase.Firestore;

[FirestoreData]
public class World
{
    [FirestoreProperty]
    public string Description { get; set; }

    [FirestoreProperty]
    public string Name { get; set; }
```

```
        [FirestoreProperty]
        public string Scene { get; set; }

        public World()
        {

        }
    }
```

2. Now that we have the `World` helper class, we will create the `TravelWindowBehaviour` script. Go to **Assets** | **_App** | **Script** and right-click to select **Create** | **C# Script**. Rename it `TravelWindowBehaviour`. Double-click on it to edit it.

   Replace all the auto-generated content with the following code:

```
using Firebase.Extensions;
using Firebase.Firestore;
using UnityEngine;
public class TravelWindowBehaviour : MonoBehaviour
{
    FirebaseFirestore db;
    public GameObject travelWindow;
    public GameObject scrollContent;
    public GameObject worldItemPrefab;

    void Start()
    {
        GetWorlds();
    }


    void GetWorlds()
    {
        if (db == null)
            db = FirebaseFirestore.DefaultInstance;
        CollectionReference docRef =
            db.Collection("Worlds");
        docRef.GetSnapshotAsync().ContinueWithOnMainTh
            read(task =>
        {
            if (task.IsCanceled || task.IsFaulted)
                return;
            QuerySnapshot worldsRef = task.Result;
            foreach (var worldDocument in
```

```
                    worldsRef.Documents)
                {
                    World world =
                        worldDocument.ConvertTo<World>();
                    AttachWorldItemToScroll(world);
                }
            });
        }


        void AttachWorldItemToScroll(World world)
        {
            GameObject worldItemInstance =
                Instantiate(worldItemPrefab,
                    scrollContent.transform) as
                        GameObject;
            worldItemInstance.transform.parent =
                scrollContent.transform;
            worldItemInstance.transform.localPosition =
                Vector3.down;
            worldItemInstance.Getcomponent
                <WorldItemBehaviour>().sceneName =
                    world.Scene;
            worldItemInstance.Getcomponent
                <WorldItemBehaviour>().worldName =
                    world.Name;
        }


        public void CloseWindow()
        {
            travelWindow.SetActive(false);
            Cursor.lockState = CursorLockMode.Locked;
        }
    }
```

The operation and flow of the script we have just developed is as follows:

I. When the script is instantiated in the scene, it calls the `GetWorlds` method and performs a query to get all the elements in the Firestore Worlds collection.

II. It converts the data from the document into a `World` class.

III. It executes the `AttachWorldItemToScroll` function, which instantiates the `WorldItem` Prefab in the Scroll View with the scene data and the world name.

We have created the `CloseWindow` function, which will be executed when the **Close** button is clicked on.

For the Close button to work correctly, don't forget to connect the GameObject **Canvas** to the **OnClick()** function.

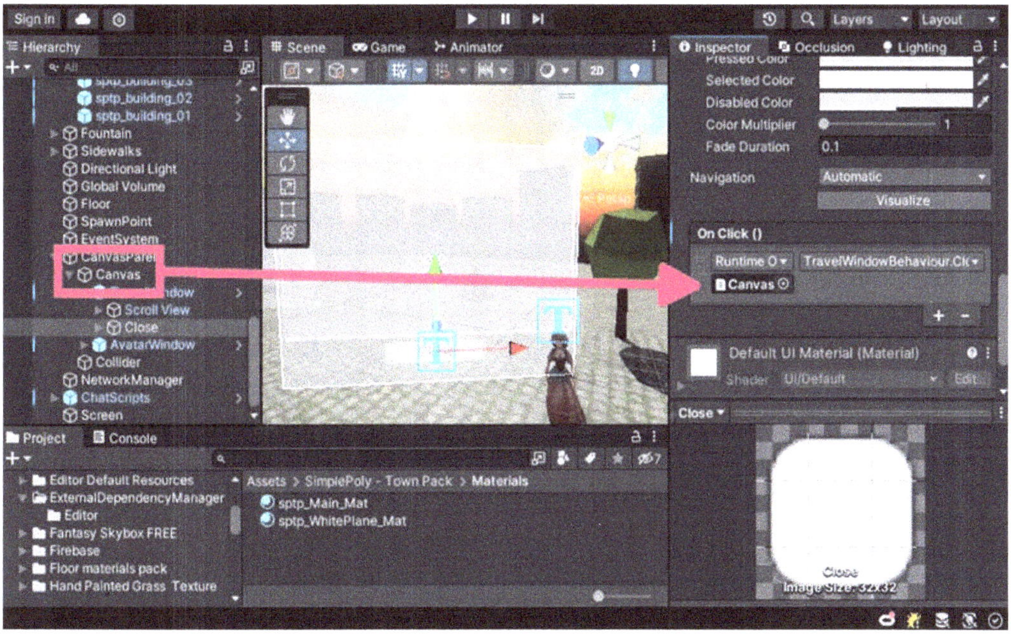

Figure 7.38 – Assigning the GameObject Canvas to the slot of OnClick()

Finally, we drag and drop the `TravelWindowBehaviour` script into the **Inspector** panel of the **Canvas** GameObject and perform the following steps to configure it correctly:

1. Drag the **Content** GameObject into the empty **Scroll Content** slot.
2. Drag the **WorldItem** Prefab found in **Assets** | **_App** | **Prefabs** into the empty **World Item Prefab** slot.
3. Drag the **Panel** GameObject into the empty **Travel Window** slot.

Figure 7.39 – Final look of the Travel Window Behaviour component

4. Well, now that we have all the functionality of the window completed, we will convert the **Panel** GameObject into a Prefab so we can easily instantiate it in other scenes. To do this, create a new Prefab called **TravelWindow** in **Assets** | **_App** | **Prefabs**.
5. Drag the **Panel** GameObject into the Prefab you just created, and finally, rename the **Panel** GameObject `TravelWindow`. Your scene will look as in the following screenshot.

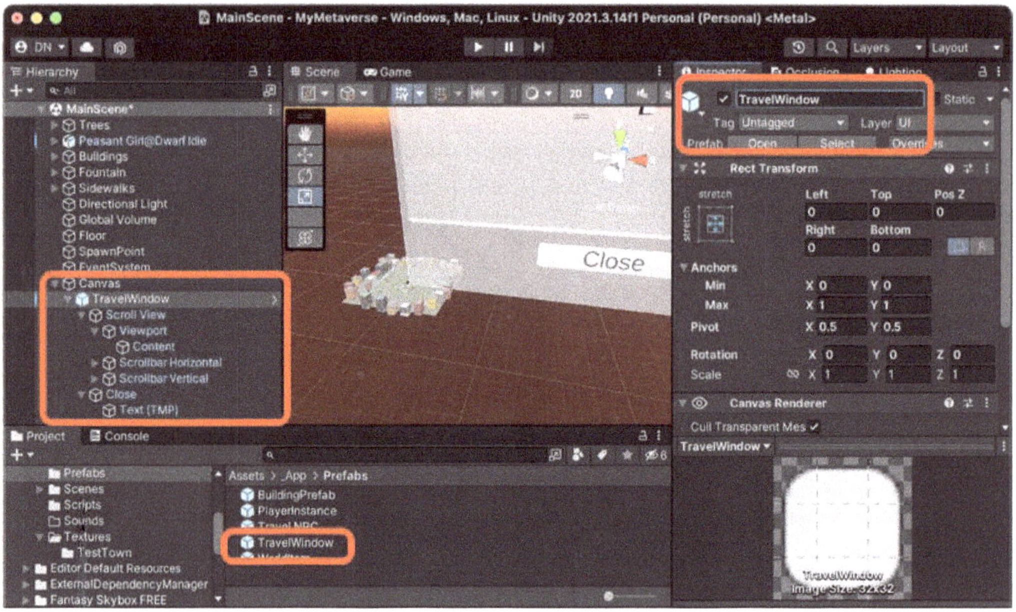

Figure 7.40 – Converting the TravelWindow GameObject into a Prefab

Good job! You have successfully completed the creation of a window that shows the available worlds to travel to, and with a button, we can change the scene.

Now, to close the loop, we need to show and hide this window depending on whether we enter or leave our NPC's area of influence.

6. To achieve this, we need to add a few lines of code to the **TravelNPCTrigger** script. Find this script by following the **Assets** | **_App** | **Script** path and double-click to edit it.

   Add the new highlighted lines of code:

```
. . . (existing code here) . . .
public class TravelNPCTrigger : MonoBehaviour
{    public GameObject travelWindow;
. . . (existing code here) . . .
}
```

The script now needs us to link the **TravelWindow** GameObject, so drag the GameObject into the empty **Travel Window** slot.

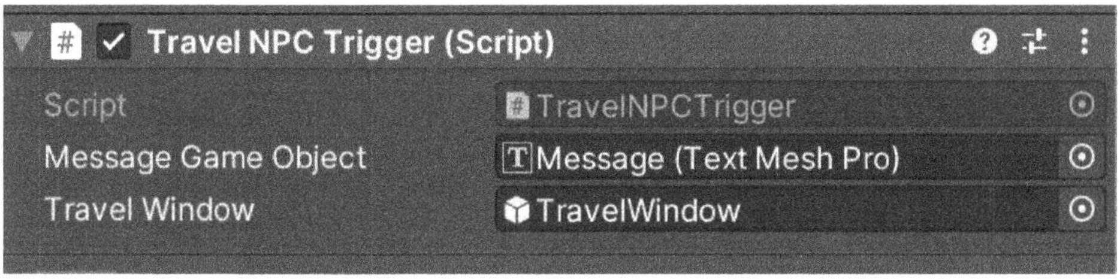

Figure 7.41 – Final look of the Travel NPC Trigger component

7. Before we continue, we need to save all the changes we have made to the NPC Prefab. To do this, select the **Peasant Girl@Dwarf Idle** Prefab found in the **Hierarchy** panel. Drag it into the **Travel NPC** Prefab found in **Assets** | _**App** | **Prefabs**.
8. In the confirmation window that appears, click on the **Replace Anyway** button, and in the second confirmation window, click on the **Original Prefab** button.

Now, everything is connected and working. If you click on the **Play** button and press the *E* key when you are near the NPC, you will see the screen we have created with the worlds that exist in the Firebase Firestore Worlds collection.

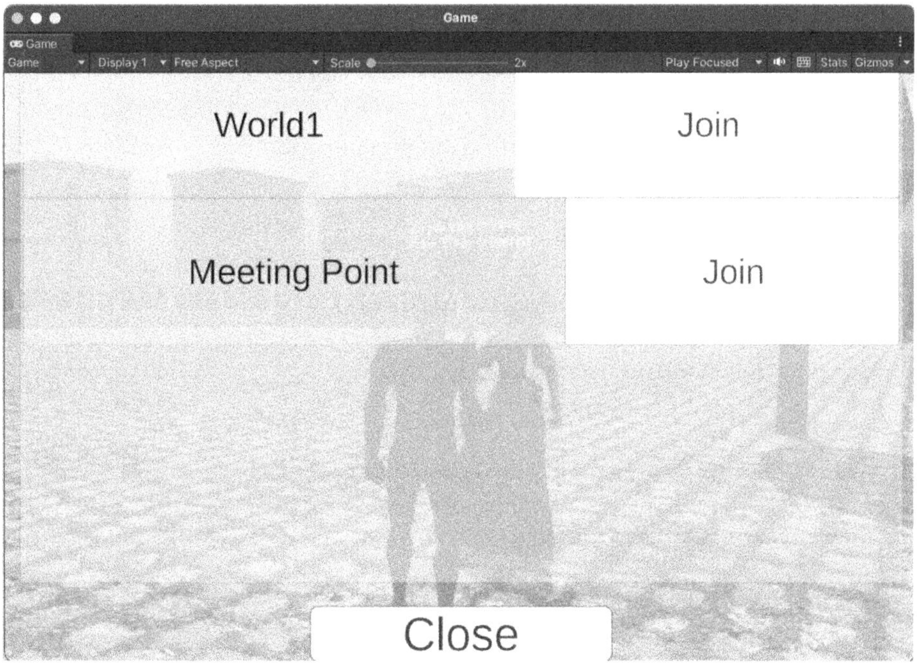

Figure 7.42 – Preview of the Travel Window operation

If you click on the **Close** button, the window will close. If you leave the area of influence as well, everything works correctly. You have done a great job; congratulations!

If you are an observant person (I'm sure you are), you will have noticed that if you click the **Join** button in **World 1**, it changes the scene – but it doesn't work entirely; the character doesn't appear. This is totally normal, as we have not configured a point of appearance in the target world. We will do so in a later section of this chapter.

Mobile support

The `TravelNPCTrigger` script we just created recognizes when the user presses the *E* key to interact, but we need to support mobile users as well. In the *Preparing our player* section of *Chapter 2, Preparing Our Player*, we included a graphical interface with touch controls for mobiles. That interface provided basic movement controls, such as walking, running and jumping. Now, we will add a new button to emulate the behavior that pressing the *E* key would have.

To add a new button to our mobile interface, follow these steps:

1. We find and edit the **StarterAssetsInputs** script located at the **Assets | StarterAssets | InputSystem** path.
2. We add only the following highlighted code snippets:

```
. . . (existing code here) . . .
namespace StarterAssets
{
public class StarterAssetsInputs : MonoBehaviour
{
. . . (existing code here) . . .
public bool interact;

. . . (existing code here) . . .

public void OnInteract(InputValue value)
{
InteractInput(value.isPressed);
}

public void InteractInput(bool newInteractState)
{
interact = newInteractState;
}
. . . (existing code here) . . .
```

OK, we have just added a new mapping to detect a new user interaction. Now, we need to link this new interaction to the code that manages the Canvas with the buttons.

3. We find and edit the **UICanvasControllerInput** script found in **Assets | StarterAssets | Mobile | Scripts | CanvasInputs**.
4. We add only the highlighted code fragment:

```
using UnityEngine;

namespace StarterAssets
{
    public class UICanvasControllerInput :
        MonoBehaviour
    {
...
        public void VirtualInteractInput(
            bool virtualInteractState)
        {
            starterAssetsInputs.InteractInput(
                virtualInteractState);
        }
...
```

Perfect; the next step will be to create a new button in the graphical interface.

5. Double-click to edit the **PlayerInstance** Prefab located in **Assets | _App | Prefabs**.
6. Let's focus on the **UI_Canvas_StarterAssetsInputs_Joysticks** GameObject. As you can see, inside it, there is a GameObject that represents each **Canvas** button. Right-click on one of them, for example, **UI_Virtual_Button_Jump**, and select **Right Click | Duplicate**. We then rename the new GameObject `UI_Virtual_Button_Interact`.

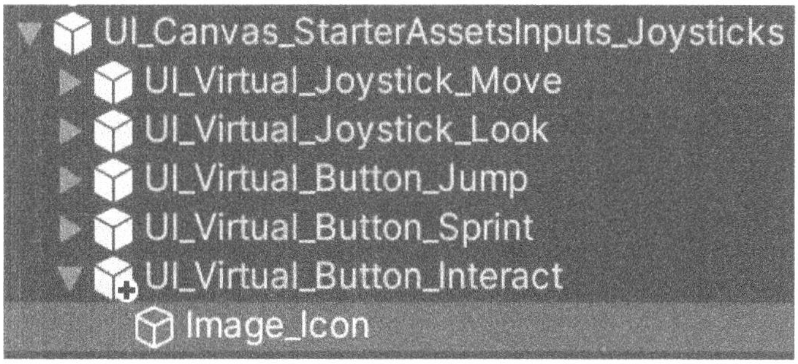

Figure 7.43 – New button added to the UI_Canvas_StarterAssetsInputs_Joysticks Prefab

7. If we deploy the **UI_Virtual_Button_Interact** GameObject, inside, we can find the GameObject that represents the icon. You can change the property to choose the icon that best represents you. In my case, I have used the following configuration:

Figure 7.44 – Image component configuration

8. Now that we have duplicated the GameObject, we must move it. Use the **Move** tool to place it where you'd like. In my case, it is as follows:

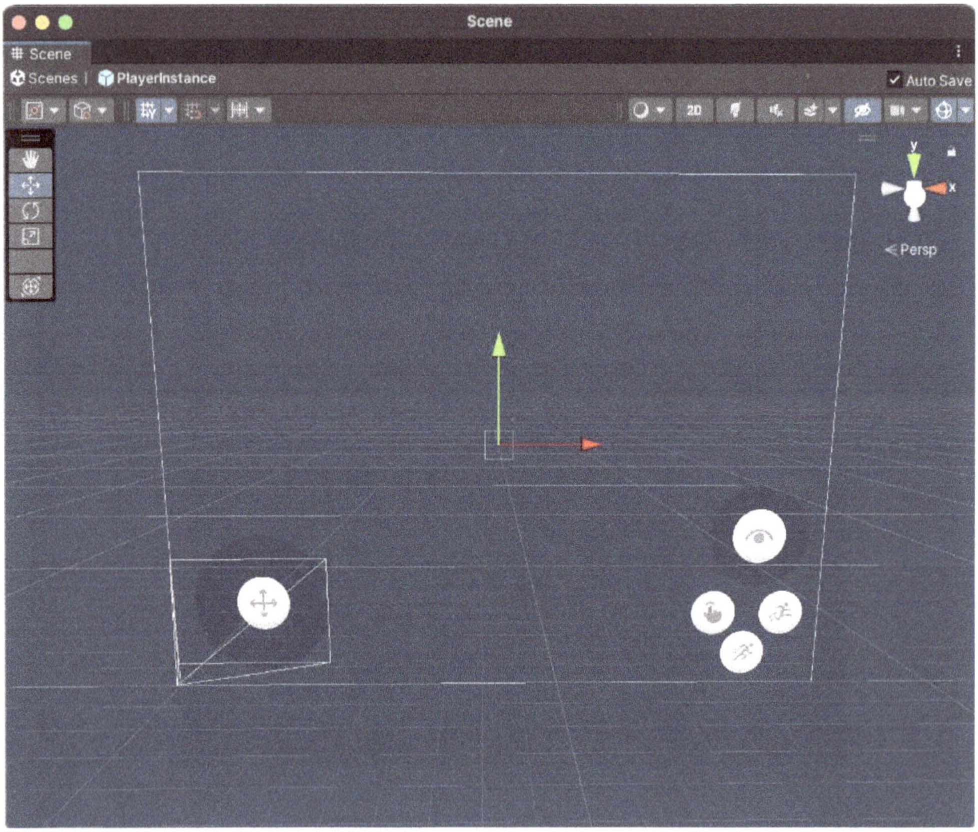

Figure 7.45 – Result after placing the new button

Good job; now you have a new button in your mobile interface. We need to do another step: link the **Touch** event to the new function we created earlier in the **StarterAssetsInputs** script.

9. Select the **UI_Virtual_Button_Interact** GameObject.
10. If we look at the **Inspector** panel, we can find a component called **UI Virtual Button**.
11. Change the **Output** property to link to the **VirtualInteractInput** function, which we created previously in the **UICanvasControllerInput** script.

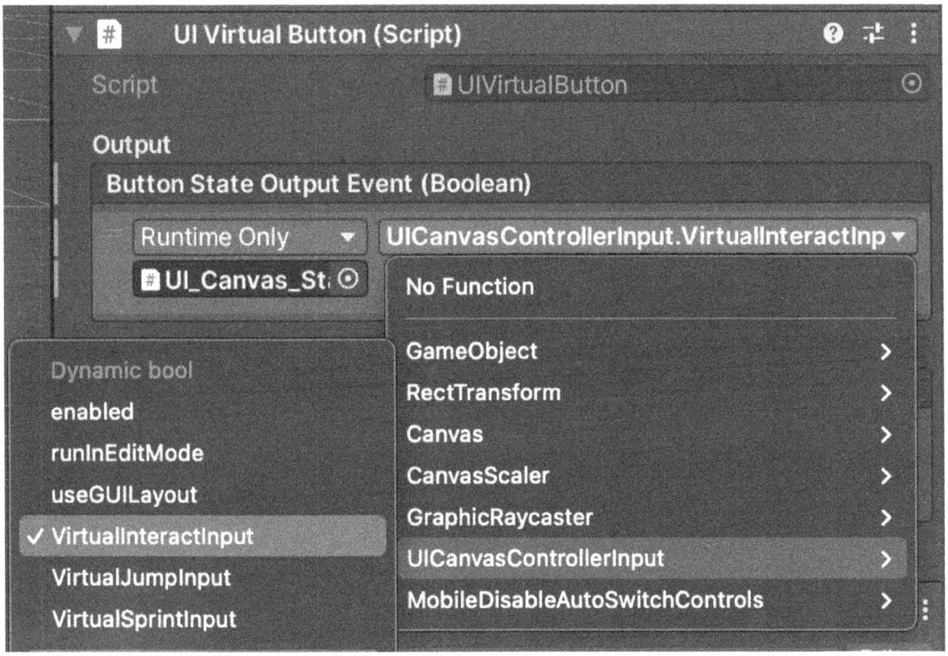

Figure 7.46 – Switching on the action of pressing the button

Now we have linked the new button with the function we previously created in the **StarterAssetsInputs** class. The next step is to create a new action that responds to the act of pressing the button. We will create what is called an action mapping.

12. Double-click on the **StarterAssets.inputactions** asset located in the **Assets** | **StarterAssets** | **InputSystem** path.
13. As you can see, there is an Action Mapping for each movement: **Jump**, **Move**, **Look**, and **Sprint**. Here, we will create a new Action Mapping called **Interact**.
14. We click the **Add New Action** button, then rename the action `Interact`. The **Action Type** property will be **Button**. Finally, we create a new **Interaction** of the **Press** type. Select the **Press Only** option in the **Trigger Behavior** property.

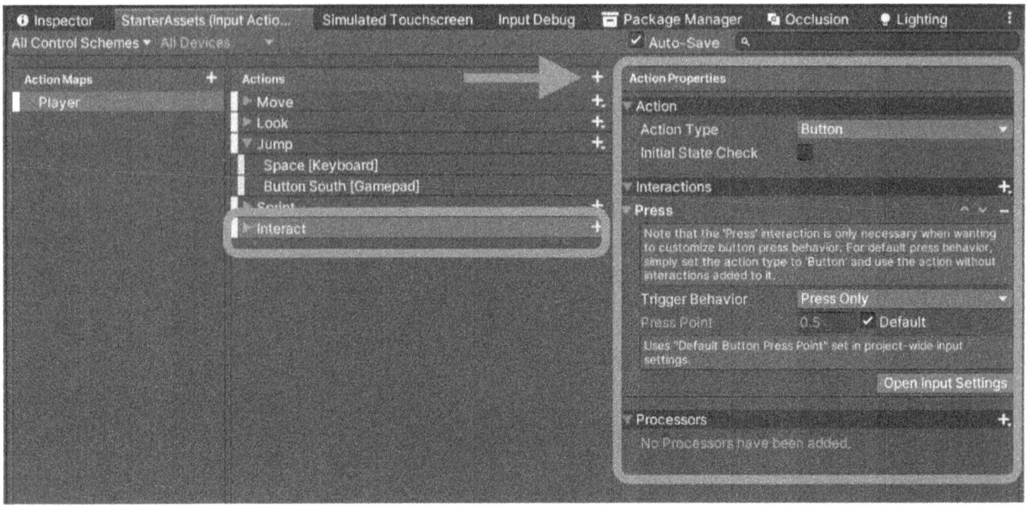

Figure 7.47 – Properties of the Interact action

15. Once the Action Mapping is created, we must configure its properties. In the **Path** property, we look for and select **Press [Touchscreen]**, which corresponds to the event of touching the mobile screen.

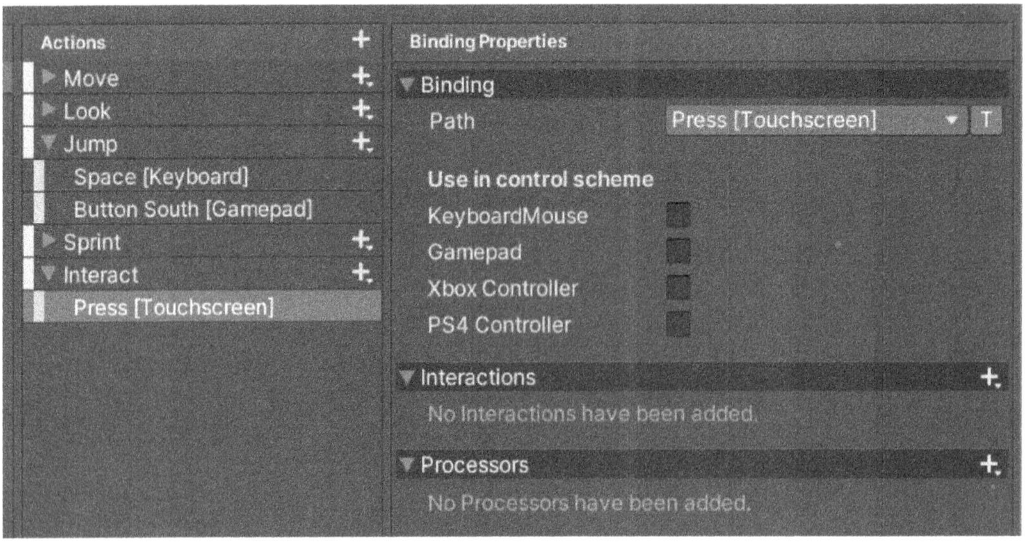

Figure 7.48 – Properties of the Press [Touchscreen] action

Good job! We already have the new Action Mapping created. Finally, we will automatically generate a class that will allow us to access and observe these Action Mappings from our scripts.

16. To do this, we look for the **StarterAssets.inputactions** asset again and select it to view its properties in the **Inspector** panel.

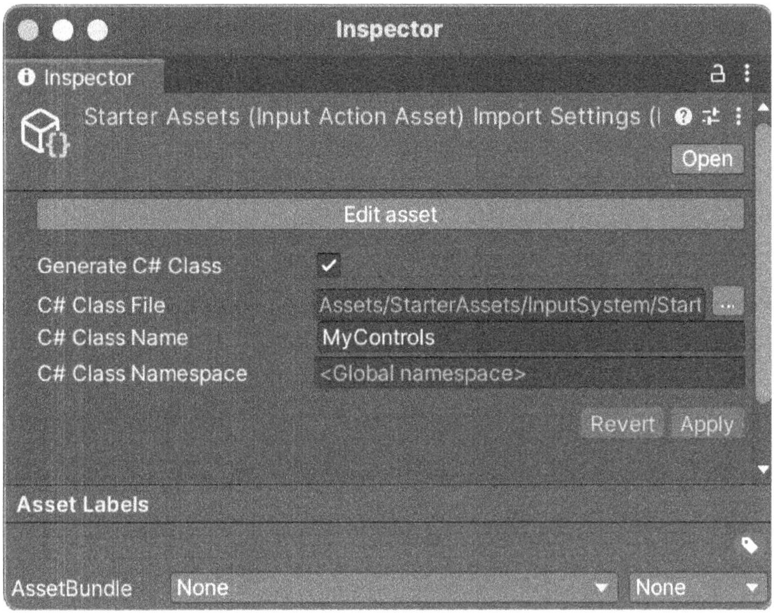

Figure 7.49 – StarterAssets.inputactions properties in the Inspector panel

17. Simply type `MyControls`, or whatever name you prefer, into the C# **Class Name** property, then check the **Generate C# Class** option, and click the **Apply** button.

    This will generate a static class called **MyControls** automatically. It will be refreshed every time we make a change or add or remove an Action Mapping. This way, we can, in our scripts, check whether the user has clicked any of our actions.

Well, we have now configured our Canvas and its new button to interact. We can update our `TravelNPCTrigger` script to add the mobile device support we have configured.

To do this, double-click on the `TravelNPCTrigger` script located in the **Assets** | **_App** | **Scripts** path and add the highlighted code blocks:

```
. . . (existing code here) . . .
using static MyControls;
using static UnityEngine.InputSystem.InputAction;

public class TravelNPCTrigger : MonoBehaviour, IPlayerActions
{
...
    MyControls controls;
```

```
    private void Awake()
    {       // We link the Interact action and enable it
               for detection in the code.
        if (controls == null)
        {
            controls = new MyControls();
            // Tell the "gameplay" action map that we want
               to get told about
            // when actions get triggered.
            controls.Player.SetCallbacks(this);
        }
        controls.Player.Enable();...
    }
. . . (existing code here) . . .
    public void OnMove(CallbackContext context)
    {
    }

    public void OnLook(CallbackContext context)
    {
    }

    public void OnJump(CallbackContext context)
    {
    }

    public void OnSprint(CallbackContext context)
    {
    }

    public void OnInteract(CallbackContext context)
    {
        if (context.action.triggered && canInteract)
            travelWindow.SetActive(true);
    }
}
```

As you can see, we have made reference to the `MyControls` class, which Unity created automatically for us, and which allows us to *listen* for user actions.

For each Action Mapping, a function is created. In this case, we have `OnMove`, `OnLook`, `OnJump`, and `OnSprint`, without functionality, as we don't need any special action when any of these events occur, but Unity forces us to have them in our script as we have inherited them from the `IPlayerActions` interface.

Finally, we have the `OnInteract` function, which is where we have applied the same logic as pressing the *E* key.

Good job! You have completed the support for mobile devices. Next, we create the functionality to be able to travel to the selected world from our menu.

Traveling to the selected world

As previously mentioned, if we travel to **World 1**, the player does not appear. This is normal. We must bear in mind that every time we create a new navigable world, we must have the same world configuration in the destination world as the one we traveled to in order to be able to *land* properly.

In this case, in **World 1**, we don't have the **SpawnPoint** Prefab we created in the preceding chapter in the scene, so the player doesn't find a place to appear. We also don't have our NPC there to travel to, so if we manage to travel there, we have no way to go back or continue navigating to other worlds.

In the following section, we will correct this problem. We will create the interface with a Canvas to allow the user to interact with the NPC again.

Canvas

To fix this, we will follow these steps and you will need to repeat them for every new scene you build in the future, as long as you want to allow travel there:

1. Open the **World 1** scene located in **Assets** | **_App** | **Scenes**.
2. Add a Canvas using the Unity **GameObject** | **UI** | **Canvas** main menu bar option.
3. Apply the same settings to the **Canvas Scaler** component that we have previously used in other scenes.

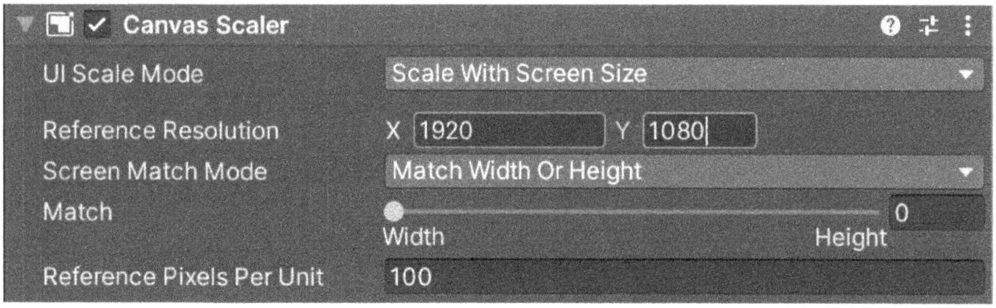

Figure 7.50 – Properties of the Canvas Scaler component

4. Drag the **TravelWindow** Prefab found in **Assets** | **_App** | **Prefabs** into the **Canvas** GameObject.
5. Add a new component to the Canvas with the **TravelWindowBehaviour** script and assign the **TravelWindow** Prefab found inside the **Canvas** GameObject to the empty **Travel Window** slot.
6. Add the **Content** GameObject to the empty slot called **Scroll Content**.
7. Add the **WorldItem** Prefab located in **Assets** | **_App** | **Prefabs** to the empty slot named **World Item Prefab**.

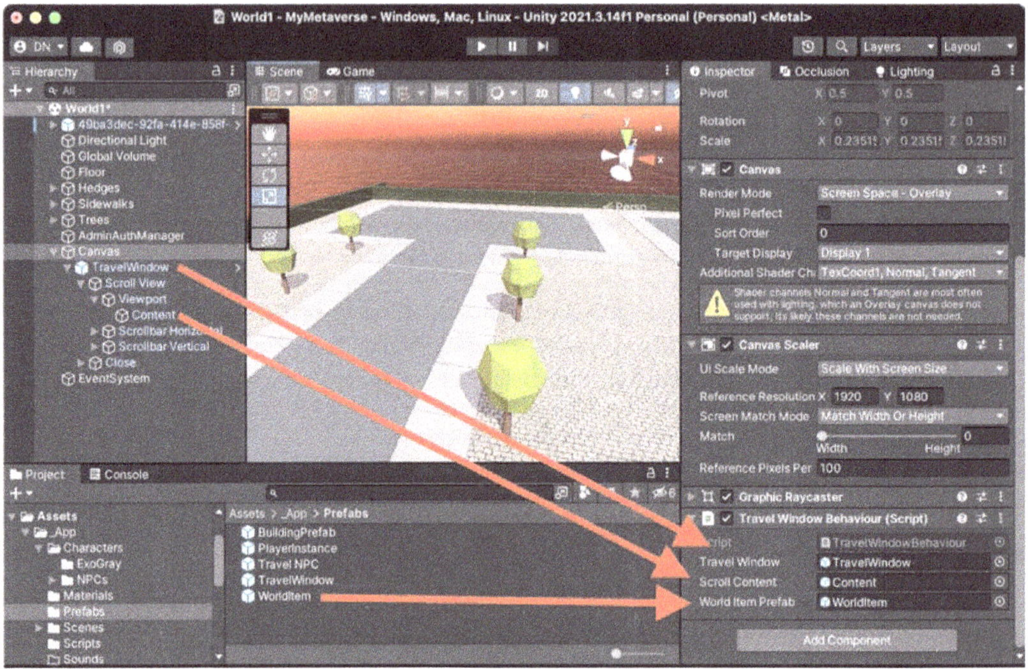

Figure 7.51 – Finalising the configuration of the Travel Window Behaviour component

Great, now we have our Canvas configured with the **World Selector** window. The next step will be to add our NPC to enable travel.

NPC

To add the NPC to the scene, simply drag the **Travel NPC** Prefab located in the **Assets** | **_App** | **Prefabs** path to any location in the scene you like.

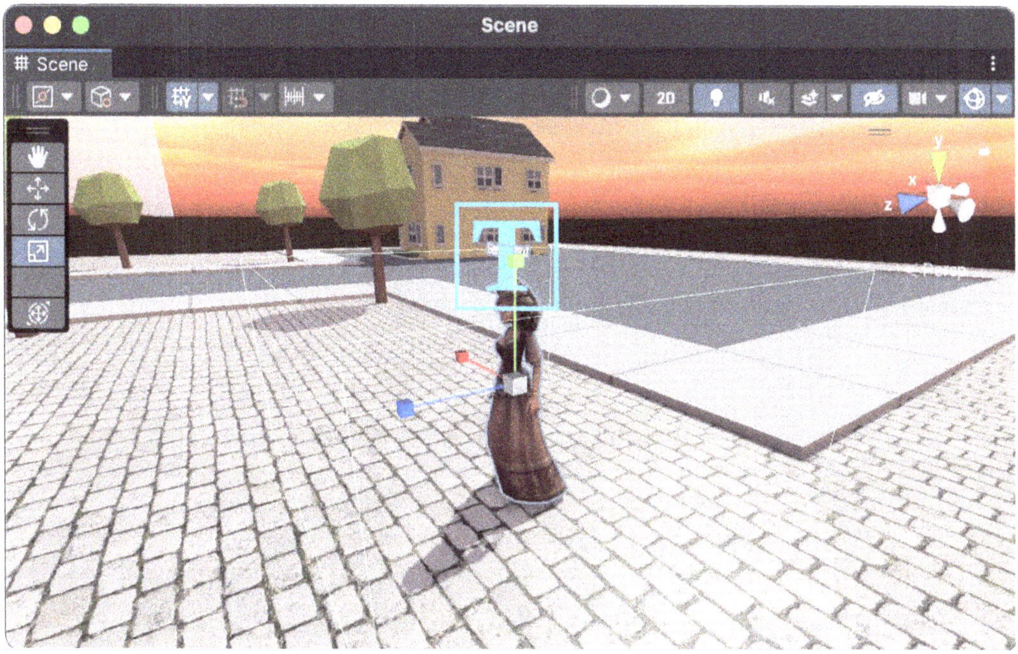

Figure 7.52 – Placing the NPC Prefab in the scene

Remember that the NPC has inside it a GameObject called **Influence** whose **Travel NPC Trigger** script needs the **TravelWindow** GameObject for it to work properly.

To correct this, drag the **TravelWindow** GameObject into the empty **Travel Window** slot.

Figure 7.53 – Final look of the Travel NPC Trigger component

With this last step, the NPC is fully functional. Finally, we will configure the **PlayerInstance** Prefab so that the player has an entry point into this world.

PlayerInstance

With this last Prefab, our **World 1** scene will be fully operational for traveling. Just drag the **PlayerInstance** Prefab found in **Assets** | **_App** | **Prefab** to any place in the scene, the one you like the most.

Great, how about trying it out?

Whether you press the **Play** button in this **World 1** scene or in the **MainScene**, you can navigate to other worlds by visiting our NPC. Impressive, isn't it? With this last step, we have reached the end of this chapter. I hope you have learned a lot and had as much fun as I did.

Summary

Throughout this chapter, we learned how to create our first NPC, a nice, animated character that will help us in our Metaverse to travel to other worlds. Our NPC has an area of influence, which we programmed to detect when the player enters or leaves. We also learned about a new Unity UI feature, Scroll View, which we linked with a query to the Firestore database to dump the Worlds collection into a scrolling view. Finally, the user can select any of these worlds to travel to another scene.

In the next chapter, we will learn how to program scripts that allow a user to get a house in ownership, link the house to the player's owner, and finally, persist the data in the Firestore database.

8
Acquiring a House

During this chapter, we will enable our metaverse to allow its users to settle in our world by acquiring a home. To do so, we will provide the Prefab of the houses we saw during *Chapter 5, Preparing a New World for Travel*, with a script that will allow a player to interact.

During this acquisition process, we will first consult the database, then, if the user has been able to obtain ownership of the building, we will update the building document in Firestore, so that it links the user's ownership information.

In this chapter, we will cover the following topics:

- The concept of home
- Enabling a user to obtain a home
- Linking a house to a player
- Converting a house to a **non-fungible token** (**NFT**)

Technical requirements

This chapter does not require any special technical requirements, but as we will start programming scripts in C#, it would be advisable to have basic notions in this programming language. We need an internet connection to browse and download an asset from Unity Asset Store. We will continue on the project we created during *Chapter 1, Getting Started with Unity and Firebase*.

Remember that we have the GitHub repository (`https://github.com/PacktPublishing/Build-Your-Own-Metaverse-with-Unity/tree/main/UnityProject`), which will contain the complete project that we will work on here.

You can also find the complete code for this chapter on GitHub at: `https://github.com/PacktPublishing/Build-Your-Own-Metaverse-with-Unity/tree/main/Chapter08`

The concept of home

Home, abode, domicile, dwelling... we all know what the terms mean, but what about in the metaverse?

The metaverse can be a reflection of reality, where we try to emulate actions that we usually do in our lives. One of the things we can do is to get a house to make it our home. If you remember in *Chapter 5, Preparing a New World for Travel*, we managed to generate houses in the scene, and they had a link – a reference in Firebase Firestore. In this chapter, we will go a step further and allow our users to own those houses.

In an ideal metaverse, a user can own a house, allow friends to enter, have a party, decorate it, and meet friends to talk – sounds fun, doesn't it? In reality, these action involves more things – for example, when a user owns a house or has permissions on other houses, we want them to be able to enter them.

The action of entering a house may sound like a very complex thing to do, but it boils down to a collision trigger. This trigger will listen if we press a specific key (as we did with the **Non-Player Character** (**NPC**) in the previous chapter), and when we press it, we redirect the trigger to a specific function, and then we change the scene that the user is in. That scene will be the inside of the house.

Impressive, isn't it? Let's start with the process that will allow the user to obtain free housing.

Enabling a user to obtain a home

The process of making a user the owner of a property is very simple, partly because we have already acquired knowledge that will provide the fundamentals to do so.

We will divide the objective into the following sections:

- Adding a trigger to the Prefab of a house
- Programming the script that allows us to acquire the house
- Linking the house to a player

To start with this task, we will make modifications on the Prefabs that we have used to build the houses in the World1 scene.

Adding a trigger to the Prefab of the house

If you remember from *Chapter 5, Preparing a New World for Travel*, we added two Prefabs that represent houses in a scene; these Prefabs are located in the `Resources` folder:

1. Go to the **Assets** | **Resources** | **Prefabs** | **Buildings** path and double-click on the **Suburb House Grey** Prefab to edit it.

2. Once opened, we will create a Cube and disable the **Mesh Renderer** component; we will only use its Collider to create the interaction zone with the player. You should see something similar to the following screenshot:

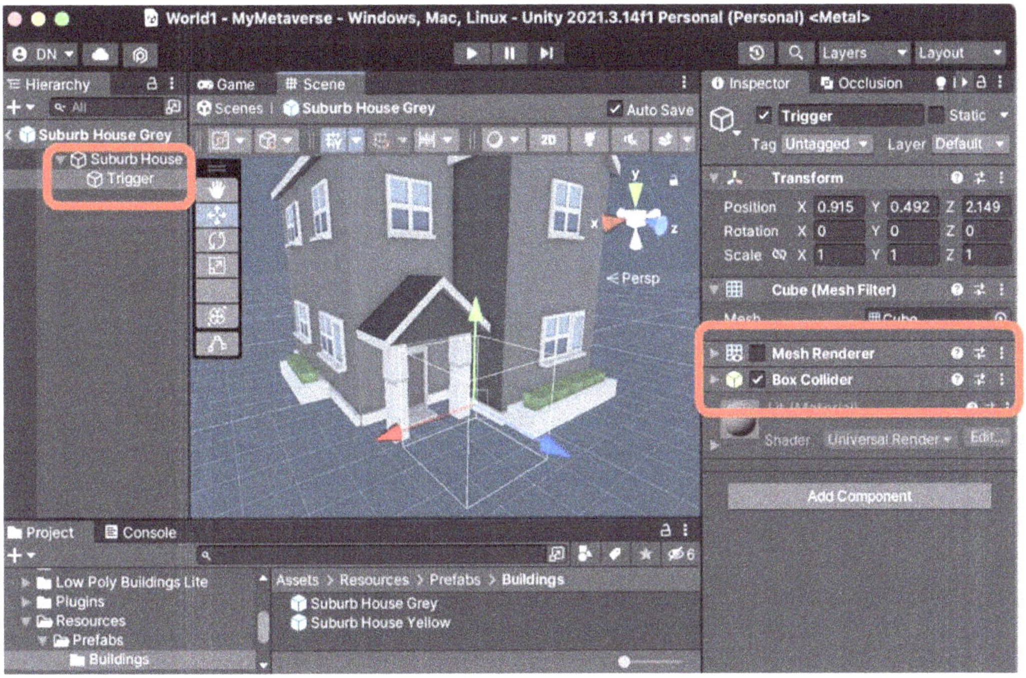

Figure 8.1 – Adding a Cube as an interaction zone

As you can see, I have renamed the Cube `Trigger`, making it more indicative of its functionality. Now, it's time to program; we will create a script that detects when a user comes in contact with the GameObject trigger.

Programming script to acquire a house

To create a new script, right-click on the **Assets** | **_App** | **Scripts** path, select the **Create** | **C# Script** option, and rename it `BuildingDoorTrigger`. Double-click to edit it.

Replace all the auto-generated code with the following:

```
using UnityEngine;
using static MyControls;
using static UnityEngine.InputSystem.InputAction;

public class BuildingDoorTrigger : MonoBehaviour,
```

```
    IPlayerActions
{
    MyControls controls;
    private void Awake()
    {
        if (controls == null)
        {
            controls = new MyControls();
            controls.Player.SetCallbacks(this);
        }
        controls.Player.Enable();
    }

    public void OnMove(CallbackContext context)
    {
    }

    public void OnLook(CallbackContext context)
    {
    }

    public void OnJump(CallbackContext context)
    {
    }

    public void OnSprint(CallbackContext context)
    {
    }

    public void OnInteract(CallbackContext context)
    {
    }
}
```

Great! We have already created the base class. Now, we will add the trigger logic to detect when a user enters and exits the collider. To do so, we will add the following highlighted code sentences to the `BuildingDoorTrigger` class that we created:

```
... (existing code here) ...
public class BuildingDoorTrigger : MonoBehaviour,
    IPlayerActions
{
... (existing code here) ...
    // If the Script is in a GameObject that has a
```

```
        Colllider with the Is Trigger property enabled, it
        will call this function when another GameObject
        comes into contact.
    private void OnTriggerEnter(Collider other)
    {
        // The player's Prefab comes with a default tag
           called Player; this is an excellent way to
           identify that it is a player and not another
           object.
        if (other.gameObject.tag == "Player")
        {
        }
    }

    // When the player leaves the area of influence, we
       will put a blank text back in.
    private void OnTriggerExit(Collider other)
    {
        if (other.gameObject.tag == "Player")
        {
        }
    }
  ... (existing code here) ...
}
```

The `OnTriggerEnter` and `OnTriggerExit` functions are key to implementing the logic of our script. These functions are responsible for knowing whether a user is within an area of influence or has left it. What we will program next is a query to Firestore that allows us to obtain information about the building that we interact with.

We want to know whether it is free to obtain it, or whether it is owned by another user. To do this, we will make further modifications to the script to add the connection to our database in Firestore. Simply add the following highlighted code blocks to the `BuildingDoorTrigger` script:

```
... (existing code here) ...
        if (db == null)
            db = FirebaseFirestore.DefaultInstance;
        if (auth == null)
            auth = FirebaseAuth.DefaultInstance;
... (existing code here) ...
    }

    private void OnTriggerEnter(Collider other)
    {
```

```
            if (other.gameObject.tag == "Player")
                GetBuildingInfo();
        }

        private void OnTriggerExit(Collider other)
        {
            if (other.gameObject.tag == "Player")
            {
                displayMessage = string.Empty;
                canInteract = false;
            }
        }

        void OnGUI()
        {
            GUI.Label(new Rect(Screen.width / 2,
                Screen.height / 2, 200f, 200f),
                    displayMessage);
        }
    ... (existing code here) ...
                else
                {
                    displayMessage = BUSY_BUILDING_TEXT;
                    canInteract = false;
                }
            });
        }
    }
```

As you can see, we have introduced a new tool in this code, **Graphical User Interface** (**GUI**), which is a native Unity class that allows you to display graphical elements on screen, such as text, buttons, shapes, and textures.

It has less customization power and more restrictions than the Canvas system we used previously, but it is still very useful when you want to display information on screen.

> **Tip: a comparison of UI systems in Unity**
>
> If you want to learn more about the differences between the two user interface systems, I recommend that you have a look at the official Unity documentation by navigating to the following web address: `https://docs.unity3d.com/2020.2/Documentation/Manual/UI-system-compare.html`.

As you can see, we launched a query to the **Buildings** collection of Firestore; we query whether the building has an owner or not, and we notify the player the status through a message on the screen with the GUI tool.

Next, we will learn about the missing functionality in the script, the possibility for a user to acquire a building, which we can link to the user in the Firestore database.

Linking a house to a player

Once we have finished the script that launches a query to the **Buildings** collection of our database in Firestore, to obtain the state of the building that we interact with, we can proceed to program the functionality to obtain it in property.

In order for a player to obtain the building, we must program in the `BuildingDoorTrigger` class the ability to detect when the user presses a key within the area of influence.

To do this, we will make a small modification to the `BuildingDoorTrigger` script to add the highlighted code block:

```
... (existing code here) ...
public class BuildingDoorTrigger : MonoBehaviour,
    IPlayerActions
{
    ... (existing code here) ...
    void Update()
    {
        // We check if the user has pressed the C key and
           is also inside the collider.
        if (Input.GetKeyDown(KeyCode.C) && canInteract)
        {
        }
    }

    public void OnInteract(CallbackContext context)
    {
        if (context.action.triggered && canInteract)
        {
        }
    }
    ... (existing code here) ...
}
```

For desktop users, we will implement the detection of the *C* key when it is pressed. For mobile users, we will use the `OnInteract` function that we implemented in *Chapter 2, Preparing Our Player*.

Finally, we will add the connection to Firebase that will perform the necessary update on the user's document for the binding. To do this, we will insert the highlighted code snippets into the `BuildingDoorTrigger` class:

```
... (existing code here) ...
public class BuildingDoorTrigger : MonoBehaviour,
    IPlayerActions
{
... (existing code here) ...
    void Update()
    {
        if (Input.GetKeyDown(KeyCode.C) && canInteract)
        {
            UpdateBuildingOwner();
        }
    }

    public void UpdateBuildingOwner()
    {
        if (building == null) return;
        DocumentReference docRef =
          db.Collection("Buildings").Document(building.Id);
        Dictionary<string, object> updatedProperties = new
          Dictionary<string, object>();
        updatedProperties["OwnerUserId"] =
          auth?.CurrentUser.UserId;

        docRef.UpdateAsync(updatedProperties)
            .ContinueWith(task =>
        {
            if (task.IsCanceled || task.IsFaulted) return;
            displayMessage = CONFIRMATION_TEXT;
            canInteract = false;
        });
    }
```

```
    public void OnInteract(CallbackContext context)
    {
        if (context.action.triggered && canInteract)
        {
            UpdateBuildingOwner();
        }
    }
    ... (existing code here) ...
}
```

Fantastic! As you can see, the `UpdateBuildingOwner` function will take care of updating the `OwnerUserId` property of the document representing the building with the player's `Id`.

What we have managed to solve with our `BuildingDoorTrigger` class is summarized in the following steps:

1. We store in different constants the texts that we will show to the user.
2. When the player enters the trigger, we call the function that gets the updated data of the building in real time in Firestore.
3. If the building has an empty `OwnerUserId` variable, it means that you can buy the building.
4. If the building has the same user ID, it means that we already own this building.
5. If it does not meet the preceding conditions, the building belongs to another player.
6. If we can acquire the building, we allow the player to do so by pressing the *C* key, and we immediately update the building in the database with our user ID.

Easy, isn't it? Now, we must add this script as a new component to the GameObject trigger and activate the **Is Trigger** property of the Collider; otherwise, it won't work.

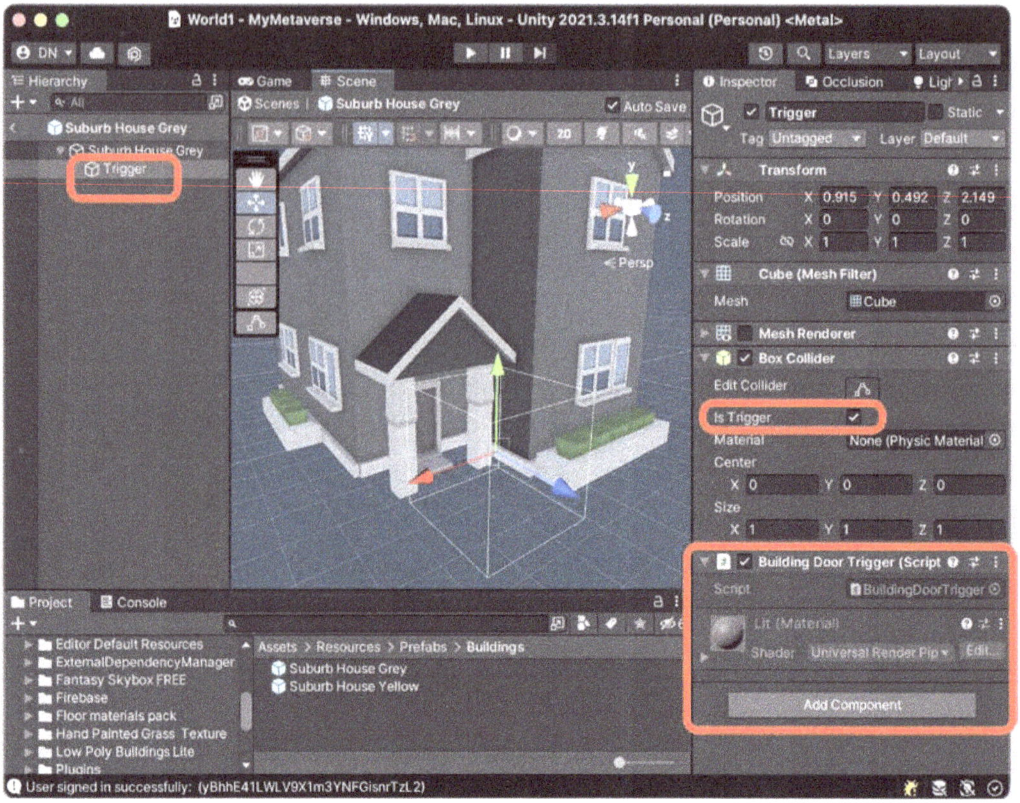

Figure 8.2 – Adding the Building Door Trigger script as a new Component

Great! Now, you just have to repeat the same steps on the other building type, the Prefab called **Suburb House Yellow**, which is also located in **Assets** | **Resources** | **Prefabs** | **Buildings**, and you will have completed the process.

Now, it's time to test the script; to do so, start the **Login** scene, **MainScene**, or directly in the scene where the buildings are located, **World1**. Navigate the character to the door of the house and see how it works. Impressive, isn't it?

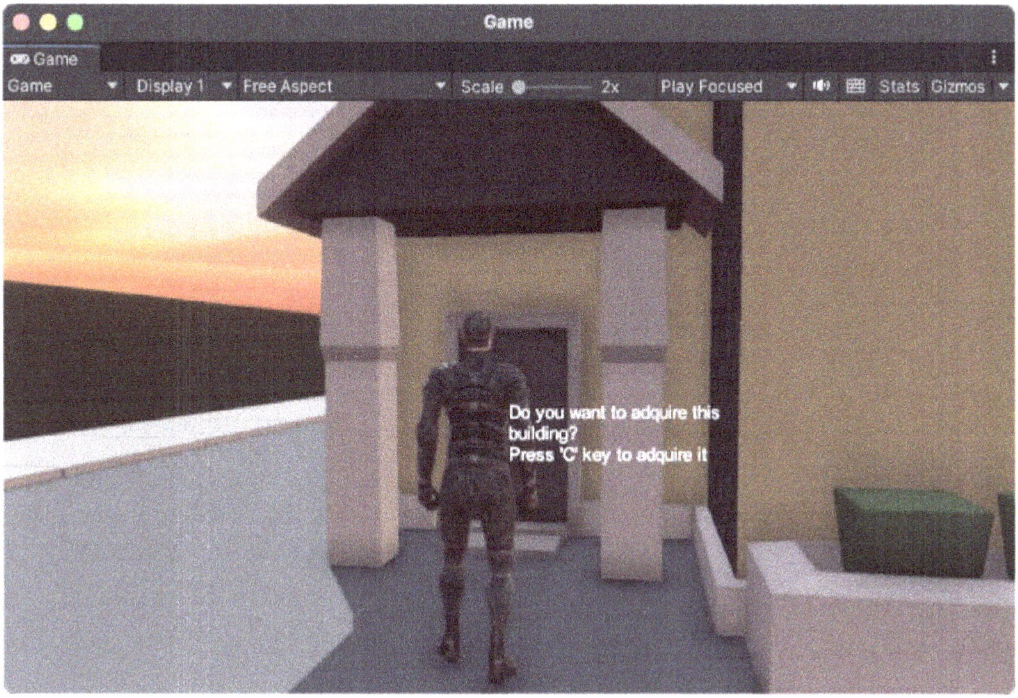

Figure 8.3 – Showing in real time the functioning of the script

Good work! Your metaverse has acquired a new functionality, and players are now able to obtain a house. The following are some ideas that you can implement in the future with the knowledge you have gained in the current playthrough of the book:

- You can extend the `BuildingDoorTrigger` script to allow a user to enter the house; the inside of the house would be a new scene.
- You can dynamically load objects inside the house, with the same logic we used in *Chapter 5*, *Preparing a New World for Travel*.
- You can search for free (or priced) assets in Unity Asset Store that represent the interior of a house and make this the target scene when you enter – for example, `https://assetstore.unity.com/packages/3d/props/interior/free-house-interior-223416`.

Fantastic! As you can see, the `UpdateBuildingOwner` function will take care of updating the `OwnerUserId` property of the document that represents the building with the player's ID.

What we have managed to solve with our `BuildingDoorTrigger` class is summarized in the following steps:

1. We make sure that we have the Firestore SDK and Firebase Authentication properly referenced.
2. We detect whether the user has entered the collider's zone of influence.
3. If the player is within the zone of influence, we display a message inviting them to interact; otherwise, we display a message with the status of the building.
4. If the player presses the *C* key within the area of influence, we proceed to update the **Buildings** collection in Firestore with the player's ID.

Now that we have learned how to allow a user to interact with a home and be able to purchase it, let's go a step further. We will learn how to generate an NFT from our code that is related to the building in some way.

There are endless possibilities when creating an NFT related to your building – for example, one that contains the ID of the building, a JSON representing the hierarchy of the GameObject, or text with a description. We will didactically create an NFT from a screenshot.

Understanding what an NFT is

Non-Fungible Token (**NFT**) technology is a type of unique and indivisible digital token used to represent the ownership or authenticity of a specific digital asset. Unlike cryptocurrencies such as Bitcoin or Ethereum, which are fungible (interchangeable with each other), NFTs are unique and cannot be replaced by other tokens.

In this book, we will discuss the creation of NFTs in a didactic and testing environment, so there will be no real monetary cost.

> **Note: NFTs in your project**
>
> The world of NFTs is very abstract; what you want to turn into an NFT in your project depends entirely and only on the goal you want to achieve. In this book, we will deal with the creation of NFT in a didactic way, with simple examples that will lay the foundations for what you want to build in the future. There are infinite possibilities in terms of creating an NFT; you can create it from a photo, a text, a file, and so on. We will learn in the following sections how to create an NFT using a screenshot of what a player is watching at that instant, which will be our NFT, but going forward, you can extend or modify the code so that the created NFT adapts to your needs.

With this short introduction to the NFT concept, we can learn about what metaverses currently exist with NFT implementation and what other related concepts need to be understood first.

Converting a house to an NFT

The term *NFT* usually relates to the world of metaverses, as if it were an indispensable part of them. You will rarely see a metaverse that does not include this technology. For example, the following metaverses include it:

- **Decentraland**: This is an Ethereum-based metaverse where users can buy, sell, and own virtual land represented as an NFT. In addition, users can create and trade unique digital assets, such as art, avatars, and 3D objects.
- **Cryptovoxels**: This is another Ethereum-based metaverse that uses NFTs to represent land parcels, avatars, wearables, and art. Users can explore the virtual world, socialize, and build on their land plots.
- **The Sandbox**: This is a metaverse powered by the Ethereum blockchain that allows users to create, buy, and sell digital assets using NFT technology. Players can build and customize their own virtual space, as well as participate in community-created games and experiences.
- **Somnium Space**: This is a metaverse where users can buy and sell virtual land using an NFT, based on the Ethereum blockchain. Users can build, explore, and socialize in an ever-expanding virtual world.

That is why during this chapter we will learn the basic pillars of this technology and how to implement its creation from our code. Before continuing, we must define some terms that we will use throughout these pages.

IPFS

In the world of NFTs, an **InterPlanetary FileSystem** (**IPFS**) is a decentralized storage and distribution protocol used to host and access files associated with NFTs. IPFS uses a peer-to-peer network to efficiently store and retrieve data in a censorship-resistant manner.

When an NFT is created, the associated digital file, such as an image, video, music, or other content, is uploaded to the IPFS network. Instead of being stored on a centralized server, the file is divided into fragments and distributed to different nodes in the IPFS network. Each fragment is identified by a unique cryptographic hash.

Goerli

Goerli is a testnet in the cryptocurrency world, specifically designed for Ethereum. Testnets, such as Goerli, allow developers and users to test and experiment with applications and smart contracts without using real tokens or interacting with Ethereum Mainnet.

Goerli is one of several testnet available for Ethereum and is characterized by its compatibility with the Ethereum 2.0 specification. It was launched in January 2019 and uses a **proof-of-authority** (**PoA**) consensus algorithm called **Clique**. Instead of relying on **proof-of-work** (**PoW**) mining, in Goerli, validators are selected by predefined authorities involved in validating transactions.

The Goerli network is primarily used by developers, enterprises, and users interested in testing and debugging **decentralized applications** (**dApps**), smart contracts, and other Ethereum-related functionality. It provides a secure and isolated environment where testing can be performed without the risk of affecting the main network or real funds.

The tokens used in Goerli have no real economic value, as they are obtained for free through faucets. These tokens are used to simulate transactions and perform functionality tests.

ChainSafe

ChainSafe is a company that specializes in the development of blockchain technologies and decentralized software solutions. It offers a wide range of blockchain-related tools and services, including infrastructure development, the creation of **Decentralized Applications** (**dApps**), and the implementation of smart contracts.

In our project, ChainSafe can benefit us in the following ways:

- **Blockchain integration**: If you want to incorporate blockchain functionalities into your Unity3D project, ChainSafe can help you integrate a blockchain network into your game or application. This will allow you to take advantage of the benefits of blockchain technology, such as digital asset ownership, transaction tracking, and decentralized security.
- **Smart contract development**: ChainSafe has expertise in developing smart contracts, which are self-executing, self-contained computer programs that run on a blockchain network. It can help you design and develop custom smart contracts for your Unity3D project, allowing you to implement business logic and specific rules in your decentralized application.
- **Consulting and advisory**: ChainSafe offers blockchain consulting and advisory services. It can provide you with strategic guidance on how to leverage blockchain technology most effectively in your Unity3D project. This includes designing blockchain architectures, selecting the right blockchain network, and implementation planning.
- **Security testing and audits**: ChainSafe also performs security testing and smart contract audits to ensure that your Unity3D project meets security and quality standards. This is especially important in decentralized environments, where the security and protection of digital assets are critical.

The creation of NFT is a complex matter, even more so when looking for a provider that has an SDK for Unity. Of all the existing options on the market, ChainSafe is undoubtedly the easiest way to implement the creation of NFTs.

With just a few lines of code, thanks to its Unity-compatible SDK, we will be able to control the creation of NFTs.

Getting started with ChainSafe

In order to use ChainSafe's technology, we must first create an account by following these steps:

1. Go to the website `https://dashboard.gaming.chainsafe.io`.
2. Enter your email; you will receive a code that you must paste into the form. Once the process is completed, you will access the ChainSafe control panel.
3. Once the registration is completed, you will have access to the control panel; now, you can click on the **Create new project** button.

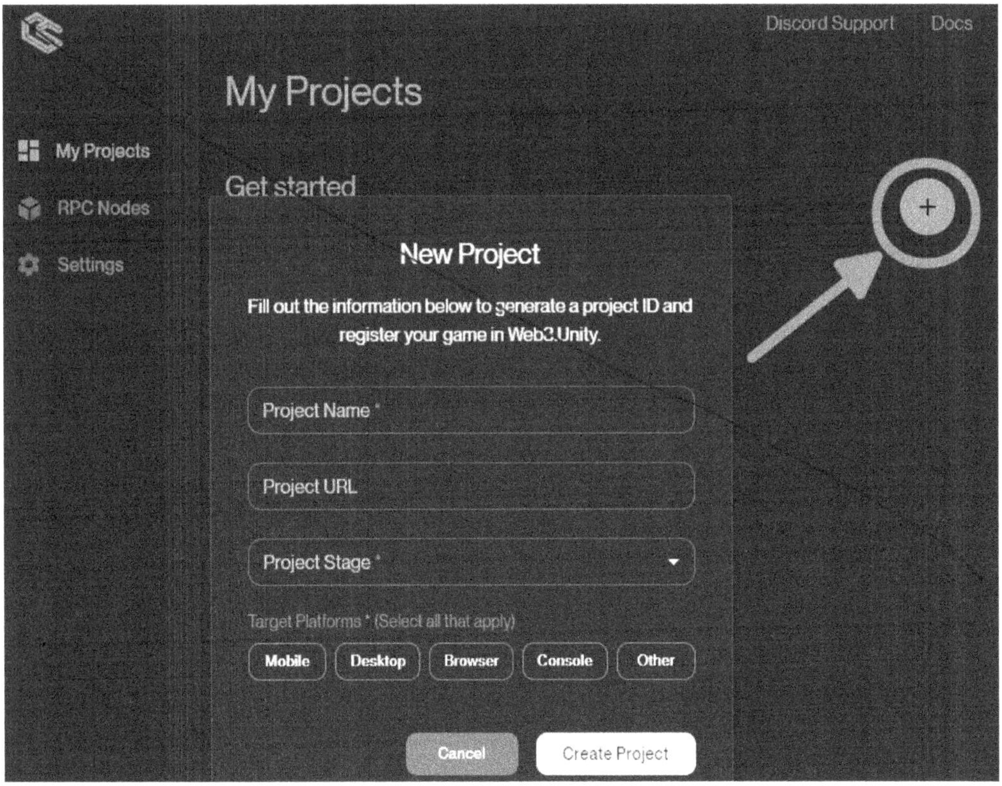

Figure 8.4 – Creating a new ChainSafe project

4. In the form, you can write the name of your choice in the **Project Name** field, then select the **Building** option in the **Project Stage** field, and finally, select the **Mobile**, **Desktop**, and **Browser** platforms.
5. Click on the **Create Project** button to finish the process. You will see your project already created in the control panel.

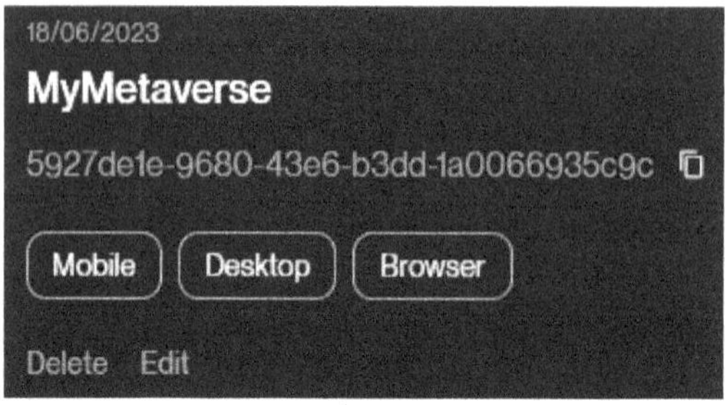

Figure 8.5 – The properties of the project created in ChainSafe

As you can see, just below the name of the project we have created, we can see the project ID. Copy it by clicking the **Copy** button, or selecting it with the mouse, right-clicking, and clicking **Copy**. We will need this ID now to complete the installation of the SDK in Unity.

Installing the ChainSafe SDK

As we mentioned previously, one of the great advantages of using ChainSafe is that it has a fully integrated SDK with Unity that we can download for free:

1. To do this, we will proceed to download it from `https://github.com/ChainSafe/web3.unity/releases/latest/download/web3.unitypackage` and then add it to our project.

   After importing the SDK, you will notice many errors warning about a dependency not being installed, specifically the `NewtonSoft-Json` library. Do not worry; installing it is really easy. Follow these steps to install the `NewtonSoft-Json` dependency:

2. Open the **Package Manager** window by clicking on **Window** | **Package Manager** in the main menu bar.

3. Click on the + button and select the **Add package by name** option.

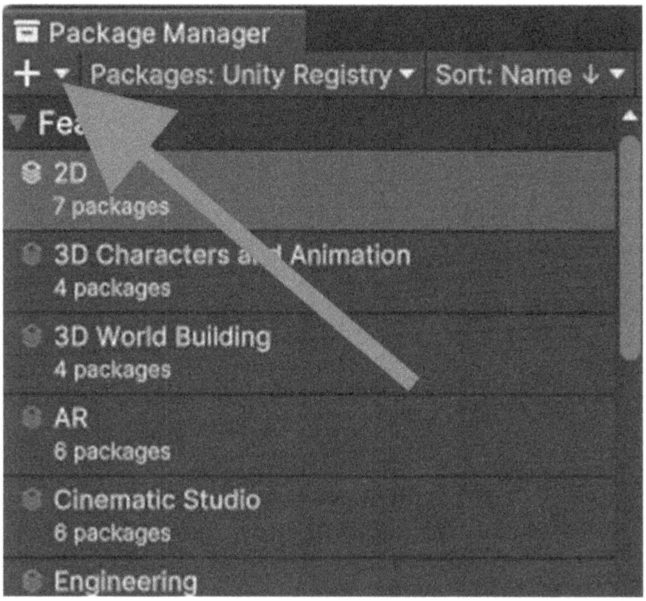

Figure 8.6 – Installing the NewtonSoft-Json package in Unity's Package Manager

4. Type `com.unity.nuget.newtonsoft-json` and press **Accept**; the package installation will start immediately.

   Once the installation of the dependency is complete, the ChainSafe SDK installation will complete, and the configuration screen will appear. If the ChainSafe configuration window does not appear automatically, you can access it in the Window | ChainSafeServerSettings option in Unity's main menu bar.

Figure 8.7 – The ChainSafe SDK properties configuration window

Now, let's proceed to configure the ChainSafe SDK to complete the installation.

5. In **Project ID**, paste the ID that we copied in the ChainSafe control panel when we created the project.
6. In the **Chain ID** property, enter the value `5`, which corresponds to the code that represents the Goerli network, the testing network we will use for Ethereum.

   The following are the values to be configured in **Chain** and **Network** to work with the Ethereum testing network:

   - In **Chain**, we enter the value `Ethereum`.
   - In **Network**, we enter the value `Goerli`.
   - The RPC server acts as a communication interface between the client and the underlying blockchain. It provides a set of methods or procedures that allow developers to interact with the blockchain protocol and perform operations related to the creation and management of NFTs. You can get an RPC server with a free layer at `https://console.chainstack.com/user/login`.

> **Tip: setting up an RPC server**
>
> During development on the testing network, the RPC server does not seem to be necessary. However, if you need one, you can get one with a free layer; on the downside, you will have to enter your credit card in the registration process. During the testing of this book, ChainSafe company did not make any charges, and they offer a fairly generous free layer. Detailed information on how to get an RPC server can be found in the guide: `https://docs.gaming.chainsafe.io/current/setting-up-an-rpc-node`.

7. Finally, click on the **Save Settings** button to save the changes and close the configuration window.

> **Tip: chains list**
>
> In this book, we will work with the Goerli test network, but if going forward you want to implement other networks in production, a list of all the networks supported by ChainSafe is available at `https://docs.gaming.chainsafe.io/current/using-the-minter/`.

We need to create an account on ChainSafe Storage, a ChainSafe service to store and generate IPFS. We will do this next.

Creating a ChainSafe Storage account

ChainSafe Storage is a set of tools and services that allows users to store and retrieve IPFS data. It offers a well-documented API, as well as an SDK for Unity, which allows developers a very easy-to-integrate system.

As we already know, NFTs are unique decentralized assets. One of the most powerful use cases of ChainSafe Storage is off-chain data storage. With ChainSafe Storage, users can rest easy knowing that the off-chain data associated with their NFTs will always be as available and decentralized as the asset itself.

Now that we know a bit about ChainSafe Storage, let's proceed to create an account and a first Bucket. To do so, follow these steps:

1. Go to `https://app.storage.chainsafe.io` and create a new account; you can easily do so with your Google or GitHub account, by email, or with your MetaMask wallet.
2. Once we have completed the registration, we will be redirected to the control panel. There, in the **Storage** section, click on the **Create Bucket** button, choose the name you want, and click on the **Create** button to finish – it's as simple as that.

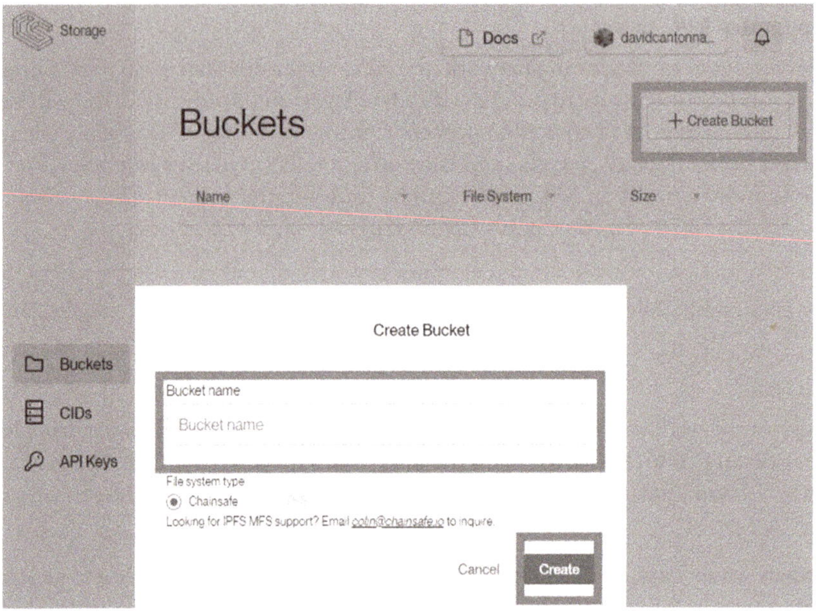

Figure 8.8 – Creating a new Bucket in ChainSafe Storage

As you can see, once we have clicked on the **Create** button, we can see the Bucket we just created. You can navigate through the Bucket in a very familiar way. The Bucket is like a root folder; if you click on it, you will access its interior, and inside it, you can create more folders and upload files.

You can, for example, create a new folder by simply clicking on the **New folder** button and choosing a name, which will automatically appear inside the Bucket.

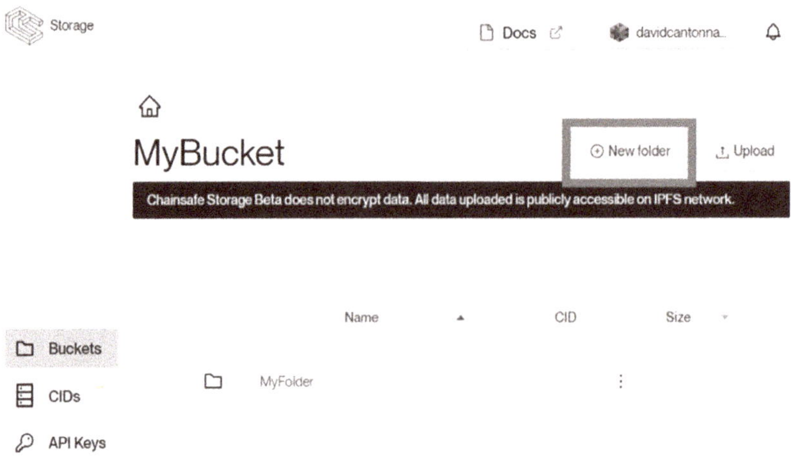

Figure 8.9 – Creating a new folder in ChainSafe Storage

Finally, we need to generate an API key that we will use in our Unity project to be able to upload files from the code.

3. In the left menu of ChainSafe Storage, select the **API Keys** option.
4. Click on the **Add API Key** button.
5. Copy and save the generated key in a safe place.

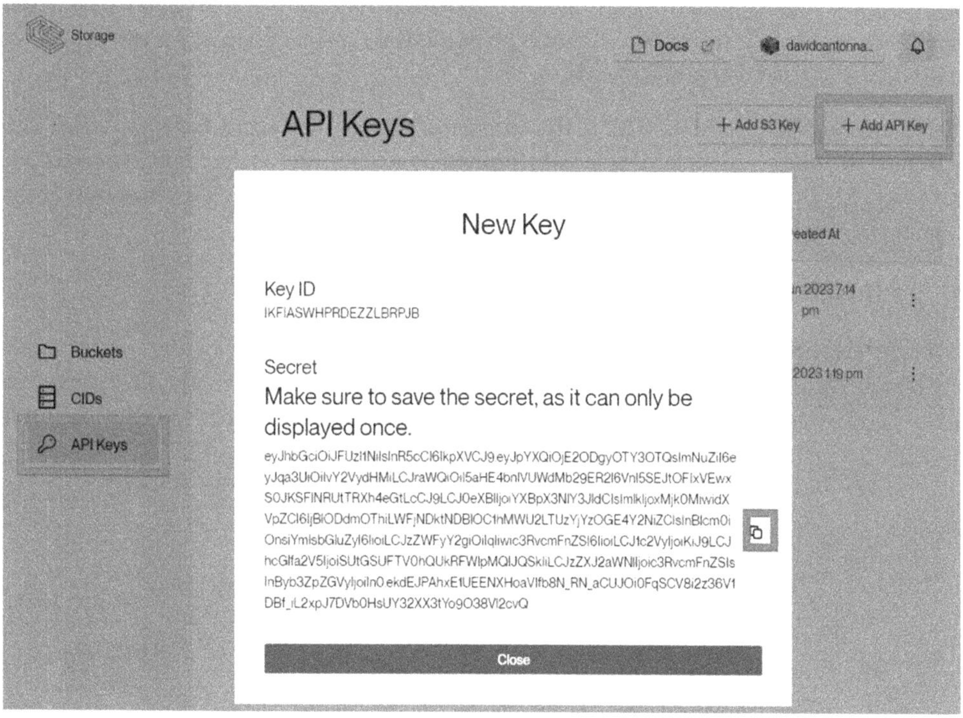

Figure 8.10 – Copying the generated API key to ChainSafe Storage

> **Tip: the generated API key**
>
> When you generate an API key in ChainSafe Storage, it is important that you store it in a safe place and copy it because once you close the window where it appears, you will not have access to it again. If you forget it or lose it, you will have to generate a new one.
>
> Once ChainSafe Storage is configured and we have an API Key available, we are ready to continue installing the necessary tools. Now, we will proceed to install the last tool, the MetaMask extension.

We already have the ChainSafe SDK fully configured, but before we go ahead and start creating code to implement the SDK, we need a couple more steps. It is strictly necessary to have the MetaMask extension installed in our browser.

MetaMask

MetaMask is a cryptocurrency wallet and browser extension that allows users to interact with dApps and access the Ethereum network. It functions as a digital wallet that allows users to store, send, and receive Ethereum tokens and other Ethereum-based assets, such as NFTs.

MetaMask integrates as an extension into popular web browsers, such as Chrome, Firefox, Brave, and Edge, and provides an intuitive user interface to manage Ethereum accounts and make secure transactions. Users can create multiple accounts on MetaMask, and each account is associated with a unique address on the Ethereum network.

In addition to being a wallet, MetaMask is also an essential tool to interact with dApps. When users visit a dApp in their browser, MetaMask automatically connects and allows the dApp to interact with the user's account. This enables actions such as signing transactions, approving token spending, and participating in smart contracts.

Installing the MetaMask extension

As previously mentioned, the MetaMask extension is absolutely necessary for any transaction on decentralized websites and applications. The extension is available for Chrome, Firefox, Brave, Edge, and Opera browsers. To install it, go to your browser's extension store and install it like any other extension.

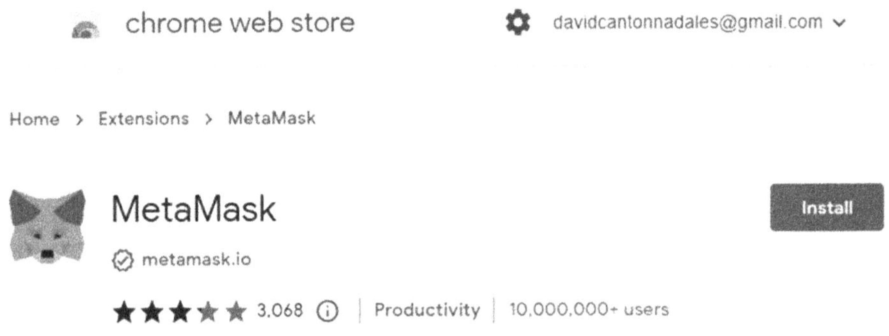

Figure 8.11 – MetaMask extension in Chrome web storage

Once the installation is finished, you will see the MetaMask welcome screen. Click on the **Create a new wallet** button and follow the instructions to complete the creation process.

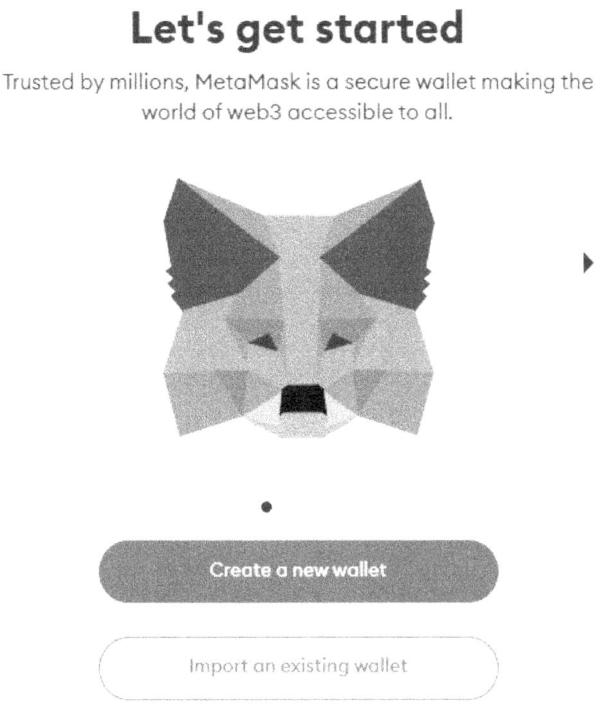

Figure 8.12 – The welcome screen of the MetaMask extension

After the installation of the MetaMask extension, we can continue with the development.

Before proceeding with further development, it should be noted that there is a cost to creating an NFT. As we will work in a testing environment, this cost is not real, but we need to obtain test coins to pay for the transactions. In the following section, we will learn how to easily generate test coins for our transactions.

Mining cryptocurrencies

Creating an NFT is not free; it has a cost – a cost associated with the gas used during the transaction. However, we will create an NFT in testing mode so that this cost is not *real*. Here is why the NFT has a cost.

The gas cost associated with creating an NFT on a blockchain is due to the nature of blockchain technology and the use of smart contracts. **The amount of gas used** is a measure of the amount of computational work required to execute a transaction or smart contract on a blockchain.

When you create an NFT on a blockchain, a smart contract is generally used to define the rules and unique properties of the token. This smart contract runs on the blockchain network and requires nodes on the network to perform calculations and validations to process and store the transaction.

The gas acts as a unit of measure to quantify the computational resources used in processing a transaction or smart contract. Each transaction that the smart contract performs, such as assigning values, storing data, or updating state, consumes a certain amount of gas.

The gas cost is determined by the gas price and the amount of gas consumed by the transaction. The price of gas is usually set by the blockchain network and can vary, according to demand and network congestion at any given time.

There are several reasons why the creation of an NFT has a gas cost:

- **Computational computations and validations**: Creating an NFT involves performing computational operations and validations on the blockchain network, which requires resources from the nodes in the network. The cost of gas covers these resources and ensures that the nodes are compensated for the work performed.
- **Prevention of spam attacks**: The cost of gas helps prevent spam and abuse attacks on the blockchain network. By requiring users to pay a fee to transact and execute smart contracts, massive and indiscriminate creation of NFTs is discouraged, ensuring more efficient and sustainable use of the network.
- **Resource balancing**: By assigning a cost to transactions, responsible use of blockchain network resources is encouraged and excessive congestion is avoided. This ensures that network resources are distributed equitably and that users who are willing to pay more for gas have priority in executing their transactions.

It is important to note that the cost of gas may vary, depending on the blockchain network used and market conditions. Therefore, it is advisable to check current gas rates before creating an NFT to understand the associated cost and make informed decisions.

When we go to create our first NFT, if we do not have enough cryptocurrencies to *pay* for the transaction, we will get the following prompt in MetaMask:

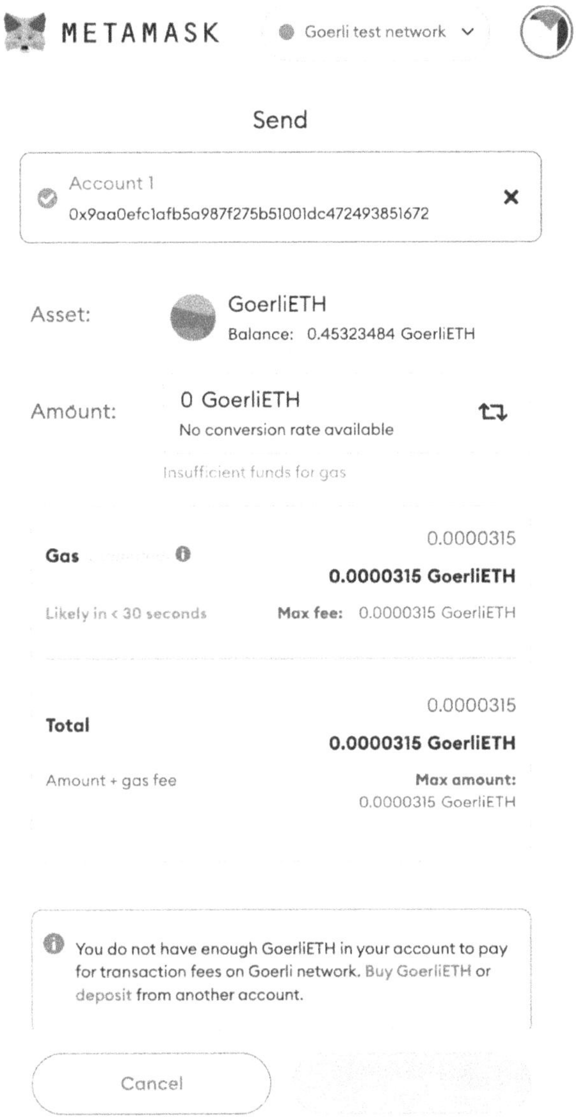

Figure 8.13 – A transaction that cannot be completed due to lack of funds

Goerli has a system for mining cryptocurrencies that can be used in its testing network, so we can pay for as many test transactions as we want. To mine-test Goerli, we will follow these steps:

1. Go to the mining website: `https://goerli-faucet.pk910.de`.
2. Enter your wallet address; you can get it by opening the MetaMask extension and clicking the **Copy** button.

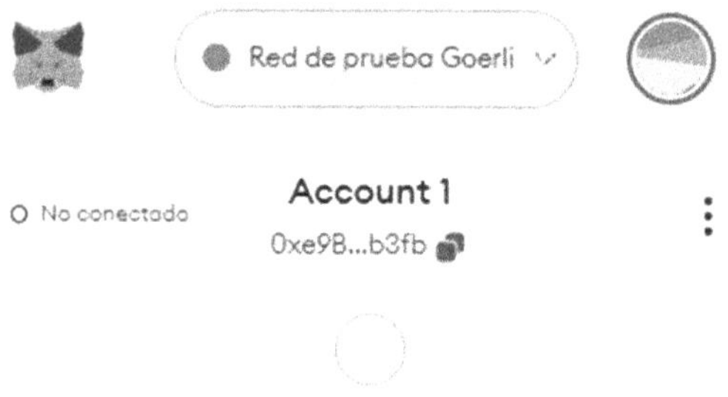

Figure 8.14 – Copying the account ID to the MetaMask extension

3. Press the **Start Mining** button and wait 30–60 minutes to mine enough Goerli. Press the **Stop Mining** button to get the reward after enough time has passed.

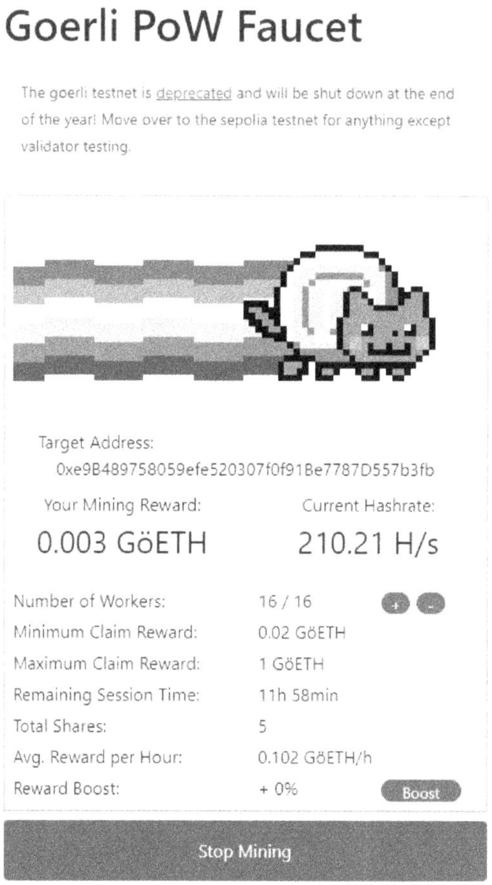

Figure 8.15 – Mining process to obtain test coins

4. After a few minutes, you will have the mined coins in your MetaMask wallet.

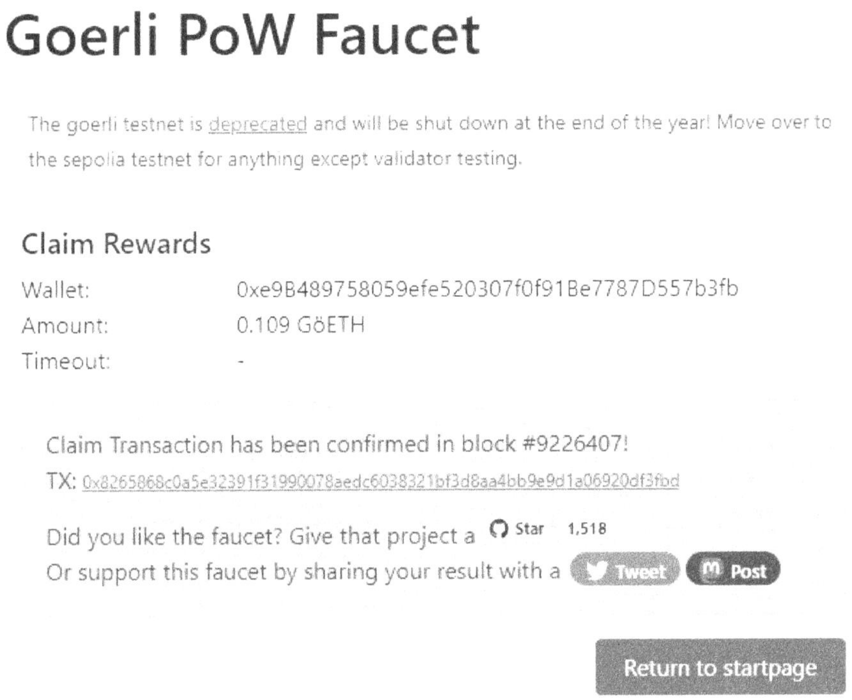

Figure 8.16 – Confirmation after obtaining the test coins

Great! You now have enough coins to do all the necessary tests. Now, we will start programming the NFT identification and creation system.

Programming the generation of the NFT

Now begins the good part – we will make some small modifications to the `BuildingDoorTrigger` script so that after the writing in the database of the new owner of the building, we can call a new function. This function will be in charge of carrying out the whole process. However, first, we need to understand the basic flow that the process should have.

The creation of the NFT can be broken down into three fundamental steps:

1. Identify the user by their MetaMask wallet, which will allow us to obtain the necessary information for the next step.

2. Create the artifact that we want to convert into NFT. The artifact will be created in two steps:

   I. In our case, we will take a screenshot of the current view of the player. This will be the object we will convert to NFT.

   II. Upload the screenshot to ChainSafe Storage to generate an ID that represents the IPFS created.

3. With the generated IPFS ID, we can execute an action called **Mint**. The Mint process can basically be broken down into two actions:

   I. The Mint process involves assigning a unique value and authenticity to a specific digital item, turning it into a unique and indivisible NFT.

   II. The NFT created can represent any type of digital asset, such as a work of art, a collectible, a virtual item, or even a piece of multimedia content.

Now that we know all the theory needed to understand the NFT ecosystem, we can start programming. The first thing we will do is to take care of the identification between our project and the MetaMask wallet, which is strictly necessary to obtain a signed session that allows us to create a transaction.

Identifying yourself in the MetaMask wallet

The primary process before starting with the generation of the NFT is the login. We must create an identification process between our project and the MetaMask wallet. For this, we will implement a few simple lines of code in our script. It is very simple; just execute the following steps:

1. Edit the `BuildingDoorTrigger` script located in the **Assets** | **_App** | **Scripts** path.
2. Add the `LoginNft` and `SignVerifySignature` methods to the end of the class:

```
public async void LoginNft ()
{
    Web3Wallet.url = "https://chainsafe.github.io
        /game-web3wallet/";
    // get current timestamp
    var timestamp = (int)
        System.DateTime.UtcNow.Subtract(
            new System.DateTime(1970, 1, 1))
                .TotalSeconds;
    // set expiration time
    var expirationTime = timestamp + 60;
    // set message
    var message = expirationTime.ToString();
    // sign message
    var signature = await
        Web3Wallet.Sign(message);
```

```
        // verify account
        var account = SignVerifySignature(signature,
            message);
        var now = (int)
            System.DateTime.UtcNow.Subtract(
                new System.DateTime(1970, 1, 1))
                    .TotalSeconds;
        // validate
        if (account.Length == 42 && expirationTime >=
            now)
        {
            print("Account: " + account);
        }
    }

    public string SignVerifySignature(string
        signatureString, string originalMessage)
    {
        var msg = "Ethereum Signed Message:\n" +
            originalMessage.Length + originalMessage;
        var msgHash = new Sha3Keccack().CalculateHash
            (Encoding.UTF8.GetBytes(msg));
        var signature = MessageSigner
            .ExtractEcdsaSignature(signatureString);
        var key = EthECKey.RecoverFromSignature
            (signature, msgHash);
        return key.GetPublicAddress();
    }
```

Well, these two methods will allow us to create a signed session to get the MetaMask wallet address.

3. Now, we will get a screenshot and upload it to ChainSafe Storage. To do this, add the following highlighted lines of code to the `BuildingDoorTrigger` class. First, let's create a variable that stores the ChainSafe Storage API key that we created, copied, and saved previously:

```
...(existing code here)...

public class BuildingDoorTrigger : MonoBehaviour,
    IPlayerActions
{
    private const string apiKey =
        "eyJhbGciOiJFUzI1NiIsInR5cCI6IkpXVCJ9.eyJpYXQi
        OjE2ODc0NTQwNTUsImNuZiI6eyJqa3UiOiIvY2VydHMiL
        CJraWQiOiI5aHE4bnlVUWdMb29ER2l6VnI5SEJtOFIxVE
        wxS0JKSFlNRUtTRXh4eGtLcCJ9LCJ0eXBlIjoiYXBpX3N
```

```
            lY3JldCIsImlkIjoxMjk0MiwidXVpZCI6IjBlODdmOThi
            LWFjNDktNDBlOC1hMWU2LTUzYjYzOGE4Y2NiZCIsInBlc
            m0iOnsiYmlsbGluZyI6IioiLCJzZWFyY2giOiIqIiwic3
            RvcmFnZSI6IioiLCJ1c2VyIjoiKiJ9LCJhcGlfa2V5Ijo
            iUEZMSVlZQ1VKT1BDVlNZQkpGU1UiLCJzZXJ2aWNlIjoi
            c3RvcmFnZSIsInByb3ZpZGVyIjoiIn0.qk6lPZqZpqc1h
            bp09OGKrWwQfJk1gYt5DkUeejKw1UXJzaSlSt9HwOL4jms
            xd8FfMF0O3uVe1YYLKYiddi9";
...(existing code here)...
```

Now, we will create a new function that will take care of taking the screenshot and uploading it to ChainSafe Storage:

```
private async Task<string> UploadIPFS()
{
    var capture =
       ScreenCapture.CaptureScreenshotAsTexture();
    byte[] bArray = capture.GetRawTextureData();
    var ipfs = new IPFS(apiKey);
    var bucketId =
       "5d9c59c9-be7a-4fbc-9f1e-b107209437be";
    var folderName = "/MyFolder";
    var fileName = "MyCapture.jpg";
    var cid = await ipfs.Upload(bucketId,
         folderName, fileName, bArray,
              "application/octet-stream");

    return $"{cid}";
}
```

The process of the function is very simple; first, we capture what the active camera is focusing on in the scene. Then, we convert the capture to an array of bytes, which is the format expected by the `ipfs.Upload` method.

As you can see, the `buckedId` variable represents the ID of the Bucket to which we are going to send the file upload. This `Id` can be obtained in the URL when you are inside the Bucket in ChainSafe Storage:

Figure 8.17 – Getting the Bucket ID in ChainSafe Storage

4. Finally, with the `folderName` and `fileName` variables, we finish configuring the path that the file we will upload will take by calling the `upload` method. This function will return the **Content Identifier** (**CID**), which represents the unique ID of the created IPFS. We will use it later to complete the creation of the NFT.

5. Now, we will create a new function that will be in charge of completing the creation of the NFT; to do so, add this new code block to the `BuildingDoorTrigger` class:

```
public async void MintNft()
{
    var voucherResponse721 = await
        EVM.Get721Voucher();
    CreateRedeemVoucherModel.CreateVoucher721
        voucher721 = new CreateRedeemVoucherModel
            .CreateVoucher721();
    voucher721.tokenId =
        voucherResponse721.tokenId;
    voucher721.signer = voucherResponse721.signer;
    voucher721.receiver =
        voucherResponse721.receiver;
    voucher721.signature =
        voucherResponse721.signature;
```

```
        string chain = "ethereum";
        string chainId = "5";
        string network = "goerli";
        string type = "721";

        string voucherArgs =
            JsonUtility.ToJson(voucher721);
        var nft = await UploadIPFS();
        // connects to user's browser wallet to call a
           transaction
        RedeemVoucherTxModel.Response voucherResponse
            = await EVM.CreateRedeemTransaction(chain,
                network, voucherArgs, type, nft,
                    voucherResponse721.receiver);
        string response = await
            Web3Wallet.SendTransaction(chainId,
                voucherResponse.tx.to,
                voucherResponse.tx.value.ToString(),
                voucherResponse.tx.data,
                voucherResponse.tx.gasLimit,
                voucherResponse.tx.gasPrice);
        print("My NFT Address: " + response);
    }
```

Fantastic! As you can see, with the ChainSafe SDK, it is really easy to create an NFT. This new function will call the previously created `UploadIPFS` function, and we will use the generated CID to bind it in the Mint process to finalize the creation of the NFT.

6. Now, to finish the circle, let's add the `UploadIPFS()` function call in `LoginNft()` by adding the highlighted line of code:

```
    public async void LoginNft()
    {
        ...(existing code here)...
        if (account.Length == 42 && expirationTime >=
            now)
        {
            PlayerPrefs.SetString("Account", account);
            MintNft();
        }
    }
```

7. Finally, before testing, we must call `LoginNft()` in the `UpdateBuildingOwner()` method; to do so, add the highlighted line of code:

```
public void UpdateBuildingOwner()
{
    ...(existing code here)...
    docRef.UpdateAsync(updatedProperties)
        .ContinueWith(task =>
    {
        ...(existing code here)...
        LoginNft();
    });
}
```

You have completed the development, and you now know how to easily create NFTs from your code with the ChainSafe SDK. Try it out!

If you run the project and go to the **World1** world, you can walk up to one of the houses and obtain it; this functionality was completed at the beginning of this chapter. Now, in addition to linking the user's ID to the Firestore Building document, the function will launch the creation of the NFT.

8. When the `LoginNft()` method is executed, a new tab will be launched in our web browser. Then, MetaMask will ask you to sign the session:

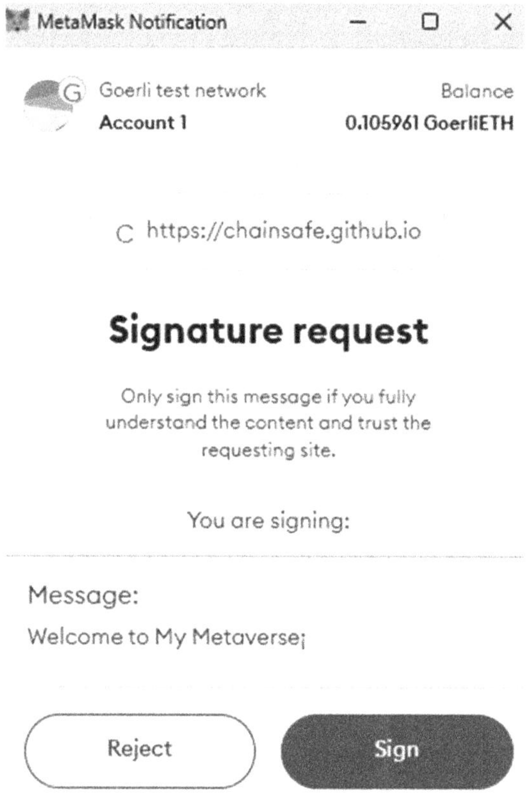

Figure 8.18 – A signature request for identification in the MetaMask wallet

9. When we click **Sign**, a message will appear on the web page that ChainSafe has opened from Unity, asking us to click **Copy** to continue the operation.

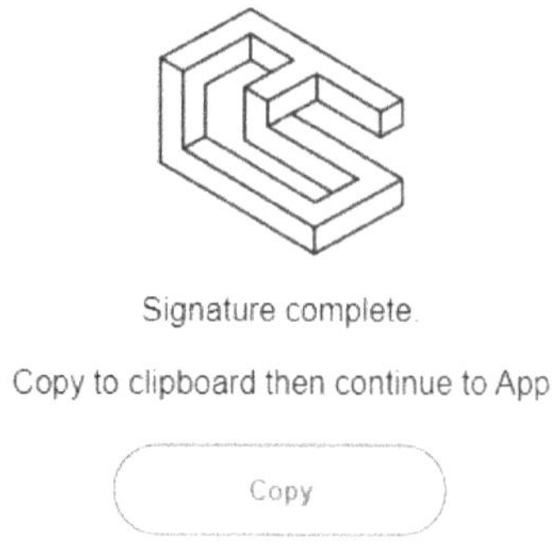

Figure 8.19 – Confirmation of the signature completed

10. Then, after clicking on the **Copy** button, we return to Unity, as indicated in the message on the web page, and a new tab will be launched in the browser – this time, to sign for the transaction and the creation of the NFT. Click on the **Confirm** button.

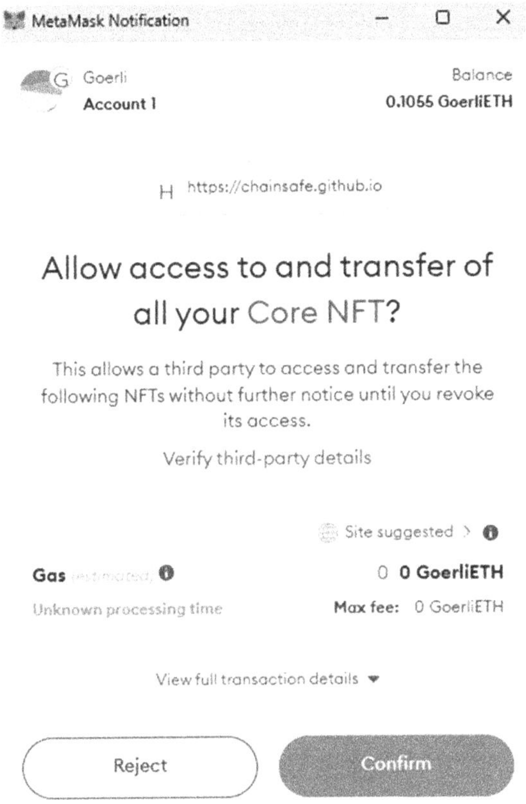

Figure 8.20 – A signature request for the transaction generating the NFT

After a few seconds, the MetaMask extension will notify us of the confirmation of the transaction.

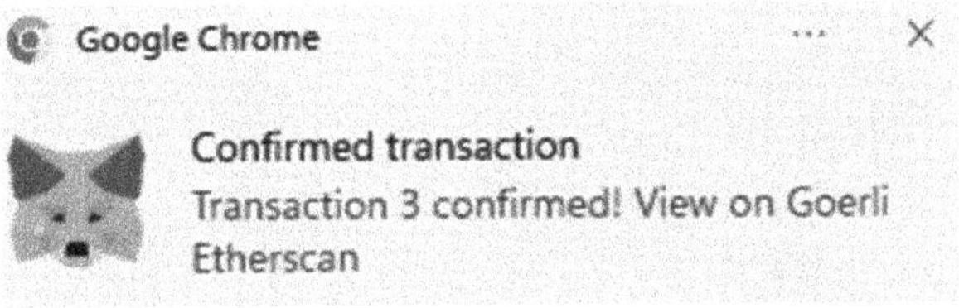

Figure 8.21 – The transaction confirmation notification in Chrome

11. If we go back to the Unity editor, we can see in the console the confirmation message that we included in the `MintNft()` function.

Figure 8.22 – The address of the NFT created in the Unity console

12. Finally, to confirm the existence of the NFT, we can check it on the website (`https://goerli.etherscan.io`) by pasting the address we obtained in the console in the search bar:

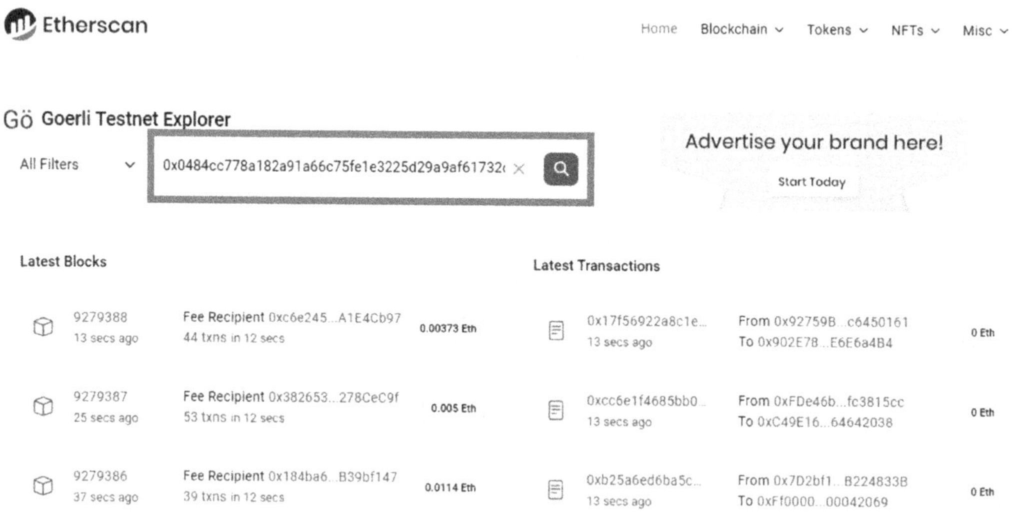

Figure 8.23 – Checking the transaction in Etherscan

In the following screenshot, we can see all the details of the transaction:

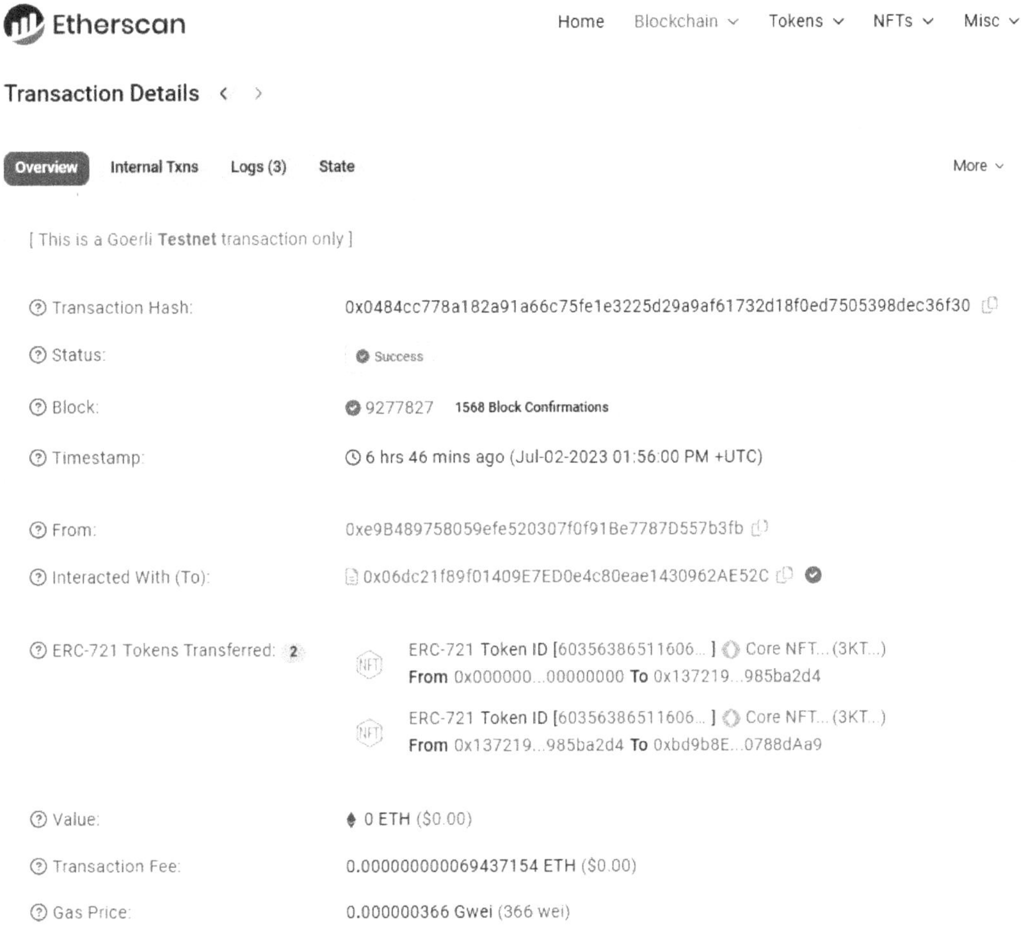

Figure 8.24 – The Etherscan transaction details

Easy, isn't it? At this point, you have learned the basics of creating an NFT from your own code.

We have reached the end of this chapter. Now, our metaverse has a new functionality – we allow a player to obtain a house. This is a great advancement in your knowledge and will open up a huge range of new possibilities.

The knowledge gained during this chapter, together with the knowledge gained previously, will allow you to create powerful combinations of new features and improvements, such as those discussed previously.

Summary

Throughout this chapter, we learned how to add a collider that serves as a trigger for a user, allowing them to interact with the houses that we dynamically load into a scene. This trigger contacts the Firestore database and queries whether a building is occupied by another player or can be acquired by the current player.

We learned a new class, GUI, allowing us to display texts on screen in a very simple way, which is ideal to display messages that incite action. We programmed a script able to query and write in the database the new owner of the building. Conversely, we learned how to use the ChainSafe SDK to create NFTs from our class.

In the next chapter, we will introduce our world to the Photon SDK, an impressive framework that will turn our metaverse into a multiplayer world. With Photon, we will be able to coordinate over a network all our player's movements; walking, jumping, turning, and running will be transmitted over the internet and accurately reproduced to other connected players.

9

Turning Our World into a Multiplayer Room

Metaverse and **multiplayer** are definitions that often go hand in hand; it is hard to imagine a single-player metaverse. One of the most fun features a virtual world can have is for it to interact with other players in real time, and for those users to interact with friends and other users in different scenarios and activities.

In this chapter, we will focus on adding this feature to our project. We will achieve this with Photon SDK, a powerful framework that offers easy integration for our Unity3D projects. We will learn how to send the movements and animations through the network so that all the users of our metaverse are synchronized in real time.

Photon offers an SDK for integrating text chat and voice chat into our project, with all the services being under the same provider. This is one of the reasons why I chose Photon for this book. We will see this in *Chapter 10, Adding Text and a Voice Chat to the Room.*

We will cover the following topics:

- Getting started with Photon SDK
- Synchronizing movements and animations in the network
- Testing synchronization with another user

Technical requirements

This chapter does not require any special technical requirements, but we will start programming scripts in C#. Therefore, it's advised that you have basic notions of this programming language.

We need an internet connection to browse and download an asset from the Unity Asset Store. We will continue using the project we created during *Chapter 1*, *Getting Started with Unity and Firebase*. Remember that we have this book's GitHub repository, which contains the complete project that we will work on here: `https://github.com/PacktPublishing/Build-Your-Own-Metaverse-with-Unity/tree/main/UnityProject`.

You can also find the complete code for this chapter on GitHub at: `https://github.com/PacktPublishing/Build-Your-Own-Metaverse-with-Unity/tree/main/Chapter09`

Getting started with Photon SDK

Photon SDK is a set of tools and services for developing networked games in Unity3D. Photon SDK allows you to create real-time multiplayer games, including online, cooperative and competitive games.

Photon SDK offers a scalable and reliable networking solution that allows you to create multiplayer online games with low latencies and a large number of concurrent players. Some of the features and benefits of Photon SDK include the following:

- **Real-time connection**: Photon SDK enables the creation of real-time, online games, which means players can interact in real time without any lag.
- **Ease of use**: Photon SDK provides an easy-to-use interface that simplifies the creation of multiplayer games. It also provides detailed documentation and technical support to help you integrate Photon SDK into your Unity project.
- **Scalability**: Photon SDK is capable of handling a large number of simultaneous players, making it ideal for massively multiplayer online games.
- **Cross-platform support**: Photon SDK supports multiple platforms, allowing you to create games that can be played on a wide variety of devices, including PCs, consoles, and mobile devices.

In short, Photon SDK can help you create multiplayer online games with scalable and reliable networking in Unity3D. With Photon SDK, you can create games that involve a large number of players, interact in real time, and support multiple platforms.

Pricing

Photon SDK offers different pricing plans to suit the needs of game developers. The following are the different pricing plans and their main features:

- **Free**: This plan is free and offers up to 20 concurrent players in an application. It also includes a limit of 100 **concurrent users** (**CCUs**) on Photon's global network. This plan does not include technical support and is only available for non-commercial use.

- **Plus**: This plan offers up to 100 CCUs on the Photon global network and technical support via email. It also includes additional features such as advanced statistics and performance analysis, access to the latest software updates, and the ability to add Photon add-ons and extensions. Pricing starts at $9 per month for 100 CCUs.
- **Pro**: This plan offers advanced features for multiplayer game development, such as the ability to customize Photon servers and custom SDK integration. It also includes priority technical support and a limit of 500 CCUs on Photon's global network. Pricing starts at $149 per month for 500 CCUs.
- **Enterprise**: This plan is designed for enterprises and offers customized features such as dedicated cloud hosting and integration with third-party services. It also includes customized technical support and a customized CCU limit on Photon's global network. Pricing varies according to specific business requirements.

In summary, Photon SDK offers different pricing plans to suit the needs of game developers, from a free plan to customized enterprise plans. Each plan offers different features and CCU limits, as well as technical support and access to software updates.

Signing up for Photon SDK

To create a new free Photon account, please execute the following steps:

1. Go to the Photon SDK website at `https://www.photonengine.com`.
2. Click the **Sign In** button on the home page.
3. Enter your credentials if you already have an account; otherwise, click on the **Create one** link.
4. Enter your email and tick the **I am not a robot** checkbox. Then, click on the **Register** button.
5. You will receive an email to confirm your new account; please follow the instructions there.
6. Complete the email verification process. An email will be sent to the email address you provided during registration. Click the verification link in the email to verify your Photon SDK account.
7. After verifying your account, log in to the Photon SDK website with your email address and password.
8. Once you are logged in, you will have access to the Photon SDK dashboard, where you can create and manage your online game applications.

With these steps, you will be able to create a Photon SDK account and start using its services and tools to develop online multiplayer games in Unity3D.

Creating a new application

To create a new application in Photon, once we have logged in, we will have access to the **Dashboard** area, where we will execute the following steps:

1. Click the **CREATE A NEW APP** button:

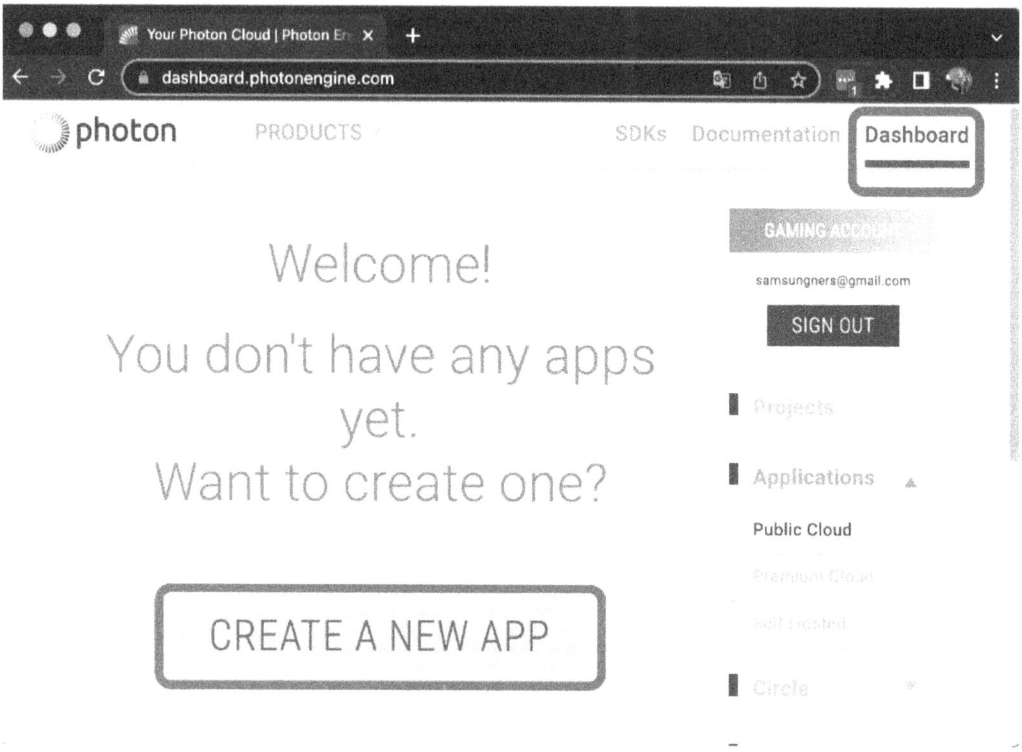

Figure 9.1 – The CREATE A NEW APP button in Photon

2. Select **Multiplayer Game** as the application type.
3. Select **PUN** as the Photon SDK type.
4. Enter a name and description of your choice.
5. Click the **Create** button.

Perfect! Once you have completed the steps, you will be redirected to the **Dashboard** area, where the application you have just created will appear:

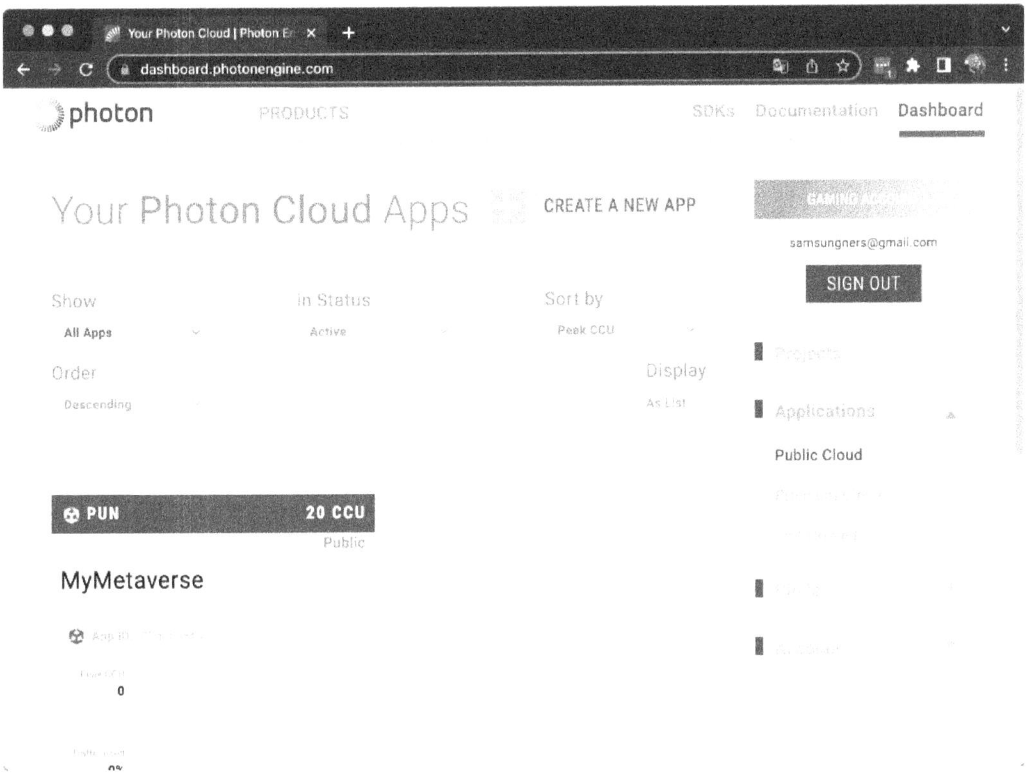

Figure 9.2 – The Photon Dashboard area

Don't close this window yet – we will need the **App ID** property of our application to configure it in Unity.

Installing Photon SDK

Finally, we have to download, install, and configure Photon SDK in our project. You can find this asset in the Unity Asset Store. To do so, follow these steps:

1. Go to the Unity Asset Store website: `https://assetstore.unity.com`.
2. In the search bar, type `PUN 2 FREE` and select the only asset that appears.
3. Click on the **Add to My Assets** button.
4. Click the **Open in Unity** button. This will open the **Package Manager** window in the Unity editor.
5. Click the **Download** button in the **Package Manager** window and then click the **Import** button.

Once the SDK import has finished, a **Photon configuration** window called **PUN Wizard** will appear. If the configuration window does not appear automatically, you can access it in the **Window | Photon Unity Networking | PUN Wizard | Setup Project** option in the Unity main menu bar. In this window, we must paste the **App ID** property of the application we created in the Photon **Dashboard** area into the text box of the window that appears in Unity.

6. In the Photon **Dashboard** area, click on the text under **App ID** to make it appear in full and copy it:

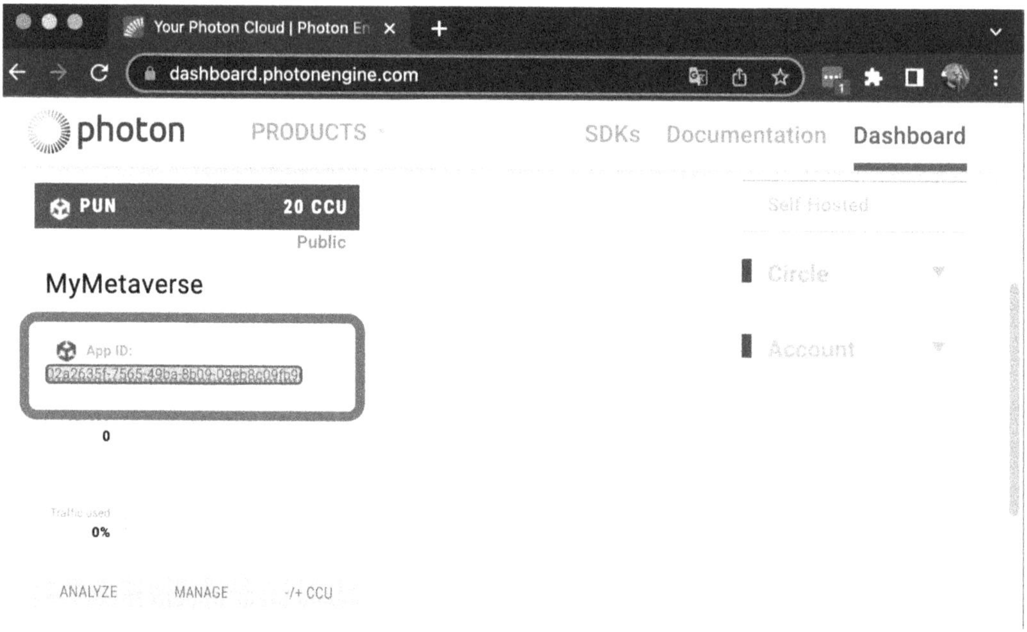

Figure 9.3 – The App ID of our project in the Photon Dashboard area

7. Paste the **App ID** property into the box in the configuration window that appears in Unity and click on the **Setup Project** button:

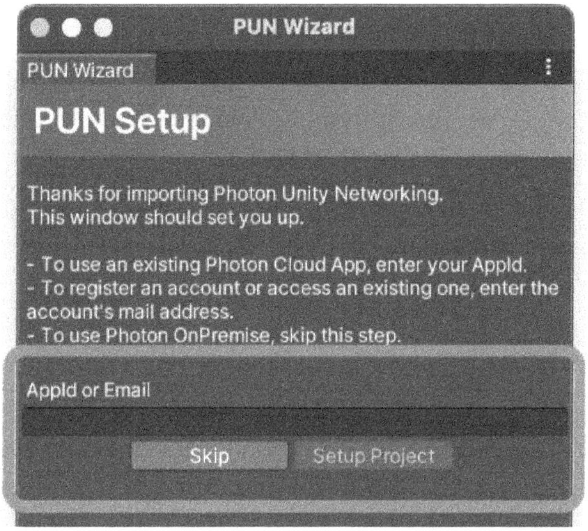

Figure 9.4 – Photon configuration window

8. You can now close this window by clicking the **Close** button.

Fantastic! You have finished setting up Photon in Unity. Now, our project is connected to the application we created in the Photon **Dashboard** area.

Photon life cycle

The life cycle of a connection in Photon SDK can vary depending on how the client and server are configured, as well as how events and game logic are handled. However, here is a general example of the life cycle of a connection in Photon SDK:

1. **Initialization**: The client initiates the connection process to the Photon SDK server. During this process, connection parameters are set, and callbacks are registered to handle events.
2. **Connection**: The client attempts to connect to the Photon SDK server. If the connection is successfully established, the `onConnectedToMaster` event occurs, indicating that the connection is active and ready to be used.
3. **Authentication**: If authentication is required, the client will send authentication data to the Photon SDK server. If authentication is successful, the `onAuthenticated` event occurs.

4. **Join/leave room**: The client joins a game room or creates a new room. If the client joins successfully, the `onJoinedRoom` event occurs. If the client leaves the room, the `onLeftRoom` event occurs.
5. **Send data**: The client sends data to the Photon SDK server, which is then transmitted to other clients in the room. If the server receives the data successfully, the `onEvent` event occurs.
6. **Disconnect**: The client disconnects from the Photon SDK server. If the disconnection is successful, the `onDisconnected` event occurs.

It is important to note that this is just a general example of the life cycle of a connection in Photon SDK and that additional or different events may occur, depending on the configuration and logic of the game.

Regions

In Photon SDK, regions are geographical locations where Photon servers are located. Developers can use regions to select a server that is closer to players and reduce latency in the game.

When a client connects to Photon SDK, it is automatically assigned the closest server based on the client's geographic location and the regions defined on the server. If no server is found in the desired region, the closest available server will be used for the connection.

Regions in Photon SDK are defined on the server and are available to clients via the SDK. Each region has a name and a set of IP addresses that correspond to the Photon servers in that region.

Developers can use the regions in Photon SDK to customize the user experience and improve game performance in different parts of the world. For example, if a game is being played primarily in North America, developers can select a server from the North American region to reduce latency and improve the user experience.

In short, regions in Photon SDK are geographic locations where Photon servers are located and are used to select the server closest to the players and reduce latency in the game.

Rooms

In Photon SDK, rooms are virtual spaces where players can interact and play together in real time. Rooms are one of the main features of Photon and allow developers to create multiplayer online games with a well-defined structure.

Rooms in Photon SDK can be of two types – **game rooms** and **chat rooms**:

- Game rooms are rooms where players can play together and interact and share information in real time. Developers can customize the features of game rooms to suit the needs of their game.

- Chat rooms, on the other hand, are rooms where users can communicate with each other via text in real time. Chat rooms in Photon SDK also allow developers to customize room features to suit the needs of their game.

Rooms in Photon SDK are created and managed on the server, which means that developers can create, join, and leave rooms through the client using the SDK. Developers can also define different parameters for the rooms, such as the maximum number of players, the type of game, and the difficulty level, among others.

In summary, rooms in Photon SDK are an important feature that allows developers to create real-time online multiplayer games and customize the room's features to meet the needs of their game.

Lobby

In the Photon SDK, the **lobby** is a key functionality that allows players to meet, interact, and join game rooms before starting a game. The lobby is essentially a space where players can view the list of available rooms and decide which one to join.

The main functions of the lobby in the Photon SDK are as follows:

- **List available rooms**: When players start the game or seek to join a multiplayer game, the lobby displays a list of available rooms that have been previously created by other players. Each room usually has a specific label or name, and may display information such as the number of players present, the type of game, the map, and so on.
- **Create and join rooms**: Players can create their own rooms or join an existing room from the lobby. By clicking on a room in the list, the player can send a request to join that room or, if configured, can enter a password for private rooms.
- **Wait and chat**: Once in the lobby, players can interact with each other via text chat. This allows them to communicate, coordinate, or simply socialize while waiting for other players to join the room or to decide which game to join.
- **Configuring room parameters**: Some implementations allow certain parameters or game rules to be configured before joining a room, such as maximum number of players, game time, difficulty, and so on.
- **Implement custom logic**: The lobby can also be a place where developers implement their own custom logic, such as advanced matchmaking, room filters, player statistics, and so on.

That's it for the *boring part* of this chapter, where we learned about the basics of Photon SDK to understand how it will help us. We also created a new account and a new application in our **Dashboard** area.

Next, we will start programming. We will use the scripts that Photon offers us to save us work and time when it comes to tedious tasks such as synchronizing movements and animations.

Synchronizing movements and animations in the network

After downloading and configuring Photon SDK, we will move on to the more technical part of this chapter. Throughout the following pages, we will learn how to connect the tools offered by this SDK to provide multiplayer technology to our project. You will be surprised how easy Photon makes this.

Photon offers a series of ready-made scripts that we only have to attach to the Prefab that represents the player. That is to say, all the *ugly* parts of programming such as synchronizing movements and animations are done for us, so we only have to know how to integrate them. That's what we will deal with in this chapter.

Preparing the player Prefab

Before we start, we must consider the details. Photon SDK will now be in charge of the scene changes and instantiating the player in the scene. So, for it to be able to perform these actions, the Prefabs that are going to be instantiated in the scene must be in a path under the Photon domain.

To do this, we must move the **PlayerInstance** player Prefab from the **Assets** | **_App** | **Prefabs** path to the new **Assets** | **Photon** | **PhotonUnityNetworking** | **Resources** path:

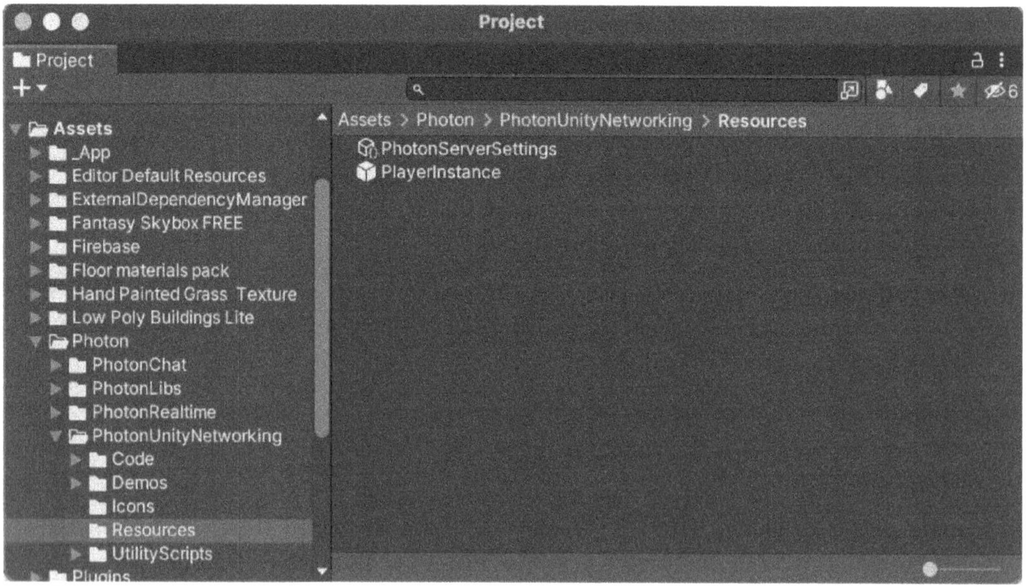

Figure 9.5 – The Resources folder

Next, we are going to add the necessary components to our player so that we can transmit movements and animations to other connected players.

To do this, we will follow these steps:

1. Double-click on the **PlayerInstance** Prefab that we have just changed folders for. It is now in **Assets** | **Photon** | **PhotonUnityNetworking** | **Resources**.
2. We will work on the **PlayerArmature** GameOject, which is where the **Animator** and **CharacterController** components are located. These are needed for the new components we are going to add:

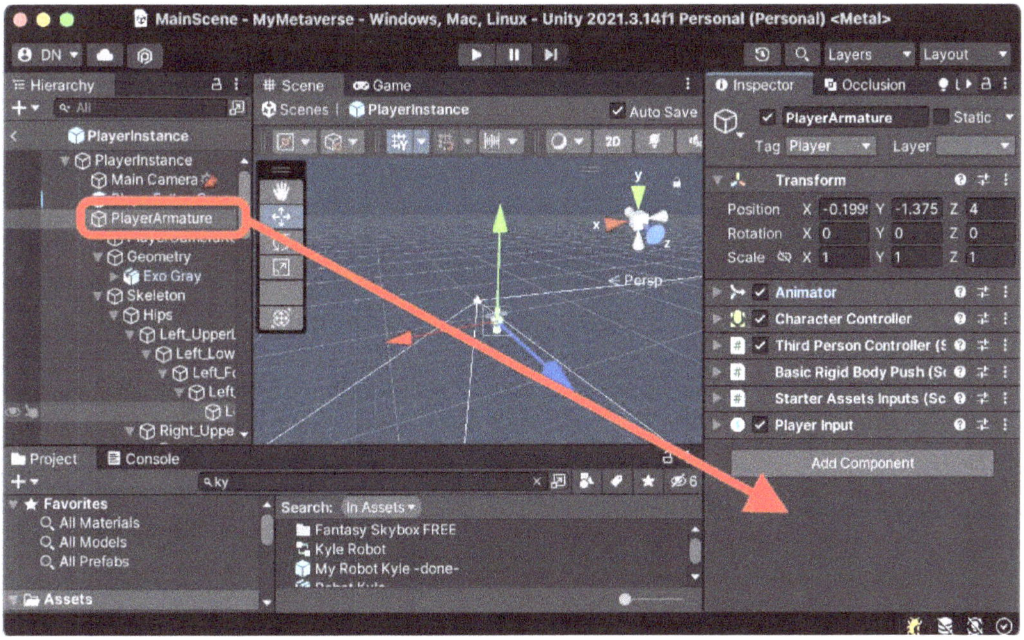

Figure 9.6 – The PlayerArmature Inspector panel

3. Click the **Add Component** button and search for and add **Photon View**. This component will be the nexus that will orchestrate movements, animations, and other parameters.
4. Click the **Add Component** button and search for and add **Photon Animator View**. This component will be in charge of sending our animations to and playing our animations for other players, automatically. Make sure it has the same configuration that's shown in the following figure:

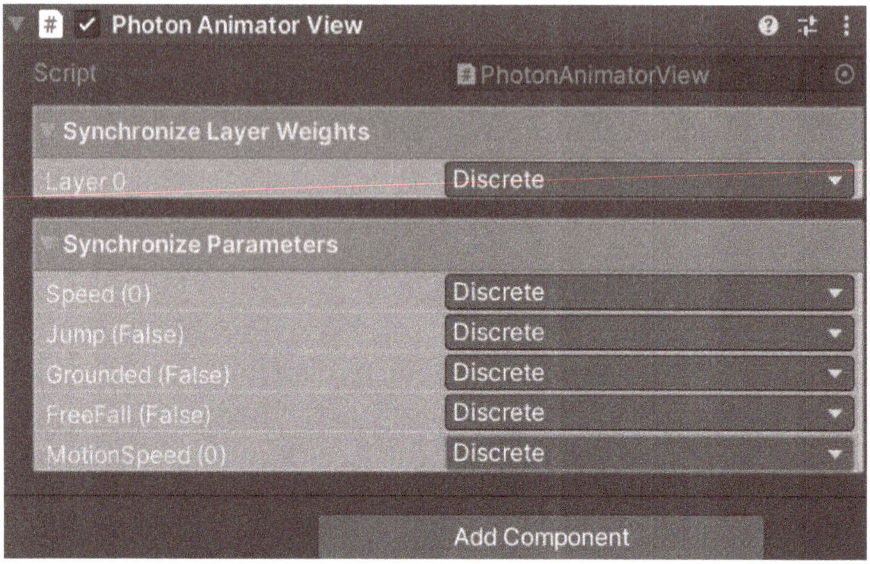

Figure 9.7 – The Photon Animator View component

5. Click on the **Add Component** button and search for and add **Photon Transform View Classic**. This component will be in charge of sending our position and rotation and playing it to other players, automatically. The configuration you must have is as follows:

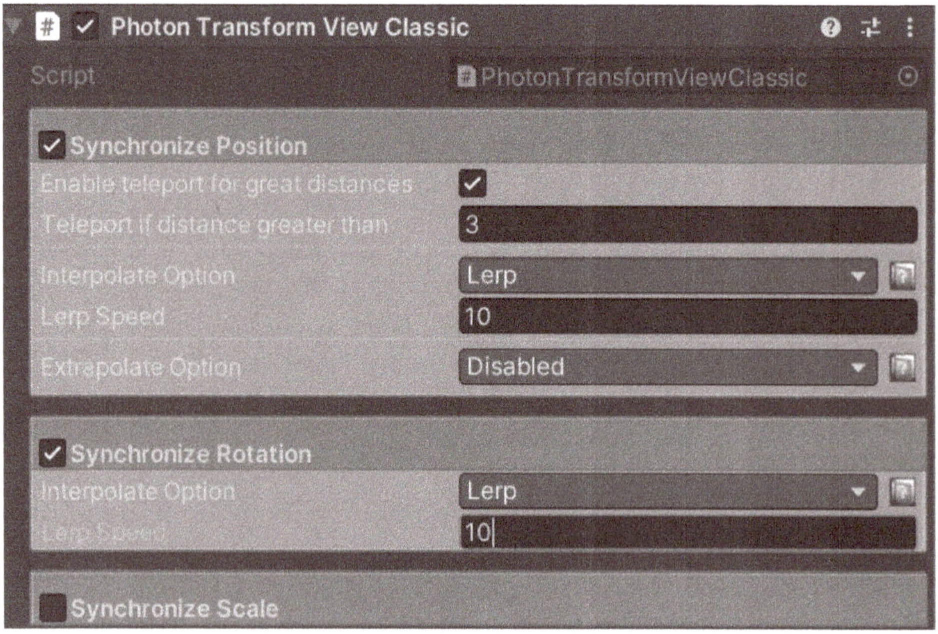

Figure 9.8 – The Photon Transform View Classic component

Excellent work! With the player Prefab set up, we just need to do one last thing.

Disabling other players' scripts

It turns out that, when multiple players enter the scene, Photon will instantiate a **PlayerInstance** Prefab for each of them. This is fine, but there will also be multiple cameras and **PlayerControllers** in the scene.

To solve this, we will create a script that will take care of cleaning our scene of any GameObject that may interfere with ours. To do this, we will follow the following steps:

1. Right-click on the **Assets | _App | Scripts** path and select **Create | C# Script**. Rename it `DisableUnneededScriptsForOtherPlayers`.
2. Double-click on the script we just created to edit it.
3. Replace all the auto-generated code with the following:

```
using System.Collections;
using System.Collections.Generic;
using Photon.Pun;
using StarterAssets;
using UnityEngine;

public class DisableUnneededScriptsForOtherPlayers :
    MonoBehaviourPun
{
    void Start()
    {
        PhotonView photonView =
           GetComponent<PhotonView>();
        CharacterController character =
          GetComponent<CharacterController>();
        ThirdPersonController controller =
          GetComponent<ThirdPersonController>();
        Transform camera =
          transform.parent.Find("Main Camera");
        Transform playerFollowCamera =
          transform.parent.Find("PlayerFollowCamera");
        Transform mobileController =
          transform.parent.Find
          ("UI_Canvas_StarterAssetsInputs_Joysticks");

        if (!photonView.IsMine)
        {
```

```
                gameObject.tag = "Untagged";
                Destroy(mobileController);
                Destroy(controller);
                Destroy(character);
                Destroy(camera.gameObject);
                Destroy(playerFollowCamera.gameObject);
                Destroy(this);
            }
            else{
                gameObject.tag = "Player";
            }
        }
    }
```

Well done! As you can see, in the script, we are looking for other components and GameObjects that may conflict with ours. With this script, we ensure that only **Camera** and **PlayerController** exist.

4. Now that we have created it, we have to add it to the **PlayerArmature** GameObject, just like we did with the other Photon scripts. Once we've added it as a component to **PlayerArmature**, the **Inspector** panel will look something like this:

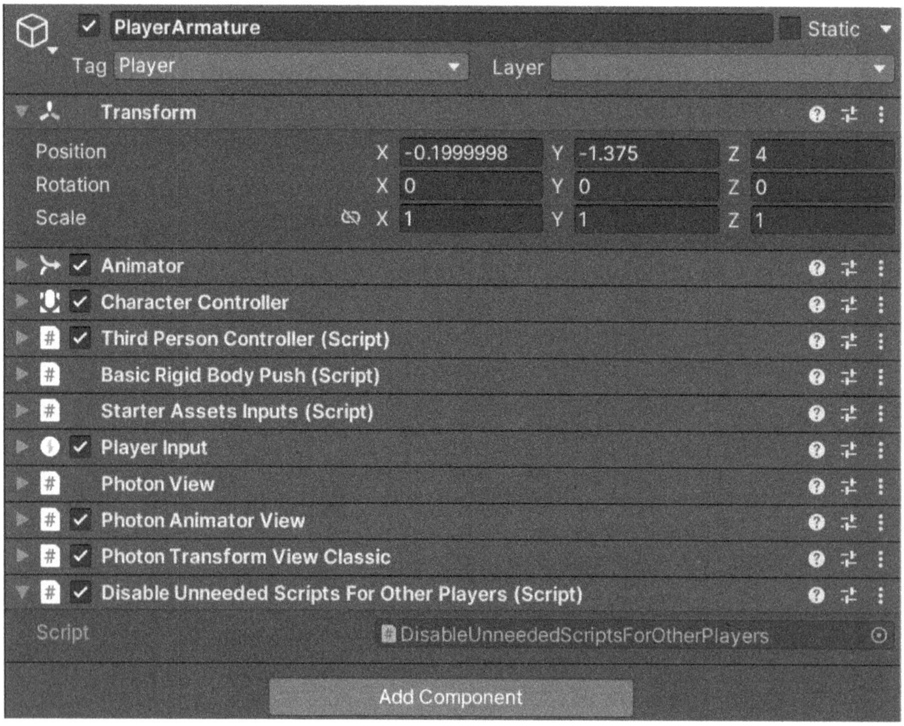

Figure 9.9 – The PlayerArmature Inspector panel

5. The next step is to modify the **InstantiatePlayer** script so that we can adapt it to Photon. To do this, locate the script in the **Assets** | **_App** | **Scripts** path and double-click it to edit it.

   Replace all the code with the following:

```
using Photon.Pun;
using UnityEngine;

public class InstantiatePlayer : MonoBehaviourPun
{
    public GameObject playerPrefab;
    void Start()
    {
        // We call Photon's Instantiate method which
           will create the player object in the scene
           and transmit the data to other connected
           users.
        PhotonNetwork.Instantiate(playerPrefab.name,
            transform.position, Quaternion.identity);
    }
}
```

With this change, we have replaced the classic `Instantiate` behavior with `PhotonNetwork.Instantiate`. With this, we get the player to instantiate in the scene but also send the event to other connected players, which means that we will be able to see when a player enters the scene.

Creating the network logic script

At this point, we have prepared the Prefab that represents the player in our project. Finally, we have to create the most important script – the one that will take care of the connection and scene changes.

To create this script, follow these steps:

1. Open the **Welcome** scene located in **Assets** | **_App** | **Scenes**, where we manage **Login** and **Registration**.
2. Create an empty GameObject in the **Hierarchy** panel and call it `NetworkManager`:

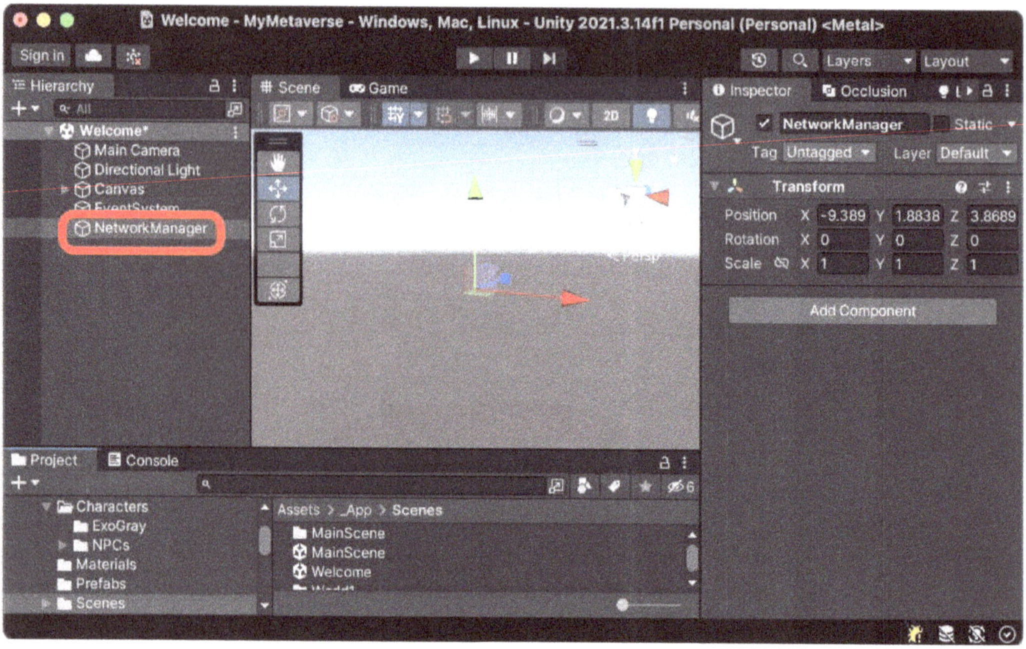

Figure 9.10 – The NetworkManager GameObject

3. Now, create the **NetworkManager** script, right-click on the **Assets** | **_App** | **Scripts** path, and select the **Create** | **C# Script** option. Rename it `NetworkManager`.
4. Double-click on the script we just created to edit it and replace all the auto-generated content with the following code:

```
. . . (existing code here) . . .
public class NetworkManager :
    MonoBehaviourPunCallbacks
{
    FirebaseAuth auth;
    static public NetworkManager Instance;
    private bool isConnecting = false;
    private const string GameVersion = "0.1";
    private const int MaxPlayersPerRoom = 100;
    public string actualRoom;


    void Awake()
    {
        FirebaseFirestore.DefaultInstance.Settings
            .PersistenceEnabled = false;
```

```
    if (auth == null)
        auth = FirebaseAuth.DefaultInstance;
    // Extended information on the configurable
       properties of the Photon SDK https://doc-
       api.photonengine.com/en/pun/v2/class_photon
       _1_1_pun_1_1_photon_network.html
    PhotonNetwork.AutomaticallySyncScene = true;
    PhotonNetwork.SendRate = 15;
    PhotonNetwork.SerializationRate = 15;
    PhotonNetwork.GameVersion = GameVersion;
    PhotonNetwork.ConnectUsingSettings();
}


private void Start()
{
    Instance = this;
}


public void JoinRoom(string sceneName)
{
    isConnecting = true;
    actualRoom = sceneName;
    if (PhotonNetwork.IsConnected)
        StartCoroutine(ChangeSceneWithDelay(
            sceneName));
}


IEnumerator ChangeSceneWithDelay(string
    sceneName)
{
    var roomOptions = new RoomOptions();
    var typedLobby = new TypedLobby(sceneName,
        LobbyType.Default);
    if(PhotonNetwork.InRoom)
        PhotonNetwork.LeaveRoom();
    while (!PhotonNetwork.IsConnectedAndReady)
    {
        yield return new WaitForSeconds(0.1f);
    }
    yield return new WaitForSeconds(1f);
```

```
            PhotonNetwork.JoinOrCreateRoom(sceneName,
                roomOptions, typedLobby);
        }


        public void Disconnect()
        {
            PhotonNetwork.Disconnect();
            SceneManager.LoadScene("Welcome");
        }


        public override void OnConnectedToMaster()
        {
            var currentUser =
                FirebaseAuth.DefaultInstance.CurrentUser;
            PhotonNetwork.LocalPlayer.NickName =
                currentUser.DisplayName;
            PhotonNetwork.JoinLobby();
        }


        public override void OnJoinedRoom()
        {
            SceneManager.LoadSceneAsync(actualRoom);
        }


        private void OnApplicationQuit()
        {
            PhotonNetwork.Disconnect();
            Application.Quit();
        }
    }
```

The **NetworkManager** script will be in charge of connecting to Photon, switching scenes, and synchronizing the player with different rooms.

5. Finally, we will add the **NetworkManager** script as a component of the **NetworkManager** GameObject:

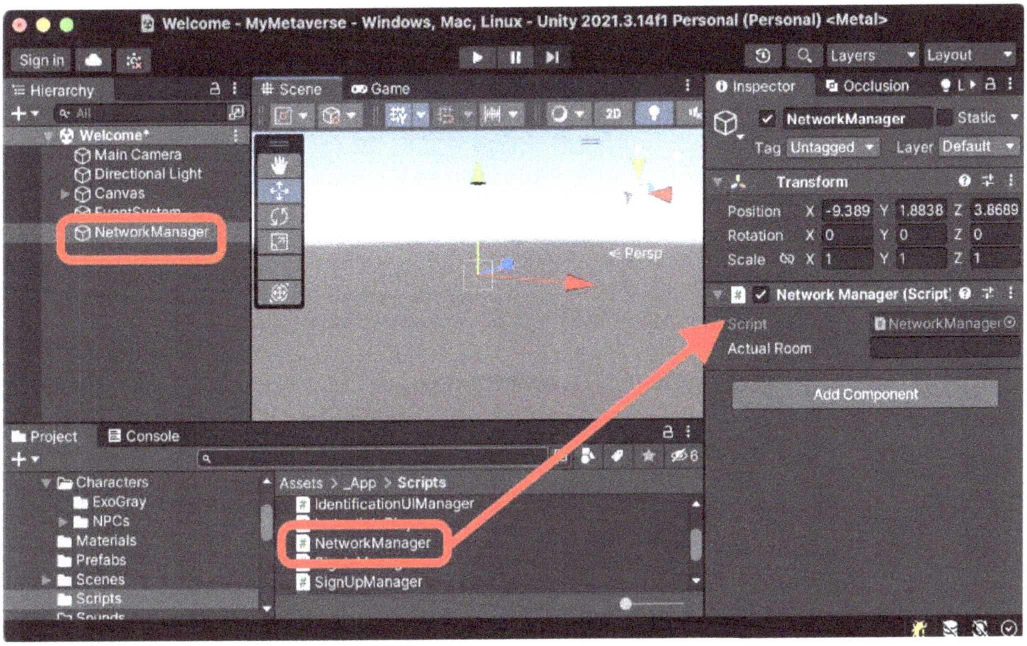

Figure 9.11 – The NetworkManager script as a component

You must copy and paste the **NetworkManager** GameObject into every scene because if we use a singleton pattern, the instance will be unique, and only one instance will be created with the first player entry, which will prevent other players from connecting and synchronizing with the server.

By removing the singleton pattern, when changing scenes, the object is destroyed and we lose the context of the multiplayer server, so the only option left is to copy and paste the object into each scene.

Adapting login for multiplayer support

Good job! We just need to make a few small changes to the **SignInManager** and **SignUpManager** scripts to close the flow.

We need to call the `JoinRoom` method of the **NetworkManager** script once we have logged in or created a new account:

1. First, open the **SignInManager** script located in the **Assets** | **_App** | **Scripts** path.
2. Next, in the `LoginUser` function, replace the `SceneManager.LoadScene("MainScene");` line with `NetworkManager.Instance.JoinRoom("MainScene");`. Your function will look similar to the following code:

```
public void LoginUser()
{
```

```
            if (string.IsNullOrEmpty(usernameInput.text)
              || string.IsNullOrEmpty(passwordInput.text))
            {
                Debug.Log("Missing information in the
                    login form");
                return;
            }
            auth.SignInWithEmailAndPasswordAsync(
                usernameInput.text, passwordInput.text)
                   .ContinueWithOnMainThread(async task =>
            {
                if (task.IsCanceled || task.IsFaulted)
                {
                    Debug.LogError("SignInWithEmailAndPass
                        wordAsync encountered an error");
                    return;
                }
                FirebaseUser existingUser = task.Result;
                Debug.LogFormat("User signed successfully:
                    {0}", existingUser.UserId);

                NetworkManager.Instance.JoinRoom(
                    "MainScene");
            });
        }
```

Great! Now, we will make the same changes to the **SignUpManager** script. Double-click to edit the script located in **Assets** | **_App** | **Scripts**.

3. In the `CreateUserDataDocument` function, replace the `SceneManager.LoadScene("MainScene");` line with `NetworkManager.Instance.JoinRoom("MainScene");`.
4. The `CreateUserDataDocument` function will look like this after the change:

```
    private async Task CreateUserDataDocument()
{
    var currentUser =
       FirebaseAuth.DefaultInstance.CurrentUser;
    DocumentReference docRef =
       db.Collection("Users").Document(currentUser.UserId);


    UserData userData = new UserData()
    {
```

```
        Uid = currentUser.UserId,
        Nickname = nicknameInput.text,
        Username = usernameInput.text,
    };


    await docRef.SetAsync(userData, SetOptions.MergeAll)
        .ContinueWithOnMainThread(task =>
    {
        if (task.IsCanceled || task.IsFaulted) return;


        NetworkManager.Instance.JoinRoom("MainScene");
    });
}
```

With that, we have made the necessary changes to enter the scene via Photon when logging in or creating a new account. If we stop to think about where else we need to make the changes, we will also remember that we make a scene change when we choose to travel with our NPC.

Adapting world navigation for multiplayer support

A change also needs to be made to the script that allows us to travel. At this point, we understand a journey as a change of scene, and also of a room. The Photon server, as we saw in the introduction, allows the server to be segmented into multiple rooms, and players can enter and leave those rooms. In our metaverse, a world, or a scene, is equivalent to a room in Photon. We will make a few changes so that changing a scene also means changing a room.

To make the change there too, we need to locate the **WorldItemBehaviour** script in the **Assets** | **_App** | **Scripts** path. Double-click to edit it.

As we have done with the previous scripts, we must replace the `SceneManager.LoadScene(sceneName);` line with `NetworkManager.Instance.JoinRoom(sceneName);`.

The code will look like this after the change:

```
using TMPro;
using UnityEngine;
using UnityEngine.UI;

public class WorldItemBehaviour : MonoBehaviour
{
    // In this variable we will connect the GameObject Name
       for later access to its properties.
    public TextMeshProUGUI worldNameGameOject;
```

```
    // In this variable we will store the name of the scene
        where we want to travel.
    public string sceneName;

    // In this variable we will store the public name of
       the world
    public string worldName;


    // When instantiating the GameObject, the script will
       assign the value of the variable worldName to the
       GameObject Name
    void Start()
    {
        worldNameGameOject.text = worldName;
    }
    // Clicking the Join button will call this function and
       change the scene.
    public void JoinWorld()
    {
        NetworkManager.Instance.JoinRoom(sceneName);
    }
}
```

Excellent! With these changes we have made to our scripts, everything should work. If you click **Play** and try the game, in the **Console** panel, you will see how Photon is recording the actions that happen in the game. You can try changing scenes and you will see that everything works as before except that now our world is multiplayer.

Testing the changes

To test the game, you must start it from the **Welcome** scene since this is the entry point where we connect for the first time in Photon:

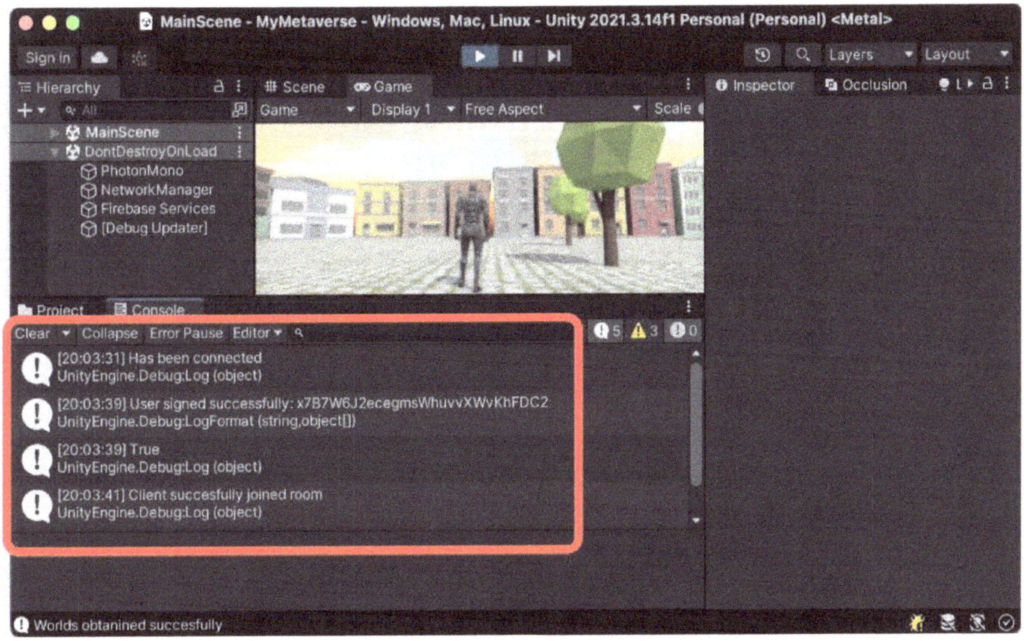

Figure 9.12 – Photon logs

Congratulations! You have done a great job and managed to turn our metaverse into a multiplayer place. Now, all the moves and scene changes we make will be shared with other players automatically.

Next, we will do some testing to make sure that everything works as it should with other players.

Testing synchronization with another user

Testing a local multiplayer game is a tedious and complicated task. We need to replicate a local development environment that allows at least two players to connect so that they can test our code.

One of the options that comes to mind is to create a build of our project and run it on other computers, or even launch it on our computer and then launch the project in the Unity editor to test two players.

But this would be terrible – we would have to compile for every test we do and, if our code fails, we would have to create a new build for each fix. Fortunately, a team of developers have developed a tool that will change our lives called **ParrelSync**.

Introducing ParrelSync

ParrelSync is an extension for the Unity editor that allows users to test a multiplayer game without building the project by having another Unity editor window open and reflecting the changes of the original project.

In short, ParrelSync will create a copy of our project that can refresh the changes we make in the original instance, so we will always have the same code in both projects. Thanks to ParrelSync, we can open two instances of Unity at the same time and in this way, test the multiplayer functionality with two users, without the need to recompile the project each time.

Installing ParrelSync

To download ParrelSync in our project, we must follow these steps:

1. Go to the ParrelSync repository page: `https://github.com/VeriorPies/ParrelSync/releases`.
2. Download the latest `.unitypackage` file. In my case, it's `ParrelSync-1.5.1.unitypackage`.
3. Once downloaded, open it so that Unity invites us to install it, and click the **Import** button:

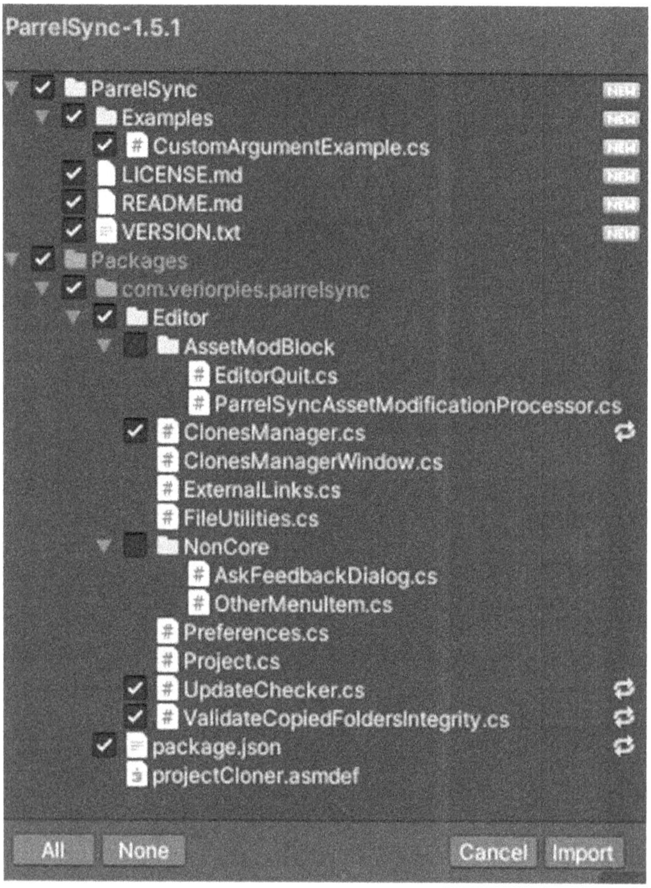

Figure 9.13 – The ParrelSync import window

Once the asset has been imported into our project, it enables a new menu in Unity's main menu bar called **ParrelSync**:

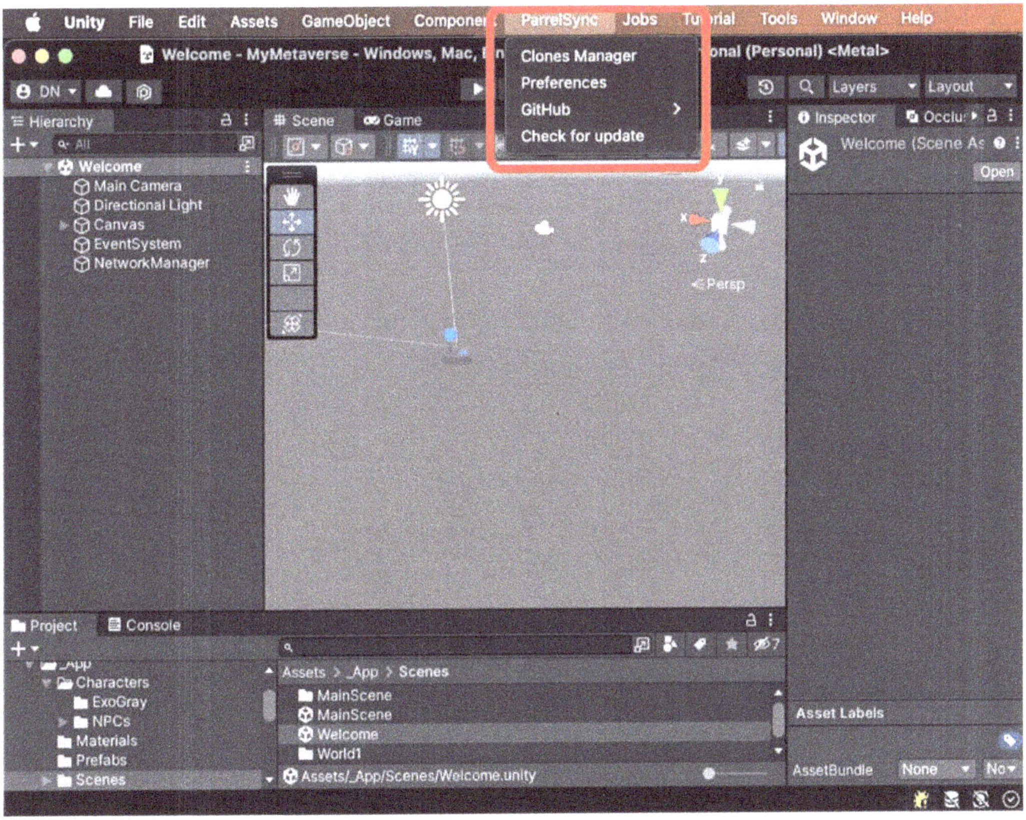

Figure 9.14 – The ParrelSync menu

To create a clone of our project, follow these steps:

1. Click on the **ParrelSync** | **Clones Manager** option in the main menu bar.
2. Click the **Create new clone** button. Once the process has finished, we will have a copy of our project in the folder where we have the original. For example, in my `Projects` folder, I have the original and the new copy:

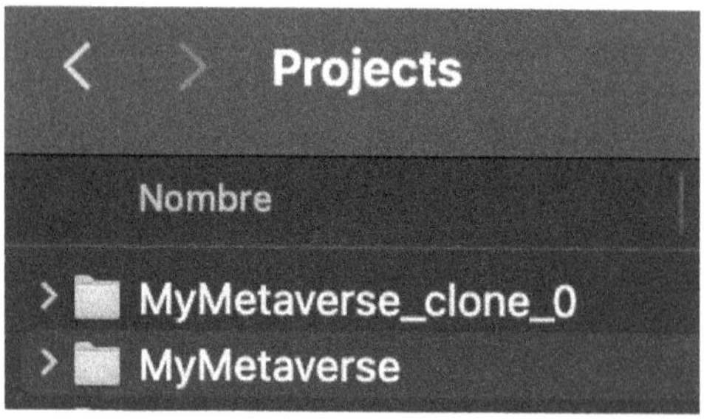

Figure 9.15 – My Projects folder

3. To open the new project, you can add it to the **Projects** list in Unity Hub:

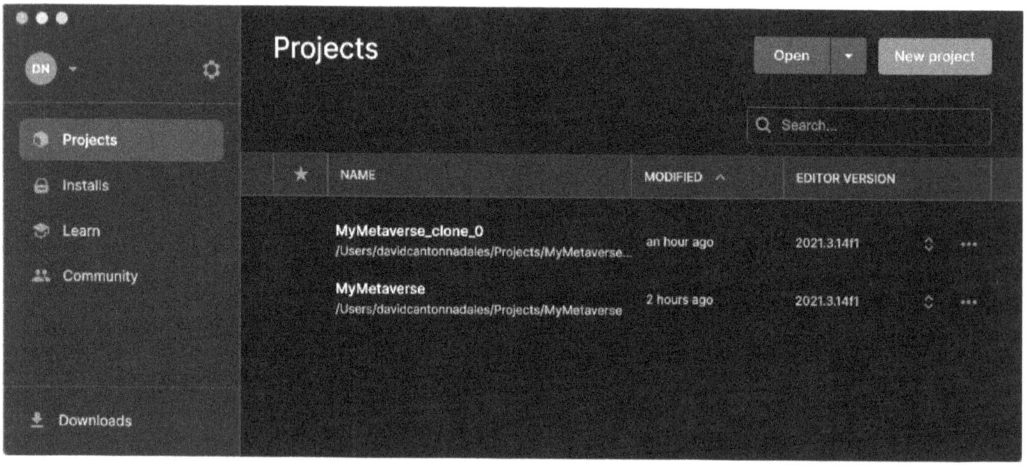

Figure 9.16 – Duplicated project in Unity Hub

4. Once the project has been imported into Unity Hub, you can open it like any other project – it will open in a new window.

Great! You now have ParrelSync installed and configured with a clone of our project. Now, it's time to test the changes we have made to the project to make it multiplayer.

Tip: more about ParrelSync

You can learn more about ParrelSync on its GitHub page: `https://github.com/VeriorPies/ParrelSync`.

Remember that any changes to code or scenes will be automatically replicated in the copy made. Before we continue, if we launch a test now, there will be a crash and we won't be able to continue on the second instance.

This is because Firebase shares the instance in a place on our hard disk, so when a second instance tries to access it, it will cause an error that prevents us from continuing. To fix this, we need to add a line of code that configures Firebase so that it doesn't create an instance with persistence. Follow these steps:

1. Locate the **NetworkManager** script located in **Assets** | **_App** | **Scripts** and open it.
2. Add the highlighted code snippet to the `Awake` function:

```
// In the Awake method we will set up a Singleton
   instance of this script, to ensure that only
   one instance exists during the whole game.
void Awake()
{
    // Prevents Firebase from blocking the
       instance when testing with the Unity Editor
       and making it inaccessible from the cloned
       project.
#if UNITY_EDITOR
    FirebaseFirestore.DefaultInstance.Settings
        .PersistenceEnabled = false;
#endif


    if (auth == null)
    {
        auth = FirebaseAuth.DefaultInstance;
    }

    // Basic configuration recommended by Photon
    PhotonNetwork.AutomaticallySyncScene = true;
    PhotonNetwork.SendRate = 15;
    PhotonNetwork.SerializationRate = 15;
    PhotonNetwork.PhotonServerSettings.AppSettings
        .AppVersion = "0.1";
    PhotonNetwork.GameVersion = "0.1";
    PhotonNetwork.GameVersion = GameVersion;

    // Launch the connection
    PhotonNetwork.ConnectUsingSettings();
}
```

Great! With this small change, we can start testing. Testing the operation is as simple as running the **Welcome** scene in each instance we have open and proceeding to the normal flow.

In my case, as I only have one user created, I proceeded to log into the first instance (the original one). In the second one, I created a new account. Now, both players are in the scene:

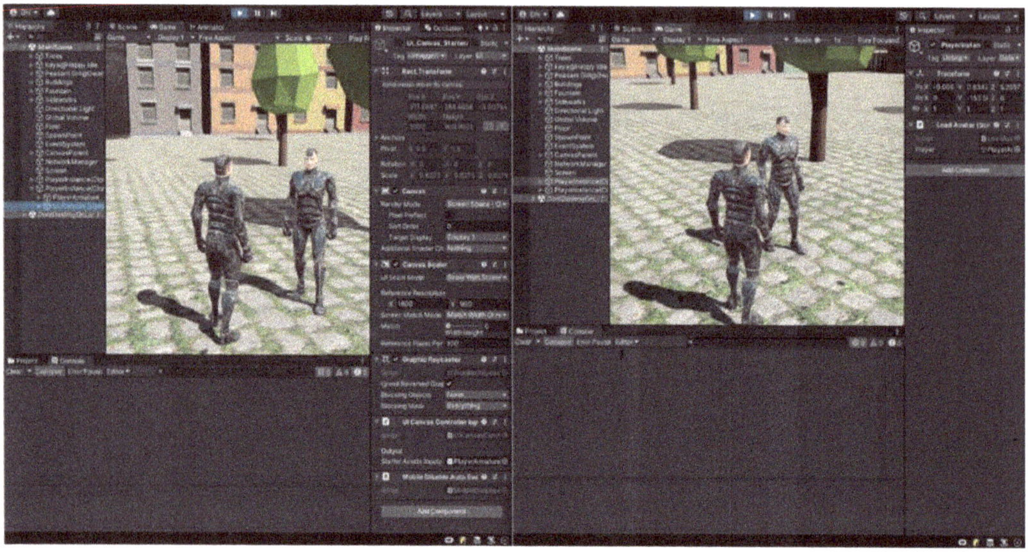

Figure 9.17 – Two players in the same scene

You can try playing with both instances; you will see that the movements and animations are perfectly coordinated. Amazing, isn't it? Great job! You have completed this amazing chapter and you are now an expert in multiplayer games with Photon SDK.

You have seen how easy it was to integrate Photon SDK into our Unity project and the quality of the result.

Summary

Throughout this chapter, we learned all about Photon and we created a new project in the **Dashboard** area. We also learned how to convert our single-player metaverse into a spectacular multiplayer metaverse. With Photon SDK, we managed to easily add new scripts that can synchronize all the players, along with their animations and movements, on the network.

We also learned how to change scenes and have them transmitted to the network to see how users appear and disappear from our scene when we change rooms in Photon.

In the next chapter, we will learn how to integrate two new services offered by Photon – Photon Chat and Photon Voice. With these services, we will add voice chat and text chat features to our metaverse.

10

Adding Text and a Voice Chat to the Room

In the previous chapter, we discovered what Photon offers us to turn our project into a multiplayer world. But the Photon SDK has much more to offer.

Integrating voice and audio chat with Unity is very easy thanks to its SDK. Throughout this chapter, we will learn how to create a voice chat and text chat project in the Photon dashboard, which we will then connect to our Unity project for further development.

By the end of this chapter, your wonderful metaverse will have two amazing new features for your users, a text chat and a voice chat, which will allow connected users to chat with each other.

We will cover the following topics:

- Getting started with Photon Chat
- Creating a chat window
- Getting started with Photon Voice
- Transmitting your voice to the network

Technical requirements

This chapter does not have any special technical requirements, but as we will work with programming scripts in C#, it would be advisable to have a basic knowledge of this programming language. We need an internet connection to browse and download an asset from the Unity Asset Store.

We will continue working on the project we created in *Chapter 1, Getting Started with Unity and Firebase*. Remember that we have a GitHub repository, `https://github.com/PacktPublishing/Build-Your-Own-Metaverse-with-Unity/tree/main/UnityProject`, which contains the complete project that we will work on here.

You can also find the complete code for this chapter on GitHub at: `https://github.com/PacktPublishing/Build-Your-Own-Metaverse-with-Unity/tree/main/Chapter10`

Getting started with Photon Chat

Now, we will discover Photon Chat; we will learn practically how to create a chat window that is connected to the room where the player is and we will test its functioning later.

Introduction to the Photon Chat SDK

The Photon Chat SDK is a real-time chat solution developed by Photon Engine. It is a set of tools and services that allows developers to easily add real-time chat functionality to their Unity 3D games.

This SDK provides essential features for real-time chat, such as public and private messaging. It is surprisingly easy to integrate into a Unity 3D project, as Photon provides a Unity 3D plugin and detailed documentation to guide developers.

By using the Photon Chat SDK in your Unity 3D project, you can enhance your users' gaming experience by allowing them to interact and communicate in real time. This can be especially useful for online multiplayer games, where communication is essential for a smooth and successful gaming experience.

In addition, the Photon Chat SDK is scalable and designed to support large volumes of concurrent users, making it ideal for popular games and large gaming communities.

Creating a Photon Chat project

As we did in the previous chapter, we must create a new project in the Photon dashboard, but this time, it will be a Chat project. To do this, we will follow these steps:

1. Go to the Photon dashboard at `https://dashboard.photonengine.com`.
2. Click on the **CREATE A NEW APP** button.

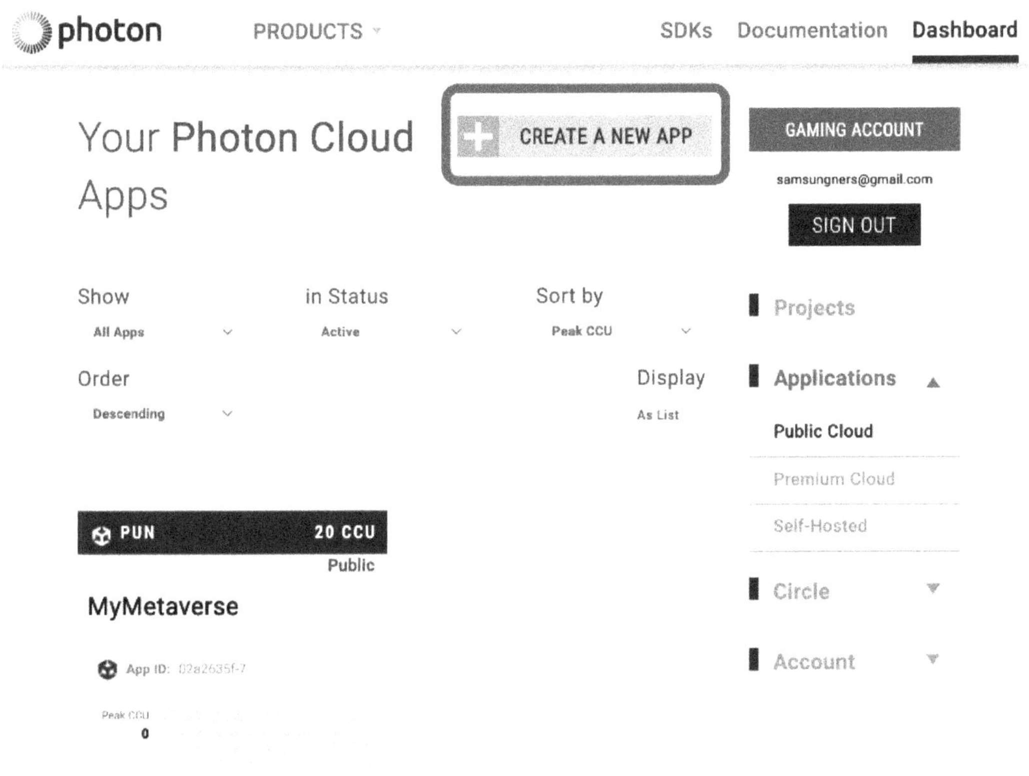

Figure 10.1 – CREATE A NEW APP button in the Photon dashboard

3. Set **Application type** to **Multiplayer Game**.
4. Select the **Chat** option in **Photon SDK Type**.
5. Fill in the **Application Name** field with the name of your choice. In my case, I have chosen `MyMetaverseChat`.
6. Finally, click on the **Create** button to finish.

Good job, you have finished creating a new project that will host our future chat. If everything went well, you should now see two projects in the Photon dashboard.

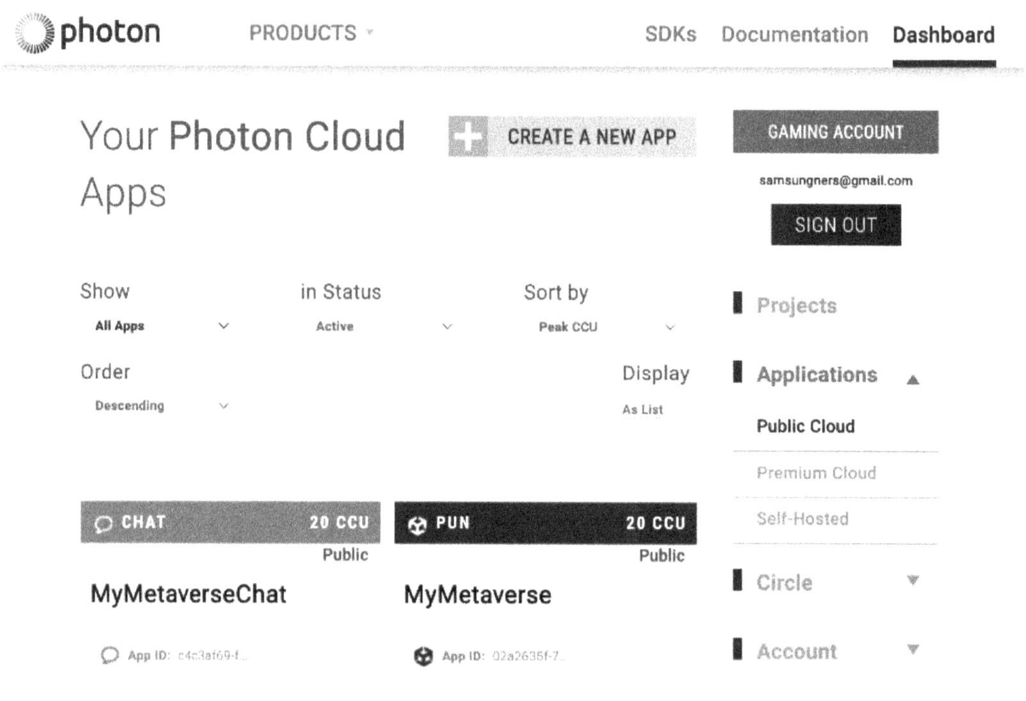

Figure 10.2 – Photon dashboard

Don't close the browser yet. We will need the app ID of the Chat project to configure it in Unity.

Configuring the project connection in Unity

Now that we have created the project in the Photon dashboard, we need to set up the connection to our project in Unity. This is really easy; just follow these steps:

1. Copy the Chat project's app ID to the Photon dashboard.
2. In Unity, in the main menu bar, select **Window** | **Photon Unity Networking** | **Highlight Server Settings**.
3. Paste the app ID you previously copied into the **App Id Chat** box.

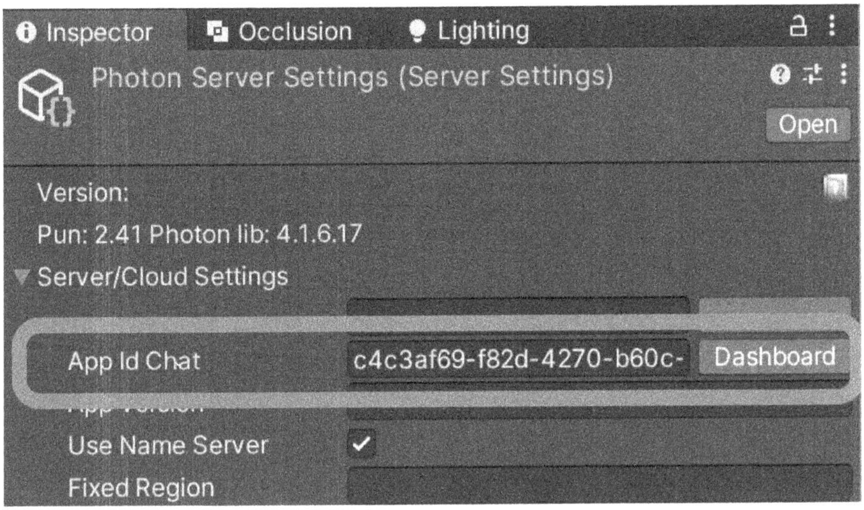

Figure 10.3 – Configuring the app ID for Photon Chat

Easy, isn't it? With these last steps, you have the Photon Chat project fully connected to the Unity project.

Next, we will focus on the chat window.

Creating a chat window

Personally, I like to optimize tasks as much as possible. A lot of the time, we waste a lot of time creating things from scratch when they already exist or we can simply reuse something else. This is the case with the chat feature we are going to implement.

We could spend more than 100 pages creating a chat interface from scratch, but what if Photon already gave us a good basis to start with?

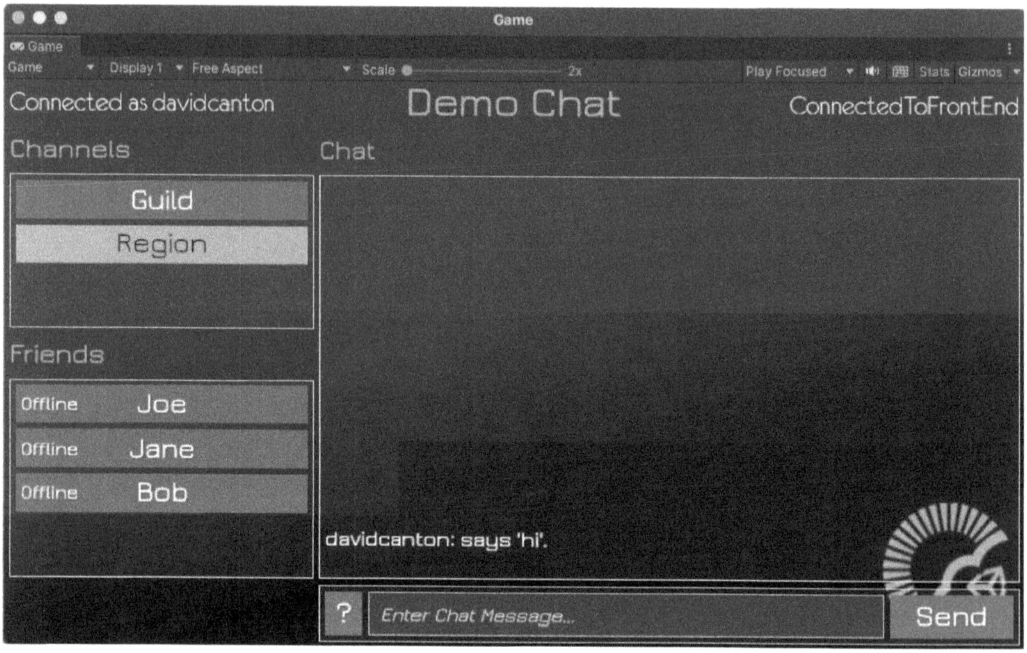

Figure 10.4 – Photon Chat example scene

As you can see in the preceding screenshot, Photon has done much of the work for us. This screenshot is of the chat interface that can be found in one of their example scenes. Specifically, you can test it by opening the **DemoChat** scene found in **Assets** | **Photon** | **Photon Chat** | **Demos** | **DemoChat**.

The strategy we are going to follow is to reuse this wonderful window that Photon offers us. We will remove the functionalities we don't need and, in this way, we will have the chat screen finished and fully functional amazingly fast. Are you ready?

Copying the essentials

As previously mentioned, we will use the chat from the Photon example scene, copy the essential elements, and from there, clean up what we don't need and adjust the elements.

To begin with, the first thing we will do is copy the elements to our scene. To do so, follow these steps:

1. Open the **DemoChat** scene found under **Assets** | **Photon** | **Photon Chat** | **Demos** | **DemoChat**.
2. Select and copy the **Scripts** and **Canvas** GameObjects.

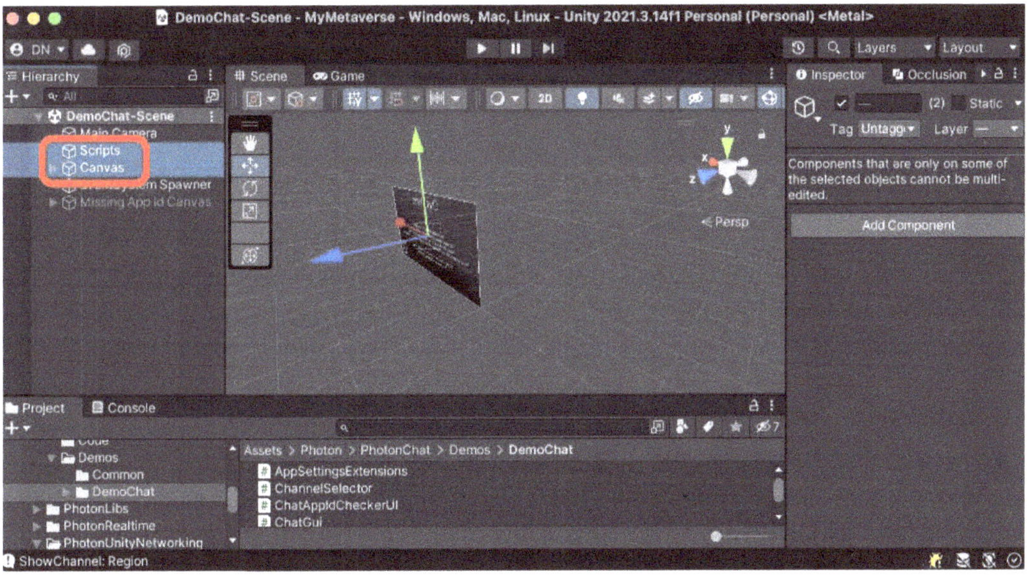

Figure 10.5 – Copying example GameObjects

3. OK, now let's paste them into our main scene. Open the **MainScene** scene, which is located in **Assets** | **_App** | **Scenes**.
4. Paste the elements into the **Hierarchy** panel.

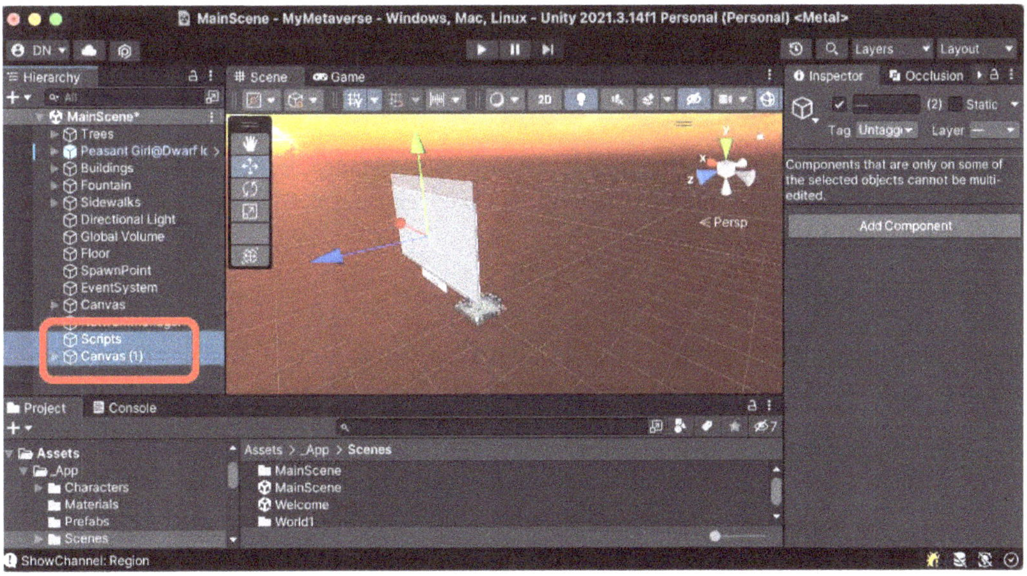

Figure 10.6 – Pasting example GameObjects

Great, now we have the necessary tools to work with. First, we will apply some order to the elements we have just glued.

5. Select the **Scripts** GameObject and give it a more descriptive name; for example, I will use the name `ChatScripts`.
6. Do the same for the **Canvas (1)** GameObject; give it a more descriptive name. I will use the name `ChatCanvas`.

You may ask, what have I copied and pasted? We will now go through a detailed explanation of each of the GameObjects that we have copied over from the Photon Chat example scene. It is important to know how the whole system works in order to be able to make future modifications to your project.

ChatScripts

The GameObject we have called **ChatScripts** carries all the logic of the communication system. If you select it, you will see in the **Inspector** panel that it has a component called **Chat GUI** and another one called **Name Pick GUI**:

- **Chat GUI**: This script implements the `IChatClientListener` interface of the Photon Chat SDK; that is, it creates and listens to different events, for example, receiving a message, entering a room, and leaving a room. In addition, it refers to the visual elements that we can find inside the **ChatCanvas** GameObject, with the intention of showing them, hiding them, or modifying their value depending on the different events that occur.
- **Name Pick GUI**: This script is in charge of showing a window for the user to enter their nickname. This window will not be needed, so we will remove it, as we will connect to the chat with the nickname the user will have previously registered with.

ChatCanvas

This element contains all the visual GameObjects that shape a chat window. Here, we will make changes to adapt it to our needs. As you may have noticed, the **ChatScripts** GameObject references elements of this canvas in its **Chat GUI** script.

Modifying ChatCanvas

Fantastic, now that we know well what we have in the scene, let's get to work. The first thing we are going to do is to clean the elements inside **ChatCanvas** that we are not going to need. To do this, follow these steps:

1. Removes the **User Id Form Panel** GameObject.
2. Remove the **Demo Name** GameObject.

3. Remove the **Watermark** GameObject.
4. Finally, it also removes the Editor **SplashScreen Panel** GameObject.

   Okay, after removing these elements, your **ChatCanvas** GameObject will look similar to as in the following screenshot.

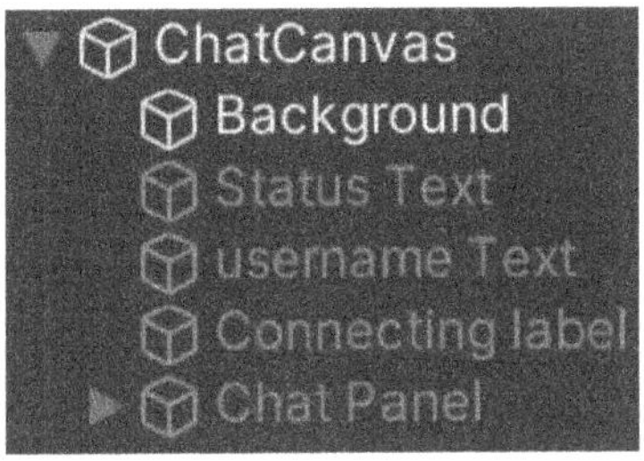

Figure 10.7 – ChatCanvas hierarchy

We have removed the GameObjects we don't need from the parent **ChatCanvas** component. Now, we need to modify the child element called **Chat Panel** as well. Well, if we deploy the **Chat Panel** GameObject, we will find the following:

Figure 10.8 – Chat Panel hierarchy

To understand what each element is and which one we are going to eliminate, let's review each of them:

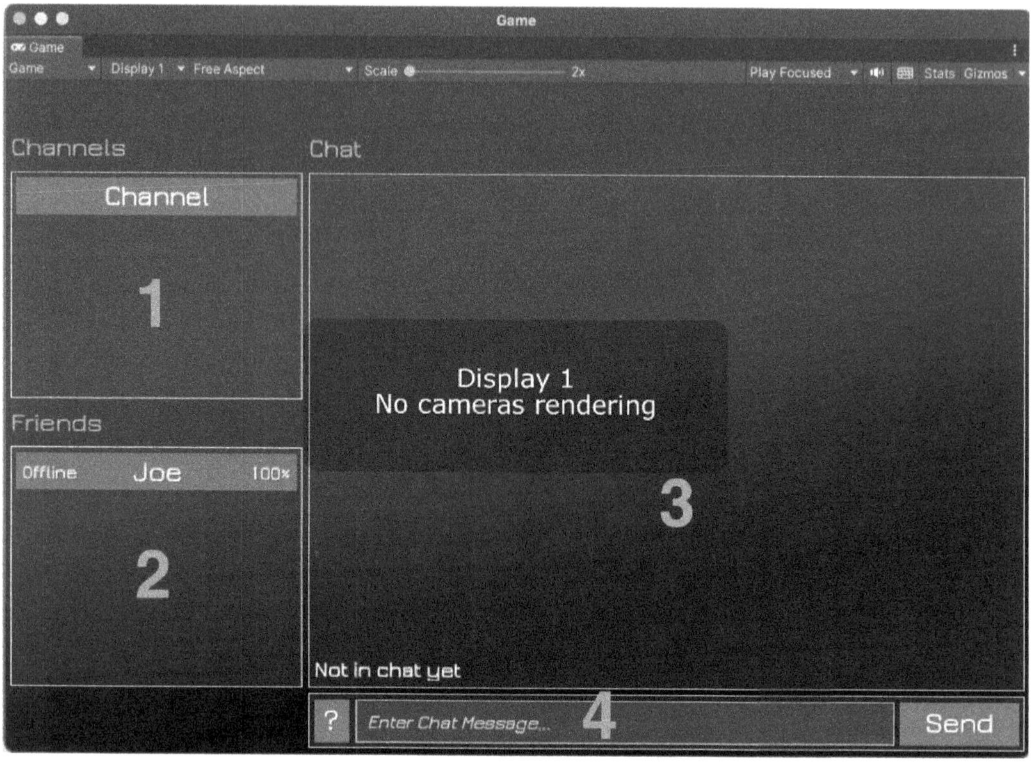

Figure 10.9 – ChatCanvas layout

The preceding figure depicts the **Chat Panel** GameObject. It is composed of four GameObjects:

1. **ChannelBar Panel**: Photon, in its demo scene, allows users to connect to receive and send messages in multiple rooms at the same time. We will take advantage of this wonderful functionality and upload our available worlds. This will allow us to communicate with other players in different rooms.
2. **Friends Panel**: In the Photon demo, this option is not finalized and simply shows fake users. We don't want to complicate the chat system too much and will end up removing this GameObject.
3. **ChatOutput Panel**: This is where the messages in the room are displayed, from both other users and us.
4. **InputBar Panel**: This is where the system allows us to write and send messages.

Great, now that you have a better understanding of the system we are setting up, let's proceed to remove the GameObject that displays the friends list. To do this, delete the item called **Friends Panel**.

For now, we will not make any more changes to the visual aspect of the chat. The next thing we will do is modify the scripts.

Modifying ChatScripts

Alright, now let's get down to business. We have already prepared the visual aspect of our chat. Now we are going to modify the necessary scripts to make it work perfectly.

Let's focus on the **ChatScripts** GameObject. Select it and look at its **Inspector** panel. As previously mentioned, we won't need the component called **Name Pick Gui**, so we are going to remove it.

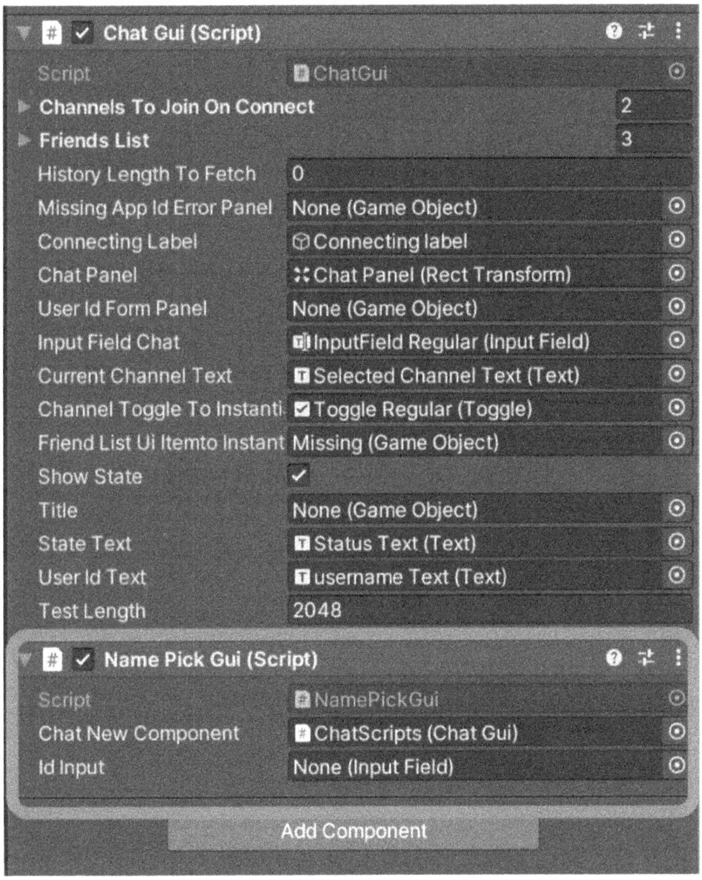

Figure 10.10 – Deleting the Name Pick Gui component

Once we have removed this component, we will only have **Chat Gui**, which is what we really need. We will also need to remove the component-dependent script; that is, we need to remove the `NamePickGui.cs` script as it will be affected by the modifications we will make to `Chat Gui` and will give an error later on.

To do this, delete the `NamePickGui.cs` file located in the **Assets** | **Photon** | **PhotonChat** | **Demos** | **DemoChat** path.

Now comes the hard work. We are going to modify the content of the `Chat Gui` script to adapt it to our new needs. To do this, double-click on the script to edit it. Wow, that's a lot of code; it's tremendous, but don't worry. Add the following highlighted lines of code to the existing code:

```
using System.Linq;
...
namespace Photon.Chat.Demo
{
    public class ChatGui : MonoBehaviour,
        IChatClientListener
    {
...
        FirebaseAuth auth;
        FirebaseFirestore db;
...
        public void Awake()
        {
            if (auth == null)
                auth = FirebaseAuth.DefaultInstance;

            if (db == null)
                db = FirebaseFirestore.DefaultInstance;
        }

        public void Start()
        {
            DontDestroyOnLoad(this.gameObject);

            GetWorlds();
...
        }

        public void Connect()
        {
...
            var currentUser =
                FirebaseAuth.DefaultInstance.CurrentUser;
            this.chatClient.AuthValues = new
                AuthenticationValues(
                    currentUser.DisplayName);
...
        }
...
        void GetWorlds()
```

```
            {
                CollectionReference docRef =
                    db.Collection("Worlds");
                docRef.GetSnapshotAsync()
                    .ContinueWithOnMainThread(task =>
                {
                    if (task.IsCanceled || task.IsFaulted)
                        return;
                    Debug.Log("Worlds obtanined succesfully");
                    QuerySnapshot worldsRef = task.Result;
                    ChannelsToJoinOnConnect =
                        worldsRef.Documents.Select(e =>
                    {
                        World world = e.ConvertTo<World>();
                        return world.Scene;
                    }).ToArray();
                    Connect();
                });
            }
        }
}
```

What we have done in this script is removed the references to the visual objects that we previously removed from the **ChatCanvas** GameObject. We have also added bold lines.

We have created a new method called `GetWorlds`, which will get all the worlds from our Firestore database and use them as chat rooms. This way, we will be able to send messages to different worlds. Sounds amazing, doesn't it?

> **Tip: Get this code on GitHub**
>
> If you want to save time, you can also copy this code from the book's GitHub repository at `https://github.com/PacktPublishing/Build-Your-Own-Metaverse-with-Unity` and navigate to the directory where it is located.

Well, we've done the hard work. Now, all we need to do to close the flow is to create a keyboard shortcut that allows us to open and close the **Chat Panel** while in our metaverse.

Opening and closing the chat window

You must have encountered a game in which to chat you must press a key. This is what we will be implementing. We are going to modify the **ChatGui** script to add a new functionality where when we press a key, for example, the *Tab* key, and the chat opens, and when we want, we can close it by pressing the *Tab* key again.

This is easy to do. Let's get on with it.

The first thing we are going to do is configure the **ChatGui** script so that it does not open the chat screen by default when starting the scene.

To implement this new functionality, follow these steps:

1. Add a new variable called `ChatCanvas` in the area where the other variables are located:

```
public GameObject ChatCanvas;
```

2. Now, after saving the changes to the script, if you select the `ChatScripts` GameObject, you will see the new empty variable appear. Drag the `ChatCanvas` GameObject to the variable.
3. Add the following line inside the `Awake` function:

```
public void Awake()
{
    ChatCanvas.SetActive(false);

    if (auth == null)
    {
        auth = FirebaseAuth.DefaultInstance;
    }

    if (db == null)
    {
        db =
            FirebaseFirestore.DefaultInstance;
    }
}
```

4. In the `Update` function, we will implement a condition that will check whether we've pressed the *Tab* key. Modify the `Update` method to add the following highlighted lines:

```
public void Update()
{
    if (this.chatClient != null)
    {
        this.chatClient.Service();
        // make sure to call this regularly!
           it limits effort internally,
           so calling often is ok!
    }

    // check if we are missing context, which
       means we got kicked out to get back to
```

```
    the Photon Demo hub.
if ( this.StateText == null)
{
    Destroy(this.gameObject);
    return;
}

this.StateText.gameObject
    .SetActive(this.ShowState);

// We check if the user has pressed the
    TAB.
if (Input.GetKeyDown(KeyCode.Tab))
{
    ChatCanvas.SetActive(
        !ChatCanvas.active);
    if(ChatCanvas.active)
    {
        // Allows you to see the mouse
            cursor
        Cursor.lockState =
            CursorLockMode.None;
    }
    else
    {
        // Hides the cursor again
        Cursor.lockState =
            CursorLockMode.Locked;
    }
}
}
```

Fantastic! You have done an outstanding job. Now we have met all the objectives to get our chat up and running. It's time to test it!

Testing our chat

Now we have everything ready to enjoy our chat system. It's time to try it out. Just open the **Welcome** scene and log in to access the main world. Once you are inside, press the *Tab* key and... magic! It works perfectly.

As you can see, you can click on the different rooms named after our existing worlds in the Firestore collection. If you press the *Tab* key again, the window will disappear, just as we programmed.

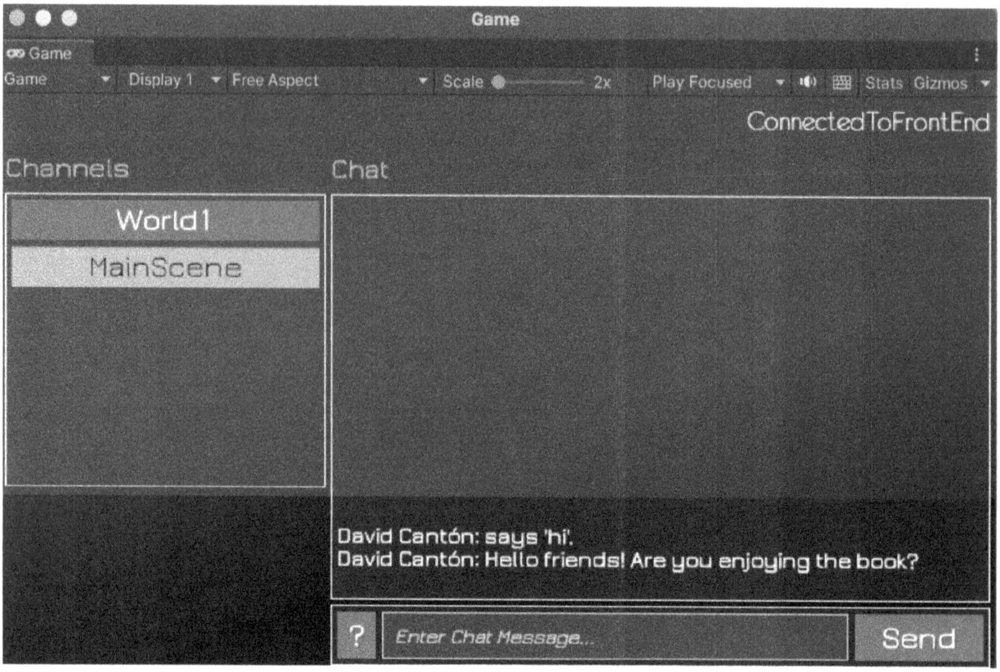

Figure 10.11 – Out chat system

To ensure that we can enjoy the chat feature in all our scenes when we travel with our player, we need to do one last configuration. Follow these steps:

1. Let's move the **ChatCanvas** GameObject into the **ChatScript** GameObject. **ChatScript** has a statement in its script that prevents it from being destroyed during a scene change. This does not happen with **ChatCanvas**, so if we place it as a child, we will also protect it from being destroyed.
2. Finally, in the **Project** panel, we follow the **Assets** | **_App** | **Prefabs** path and drag the **ChatScript** GameObject into the folder to generate a new Prefab.

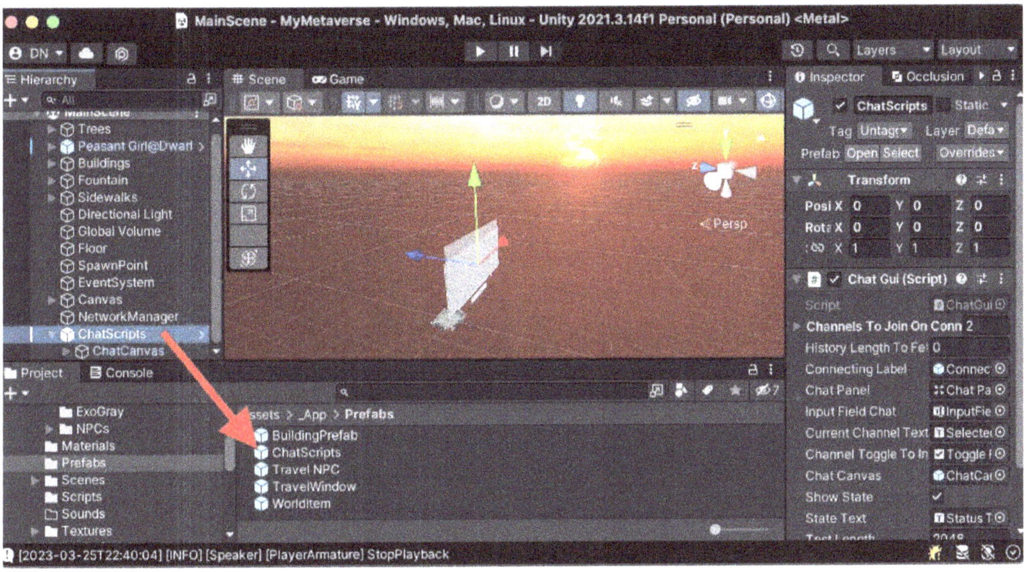

Figure 10.12 – Creating a Prefab with ChatScripts GameObject

Your metaverse already has an impressive chat system, congratulations!

Okay, so far, we have managed to create a chat system that works perfectly. Now we are going to develop another type of chat, the voice chat. You will be surprised how easy Photon makes it to implement voice chat in your project.

Getting started with Photon Voice

We have managed to integrate a chat system into our project, but in this book, we want to go further. Now, we also want our users to be able to communicate by voice, which is also provided by Photon with its Photon Voice SDK.

As with text chat, we will enable a special key to activate the microphone and be able to talk. Are you ready? First of all, let's have a short introduction to the Photon Voice SDK.

Introduction to the Photon Voice SDK

The Photon Voice SDK is a Unity tool that enables real-time voice integration in multi-user applications and games. With the Photon Voice SDK, developers can add voice communication to their Unity projects with ease, allowing users to talk to each other during gameplay.

Some of the advantages of using the Photon Voice SDK in a Unity 3D project are as follows:

- **Easy integration**: The Photon Voice SDK is easy to integrate into any Unity 3D project. Developers can quickly add real-time voice communication to their games or applications.
- **Excellent voice quality**: The Photon Voice SDK uses high-quality voice compression technology to ensure excellent sound quality, even on poor network connections.
- **Reduced latency**: The Photon Voice SDK uses Photon's high-speed network to minimize latency, ensuring smooth and uninterrupted communication.
- **Customization**: The Photon Voice SDK can be customized to suit the specific needs of your project. Developers can control the voice quality, volume, and other aspects of voice communication.

In summary, the Photon Voice SDK is a powerful tool that can benefit any Unity 3D project that requires real-time voice communication. With its ease of use, high sound quality, and reduced latency, the Photon Voice SDK is an excellent choice for any developer looking to add voice communication to their project.

Creating a Photon Voice project

As we did before for Chat, we need to create a new project in the Photon dashboard, but this time, it will be a Voice project. To do this, we will follow these steps:

1. Go to the Photon dashboard at `https://dashboard.photonengine.com`.
2. Click on the **Create a new App** button.
3. Set **Application type** to **Multiplayer Game**.
4. Select the **Voice** option in **Photon SDK** type.
5. Fill in the **Application Name** field with the name of your choice. In my case, I have chosen `MyMetaverseVoice`.
6. Finally, click on the **Create** button to finish.

Excellent, you have finished creating a new project that will host our future voice chat. If all goes well, you should now see three projects on the Photon dashboard.

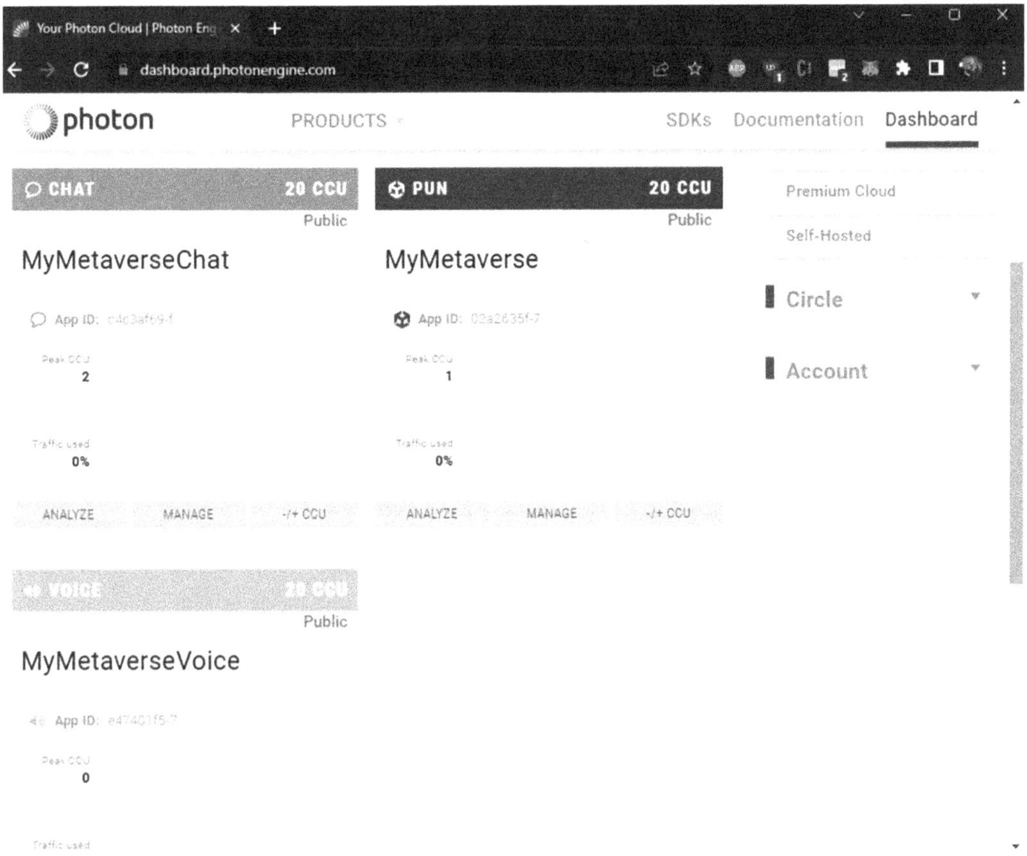

Figure 10.13 – Photon dashboard

As you know, we are going to need the app ID of the new voice chat project to configure Unity, so ensure you copy it.

Installing the Photon Voice SDK

We have to download, install, and configure the Photon Voice SDK in our project. You can find this asset in the Unity Asset Store. To do so, follow these steps:

1. Go to the Unity Asset Store website: `https://assetstore.unity.com`.
2. In the search bar, type `PHOTON VOICE 2` and select the first asset that appears.

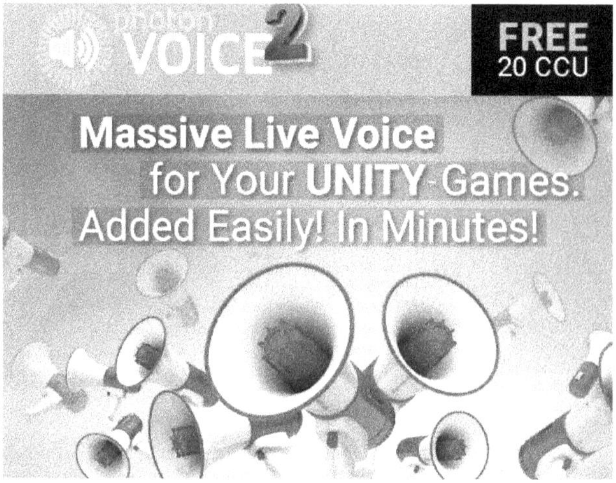

Figure 10.14 – Photon Voice 2 asset preview

3. Click on the **Add to My Assets** button.
4. Click the **Open in Unity** button. This will open the **Package Manager** window in the Unity editor.
5. Click the **Download** button in the **Package Manager** window and then click the **Import** button.

> **Tip: Configuration notice**
> It is possible that when importing the Photon SDK, the Photon configuration file where we have configured the different app IDs is reset. If this happens, don't worry; follow the steps in the *Configuring the project connection in Unity* section to re-insert the necessary app IDs.

Once the SDK has been downloaded and imported, we can continue with the app ID configuration in Unity.

Configuring the project connection in Unity

As we did previously for text chat, we need to set up the connection to our project in Unity. Just follow these steps:

1. In Unity, in the main menu bar, select the **Window** | **Photon Unity Networking** | **Highlight Server Settings** option.
2. Paste the app ID you previously copied into the **App Id Voice** box.

Well, with these simple steps, we now have the Photon Voice SDK configured and connected to our project. We can continue adding voice chat functionality to our project.

Transmitting your voice to the network

Transmitting our voice to the multiplayer server of our metaverse is really easy with Photon. The SDK has ready-to-use components that will make all the hard work easier.

To start, we need to find the Prefab we've used for our character. Find the **PlayerInstance** Prefab, which is located in the **Assets** | **_App** | **Photon** | **PhotonUnityNetworking** | **Resources** path, and double-click to edit it.

With the **PlayerInstance** Prefab in **Edit Mode**, we will follow these steps to add voice functionality:

1. Select the **PlayerArmature** GameObject and display its **Inspector** panel.
2. Add a new component called **Audio Source**.
3. Add another component, called **Photon Voice View**.
4. We will also need another component, called **Recorder**.
5. Finally, add the **Speaker** component.

Great, now the **Inspector** panel of the **PlayerArmature** GameObject will look as in the following screenshot:

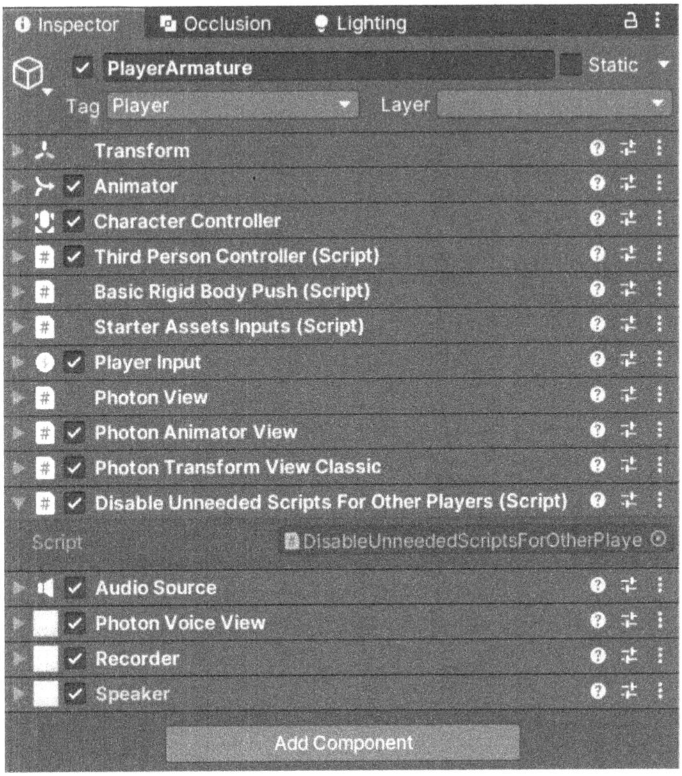

Figure 10.15 – PlayerArmature Inspector panel

Perfect. The next thing we are going to do is to create a script that we will attach to the player's Prefab, which will be in charge of transmitting the voice only when we press a key. Before we go any further, we need to make an adjustment to the **Recorder** component, which we recently added.

We have to deactivate the **Transmit Enabled** option; otherwise, it will send voice recordings constantly, and we only want it to do so when we press a button.

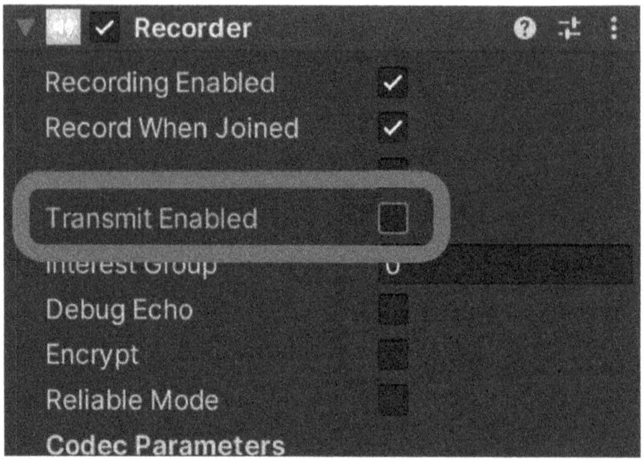

Figure 10.16 – Transmit Enabled option

Great, with this change, we can continue. Now, we will create a simple script that will be in charge of activating the voice transmission when we press a button. To do this, follow these steps:

1. Following the **Assets** | **_App** | **Scripts** path, right-click and click on the **Create** | **C# Script** option. Rename it `PushToTalk`.
2. Double-click to edit the script and replace all the auto-generated code with the following:

```
using System.Collections;
using System.Collections.Generic;
using Photon.Pun;
using Photon.Voice.Unity;
using UnityEngine;

public class PushToTalk : MonoBehaviourPun
{
    public KeyCode PushButton = KeyCode.V;
    public Recorder VoiceRecorder;
    private PhotonView view;
    private bool pushtotalk = true;
    public GameObject localspeaker;

    void Start()
```

```
        {
            view = photonView;
            VoiceRecorder.TransmitEnabled = false;
        }

        void Update()
        {
            if (Input.GetKeyDown(PushButton) &&
                view.IsMine)
            {
                VoiceRecorder.TransmitEnabled = true;
            }
            else if (Input.GetKeyUp(PushButton) &&
                view.IsMine)
            {
                VoiceRecorder.TransmitEnabled = false;
            }
        }
    }
```

The script is very simple. In the `Update` method, we check whether the user has pressed the *V* key to activate the transmission by means of `VoiceRecorder.TransmitEnabled = true;` if, on the other hand, we set the value to `false`, we will stop the voice emission, easy, isn't it?

There is only one step left, and that is to add this script we just created as a component in the **PlayerArmature** GameObject.

3. Finally, before we start testing, we will drag the **PlayerArmature** GameObject to the **Voice Recorder** and **Localspeaker** boxes to complete the configuration.

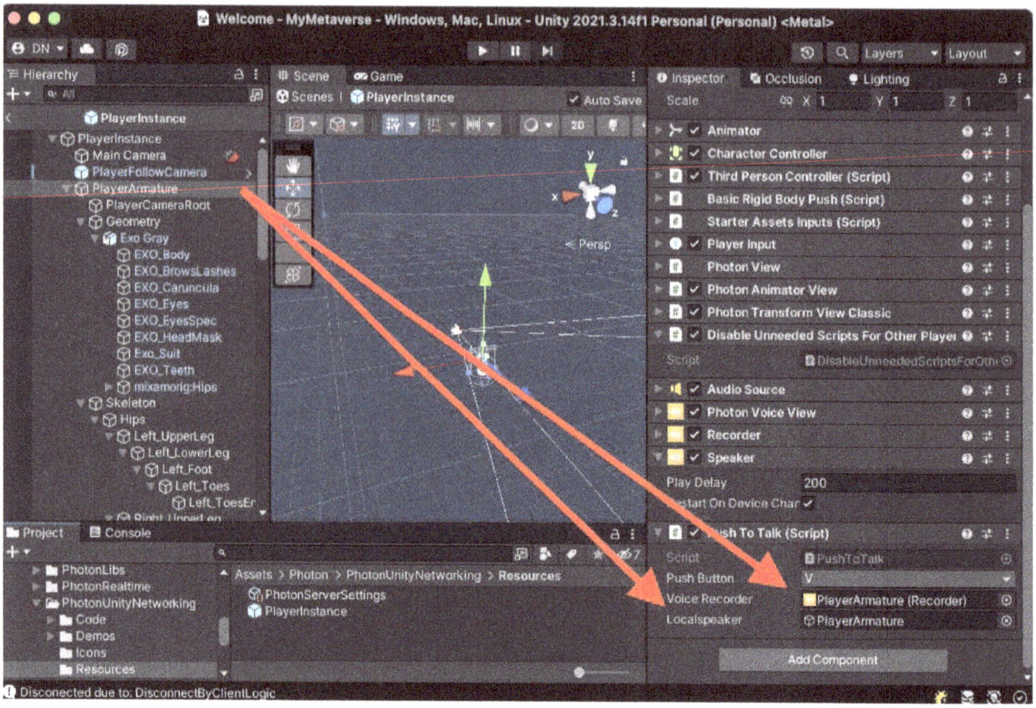

Figure 10.17 – Configuring the Push To Talk component

Good job! You have finished creating the voice chat. You can now try it out and enjoy how much you have learned and achieved during this chapter.

To test it, perform the same actions we learned in *Chapter 9*, *Turning Our World into a Multiplayer Room*. With the cloned project, you can launch two instances of the same project, and in each of the instances, with an identified player, you can test both voice chat and text chat. I promise you that the result will surprise you.

Summary

Throughout this exciting chapter, we learned a lot about what Photon can offer us in terms of multiplayer communication. With Photon Chat, we discovered how easy it is to provide a professional chat service in our project, from downloading the SDK to configuring and programming the scripts needed for operation. We also learned a lot about the Photon Voice SDK and added, in a very few steps, fast and effective communication functionality to our project.

In the next chapter, we will continue to add exciting features to our metaverse, expanding its characteristics. We will add another NPC that will allow us to modify our appearance, that is, choose another avatar.

Part 3: Adding Fun Features Before Compiling

In this last part, we will learn how to create new and fun functionalities in our project, which will provide us with a wide creative spectrum and give us new ideas. We will create a new NPC, this time helping us to change the look of our character by exploring the free Mixamo tool to get new looks and animations. Also, we will learn how to play streaming videos on any surface of our scene. In this case, we will design a cinema screen.

Finally, before compiling our project, we will learn, in a very practical and detailed way, how to prepare and convert our project to be able to run it on our Meta Quest 2 glasses. Finally, we will review new optimization tricks, which will help us to export our Metaverse, for all platforms, that is, Windows, Mac, Linux, Android, and iOS.

This part has the following chapters:

11
Creating an NPC That Allows Us to Change Our Appearance

A fun feature for our metaverse is to allow our users to choose an alternative appearance. This will be very easy as we will follow most of the steps in *Chapter 7, Building an NPC That Allows Us to Travel*. The mechanism we will build in this chapter will allow you to add multiple avatars that your users will be able to select from a custom window.

We will also modify some existing scripts to allow avatar changes to be saved to `PlayerPrefs`, a new concept we will learn about in this chapter, so that when a user logs in again, the previously selected avatar will be loaded.

In the following figure, you can see the final result of the NPC that we are going to build throughout this chapter. It is spectacular, isn't it?

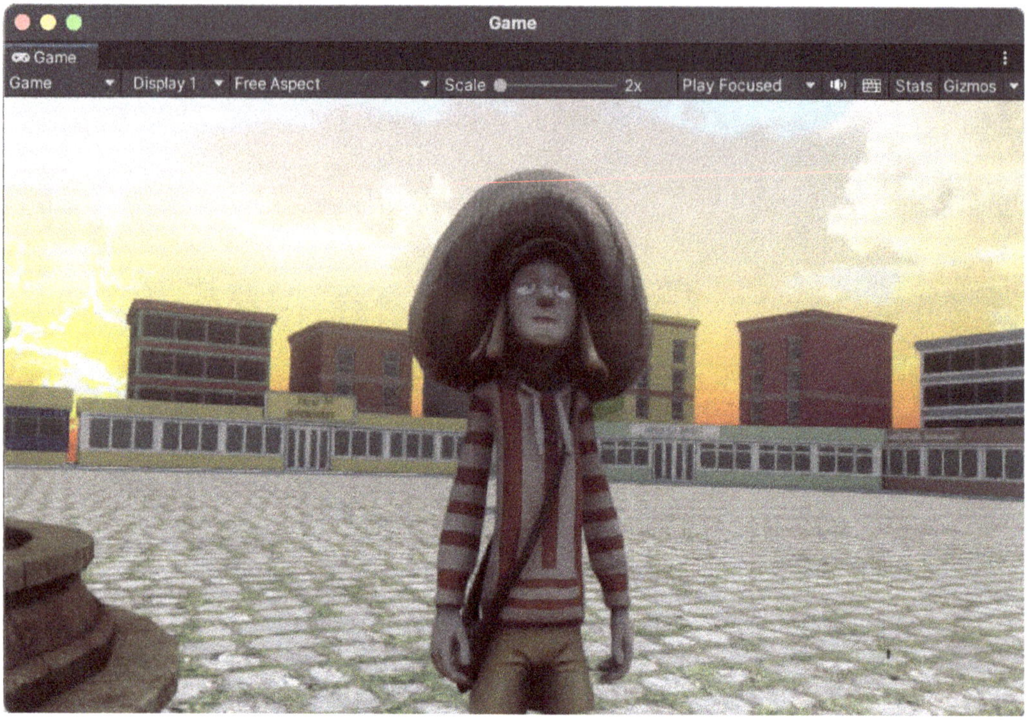

Figure 11.1 – Final result

We will cover the following topics in this chapter:

- Choosing an aspect
- Bringing the NPC to life
- Triggering the NPC when we are close
- Showing the available avatars in a window
- Persisting the new appearance

Technical requirements

This chapter does not require any special technical requirements, but as we will start programming scripts in C#, it would be advisable to have basic knowledge of this programming language. You will need an internet connection to browse and download an asset from the Unity Asset Store.

We will continue on the project we created during *Chapter 1*, *Getting Started with Unity and Firebase*. Remember that we have the GitHub repository (`https://github.com/PacktPublishing/Build-Your-Own-Metaverse-with-Unity/tree/main/UnityProject`), which will contain the complete project that we will work on here.

You can also find the complete code for this chapter on GitHub at: `https://github.com/PacktPublishing/Build-Your-Own-Metaverse-with-Unity/tree/main/Chapter11`

Choosing an aspect

To get a 3D character, we will go back to the Mixamo website, just like we did in *Chapter 2* for our main character. The steps we will take during this section are the same for all the characters we will find on the Mixamo website; what I mean by this is you are totally free to choose the character with the look you like the most from the list.

Without further ado, let's get down to business.

The first thing we will do is choose a character in Mixamo. To do so, follow these steps:

1. Go to the following website: `https://www.mixamo.com`.
2. Sign in with your username or create a new account; it's free.
3. Click on the **Characters** tab and choose the one you like the most. In my case, I have chosen **Kaya**, who has a look that really brings out his clothes, which invites us to think that he can help us with our appearance, don't you think?

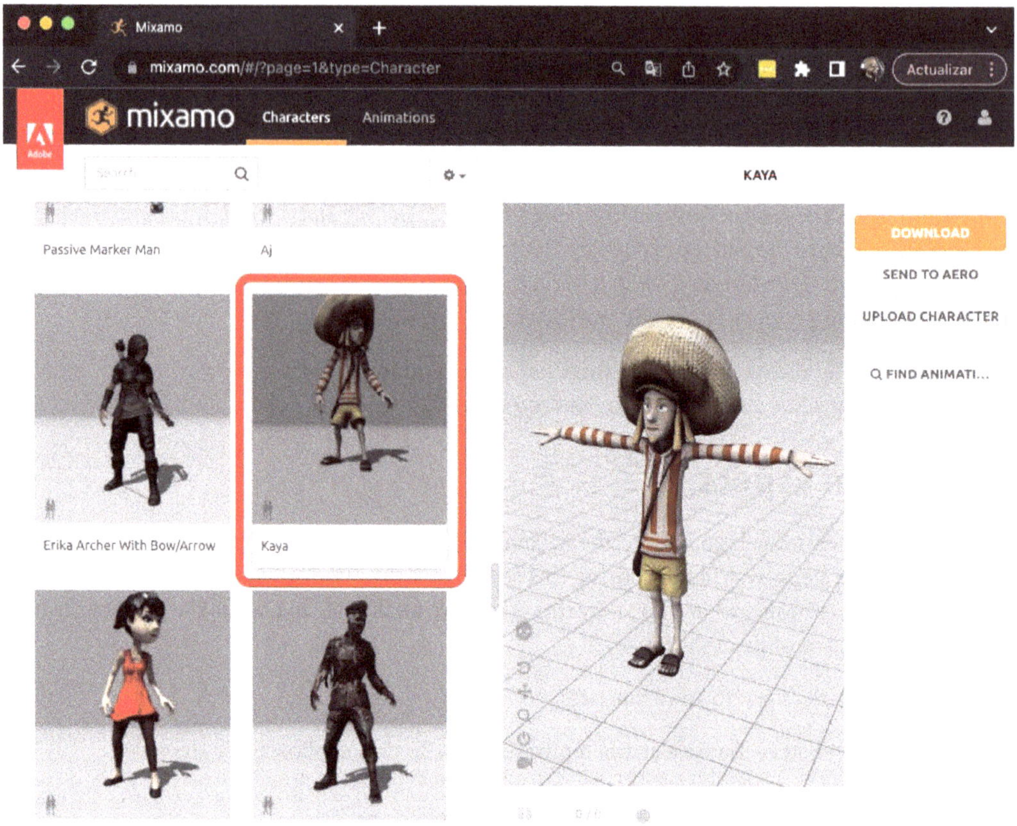

Figure 11.2 – Selecting a character in Mixamo

4. We will take the opportunity to implement a default animation to give the character some life. Click on the button on the right called **FIND ANIMATIONS**.
5. As you can see, there is a long list of animations available. The idea is to implement an animation with a closed loop. The *idle* type animations are perfect for this. Type the word `idle` in the search box and press the *Enter* key.

6. I have selected the animation called **Happy Idle**. Click on it in the list; you will see how it is applied to the character preview.

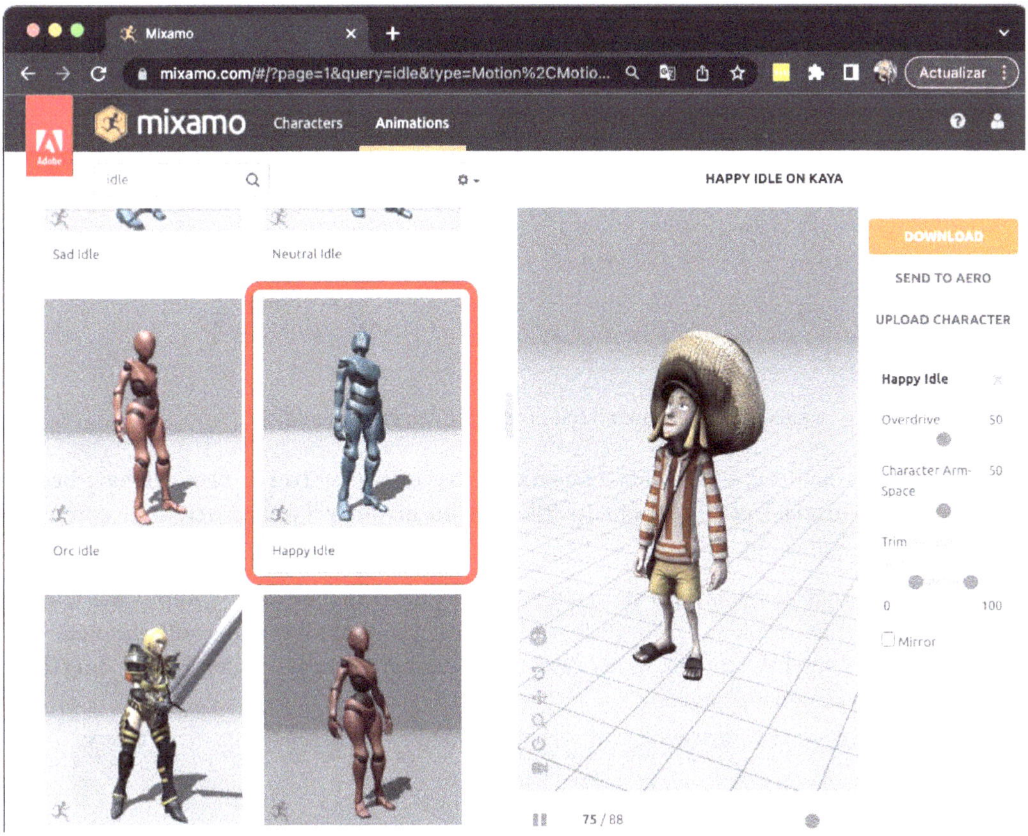

Figure 11.3 – Selecting an animation in Mixamo

7. Now, click on the **DOWNLOAD** button and a modal window will open with configuration options. Set **Format** to **FBX for Unity(.fbx)** and **Skin** to **With Skin**, and click again on the **DOWNLOAD** button.

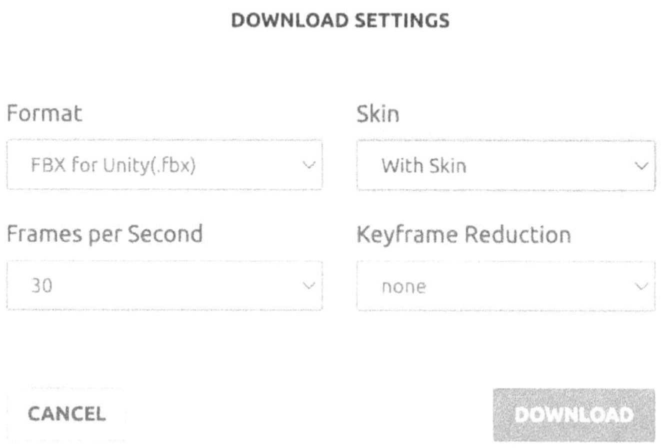

Figure 11.4 – Download options for the selected animation

Great; if all went well, a file called `Peasant kaya@HappyIdle.fbx` will have been downloaded to your `Downloads` folder. The filename may vary if you have chosen a different character or animation.

8. Now, we are going to import this file to our project. To do so, in the **Assets** | **_App** | **Characters** path, create another folder, called `NPCs`, and inside it, create another folder with the name of the character you have chosen. In my case, I have created the `Kaya` folder. Finally, drag the downloaded file into this folder:

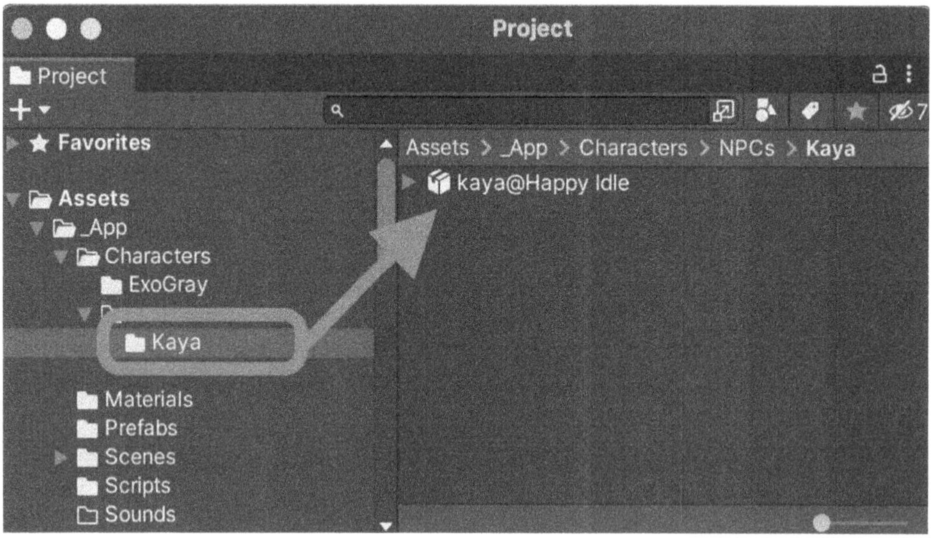

Figure 11.5 – Importing the FBX of the selected character

Good work, we have successfully imported the character file into the project; now we need to apply a bit of configuration to make it work properly.

9. If you select the `kaya@Happy Idle` file in the **Project** panel, you will see the preview on the right, but the character appears completely white. To fix this, at the top of the **Inspector** panel, select the **Materials** tab and click on the **Extract Textures...** button.

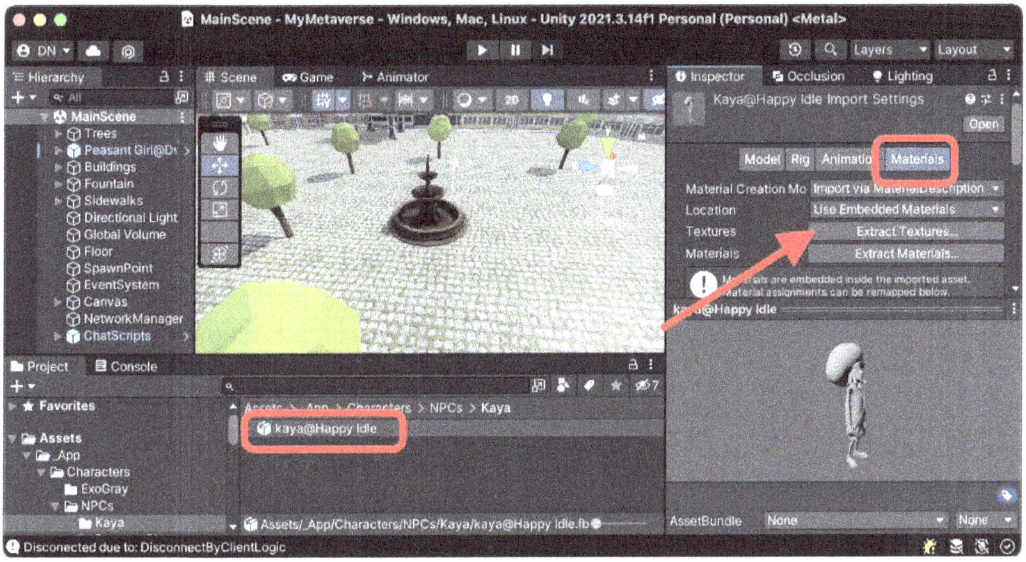

Figure 11.6 – Extracting textures from the FBX

10. Unity will ask you for a destination folder to extract the textures. By default, the same folder as the FBX file will be selected; click **Accept**.

If everything went well, you will see in the `Kaya` folder the textures and the preview with character colors.

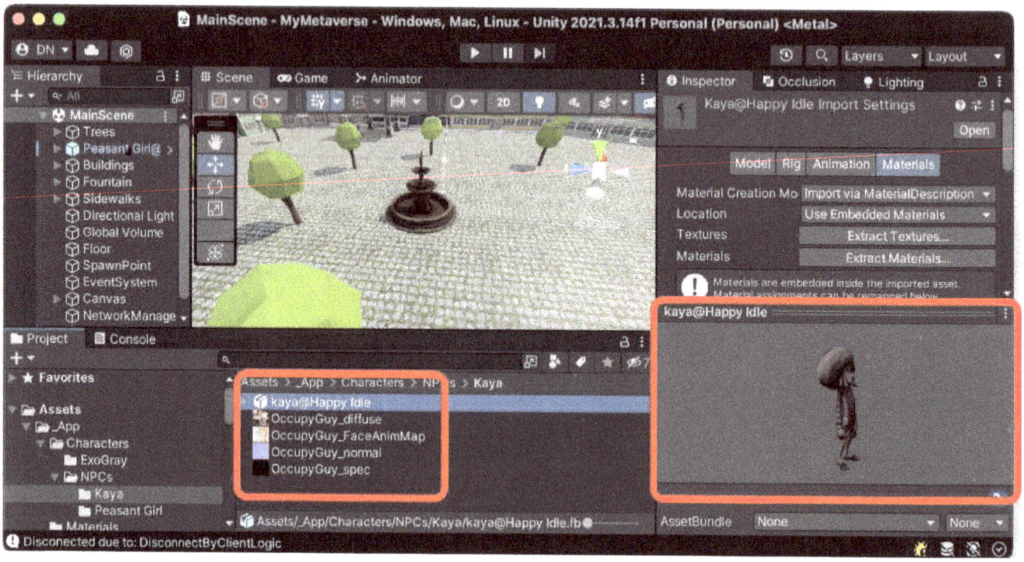

Figure 11.7 – Preview of the FBX with the applied textures

Our character is now set up so we can continue working with it. Next, we will create a new Prefab to house this character and we will learn how to create an Animation Controller and an Animator component that will animate the character once the scene is executed.

Bringing the NPC to life

This NPC will be present in our main world, and when the players want to change their appearance, they will have to come back to our nice village; endearing, isn't it? We will create a Prefab that houses the character and its functionalities.

To achieve this, we will follow these steps:

1. Anywhere in **MainScene**, drag and drop the **Kaya@Happy Idle** asset. If we click **Play**, we will see that the character is static: it has no life, and it doesn't move with the animation we selected in Mixamo. This happens because we need to add an **Animator** component and link the animation.

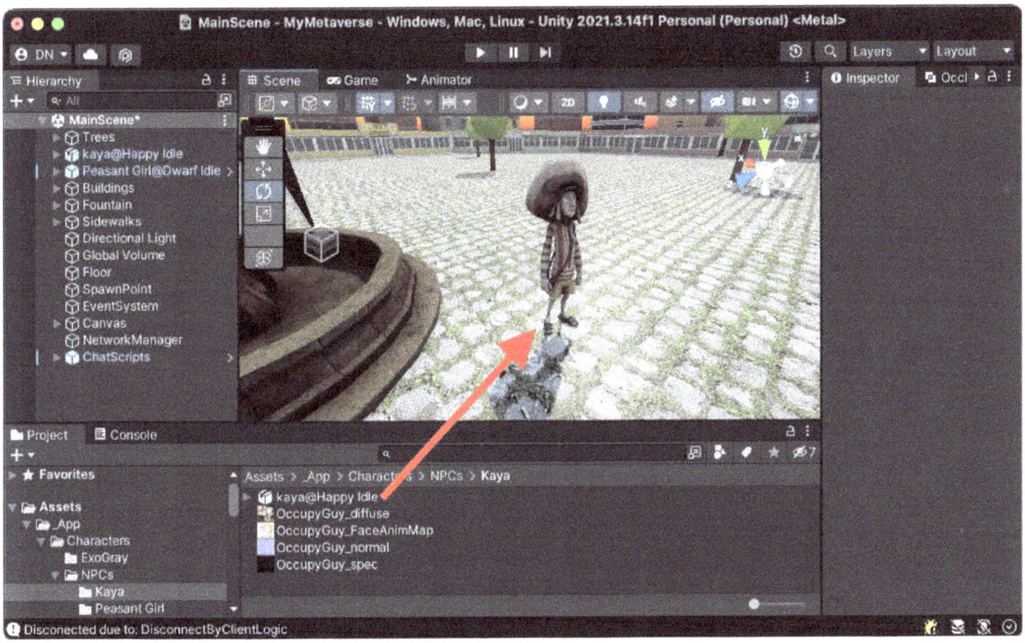

Figure 11.8 – Placing the character in the scene

2. With the **Kaya@Happy Idle** GameObject, click on the **Add Component** button in the **Inspector** panel and add the **Animator** component.

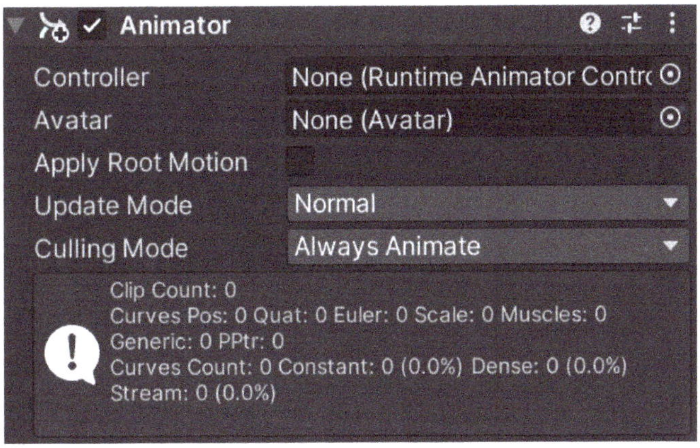

Figure 11.9 – Animator component

3. The **Animator** component needs an Animator Controller. To create an Animator Controller, in the **Assets** | **_App** | **Characters** | **Kaya** folder, right-click and select the **Create** | **Animation Controller** option and rename it `Kaya Animation Controller`.

4. Double-click on the Animation Controller we just created; this will open the **Animator** window, which manages the character states and animations. Drag the `Kaya@HappyIdle` file to any part of the graph.

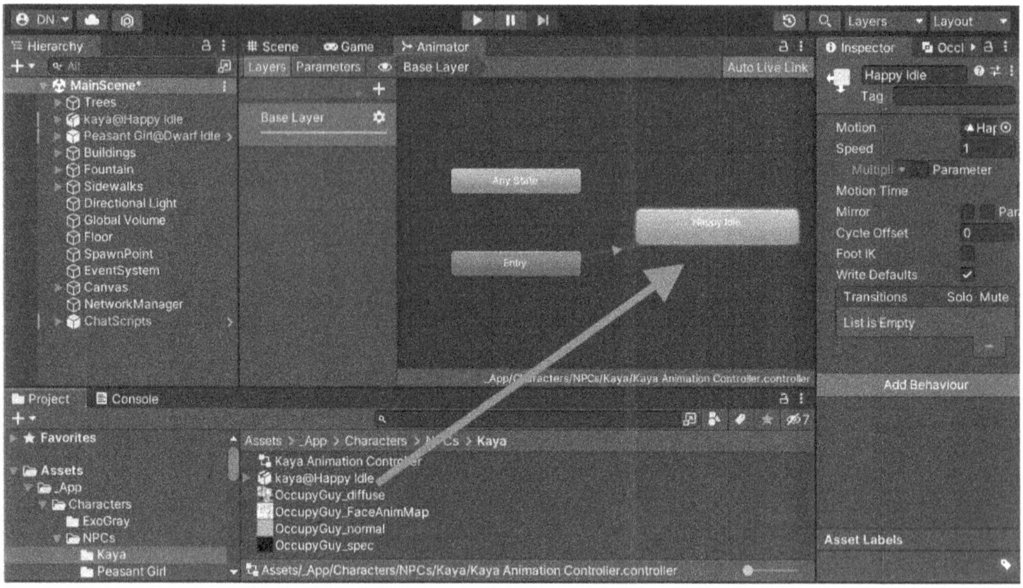

Figure 11.10 – Animator component state diagram

What we have just done is tell the Animation Controller, when it runs, to launch the **Happy Idle** animation. By default, the animations will run only once, but we want it to be an infinite loop.

5. To achieve an infinite loop in our animation, we select the **Kaya@Happy Idle** asset in the **Project** panel and, in its **Inspector** panel, we select the **Animation** tab, activate the **Loop Time** checkbox, and click on **Apply**.

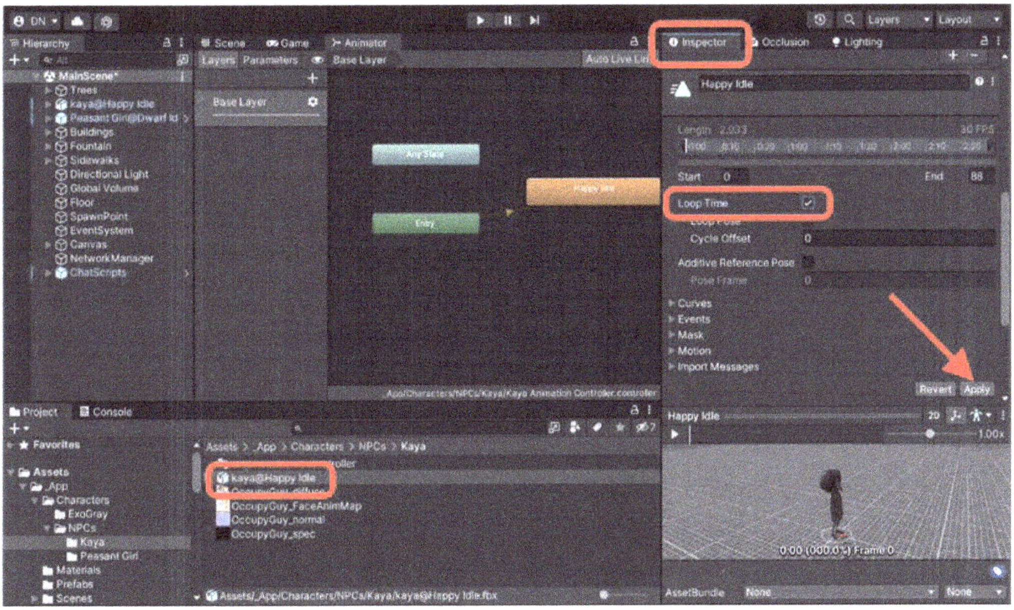

Figure 11.11 – Applying a loop to the animation

6. Finally, to finish configuring the animation of our character, with the **Kaya@Happy Idle** GameObject selected in the **Hierarchy** panel, we drag **Kaya Animation Controller** to the empty slot of the **Animator** component.\

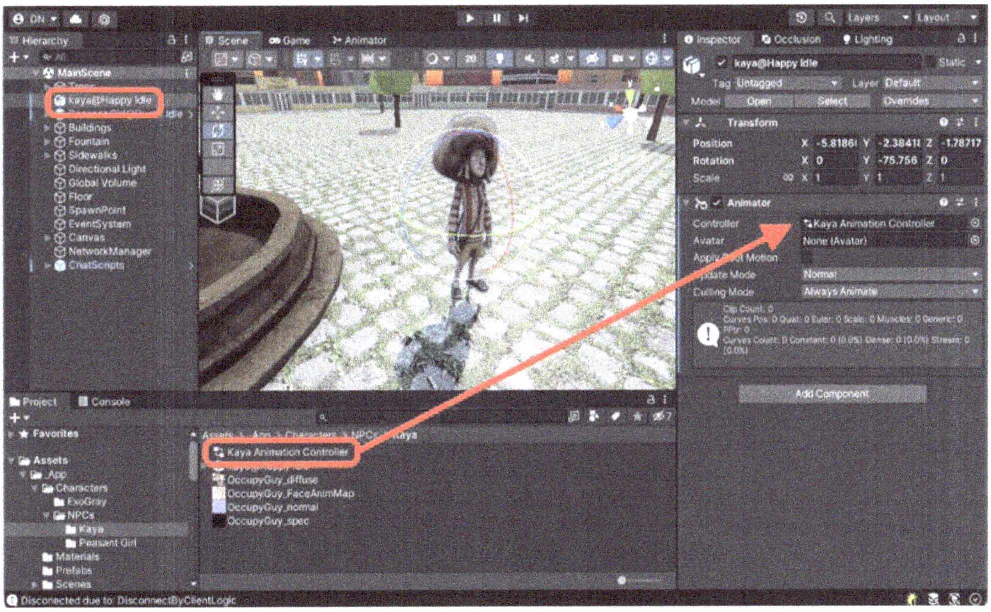

Figure 11.12 – Assigning the controller to the Animator component

Good job! Now you can see the NPC with movement. Click the **Play** button and enjoy your creation.

Finally, we will encapsulate our NPC with the animation in a new Prefab.

7. In the **Assets** | **_App** | **Prefabs** path, right-click and select the **Create Prefab** option, and rename it `Avatar NPC`.
8. Select the **Kaya@Happy Idle** GameObject from the **Hierarchy** panel and drag it to the **Avatar NPC** Prefab. Another way to create a Prefab quickly is to directly drag the object from the **Hierarchy** panel to the folder.
9. A confirmation modal will appear; click on the **Replace Anyway** button and then another modal window will appear asking whether you want to create an original Prefab or a variant. Click on the **Prefab Original** button.

Great job! You have successfully completed the creation of a Prefab that encapsulates the character in a single piece, with its **Animator** component. Don't forget to save your changes.

Next, we are going to work on programming its functionality. We will create a C# script that, when the player is close, will show us a text on the screen to invite us to interact with the NPC. Sounds interesting, doesn't it?

Triggering the NPC when we are close

Do you remember a game with an NPC in which, when you approach them, the typical text **Press the 'E' key to interact** appears on the screen? That's exactly what we're going to program. We will create a radius of influence for the NPC; when we enter this radius, we will show the message, and if we leave, we will hide it.

The radius of influence will be a Collider. In Unity, Colliders emit a trigger when an object containing a **Rigidbody** component comes into contact; the same happens when the object leaves the Collider. We will use these events to create our dialog flow.

To create our area of influence, we will first add a Collider:

1. To do this, right-click on the GameObject of our NPC called **Kaya@Happy Idle** and select **3D Object** | **Cube**, then rename it `Influence`.
2. Change the size of the cube to your own liking.

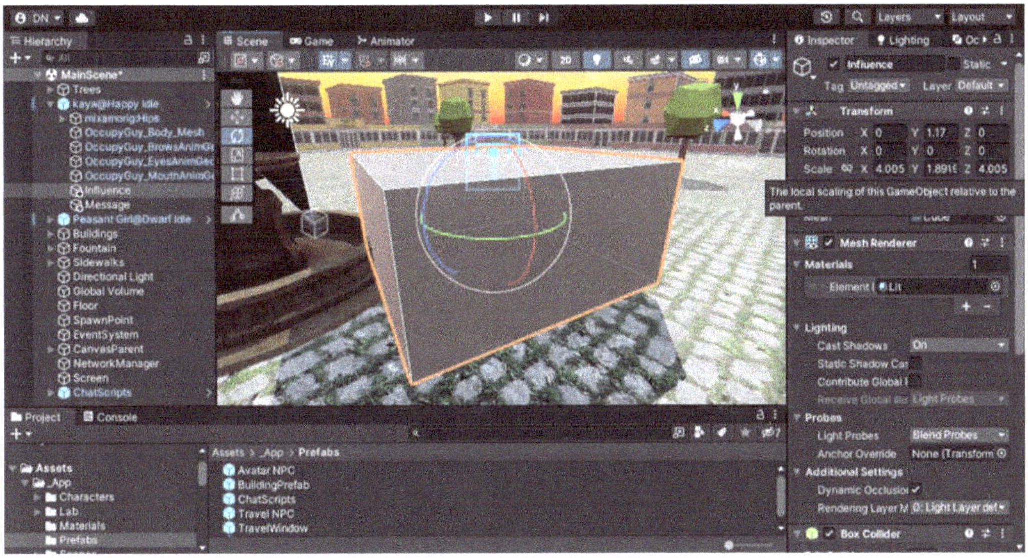

Figure 11.13 – Adding a cube to the scene

3. Finally, with the **Influence** GameObject selected, in the **Inspector** panel, deactivate the **Mesh Renderer** component; this will make the **Cube** GameObject invisible but we will keep its Collider, which is what we really need.

   It's important to know that the size of the Cube doesn't have to be exactly the same size as the image; you can use whatever you want, as it won't affect the performance we're looking for in this chapter.

4. Finally, an important setting that we should always apply when we want to use a Collider as a *tip-off* and not as a physical barrier that prevents the player from crossing it is to activate the **Is Trigger** checkbox in the **Collider** component. This will completely change the behavior of the Collider and allow it to trigger events when an object enters or exits it.

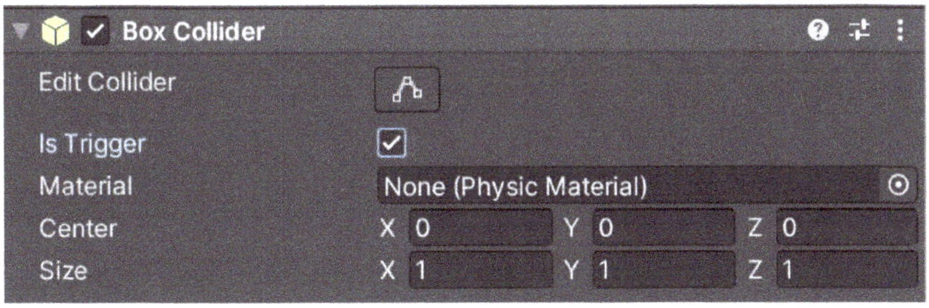

Figure 11.14 – Box Collider component

Great, we have created an area of influence. Before we start programming, we need to add a modification to our NPC, which is to place a text over his head to show a warning when the player can interact.

5. We'll use a 3D **Text** object. To add it, right-click on the NPC's GameObject and select **3D Object | Text - TextMeshPro**, and rename it `Message`.

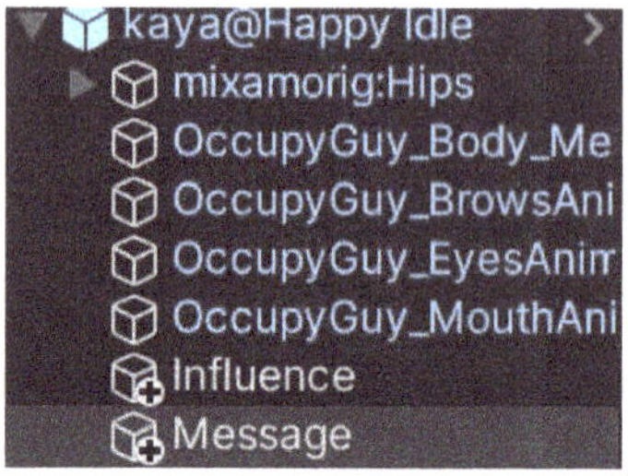

Figure 11.15 – Newly added GameObjects, Influence and Message

6. As you will have seen, our new **Message** GameObject has a **Scale** tool and **Rotation** tool that don't match the orientation of our NPC character. Use the **Scale**, **Move**, and **Rotate** tools to make it look similar to the following figure.

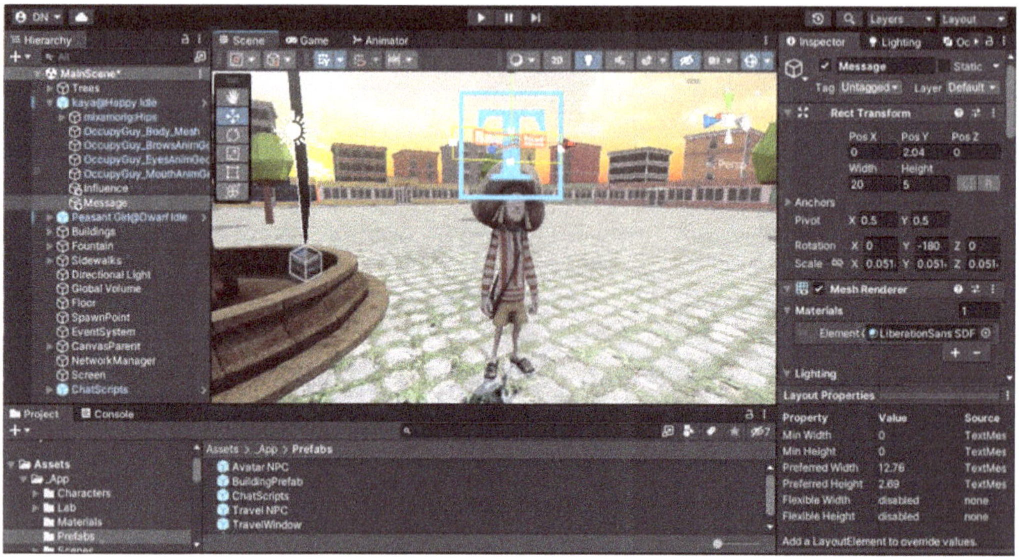

Figure 11.16 – Final preview of the NPC with the Message GameObject completed

Finally, we must know that the text **Sample Text** that comes by default in the component will be modified dynamically from the script that we create next; that's why we must foresee that it looks good if the text is longer.

7. To do this, we will modify some properties in the **TextMeshPro** component of the **Message** GameObject.
8. Select the option to align text to the center.

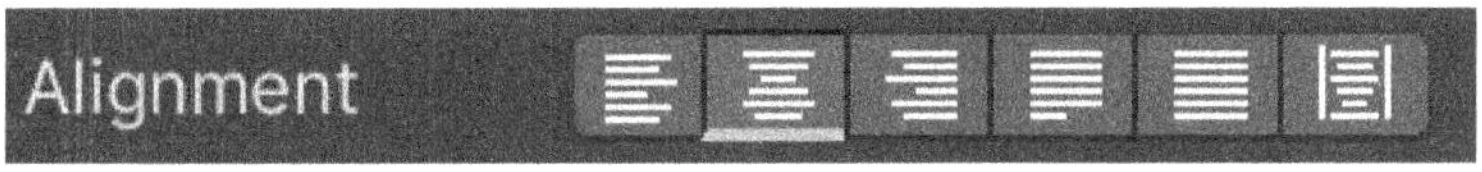

Figure 11.17 – Text alignment options

9. Check the **Auto Size** box and change the minimum font size to `14` and the maximum to `24`.

Figure 11.18 – Text size options

10. Great, now we have all the ingredients to prepare our script. To create a new script, navigate to the **Assets** | **_App** | **Scripts** path and right-click to select **Create** | **C# Script**. Finally, rename it `AvatarNPCTrigger` and double-click to edit it.

    Replace all the default code with the following:

```
using TMPro;
using UnityEditor.ShaderGraph;
using UnityEngine;
using static MyControls;
using static UnityEngine.InputSystem.InputAction;

public class AvatarNPCTrigger : MonoBehaviour,
    IPlayerActions
{
... (existing code here) ...

    {
        if (other.gameObject.tag == "Player")
        {
            messageGameObject.text = NPC_TEXT;
            canInteract = true;
        }
    }

    private void OnTriggerExit(Collider other)
    {
```

```
            if (other.gameObject.tag == "Player")
            {
                messageGameObject.text = string.Empty;
                canInteract = false;
                avatarWindow.SetActive(false);
            }
        }

        void Update()
        {
            if (Input.GetKeyDown(KeyCode.E) ||
                OVRInput.Get(OVRInput.RawButton.B) &&
                    canInteract)
            {
                avatarWindow.SetActive(true);
    ... (existing code here) ...

        public void OnInteract(CallbackContext context)
        {
            if (context.action.triggered && canInteract)
                avatarWindow.SetActive(true);
        }
        public void OnMove(CallbackContext context)
        {
        }
        public void OnLook(CallbackContext context)
        {
        }
        public void OnJump(CallbackContext context)
        {
        }
        public void OnSprint(CallbackContext context)
        {
        }
    }
```

11. Now that we have the script programmed, we will assign it to the **Influence** GameObject and drag the **Message** GameObject to the empty **Message** GameObject slot.

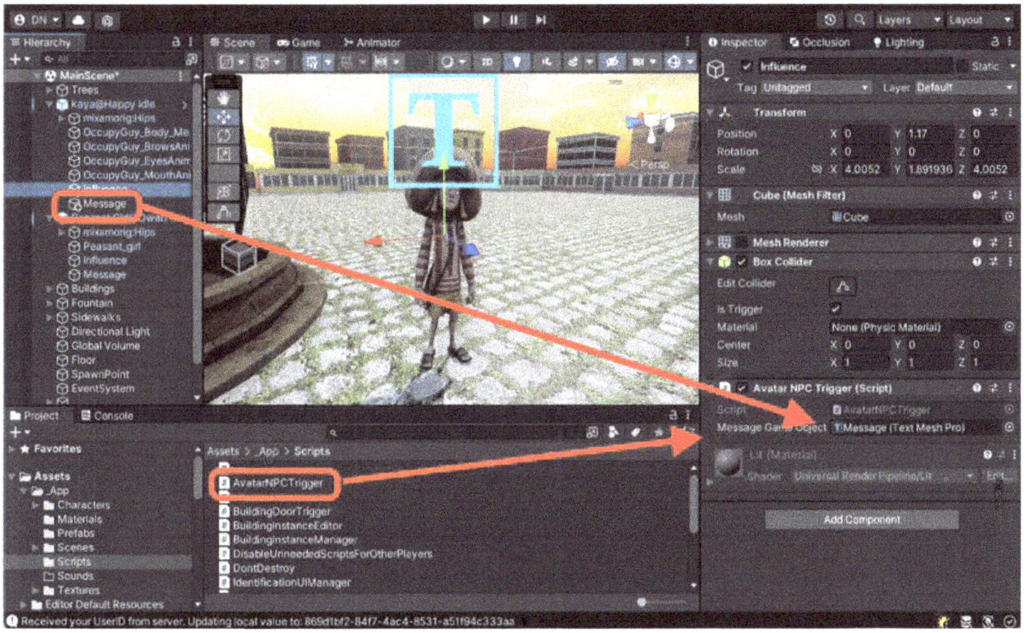

Figure 11.19 – Assigning the Message GameObject to the Avatar NPC Trigger component

Good work! Before testing it, let's briefly explain how the script works.

12. By placing the script on a GameObject that has a Collider with the **Is Trigger** property enabled, this enabled property will execute the `OnTriggerEnter` and `OnTriggerExit` functions automatically every time a GameObject comes into contact, or stops being in contact, with the Collider.
13. When our player comes into contact, we will know it is him by comparing the `Player` tag that is assigned by default to the **PlayerArmature** Prefab.
14. We will show a message in the **Message** GameObject and we will delete it when the player exits the GameObject.

Now we can see it in action. Click the **Play** button and approach the NPC, then walk away.

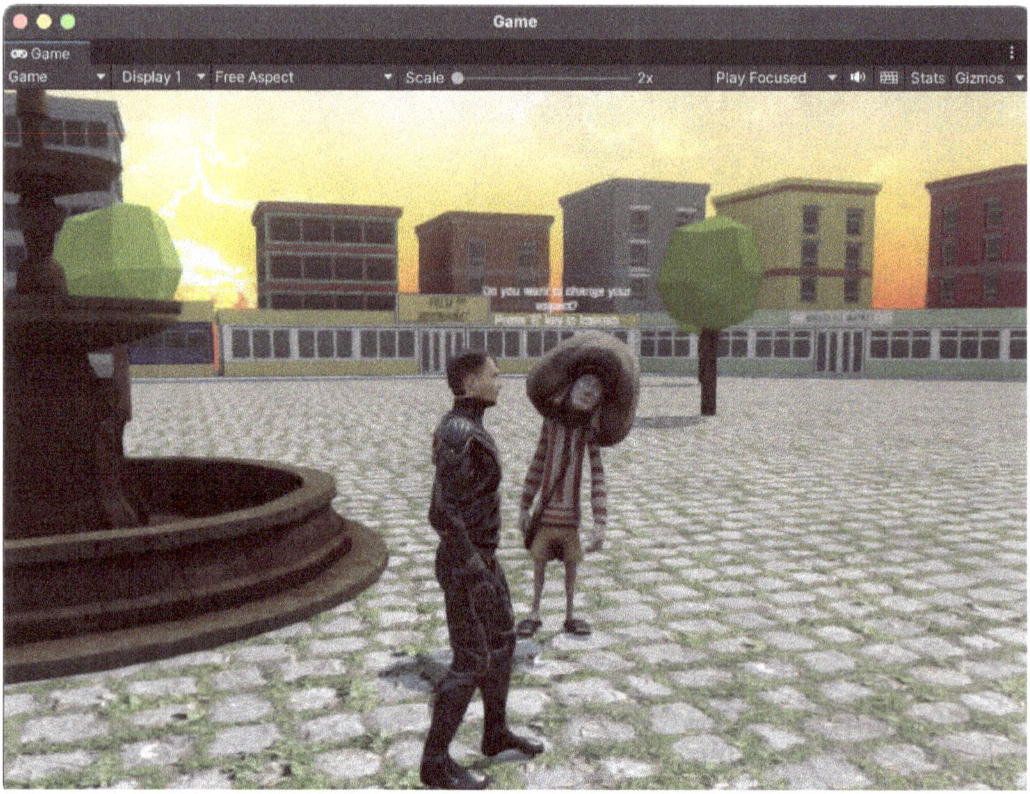

Figure 11.20 – Preview of the final result of the NPC's performance

> **Tip: Play**
>
> Remember that since we included the login system, it is necessary to start from the **Welcome** scene in order to follow the correct flow of identification and subsequent entry into the scene.

Excellent work, you have finished and implemented a system that allows our NPC to detect the player and display a message to invite them to interact. Learning how a trigger Collider works will open up an infinite range of possibilities for future improvements and functionalities in this and other projects.

In the next section, we will expand and improve our script to detect when the player presses the *E* key and display a window with the avatars we have available.

Showing the available avatars in a window

Unlike in *Chapter 7, Building an NPC That Allows Us to Travel* (where we created a window with a **Scroll Rect** component that displayed a list with text and a button, allowing us to navigate to other worlds), here we will now see how we can display a list with images. These images will be the visual representations of the available avatars; sounds interesting, doesn't it? For now, we will use the **Scroll Rect** component to host the list of available worlds we want to show to the user.

Now, we will start the creation of our popup with scrollable content. To do so, follow these steps:

1. We currently have a **Canvas** element in the scene as a child. In *Chapter 7, Building an NPC That Allows Us to Travel*, we created a GameObject with a panel to group everything related to the window that showed the available worlds.

Figure 11.21 – GameObject Canvas in the Hierarchy panel

We will follow the same pattern; this time, for our **Aspect selection** window, we can reuse the **TravelWindow** GameObject to make it easier for us to work with this new element we have to build.

2. Duplicate the **TravelWindow** GameObject. To do so, right-click and select the **Duplicate** option and rename it `AvatarWindow`.

Figure 11.22 – Newly renamed AvatarWindow GameObject

Good work; we have saved a lot of time by duplicating the **TravelWindow** GameObject, and now we need to keep working to make it work.

The next step is to create a Prefab that represents an element that we want to appear in the scroll and that contains a previsualization of the avatar we want to change to, as well as a button to select it.

3. Right-click on the **Content** GameObject, which is located inside **AvatarWindow** | **Scroll View** | **Viewport**, select **UI** | **Panel**, and rename it `AvatarItem`.
4. With the **AvatarItem** GameObject selected, resize it in the **Inspector** panel: change the **Width** value to `500` and **Height** to `500`. It will look similar to the following figure:

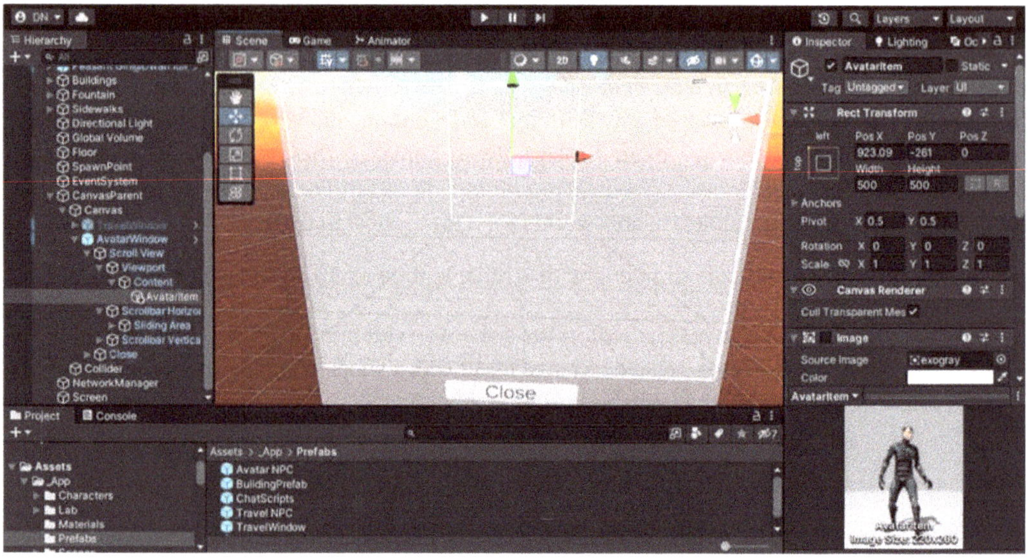

Figure 11.23 –AvatarItem GameObject preview

5. The **AvatarItem** GameObject will contain an image that will represent the avatar. This same image will be a button that will allow us to select it, so we will need to add a **Grid Layout Group** component to organize the content by columns and rows. To add it, add the **Grid Layout Group** component to the **Inspector** panel of the **Content** GameObject.

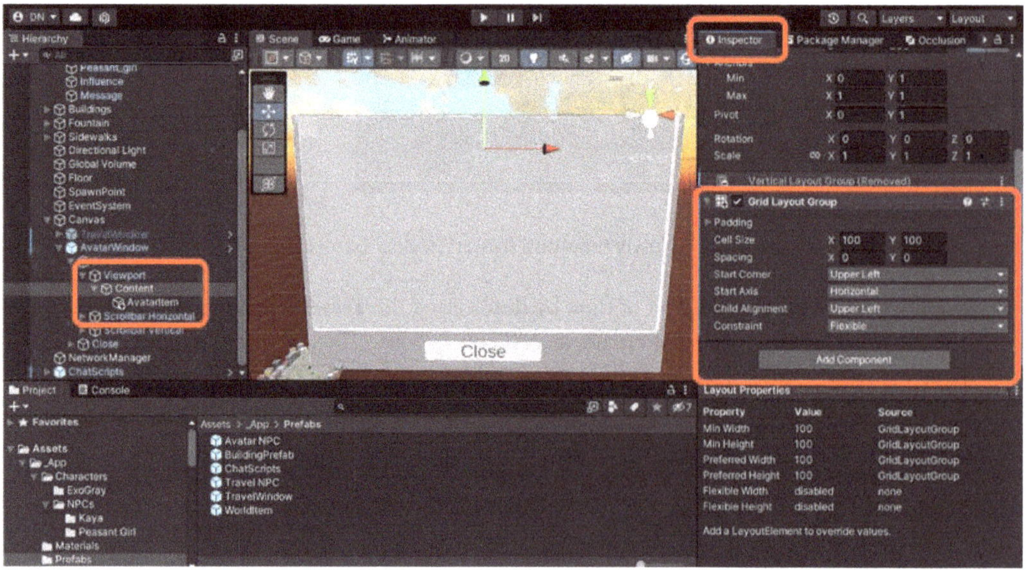

Figure 11.24 – Adding the Grid Layout Group component

6. We are going to make a modification to the **Grid Layout Group** component to change the size of the elements that will be inside this layout. For this, change the **Grid Size** property to **X**: `500` and **Y**: `500`, and finally, change **Spacing** to **X**: `5` and **Y**: `5`.

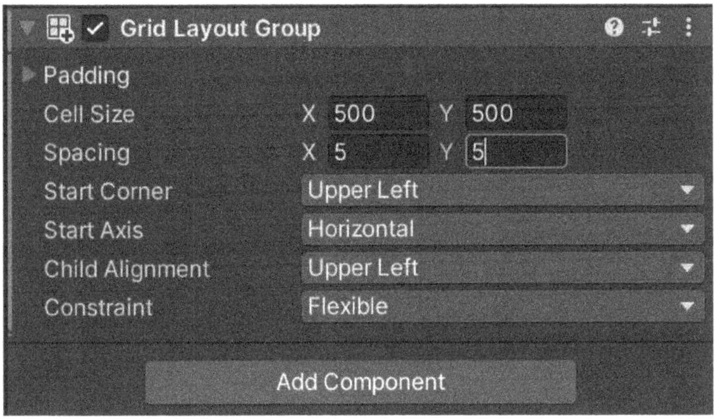

Figure 11.25 – Grid Layout Group component configuration

7. You will notice that the **AvatarItem** GameObject is a panel element that already comes with an **Image** component added. This is perfect; we just need to create a **Button** component to recognize our clicks. To add it, select the **AvatarItem** GameObject and add the **Button** component.

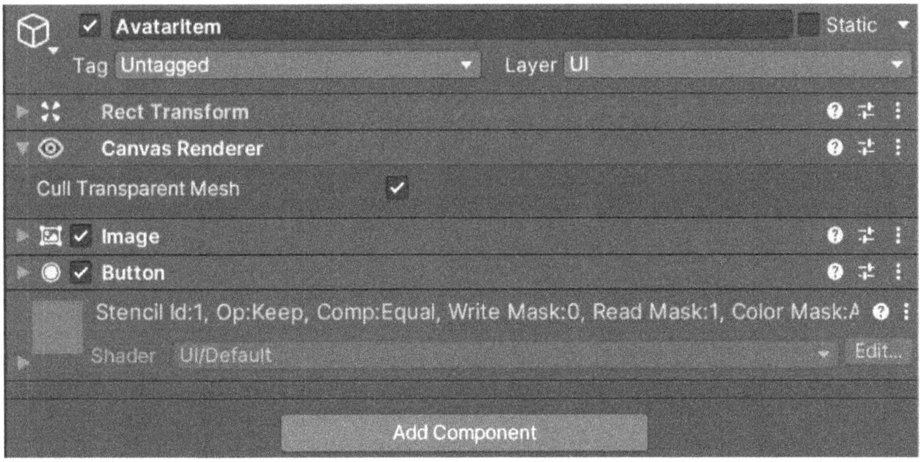

Figure 11.26 – AvatarItem GameObject Inspector panel

You're doing great! Now, to finalize the behavior of the **AvatarItem** GameObject, we'll create a script to give it functionality before turning it into a Prefab.

8. To do this, go to **Assets** | **_App** | **Scripts**, right-click to select **Create** | **C# Script**, and rename it `AvatarItemBehaviour`. Double-click on it to edit it.

   Replace all the auto-generated code with the following:

```
using UnityEngine;
using UnityEngine.UI;

public class AvatarItemBehaviour : MonoBehaviour
{
    public GameObject avatarPrefab;

    public void ChangeAvatar()
    {
        var player = GameObject.FindWithTag("Player");
        var content =
            player.transform.Find("Geometry");
        var oldAvatar = content.GetChild(0);
        var newAvatar = Instantiate(avatarPrefab,
            oldAvatar.transform.position,
                oldAvatar.transform.rotation);

        Destroy(oldAvatar.gameObject);
        newAvatar.transform.parent = content;
        StartCoroutine(ResetAnimator(player));
    }

    IEnumerator ResetAnimator(GameObject player)
    {
        player.SetActive(false);
        yield return new WaitForSeconds(1f);
        player.SetActive(true);
    }
```

   Save the changes and add this script to the **AvatarItem** GameObject, then we will follow a few simple steps to configure this component we have just created.

9. The first thing to do is to connect the Prefab of the avatar we want to select. For this first option, we will use the one we already have downloaded and currently use for our player; drag the **Exo Gray** Prefab found in **Assets** | **_App** | **Characters** | **Exo Gray** and drop it into the **Avatar** Prefab box of the **Avatar Item Behavior** component.

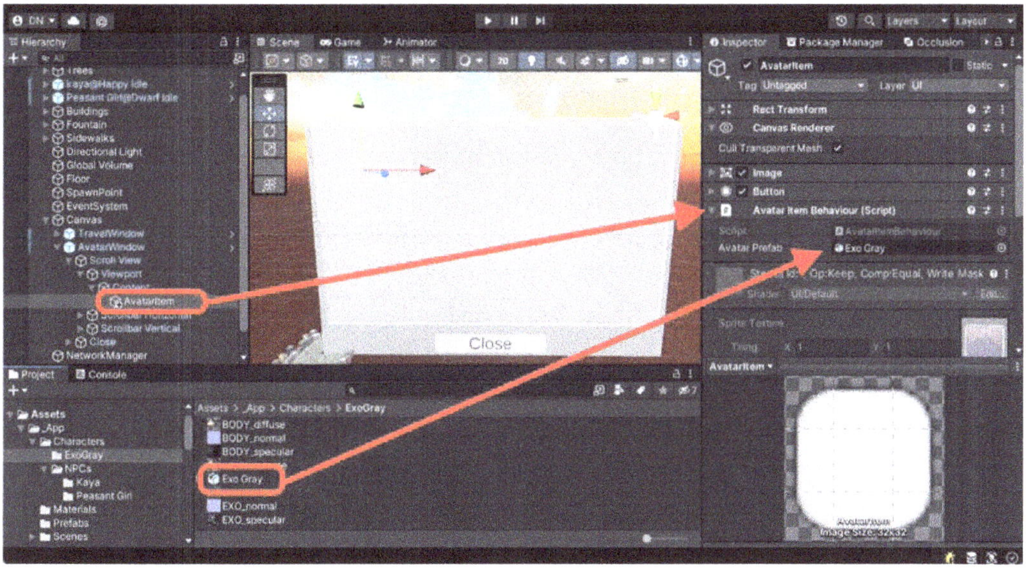

Figure 11.27 – Linking the Prefab Exo Gray to the Avatar Prefab slot

10. Now, we will create the `OnClick` event on the button of the **AvatarItem** GameObject and connect it to the `ChangeAvatar` function of the `AvatarItemBehaviour` script.

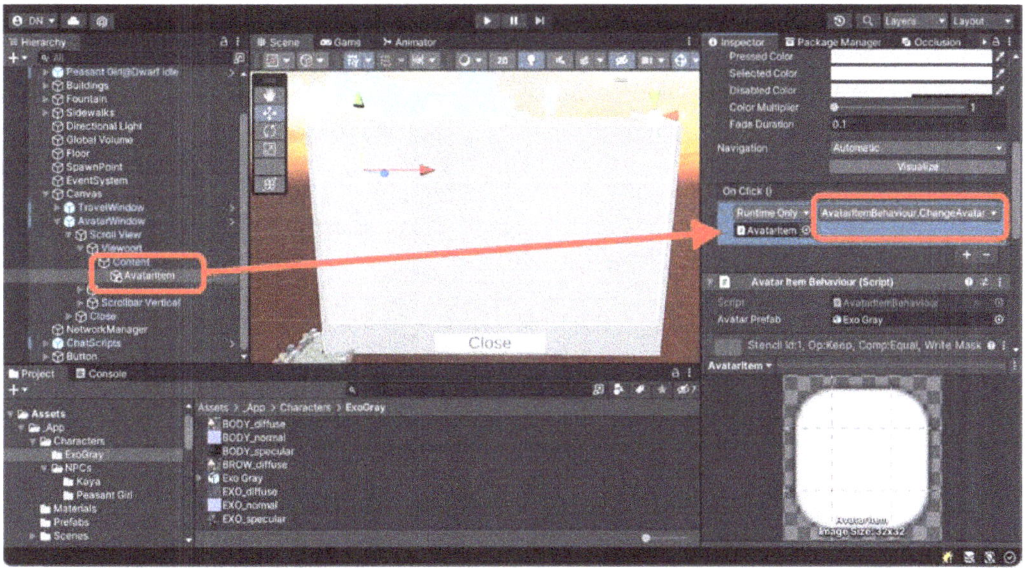

Figure 11.28 – Binding the ChangeAvatar function to the Click event of the button

Perfect; with these two steps, we have already configured the behavior of our button. Finally, we are going to configure a representative image of the avatar that we will obtain when clicking this button.

11. Go to the Mixamo website (`https://www.mixamo.com/`) and log in, if you are not already logged in.
12. In the **Characters** tab, look for **Exo Gray**, which is the character we used in *Chapter 2, Preparing Our Player*, for our player; if you used another one, search for it by name.
13. We will need the representative image of this model. Right-click on the image in the list of results and download it.

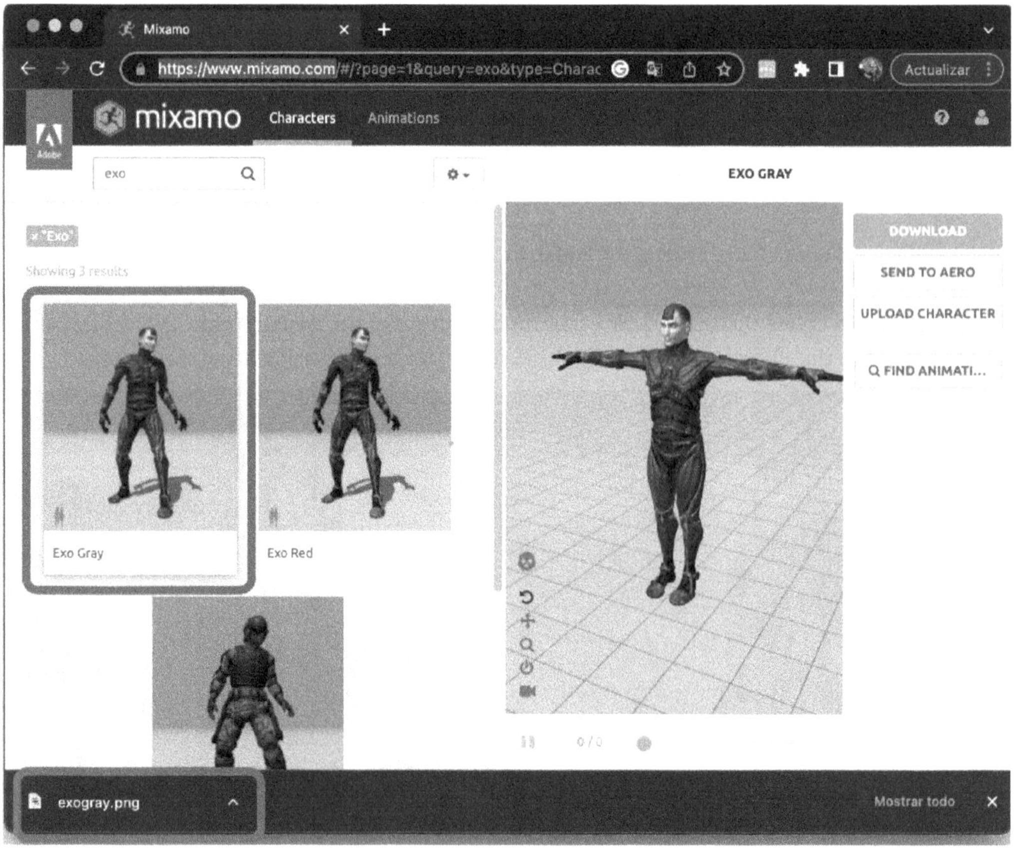

Figure 11.29 – Downloading the preview image of the character

14. Great, now we will set up the image for our button. First, we need to add the downloaded image to the **Assets** | **_App** | **Textures** folder; just drag and drop it there.

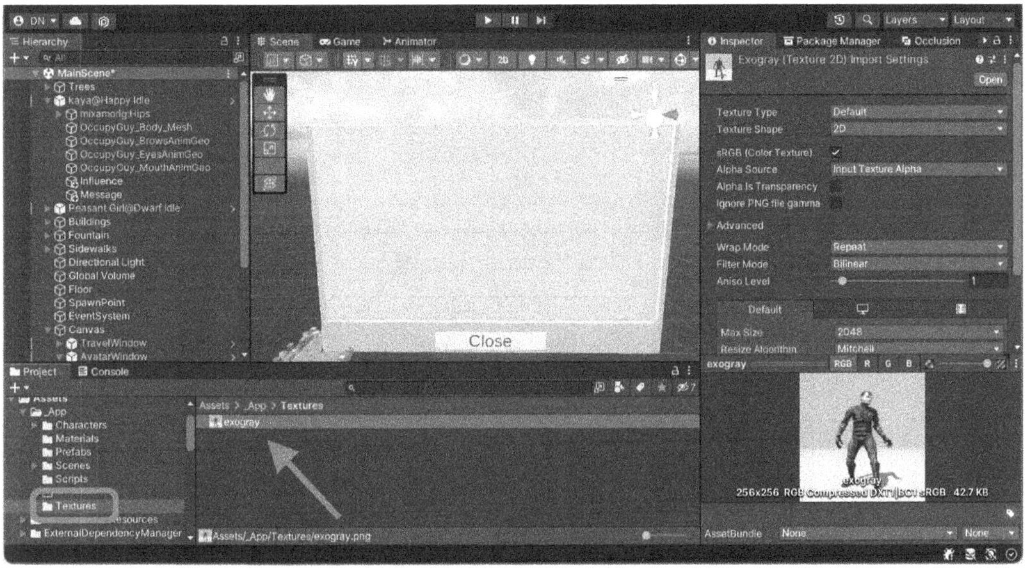

Figure 11.30 – Importing the image into our project

15. In order to use this image in a UI type component, we need to configure it, so if you select the image and look at the **Inspector** panel, you must select **Sprite (2D and UI)** in the **Texture Type** option and **Single** in the **Sprite Mode** option.

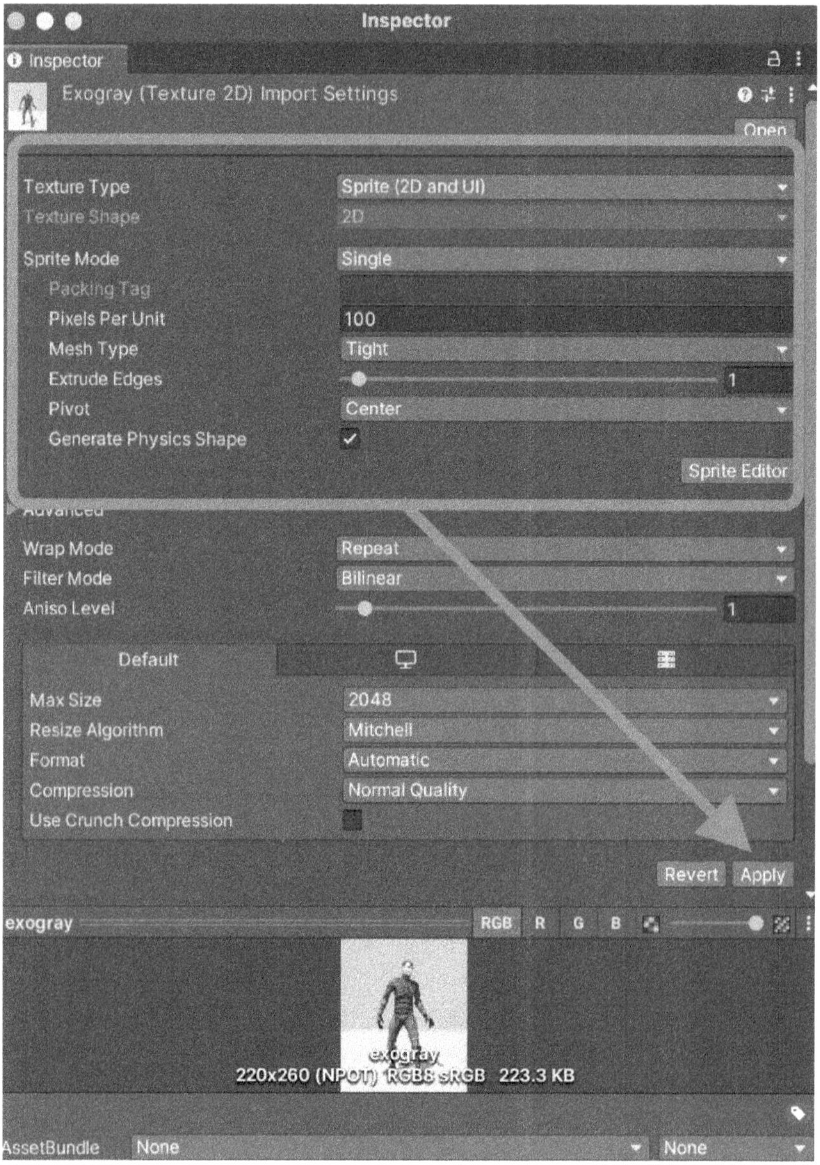

Figure 11.31 – Applying settings to the image

16. Finally, click on the **Apply** button to make the changes effective.

17. Now that we have it ready, we go to the **AvatarItem** GameObject and, in the **Image** component, we drag and drop our image into the **Source Image** property.

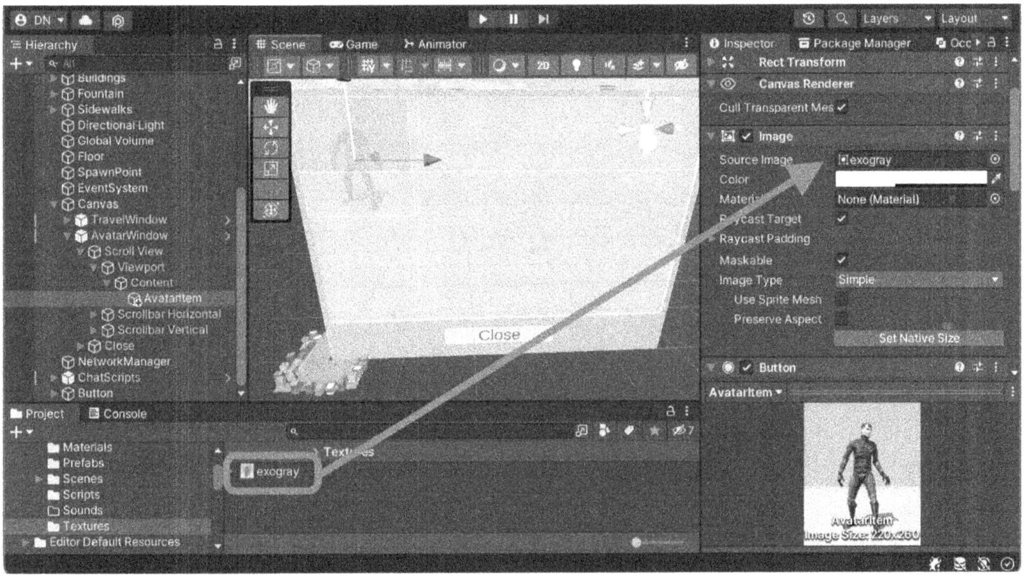

Figure 11.32 – Adding the imported image to the Image component

As you will have been able to observe, now an item appears with the image in the window, but it comes out with little opacity. This is because, by default, the **Image** component of the **Panel** element brings a defined opacity.

18. To be able to set it to 100%, simply click on the white bar in the **Color** property; a window with the color palette will be displayed. The last bar that appears is in reference to **Alpha**, to the transparency, and by default, it comes with a value of `100`. Change that value to `255` so that it is at maximum opacity.

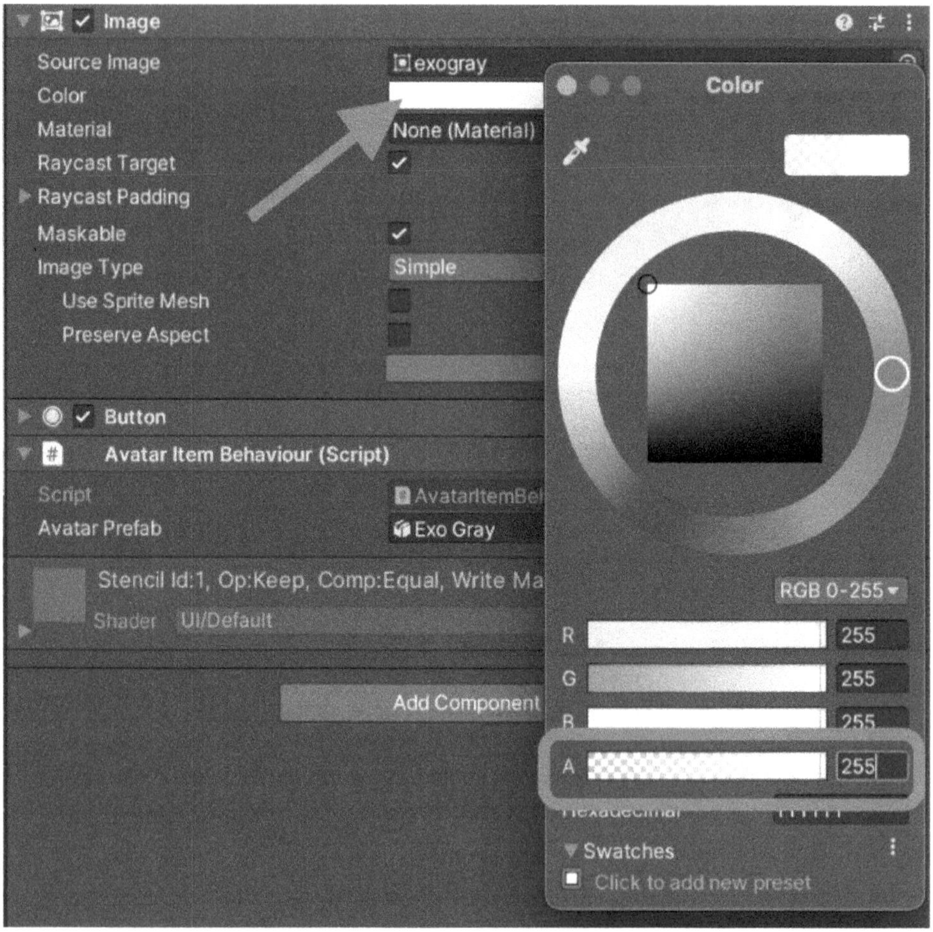

Figure 11.33 – Setting the image opacity

Fantastic! We have finalized the button that allows the user to select an avatar, but right now, it doesn't make much sense as it is the same avatar we are using on our player. To make this more fun, we need another avatar to change to.

Next, we're going to include a second avatar with its respective button; we'll make sure you learn everything you need to know to add a third, a fourth, and so on. The first thing we are going to do is to download a second FBX model from Mixamo.

19. On the Mixamo website (`https://www.mixamo.com/`), go to the **Characters** tab and look for another model. I will choose **Exo Red**, which is similar to the one we have now but with red clothes; I like it. Select it and click on the **DOWNLOAD** button.

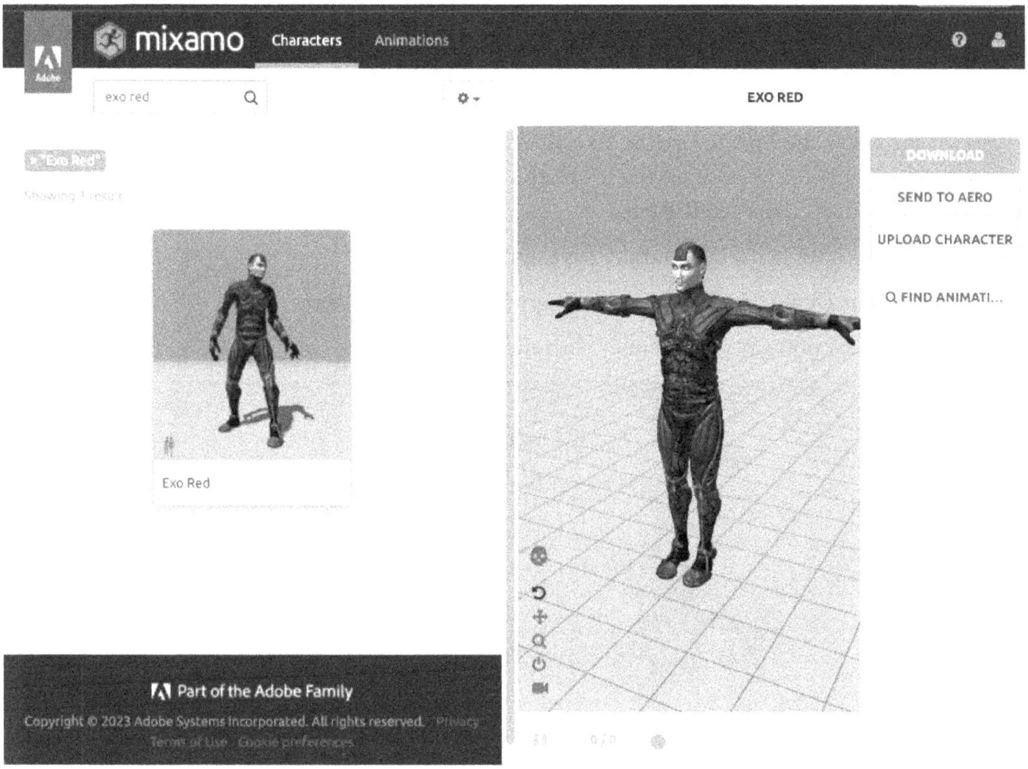

Figure 11.34 – Preview of the selected character in Mixamo

20. In **DOWNLOAD SETTINGS**, select the **FBX for Unity(.fbx)** format, set **Pose** to **T-pose**, and click the **DOWNLOAD** button again.

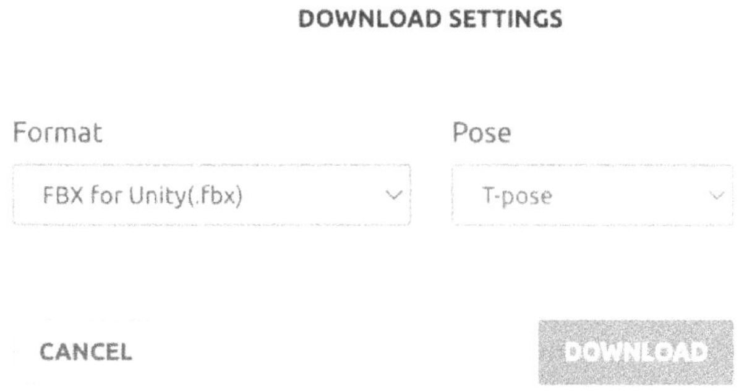

Figure 11.35 – FBX download options

21. Once downloaded, create a new folder called `ExoRed` in the **Assets** | **_App** | **Characters** path and drag the FBX there.
22. Great! Now, we'll do a little configuration to make it easy to change the avatar using our script; we'll configure the FBX to use the **Exo Gray** avatar, our main model. This way, we can use the same Animator without having to create a new one. Select the `exo_red` FBX and, in the **Inspector** panel, select the **Rig** tab.
23. Select **Humanoid** in the **Animation Type** option.
24. In the **Avatar Definition** option, select **Copy From Other Avatar**.
25. In the **Source** option, search for and select the avatar.

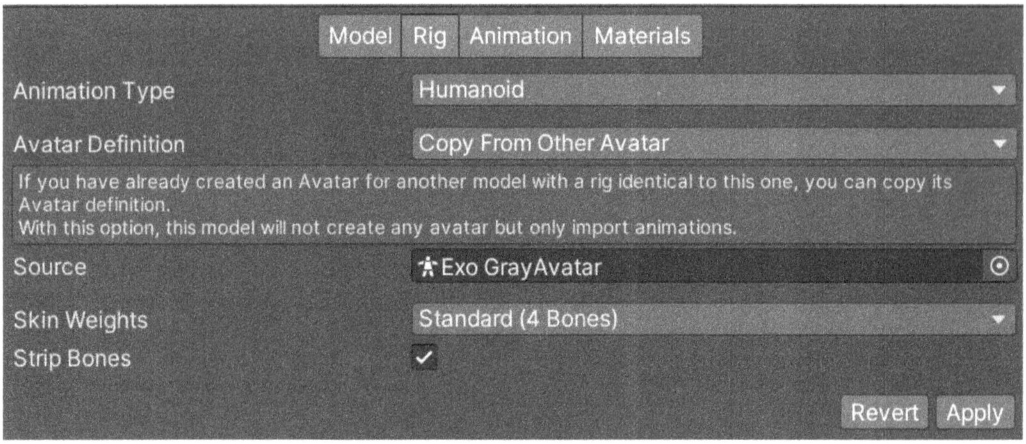

Figure 11.36 – Configuring the Rig options of the FBX model

26. Click on the **Apply** button to confirm the changes.
27. Finally, we need to extract the textures. Simply select the **Materials** tab and click the **Extract Textures** button. A window will appear to select the destination folder; keep the same one selected and click **OK**.

    We have just finished setting up the model for a second avatar; now we will create another button in our selection window, but this time, it will be easier, as we can duplicate the previous one and change the necessary settings.

28. In the **Hierarchy** panel, find the **AvatarItem** GameObject found in **Canvas** | **AvatarWindow** | **ScrollView** | **Viewport** | **Content,** right-click on it, and then click on the **Duplicate** option.

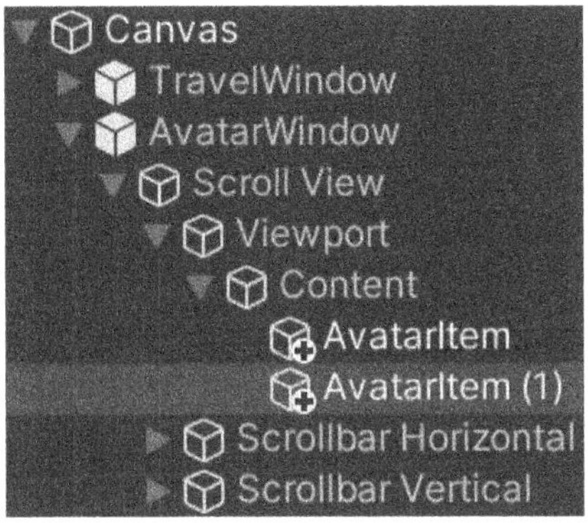

Figure 11.37 – Duplicating the AvatarItem GameObject

Perfect, we now have a second item in our avatar selection window. You will have noticed how the **Grid Layout Group** component we added arranges the images in a grid, by columns and rows.

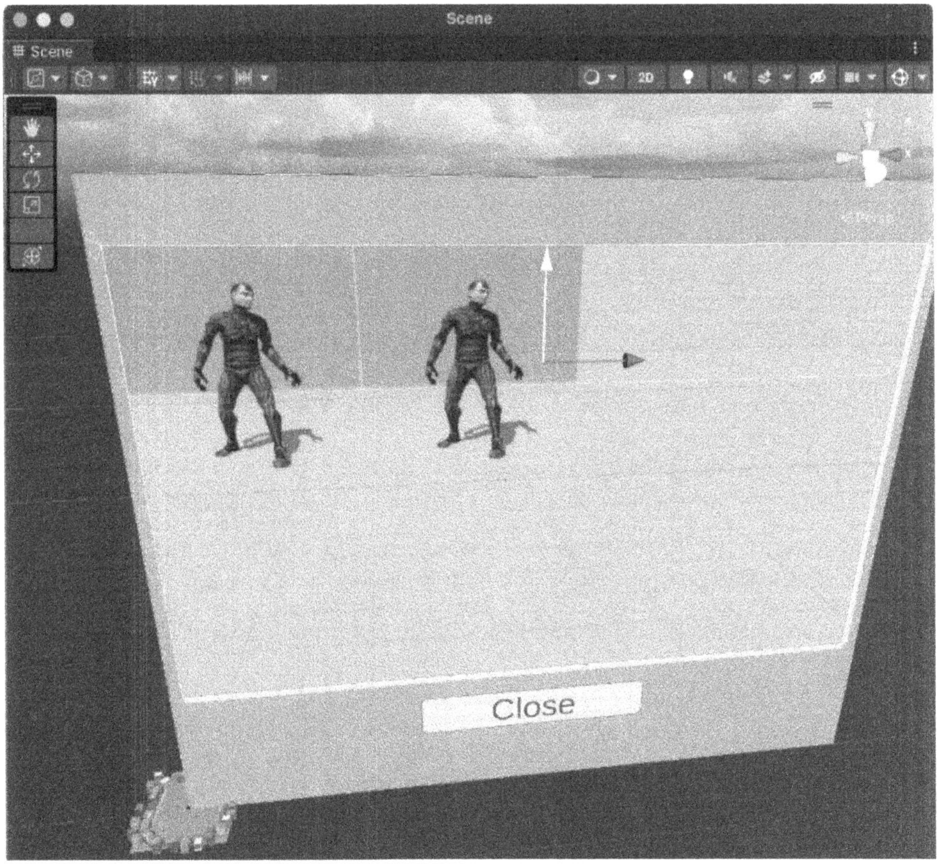

Figure 11.38 – Preview of the sorted AvatarItems GameObjects

We have already downloaded the new FBX and we have also created the new button. We need the image; for that, we again go to the Mixamo website and download the image as we did for the previous one.

29. Drag the downloaded image to the **Assets** | **_App** | **Textures** path and apply the configuration we did before to the **Texture Type** and **Sprite Mode** properties.
30. Finally, we assign the new image and the FBX of the new character to the properties of the **Avatar Item Behavior** component.

Figure 11.39 – Setting up the second AvatarItem GameObject

Good job! We have completed the configuration of the two buttons that will allow us to change our look, but before we test it, we need to do a few last steps to tidy up what we have just done.

Now that we have more elements inside our Canvas, we need to modify the `AvatarNPCTrigger` script to add a function to close the window when we click the **Close** button.

31. Open the `AvatarNPCTrigger` script and add the following function:

```
public void Close()
{
    avatarWindow.SetActive(false);
    Cursor.lockState = CursorLockMode.Locked;
}
```

Great, now we just need to link the function to the **Close** button.

32. Select the **Close** button found in **Canvas** | **AvatarWindow**.
33. In the **Button** component, drag the **Influence** GameObject, which is inside the GameObject that represents our NPC, to the **OnClick** property and select the **Close** function.

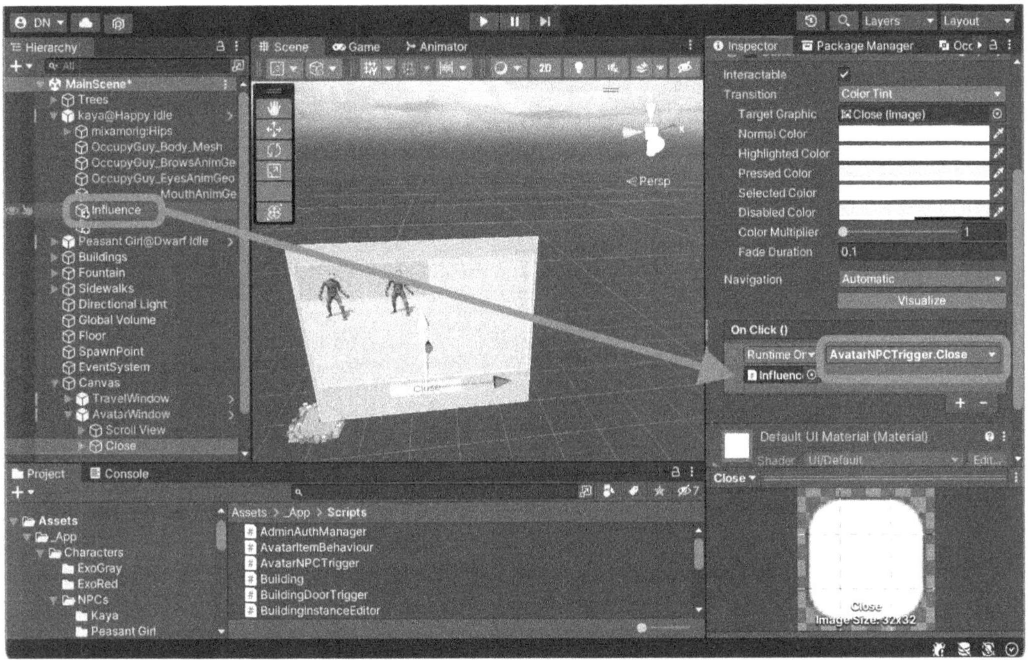

Figure 11.40 – Binding the Close function to the button's Click event

34. Finally, to close the functionality completely, we need to link the **AvatarWindow** GameObject with the `AvatarNpcTrigger` script. Simply drag the **AvatarWindow** GameObject to the script property attached to the **Influence** GameObject.

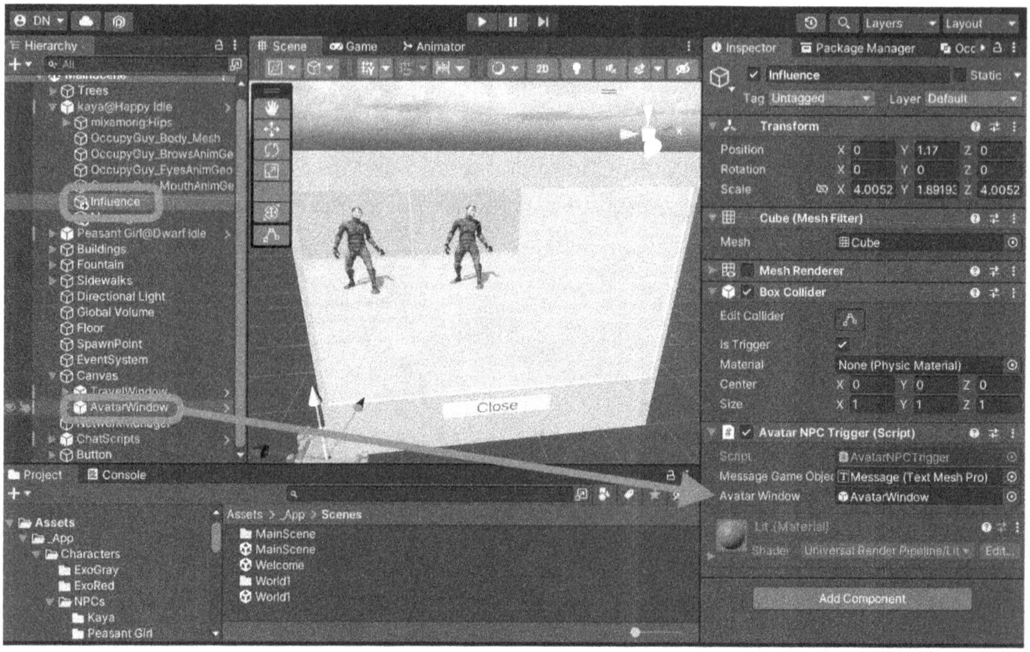

Figure 11.41 – Assigning the AvatarWindow GameObject to the Avatar NPC Trigger avatar component

Great, now everything is connected and working, it's time to try it if you dare. I certainly do! Start the project and go to our new NPC and interact with him; you will be surprised by the result.

Figure 11.42 – Final preview of the functioning of the AvatarWindow screen

If I choose the **Exo Red** character... voila! It works perfectly; excellent work!

Figure 11.43 – Preview of the change of avatar for our character

So far, we have done the hardest work, and we have managed to implement a nice system that allows our users to choose an alternative appearance. Now, we are going to go one step further and introduce the `PlayerPrefs` concept, which, if you didn't know about previously, will open up a whole new range of possibilities.

We will use this local data persistence system to save the user's avatar selection so that when they log in again in the future, they will have their avatar available.

Persisting the new appearance

PlayerPrefs is a class in Unity3D that is used to store player configuration data. It allows developers to save and load persistent values on the user's device, meaning that the values will be retained after closing and reopening the game.

It is useful for saving information such as audio settings, player high score, game progress, and other player data. It can also be used to create a personalized experience for the player, for example, allowing them to choose their own username or the color of their character; in our case, it will be the selected avatar.

To use `PlayerPrefs` in your Unity3D project, simply call its static methods to save and load data. For example, to save the player's high score, you can use the following code:

```
int highScore = 1000;
PlayerPrefs.SetInt("HighScore", highScore);
```

And to load the previously saved high score, you can use the following code:

```
int savedHighScore = PlayerPrefs.GetInt("HighScore");
```

It is important to note that `PlayerPrefs` is not suitable for storing large amounts of data, as the values are stored in a file on the user's device, and overloading that file can slow down game performance.

OK, now that we know what `PlayerPrefs` is and how it can help us, let's get started.

The first thing we're going to do is modify the `AvatarItemBehavior` script, and we're going to add the highlighted lines:

```
...
public class AvatarItemBehavior : MonoBehaviour
{
...
    // Pressing the Join button will call this function and
       change the scene.
    public void ChangeAvatar()
    {
...
        var newAvatar = Instantiate(avatarPrefab,
            oldAvatar.transform.position,
                oldAvatar.transform.rotation);

// Save the name of the selected prefab in the Avatar key.
        PlayerPrefs.SetString("Avatar", avatarPrefab.name);

        // We destroyed the old FBX model.
```

```
            Destroy(oldAvatar.gameObject);
...
        }
...
    }
```

With this modification, we guarantee that the last avatar selection will be saved in the `Avatar` key. The next step would be to load the avatar when the player starts the game. To do this, we need to make a series of modifications.

As we already know from previous chapters, in order to instantiate a GameObject through code, we need it to be in the `Resources` folder; otherwise, it won't allow us to do so:

1. Create an `Avatars` folder in the **Assets** | **Resources** | **Prefabs** folder.
2. Move the two character Prefabs found in **Assets** | **_App** | **Characters** | **Exo Gray** and **exo_red** to the folder we just created.

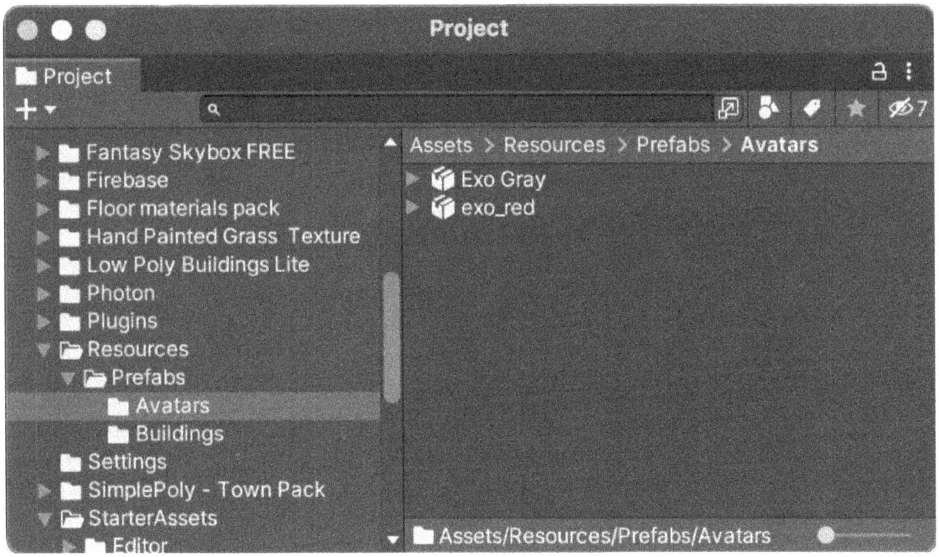

Figure 11.44 – Moving the Prefabs to the Avatars folder

Perfect, now we have both Prefabs ready to be manipulated by code. The next step will be a new script to manage and load the avatar we have saved in `PlayerPrefs`.

3. Right-click on the **Assets** | **_App** | **Scripts** path and select the **Create** | **C# Script** option, rename it `LoadAvatar`, and replace all the auto-generated code with the following:

```
using System.Collections;
using System.Collections.Generic;
using Photon.Pun;
using StarterAssets;
using UnityEngine;

public class LoadAvatar : MonoBehaviour
{
    public GameObject Player;

    void Awake()
    {
        if(Player.GetComponent<PhotonView>().IsMine)
            ChangeAvatar();
    }

    public void ChangeAvatar()
    {
        string avatarName =
            PlayerPrefs.GetString("Avatar") ??
                "Exo Gray";
        var content =
            Player.transform.Find("Geometry");
        var oldAvatar = content.GetChild(0);
        GameObject avatarInstance =
            Instantiate(Resources.Load
            ("Prefabs/Avatars/" + avatarName),
            oldAvatar.transform.position,
            oldAvatar.transform.rotation) as
            GameObject;

        Destroy(oldAvatar.gameObject);
        avatarInstance.transform.parent = content;
        StartCoroutine(ResetAnimator());
    }

    IEnumerator ResetAnimator()
    {
        Player.SetActive(false);
        yield return new WaitForSeconds(1f);
        Player.SetActive(true);
```

```
        }
    }
```

4. Double-click on the **PlayerInstance** Prefab found in **Assets | _App | Photon | PhotonUnityNetworking | Resources** to edit it.
5. Add the `LoadAvatar` script as a component to the root **PlayerInstance** GameObject.

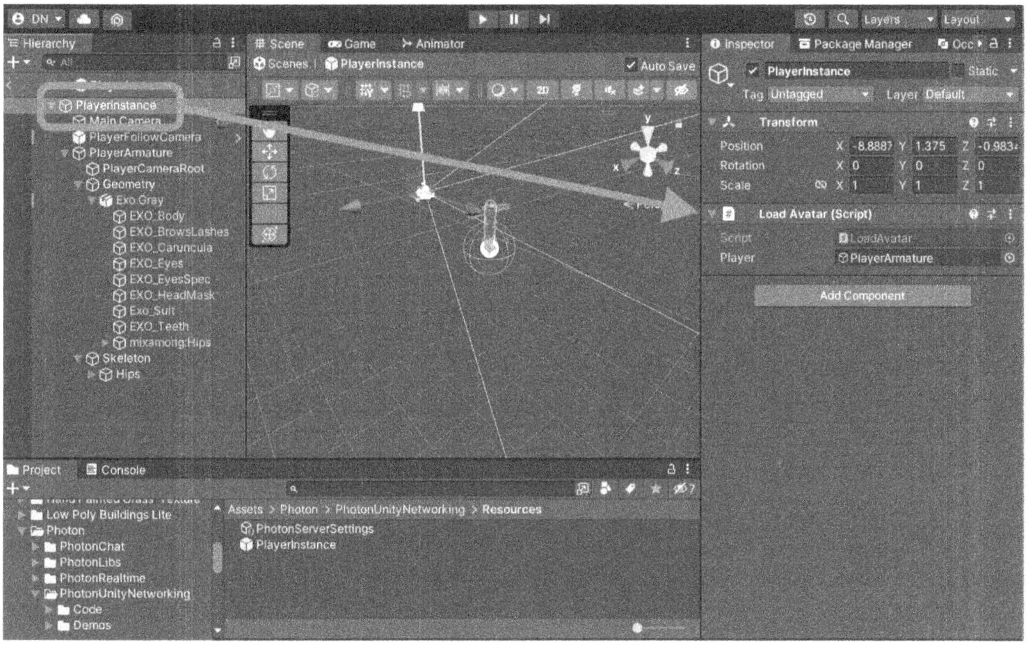

Figure 11.45 – Adding the LoadAvatar script as a Component to the PlayerInstance GameObject

6. Finally, drag the **PlayerArmature** GameObject into the empty **Player** slot of the **Load Avatar** component.

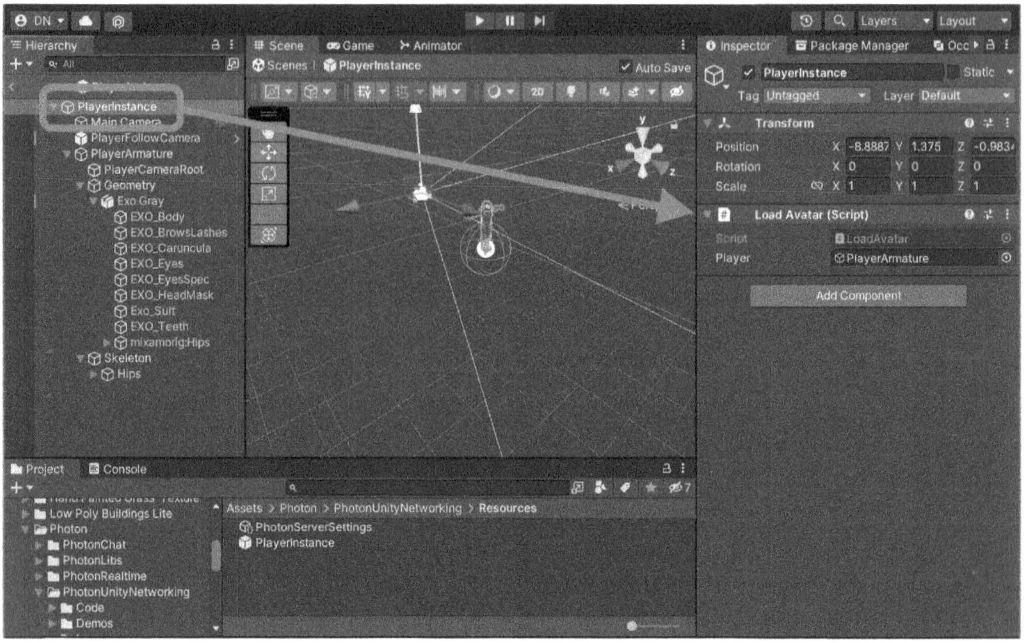

Figure 11.46 – Assigning the PlayerArmature GameObject to the Load Avatar Component

With this last step that we have just taken, we close the circle of choosing a new avatar. Now, if you run the project, you will see that the player is loaded with the last avatar you selected; it is impressive, isn't it?

As another new and incredible functionality that takes your metaverse to the next level, I invite you to browse the Mixamo website looking for new avatars. With the topics that we have learned about in this chapter, you can add infinite variants for your users.

Summary

Throughout this exciting chapter, we learned how to create a new NPC for our virtual world. This time, our NPC offers new functionality to your metaverse, allowing users to choose an alternative avatar. We have learned about a new UI component called **Grid Layout Grid**, which allows for the row and column layout of a multitude of child components and offers scrolling for easy navigation.

We also learned how to replace one Prefab with another, being now able to customize a character at runtime through code. To save the user's choice, we learned about the `PlayerPrefs` class, a fast and efficient way to save simple information in the user's session.

In the next chapter, we will deal with a very interesting functionality: we will create screens that allow you to play streaming videos in your scene, and you will be able to play videos hosted on a server or hosting. This is very interesting if you want to make, for example, a cinema-type scene, advertising videos, and so on. If you find this as interesting as I do, see you in the next chapter!

12
Streaming Video Like a Cinema

We have reached the final stretch of the book and the construction of our metaverse. It is time to fill our virtual world with fun functionalities to leave a mark on our future users. In the metaverse, you can do an infinite number of activities with your friends, such as watching a video or a film. In this chapter, we are going to discover a new Unity class, called `VideoPlayer`.

With `VideoPlayer`, we can play local videos or videos from an internet address as if they were textures, on any 3D object – fun, right? Put simply, we integrate this feature, which gives us endless new ideas. For example, you could program a cinema in which to play movies or commercials anywhere in your world. Right now, your mind probably cannot stop generating new ideas.

In this chapter, I will give you the keys to set up any future video playback system on 3D objects.

In the following screenshot, you can see the final result of the magnificent cinema screen, streaming video playback, that we are going to build throughout this chapter.

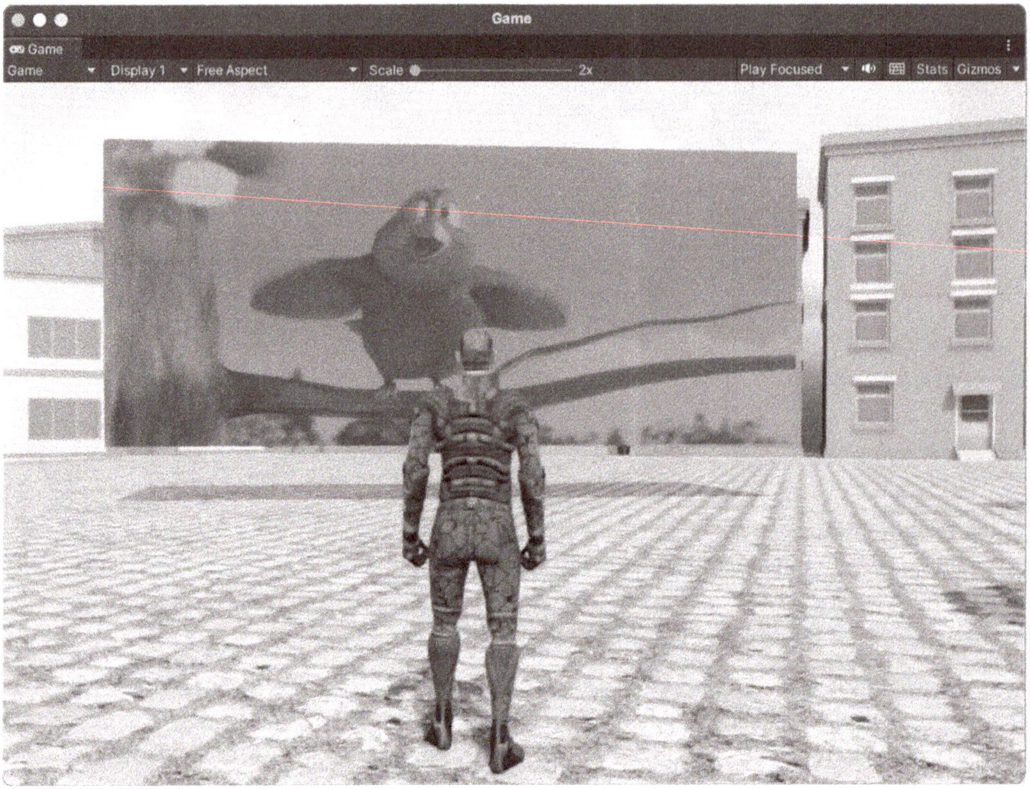

Figure 12.1 – Final result

We will cover the following topics:

- Introducing the `VideoPlayer` class
- Designing our screen
- Playing videos

Technical requirements

This chapter does not have any special technical requirements, but we will start programming scripts in C#. It would be advisable to have basic knowledge of this programming language. You'll need an internet connection to browse and download an asset from the Unity Asset Store.

We will continue with the project we created in *Chapter 1*. Remember that we have the GitHub repository, `https://github.com/PacktPublishing/Build-Your-Own-Metaverse-with-Unity/tree/main/UnityProject`, which contains the complete project that we will work on here.

Introducing the VideoPlayer class

Unity 3D's `VideoPlayer` class is a useful tool to play and control videos in real time within a game or application. This class allows you to load and play videos in different format, and offers several configuration options, such as audio track selection and image quality settings.

The `VideoPlayer` class can be used in a wide variety of projects, from games to educational applications, to multimedia presentations. Here are some ways in which the `VideoPlayer` class could help your project:

- **Integrate videos into your game**: If your game needs to include videos, either to introduce the plot or to show cinematic sequences, the `VideoPlayer` class will allow you to play videos within the game and control their playback.
- **Multimedia presentations**: If you are creating an application that includes multimedia presentations or video tutorials, the `VideoPlayer` class can be a useful tool to play and control the videos.
- **Educational applications**: If you are creating an educational application, you could use the `VideoPlayer` class to play videos that explain complex concepts or illustrate real-life situations.

In summary, the `VideoPlayer` class is a useful tool for playing and controlling real-time video in Unity3D projects and can be used in a wide variety of applications, from games to multimedia presentations, to educational applications.

Designing our screen

In our practical example, we will create a screen similar in shape and proportion to a cinema screen or a billboard – elongated, in full HD style. The screen that we will use to play videos is nothing more than a plane. Follow these steps:

1. Go to our main scene, **MainScene**.
2. Select the **GameObject** | **3D Object** | **Plane** option from the main menu bar and rename it `Screen`.

3. Use the **Move**, **Rotate**, and **Scale** tools to shape the plane until it looks like this:

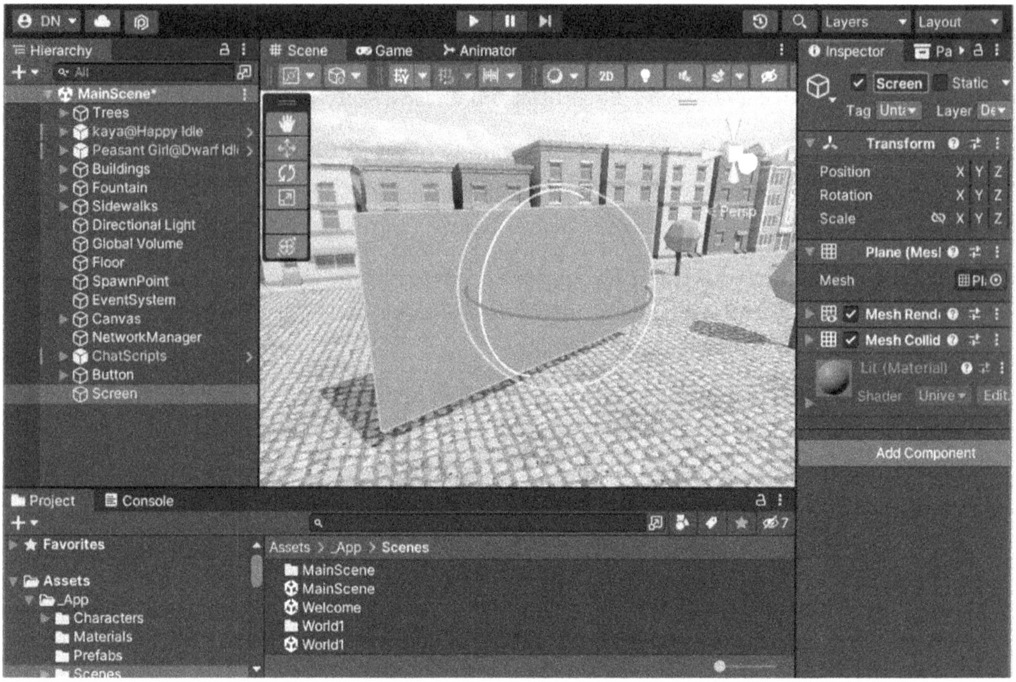

Figure 12.2 – Adding a plane to the scene

Perfect – easy, isn't it? You have created the sketch of the video screen. Don't worry if, later on, when you play videos it looks distorted. We will be able to modify the proportion of the screen later on.

Now that we have the GameObject that physically represents our screen, we need to add the **VideoPlayer** component to give it the functionality we are looking for. To do this, simply select the **Screen** GameObject in the **Hierarchy** panel and click the **Add Component** button in the **Inspector** panel, then type in the search box and add the **Video Player** component.

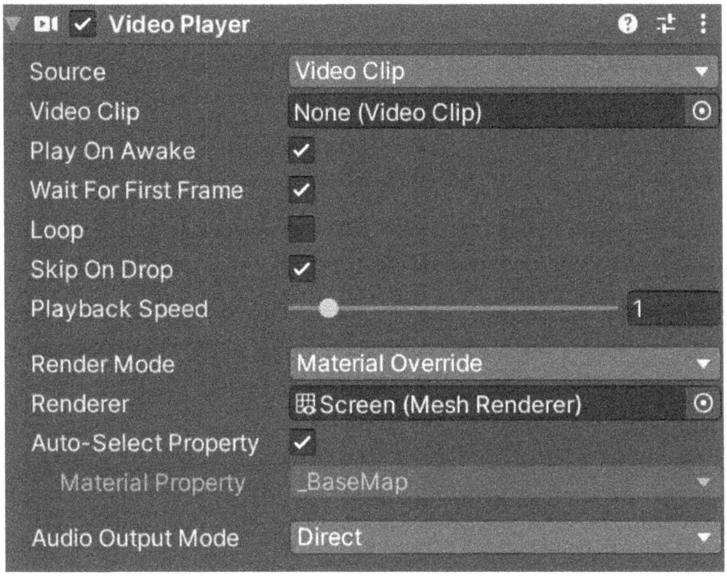

Figure 12.3 – Video Player component

We will now explain a little more about this new component we are discovering here. As you can see, its properties are very intuitive:

- **Source**: Allows us to select whether we are going to play a local video in a file or from a URL.
- **Video Clip**: Links the physical file or the URL depending on the option we have chosen in **Source**.
- **Play On Awake**: The video will play automatically when the scene starts if this option is activated.
- **Wait For First Frame**: When enabled, the Video Player will wait for the first frame of the video to be rendered before starting the actual playback. This can be useful to prevent the video from displaying a brief flicker or distorted image when playback starts, as it ensures that the first frame has fully loaded before displaying it.

  This feature is especially beneficial when working with videos that are high resolution or require considerable time to load, as it ensures a smoother and more fluid playback experience by eliminating potential initial display problems. I recommend having it always activated.
- **Loop**: If enabled, the video will play again once it ends. I recommend having this option checked for our demo.
- **Skip On Drop**: If this option is enabled, when the game performance is not fast enough to play the video at the desired speed, Unity may choose to skip some frames of the video to maintain synchronisation with the audio. This means that, instead of slowing down video playback to match the game's framerate, frames will be skipped to maintain consistency between video and sound.

While this can help to avoid slow playback or stuttering, it can also result in a less smooth visual experience and loss of detail in the video if there are many frame drops. Therefore, it is important to balance this setting based on the specific needs of the project and the performance capabilities of the target device.

- **Playback Speed**: This displays the playback speed.
- **Renderer**: This is the mesh where the video will be played. By default, it takes the mesh of the object where the script is placed (if available, otherwise, we should link one).
- **Auto-Select Property**: When you enable this option, the component itself will automatically choose the main texture of the **Renderer** object to be used as the material for the video. If you deactivate this option, you will have the possibility to manually choose which texture of the Renderer object will be used as the material for the video.
- **Audio Output Mode**: This allows you to control where to play the audio coming from the video. I recommend leaving it as the default.

> **Tip: extended VideoPlayer class information**
>
> In this chapter, we are going to cover the fundamental, basic aspects of the `VideoPlayer` class. If you need to learn more about its properties and find a more advanced example, you can consult the official documentation at `https://docs.unity3d.com/ScriptReference/Video.VideoPlayer.html`.

Perfect – now we know in detail about the **VideoPlayer** component and its properties. We are now ready to start the fun (more fun) part of the chapter. Next, we will look for a URL that offers a free video for our demo and we will also test the option of playback via a local file. Are you ready?

Playing videos

Earlier, in the introduction of this chapter, we saw a spoiler of what we are going to cover in this section. We will cover both local and online video playback. For this, we will need a valid URL and a valid file, which we will look at shortly.

If you're wondering about the video formats supported by this component, here's a look at some of the wide variety of video formats supported:

- **MP4**: This is one of the most common video formats and is widely compatible with most devices
- **WEBM**: This video format is especially useful for online video streaming and is often used in web applications
- **OGV**: This is an open source video format that is often used in video editing software and web browsers

- **MOV**: This is a video format developed by Apple and is widely supported by Apple devices
- **AVI**: This is one of the oldest video formats and is compatible with most media players
- **FLV**: This is a popular video format for online video streaming and is often used on video hosting websites

In general, Unity3D's `VideoPlayer` class supports most popular video formats. However, it is important to note that some formats may require additional plugins to work properly in Unity3D.

> **Tip: advanced information on video formats**
>
> It is true that the `VideoPlayer` class accepts a wide variety of formats, but it is possible that there will be incompatibilities between some of them and the target platform. I recommend the following URL of the official documentation, which offers clarification: `https://docs.unity3d.com/Manual/VideoSources-FileCompatibility.html`.

Next, we will learn how to play streaming videos, that is, videos hosted externally, for example, on a server.

Online reproduction

For the practical example of online playback, we will use a video in `.mp4` format hosted on the internet. Don't worry, I have found the perfect one for you – it can be found at `http://commondatastorage.googleapis.com/gtv-videos-bucket/sample/BigBuckBunny.mp4`.

To test our video playback system, please follow these steps:

1. Select the **URL** option in the **Source** property.

Figure 12.4 – Configuring the Video Player component with a URL

2. Paste the preceding URL into the **URL** property.

   The **VideoPlayer** component is designed to make your life easier when it comes to automatic video playback. To test it out, follow the usual flow of execution and look at the wonderful cinema screen you have just designed.

Figure 12.5 – Preview of the cinema screen playing an external video

Really impressive, isn't it? The result is truly amazing, and we have only taken a couple of steps. The next practical exercise we will do will be the same but choosing a physical file that we import into our project instead of a URL.

Local reproduction

This step is optional – I personally recommend using the URL method, as it will allow for more versatility. In the future, for example, in an update to your metaverse, you can create an API that provides a custom URL, with the video chosen dynamically, with some logic that you may find very interesting.

Using local files, as videos are often heavy, can compromise the final size of our project. For didactic purposes, we will also see how to make a reproduction by means of a local file. To do so, follow these steps:

1. Download a sample `.mp4` file. The preceding URL may be useful. If you open the URL in your web browser, you have the option to download by clicking on the button with the three little dots in the bottom-right corner of the video.

Figure 12.6 – Downloading a video to our computer

2. Create a new folder, in the **Assets** | **_App** | **Videos** path.
3. Drag the downloaded `.mp4` file to the new folder.

   Great, we have the file imported into our project. Now we need to configure the **VideoPlayer** component to be able to read it.
4. Change the **Source** property to **Video Clip**.

5. Drag the imported video into the **Video Clip** slot.

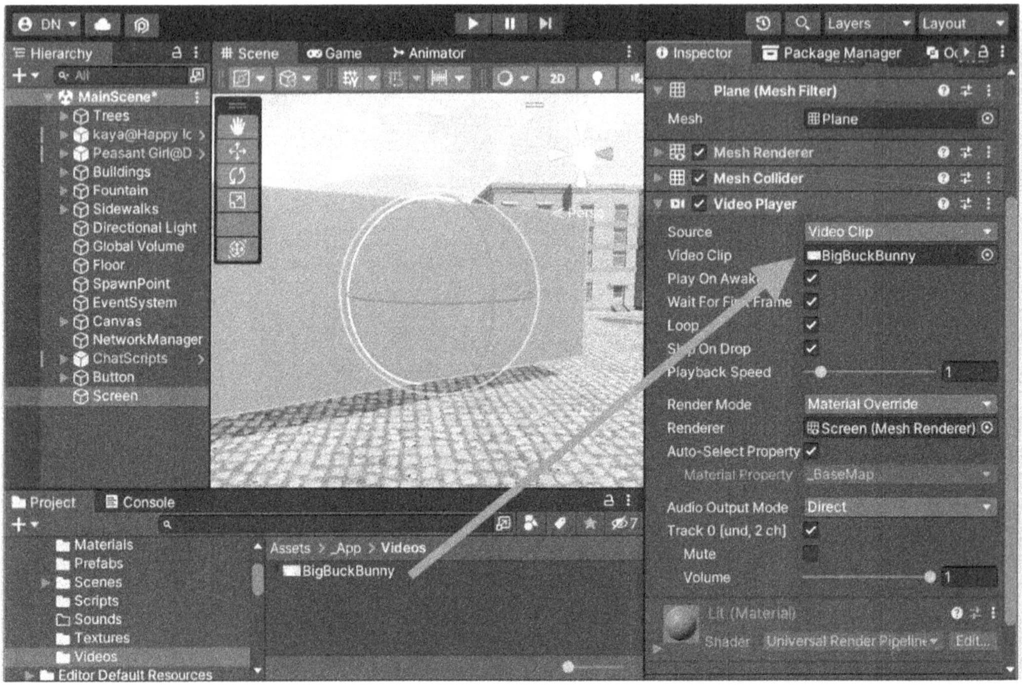

Figure 12.7 – Linking the imported video in our project to the Video Player component

Great – that should do it. You can check it again to make sure everything is working as it should.

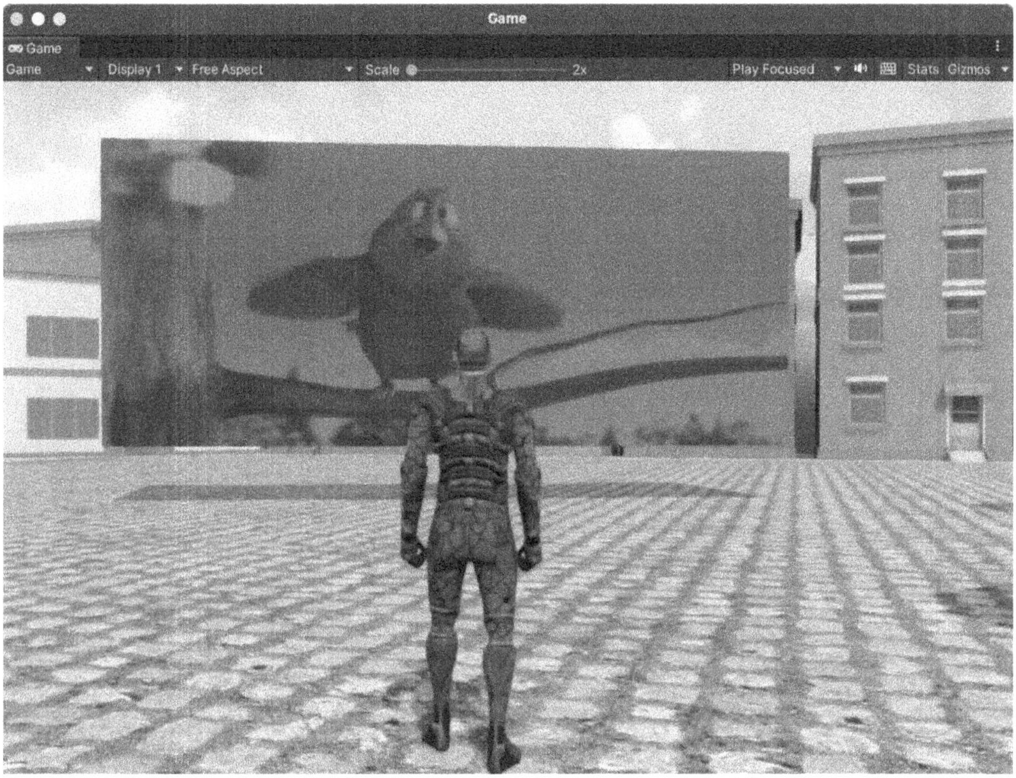

Figure 12.8 – Cinema screen preview playing a local video

As you can see, it works just as well as the first option of playing via a URL, with the difference that now our project will be heavier. That's why you should avoid importing heavy files in your project. So, if you must play videos, try to do it via URLs as much as possible.

Congratulations, you have completed this short but exciting chapter! Now your metaverse has new functionality and is becoming a project that will leave your friends and users open-mouthed.

Summary

In this chapter, we have discovered a new Unity3D class, the `VideoPlayer` class, which allows us to easily play videos both online via URLs and locally via video files. We have reviewed which formats this class accepts and analyzed in detail the properties of the class.

In the next chapter, we will learn how to integrate the Oculus Quest 2 with its respective SDK into our project.

13

Adding Compatibility for the Meta Quest 2

We have reached the chapter where we will create and implement the last functionality that defines a metaverse, and that's **Virtual Reality**.

We will use the Meta Quest 2 virtual reality goggles. The main advantage of these goggles is their *standalone* mode, which allows autonomous operation without depending on a mobile phone or PC connection.

Of course, this chapter is optional, if you do not want to implement this functionality in your metaverse or simply do not have the goggles to follow the content of these pages, you can continue to the next chapter; its completion will not affect the remaining chapters.

Our mission in this chapter is to understand what the Meta Quest 2 goggles are and how they can help us to offer a fun experience in our project, and to develop a simple implementation that allows the player to choose whether to participate in virtual reality or normal mode.

In the following screenshot, you can see the final result of this chapter. The Meta Quest 2 controllers appear in our scene. We can move them and interact with our environment. It's really exciting, isn't it?

Figure 13.1 – Final result

We will cover the following topics:

- What is Meta Quest?
- Get started with the Meta Quest SDK

Technical requirements

To carry out the tasks in this chapter, it is necessary to have the Meta Quest 2 virtual reality goggles. This is an optional chapter. Its completion and follow-up will not affect the following chapters. You can skip it if you do not have this type of goggles or you are simply not interested in including this functionality in your project.

Otherwise, this chapter does not have any special technical requirements, but we will start programming scripts in C#, so it would be advisable to have basic knowledge of this programming language. You'll need an internet connection to browse and download an asset from the Unity Asset Store.

We will continue with the project we created in *Chapter 1*. Remember that we have the GitHub repository, `https://github.com/PacktPublishing/Build-Your-Own-Metaverse-with-Unity/tree/main/UnityProject`, which contains the complete project that we will work on here.

You can also find the complete code for this chapter on GitHub at: `https://github.com/PacktPublishing/Build-Your-Own-Metaverse-with-Unity/tree/main/Chapter13`

Understanding what Meta Quest is

The **Meta Quest 2** virtual reality goggles are a high-end virtual reality device developed by the Meta company, formerly known as Oculus. The goggles were launched in October 2020 and are the sequel to the popular Meta Quest device.

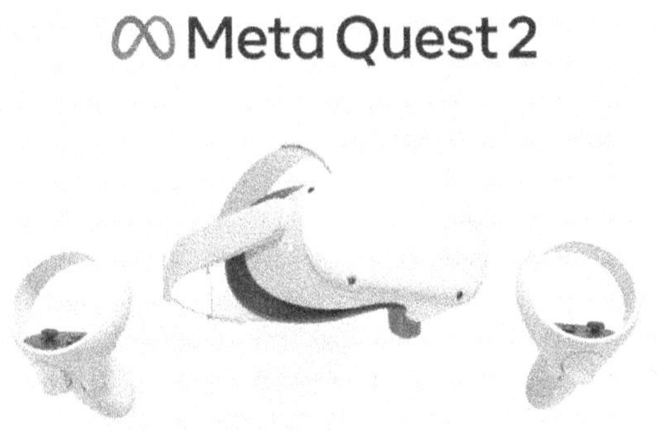

Figure 13.2 – Official Meta Quest 2 product

The Meta Quest 2 features an 1832 x 1920 resolution display per eye and a 90 Hz refresh rate, providing a very immersive and fluid virtual reality experience. It also has a Qualcomm Snapdragon XR2 processor, 6 GB of RAM, and internal storage of up to 256 GB.

The Meta Quest 2 VR goggles come with handheld controllers that allow users to interact with virtual objects in a natural way. They also feature motion-tracking technology that allows the user to move around the virtual space without the need for external sensors.

The Meta Quest 2 is a high-quality virtual reality device that offers an immersive and realistic experience for users who want to enjoy virtual reality games and experiences.

Technical requirements for the VR goggles

To use the Meta Quest 2 VR goggles, the following technical requirements are necessary:

- A personal computer with Windows 10 or higher or a Mac computer with MacOS 10.13 or higher
- A USB 3.0 port for connecting the Meta Quest 2 charging and data cable
- The Meta Quest application installed on a computer to manage the device's content and settings
- A Meta Quest account with access to the VR app store
- Access to a high-speed Wi-Fi network to download and update content

In addition to the preceding requirements, it is advisable to have a large enough space to move around while using the device, and additional hardware such as a good set of headphones or speakers to enhance the listening experience.

Importantly, one of the advantages of the Meta Quest 2 is that it does not require the installation of external sensors or tracking devices, which makes its setup and use very simple and accessible to users of all technical levels.

Cost of the Meta Quest 2 goggles

The price of the Meta Quest 2 virtual reality goggles depends on the version you purchase and the amount of internal storage you want. Currently, on Meta's official website, the prices are as follows:

- Meta Quest 2 with 64 GB internal storage: $299 USD
- Meta Quest 2 with 256 GB internal storage: $399 USD

Please note that these prices may vary depending on geographic region and local currency, as well as promotions and offers available at the time of purchase. In addition, some authorized retailers may offer different prices and accessory packages with the glasses.

Getting started with the Meta Quest SDK

In order to implement the headset in our project, we must first carry out preliminary tasks to prepare the environment. The roadmap for this section can be summarized in five phases:

1. Preparation of the glasses
2. Developer Mode
3. Configuring our project
4. Adding VR compatibility to our project
5. Testing in the Meta Quest 2

Once you have finished the roadmap, you will be able to run the project in the virtual reality glasses. Are you ready? Let's start!

Preparation of the glasses

If you already have the Meta Quest 2 and have it configured, skip this part. Here, we will look at the instructions to configure a new pair of glasses. As you will see, setting up a new pair of glasses is really easy. You just need to download the Meta Quest app on your mobile device.

In the Apple App Store, you can find it under the name **Meta Quest**.

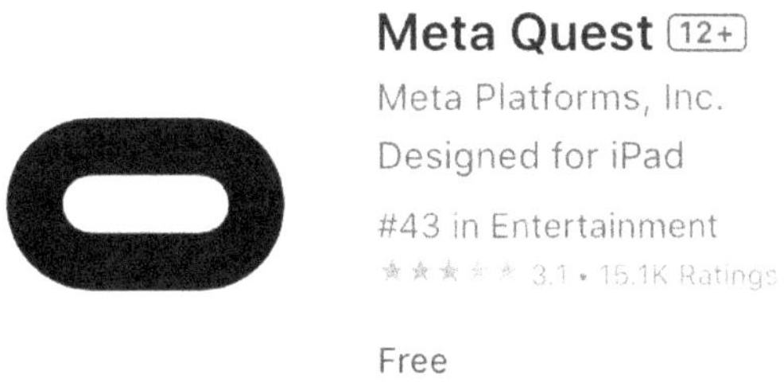

Figure 13.3 – Meta Quest application on Apple's App Store

On the Google Play Store, however, you can find it under the name **Oculus**.

Figure 13.4 – The Oculus app on the Android Play Store

Once you have downloaded the app on your mobile device, all you have to do is follow the interactive steps:

1. Create a new account or log in if you already have one.
2. Pair the glasses, following a few simple steps.
3. Connect to a Wi-Fi network.
4. Update the firmware.

> **Tip: complete installation guide**
>
> The process of setting up the Meta Quest 2 glasses is really simple by following the steps that the app interactively shows you. But in any case, I'll share with you a very complete guide that explains the whole process step by step: `https://howchoo.com/vr/how-to-setup-oculus-quest-2`. You can also have a look at the official guide, which also includes a section on troubleshooting that you might encounter in the process: `https://www.meta.com/en-us/help/quest/articles/getting-started/getting-started-with-quest-2`.

Before proceeding to the next step, I recommend that you first have some fun with your new glasses. It will help you to familiarize yourself with the buttons, interface, and movements. It will also serve as a knowledge base to continue with the development of the book.

Developer Mode

By default, the Meta Quest 2 virtual reality goggles do not have the Developer Mode activated. Without this mode active, it will be impossible to make a connection between our computer and the hardware of the goggles.

There are two ways to make the connection, a basic mode via USB cable and another mode using the Meta Quest Developer Hub software, which is more intuitive and also allows us to connect and send our project to the goggles without the need for a USB connection.

Meta Quest Developer Hub

Meta Quest Developer Hub is a set of online tools and resources for developers working on Meta's virtual reality platform (formerly known as **Facebook Reality Labs**). It is designed to provide developers with a comprehensive set of tools for creating and publishing applications and games on the Meta platform.

Tools offered by the Meta Quest Developer Hub include the following:

- Documentation and guides for developing applications and games on the Meta platform
- Development tools such as Meta Software Development Kit and the Meta Quest Emulator, which allow developers to create and test their applications without the need for a physical device
- Access to Meta's publishing platform, where developers can submit their applications and games for review and approval before they are published on the Meta Store
- A developer community, where developers can interact with each other, share knowledge, and receive support from other Meta developers and experts

In short, Meta Quest Developer Hub is a comprehensive platform for developers who want to create and publish applications and games on Meta's virtual reality platform.

The software is available for Mac and Windows computers and can be downloaded by visiting `https://developer.oculus.com/downloads/package/oculus-developer-hub-mac` if you are a Mac user or `https://developer.oculus.com/downloads/package/oculus-developer-hub-win` if you are a Windows user.

Once downloaded, follow the usual installation process. Once finished, open the application to proceed with the configuration.

In the following screenshot, we can see the welcome screen when we open the hub for the first time.

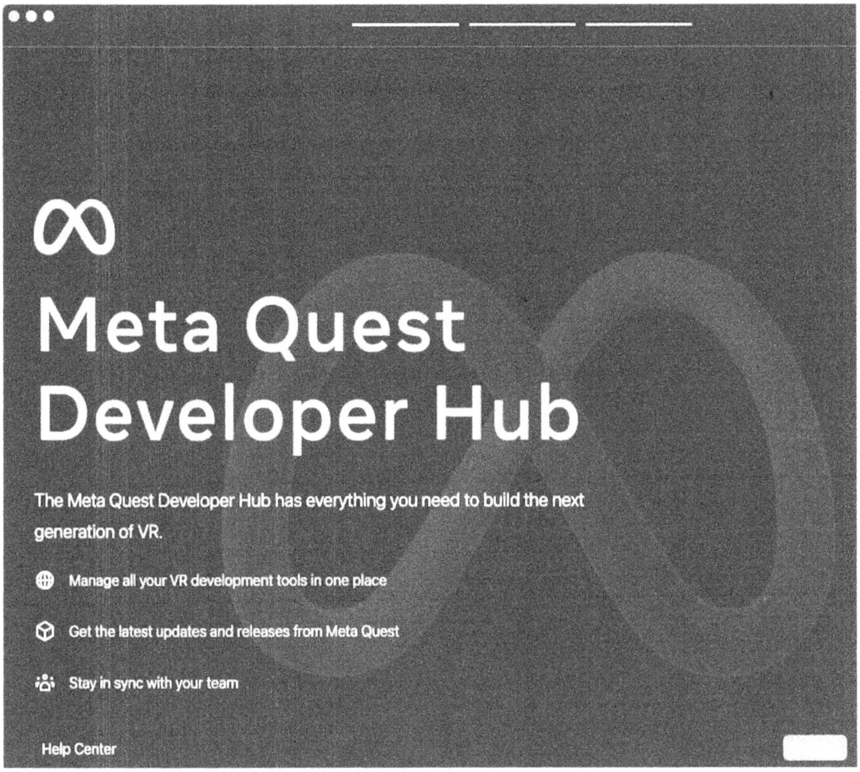

Figure 13.5 – Meta Quest Developer Hub welcome screen

Once open, click the **Continue** button and follow the initial setup process. The initial configuration is mainly to request access to Bluetooth and to log in. It is important that you log in with the same account that you used in the configuration of the Meta Quest 2.

The next step will be to pair the glasses. To do this, click on the **Device Manager** option. In my case, it has asked me to verify myself as a new developer.

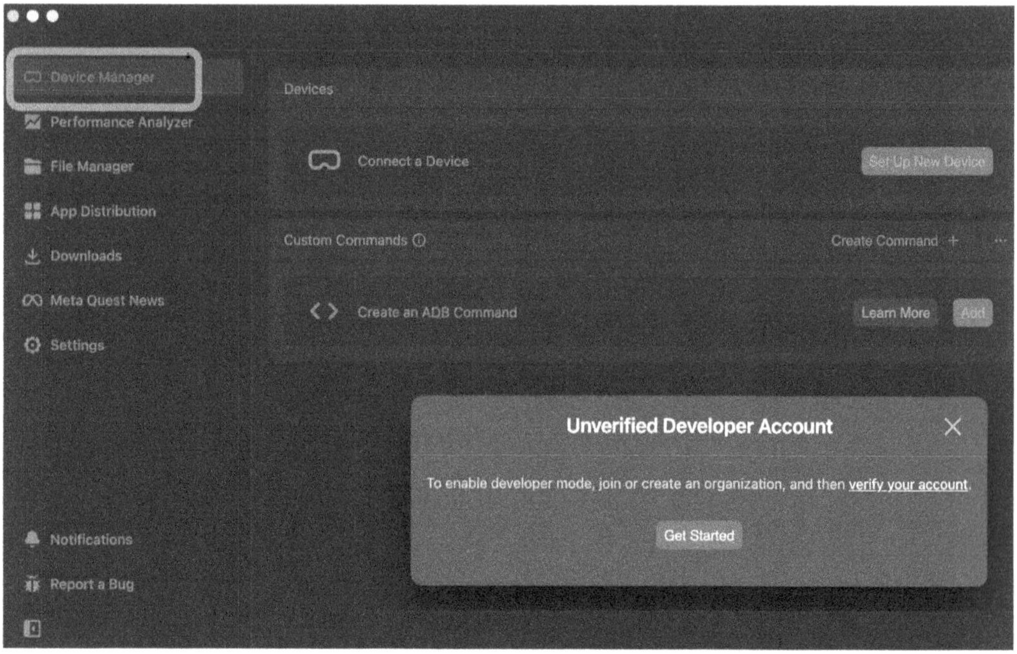

Figure 13.6 – Unverified developer warning

Verification as a developer basically consists of two steps:

1. Verify the account.
2. Create a new organization.

First, we will verify the account by logging in at `https://developer.oculus.com/manage/verify`. If necessary, log in again with the same credentials we have been using.

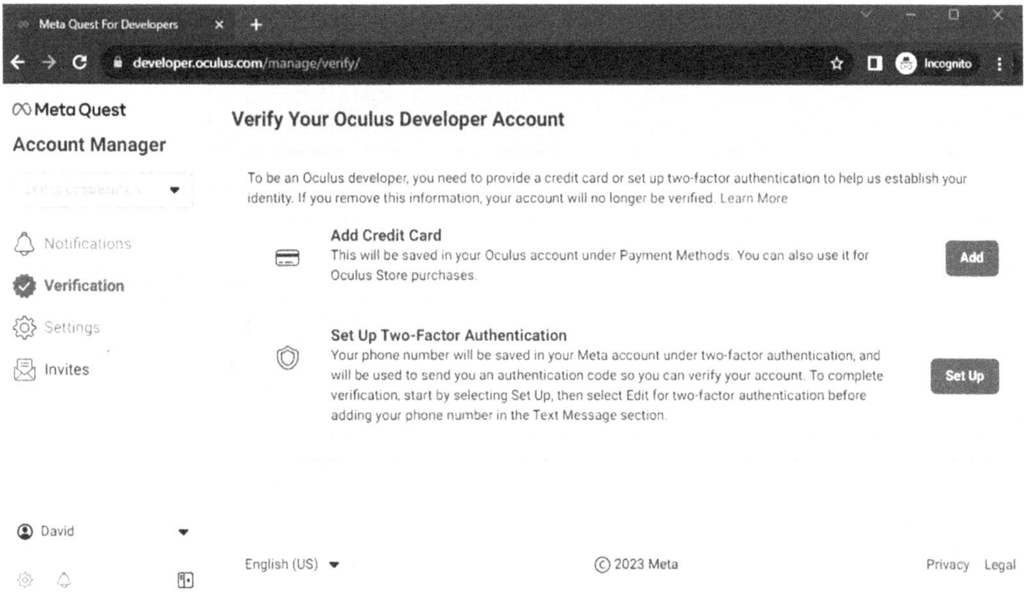

Figure 13.7 – Necessary steps for developer verification

Once we have been able to access the user verification panel, we can see that Meta asks us to add a credit card and activate two-step verification.

> **Tip: credit card information**
>
> In order to activate the Developer Mode, it is strictly necessary that you enter a valid credit or debit card. It is important to know that Meta will not charge your account.

Finally, once we have added the card, we have to configure the two-step identification, which will invite you to create a secure password and add a phone number to receive an SMS when you want to log in.

Once this second step has been completed, we can return to the verification control panel, and we will see that all the steps have been completed.

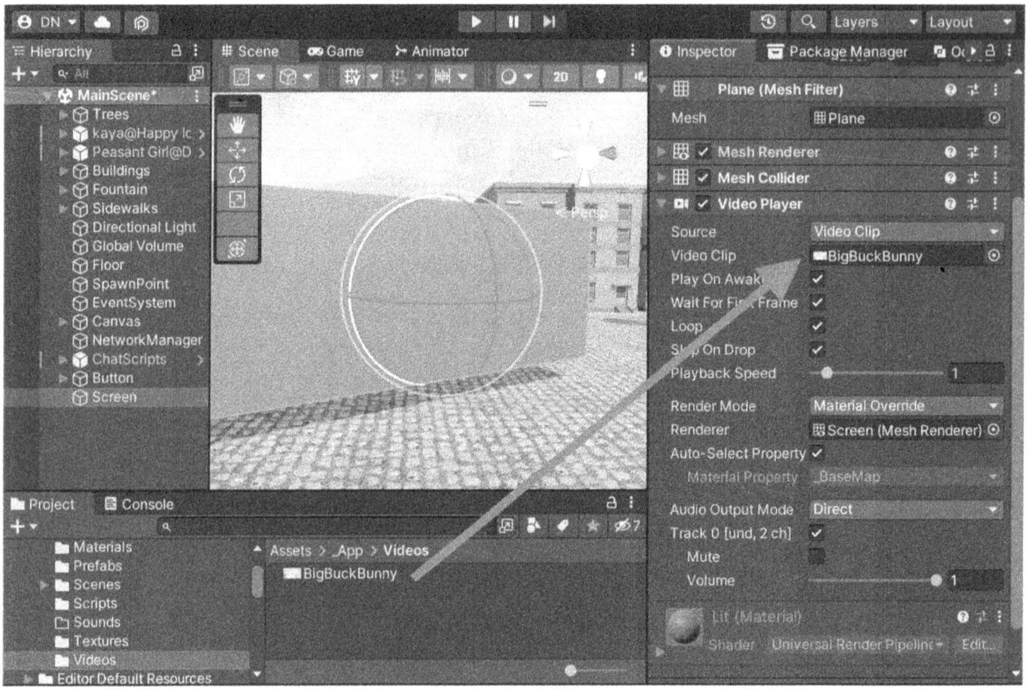

Figure 13.8 – Successful bank card verification

Good job! Now, we just need to create a new organization to finalize the activation of the Developer Mode.

In the menu on the left, click on the **Select Organization** button and click on the **Create** button.

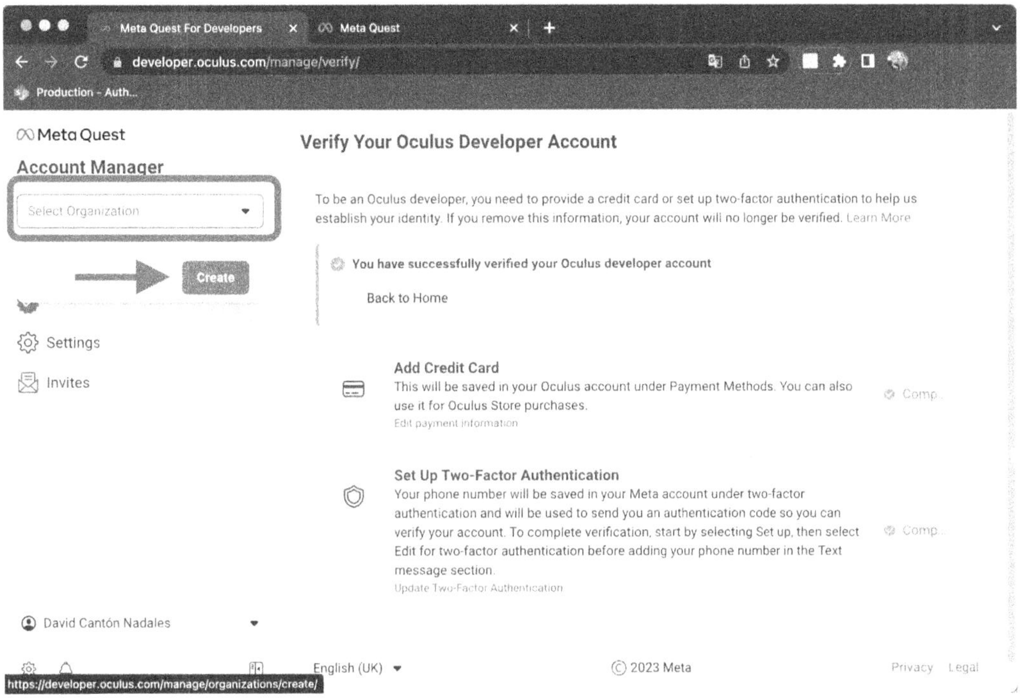

Figure 13.9 – Creating a new organization

The process of creating an organization will simply ask for a name that represents *our company*. This name will be the one that will appear in the Meta Store in the future, when we have published our app.

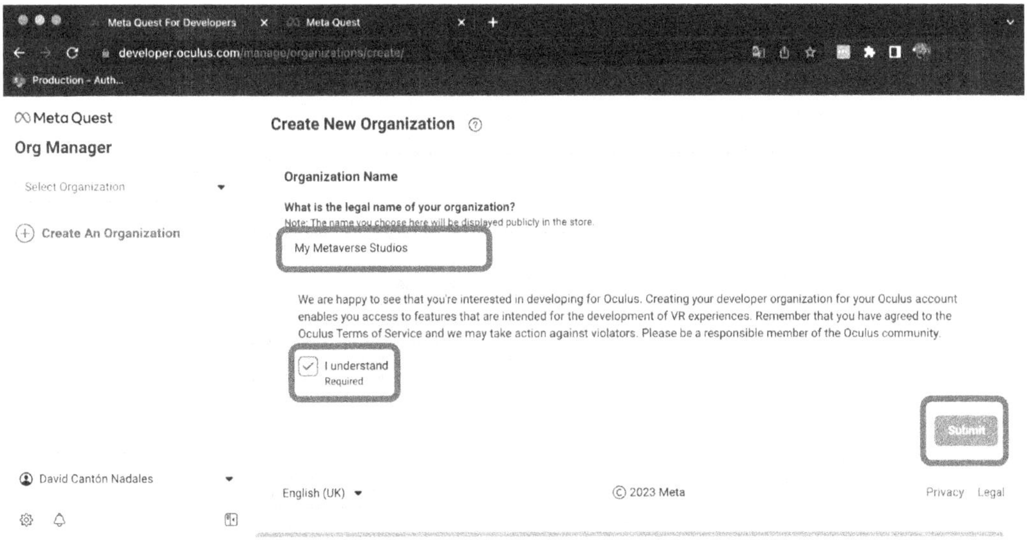

Figure 13.10 – Final step of the creation of an organization

Great – we have completed the developer verification. We can go back to the Meta Quest Developer Hub application to continue with the pairing of the glasses.

Click on the **Set up New Device** button and execute the steps that follow. As can be seen in the following screenshot, the hub warns us about the things we must take into account before continuing with the pairing of a new Meta Quest 2.

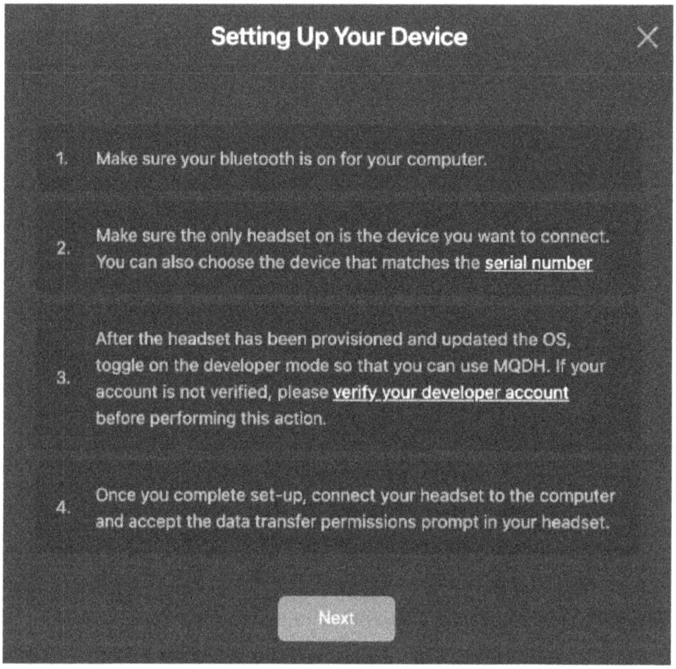

Figure 13.11 – Checklist to consider before proceeding with the pairing

The software will try to find our glasses. It is important that they are on and close by.

Figure 13.12 – Meta Quest 2 detected

The configuration wizard will take care of everything, and finally, if everything has gone well, it will ask us to activate Developer Mode.

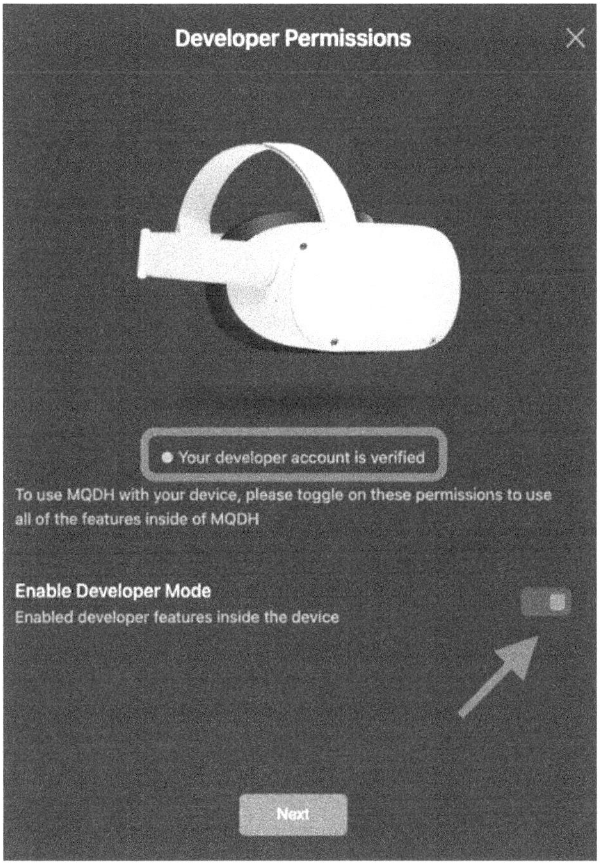

Figure 13.13 – Button to enable Developer Mode

Simply activate the toggle button and click on the **Next** button again to finish. The last step is to connect the glasses with the cable to finish assigning permissions and the configuration.

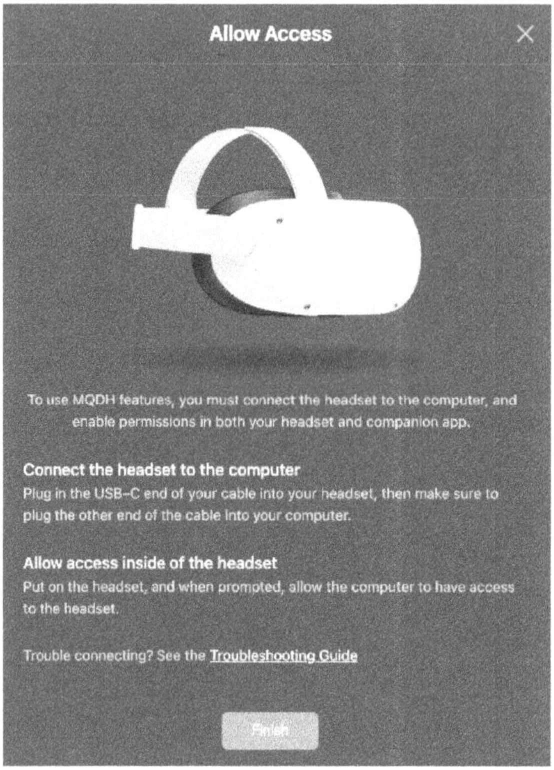

Figure 13.14 – Request for permission

Connect the glasses to your computer with the USB cable and click the **Finish** button.

When we connect the glasses, we will see a warning that the glasses have been detected but we have not given permission.

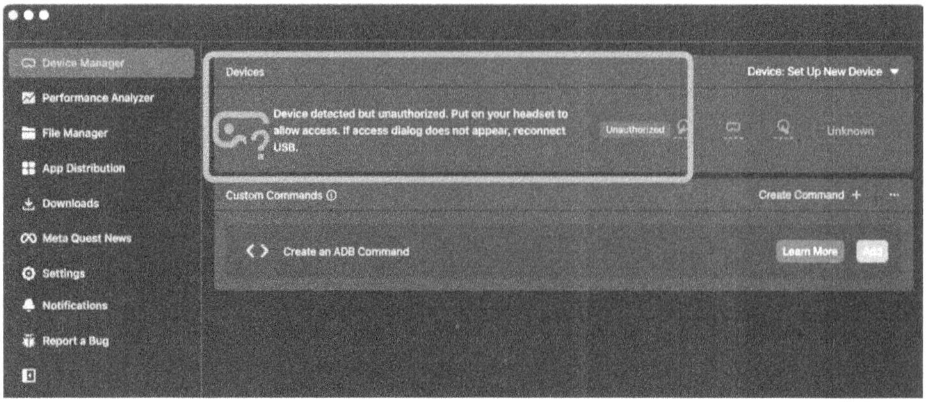

Figure 13.15 – Unauthorized device warning on the computer

The message we get invites us to put on the glasses to accept permissions. It is necessary to use one of the controls to accept several permissions that will be requested.

Once we have accepted all the permissions, we can see in Meta Quest Developer Hub that the glasses appear correctly.

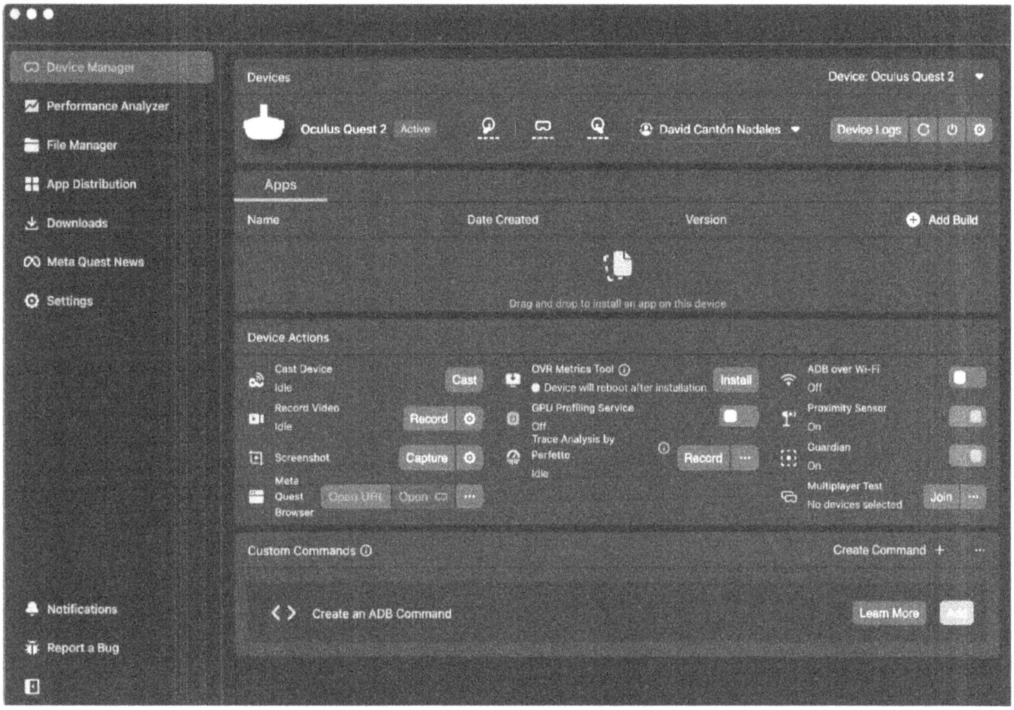

Figure 13.16 – Oculus Meta Quest 2 control panel

Tip: about permissions

It is important that you click the **Always Allow this Computer** button instead of the **Allow** button. This will help you to avoid being prompted for permissions every time you make a connection to the glasses and modify a setting.

Finally, to be able to transmit builds to the glasses without the need to be connected with the USB cable, we are going to enable the **ADB over Wi-Fi** option.

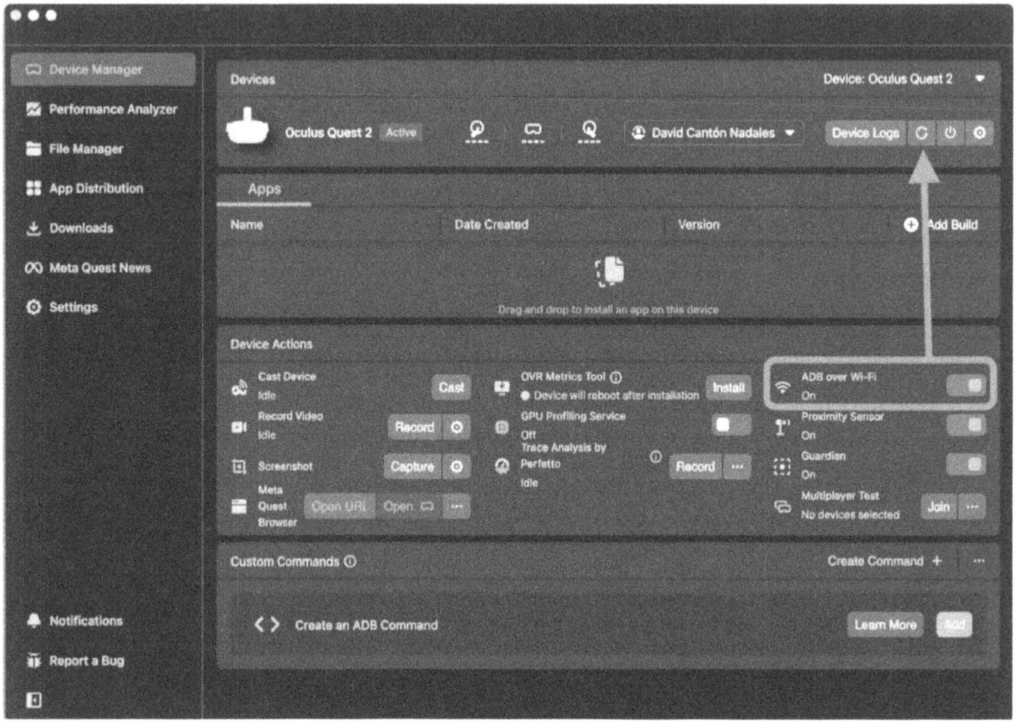

Figure 13.17 – Enabling Wi-Fi connection

If you get an error message when activating the option, it is recommended that you click the **Reset** button to try again once the goggles are turned back on.

Good work! You have just finished configuring your Meta Quest 2 and it is now ready for development.

> **Tip: advanced information about MQDH (Meta Quest Developer Hub)**
>
> If you need more information or a solution to a problem not covered in this chapter, the official documentation is available at `https://developer.oculus.com/documentation/unity/unity-quickstart-mqdh`.

Next, once we've got our device and Meta Quest Developer Hub software set up correctly, we're going to get down to the practicalities of integrating the development SDK for a first-person VR experience in the metaverse that we've been building throughout these chapters.

Configuring our project

Meta has an SDK that allows full integration into Unity 3D. This SDK, called the **Oculus Integration SDK**, is a software development toolkit that helps developers create virtual reality applications and games for Oculus devices, including Oculus Quest, Oculus Rift, and Oculus Go.

This SDK offers a wide range of features and tools, including the following:

- Access to Oculus device sensors and tracking features, such as user position and orientation, hand tracking, and eye tracking
- Integration with the Oculus Touch controller to enable user interaction with the app or game
- 3D audio integration to provide an immersive and realistic audio experience
- Oculus Store integration, allowing developers to implement in-app purchases and in-app publishing in the store
- Debugging and profiling tools to help developers detect and fix problems in their applications and games

This SDK is available for free from the Unity Asset Store or from the official website: `https://developer.oculus.com/downloads/package/unity-integration/`.

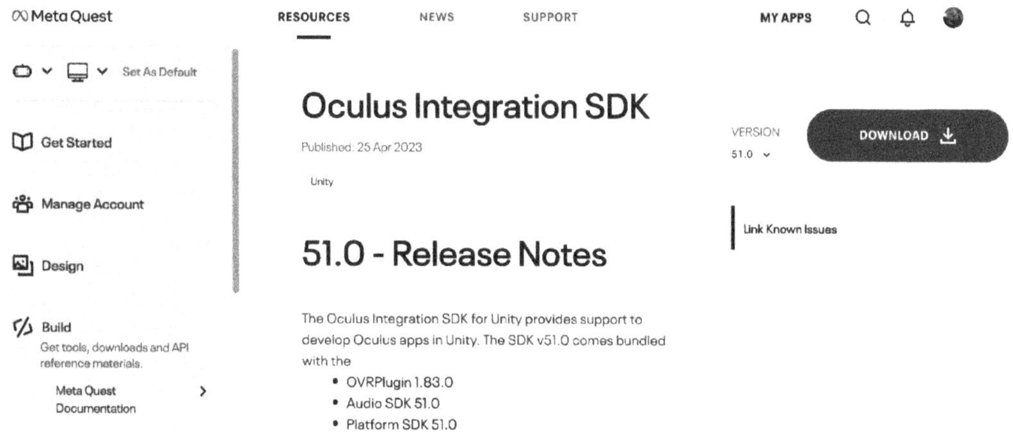

Figure 13.18 – Oculus Integration SDK download page

In both places, you can find the latest version. I invite you to download it from the place that is most convenient for you. Once downloaded, import it into your project like any other asset and wait for the process to finish.

It is possible that during the import process, Unity will launch a warning about the **OVRPlugin** plugin update. If this happens, click on the **Yes** button to update it.

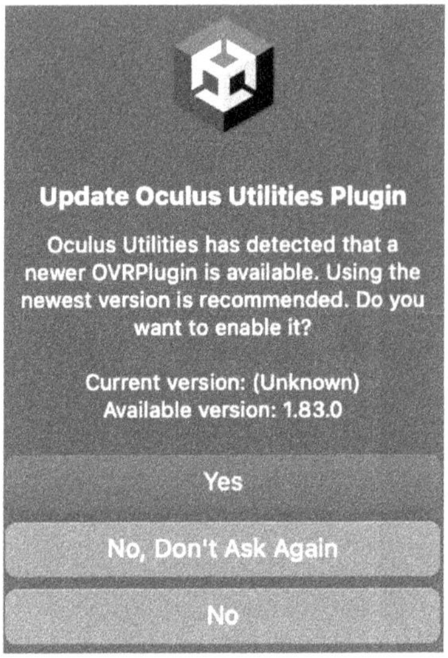

Figure 13.19 – Plugin update notice

After the plugin update, you will get another window inviting you to restart the Unity editor to apply the changes. Click the **Yes** button again.

One of the features of the plugin that we have just updated is to offer a cleanup of VR-related assets that become obsolete after the update. You will receive a notification inviting you to perform the cleanup as a recommended task before finalizing the configuration. Click the **Show Assets** button.

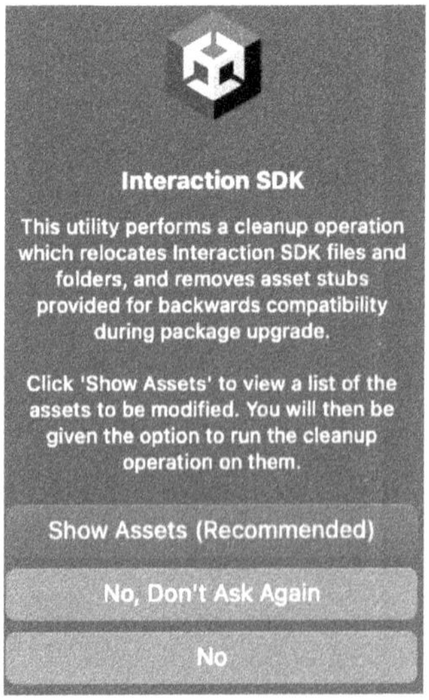

Figure 13.20 – Warning to clean up unneeded files

We will then be shown a window with a list of assets related to the Oculus SDK that we no longer need and that we should (as a recommended option) clean up.

Figure 13.21 – Warning to clean up unneeded files to be deleted

Click on the **Clean Up** button and wait for the process to finish before continuing.

Oops! It is possible that, again, a window will appear warning us about the update of another plugin. This time, click on **Upgrade** and then click on **Restart** when you are notified that it is necessary to restart the editor to continue.

Fantastic – we now have the SDK installed in our project to include VR functionality. We will now see a new folder called `Oculus` has appeared in our folder structure. Everything is ready to continue.

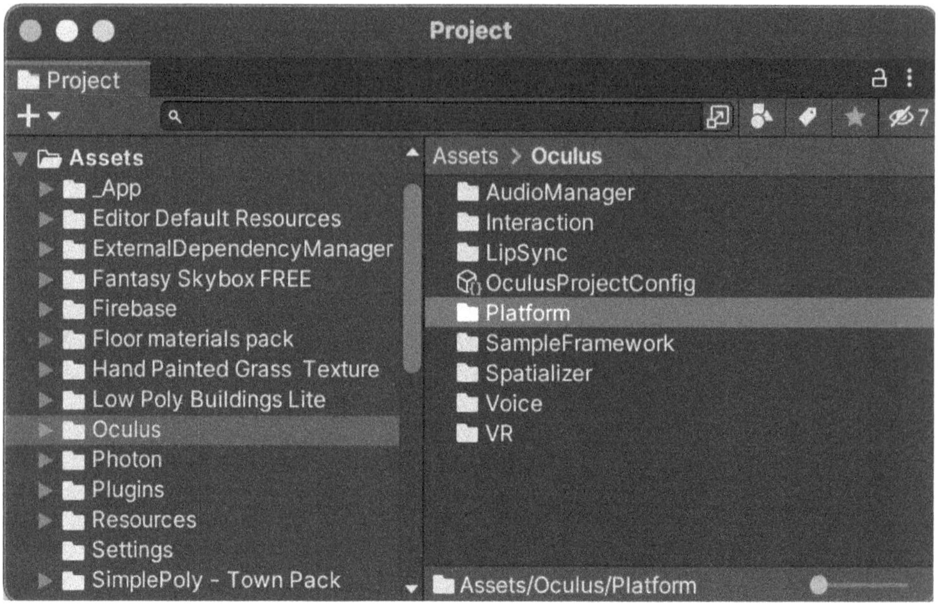

Figure 13.22 – Folder structure

Excellent work – you just finished the project setup and now we have the Oculus SDK integrated. Once we have finished installing the SDK, we need to perform some extra steps in our project to enable VR mode. We'll look at those next.

Enabling VR support

Next, we are going to go over the steps we need to follow in order to compile and send our project to the Meta Quest 2 goggles. It is important to know that VR applications are compiled for the Android operating system, so the first thing we will do is configure the project for Android:

1. Click on the **File | Build Settings** option in the main menu bar.
2. Select **Android** from the list.

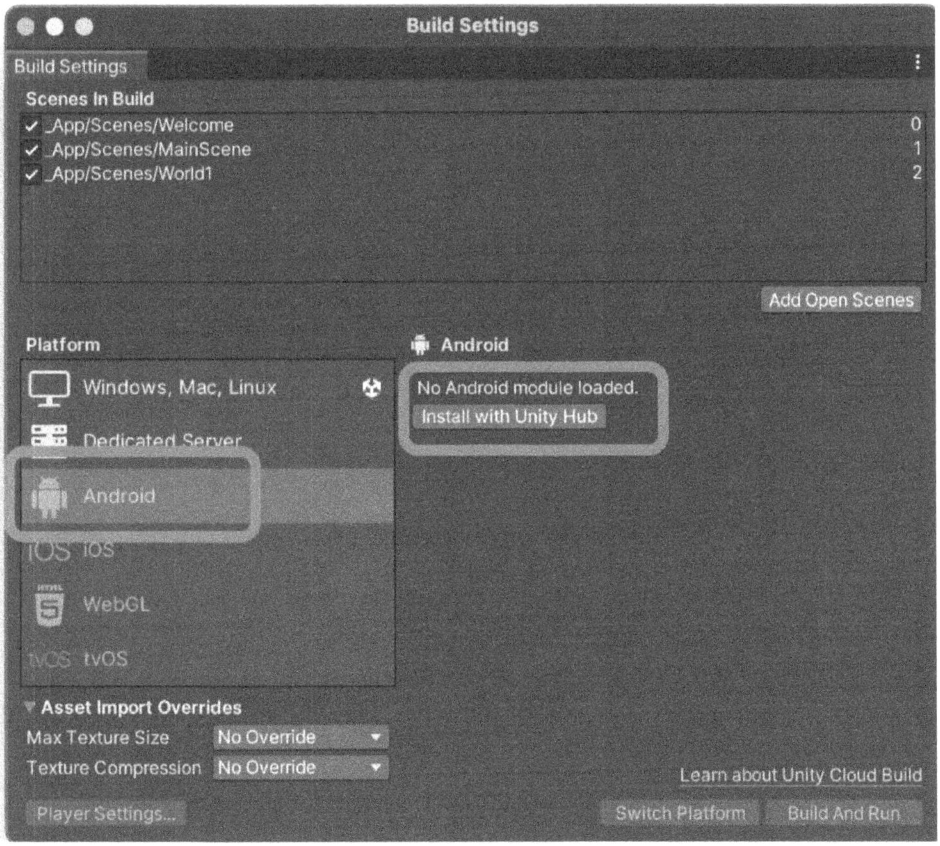

Figure 13.23 – Android platform not installed in the Build Settings window

3. In my case, I didn't have the Android module previously installed, so if the same thing happens to you, click on the **Install with Unity Hub** button. Otherwise, if you already have it installed, skip this step. In the following screenshot, we can see the installation window of the Android platform in Unity Hub. The marked options are those strictly necessary for correct operation.

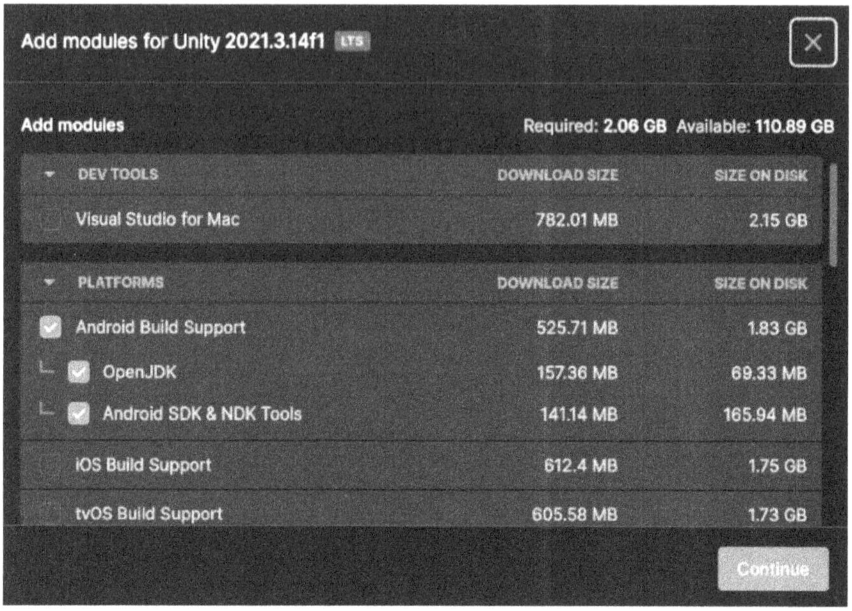

Figure 13.24 – Android platform installation window in Unity Hub

Unity Hub will show a confirmation window to download the Android module. This process can take several minutes, depending on your internet connection. It is a good time to have a coffee.

4. Once the Android module is installed, we must restart the Unity editor to refresh the list of available modules, and then we can select **Android** and click the **Switch Platform** button. This process can take several minutes.

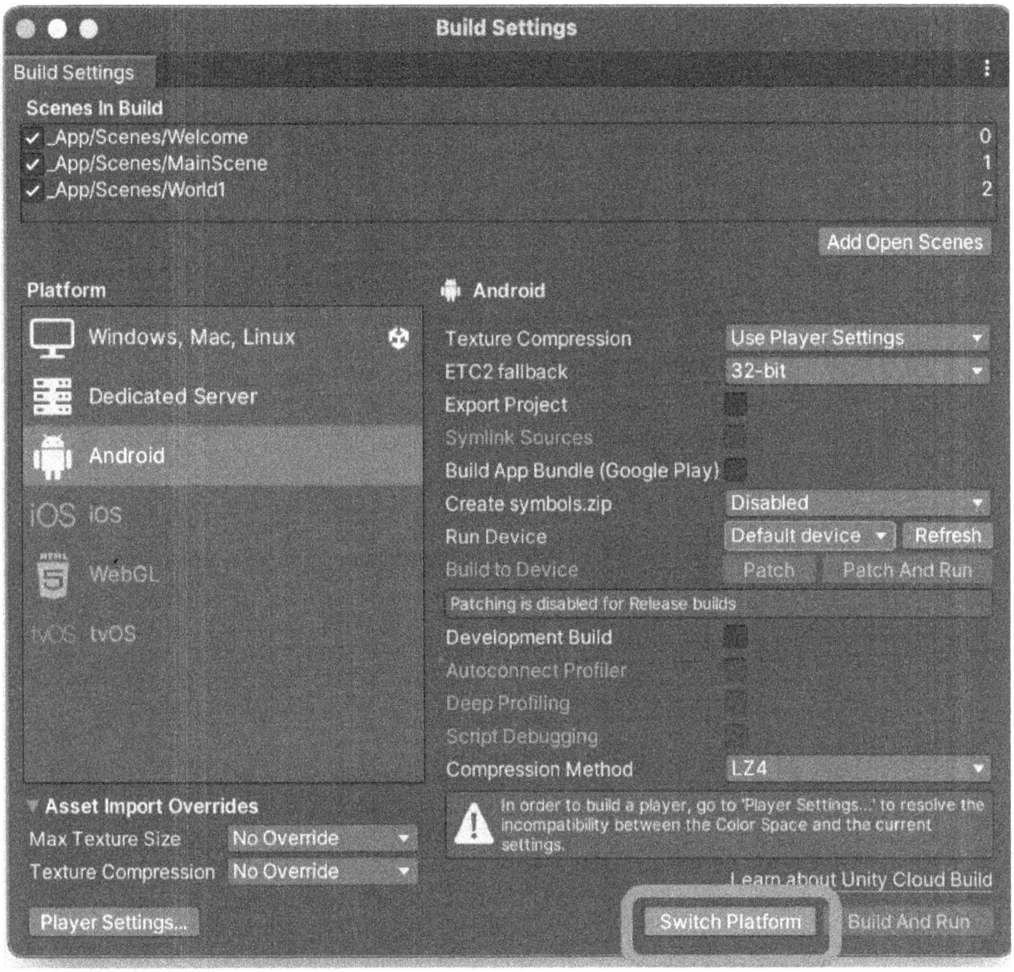

Figure 13.25 – Android platform already installed in Build Settings

Once the platform change process has finished, you may encounter two notices from Unity, one of which is to warn you about a difference in the bundle app ID we set up in Firebase and the one the project currently has.

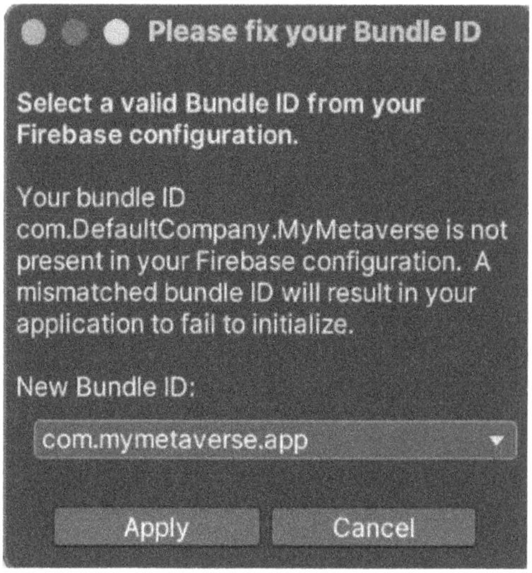

Figure 13.26 – Warning about bundle ID inconsistency with Firebase

5. The correction of this problem is simple. Select the bundle ID from the list that we configured in Firebase and click the **Apply** button.

   The second warning is to activate the plugin that allows auto-resolution versions in the plugins that we have installed in our project.

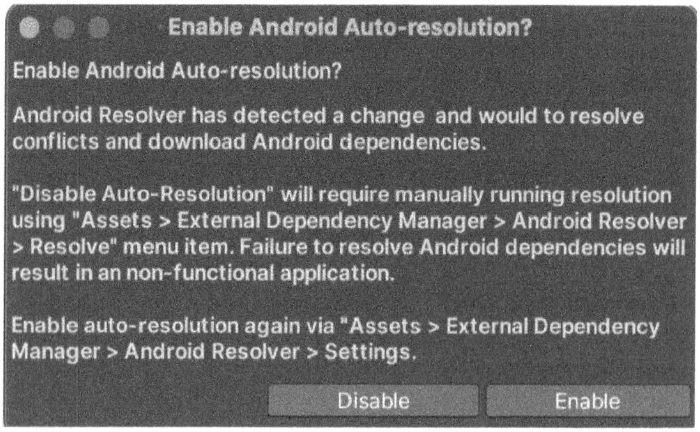

Figure 13.27 – Notice to enable dependency auto-resolution mode in Android

6. This task is very useful as it automatically solves problems with version dependencies of Android plugins. I recommend clicking the **Enable** button. It will automatically start a review and download of fixes. This process can take several minutes.

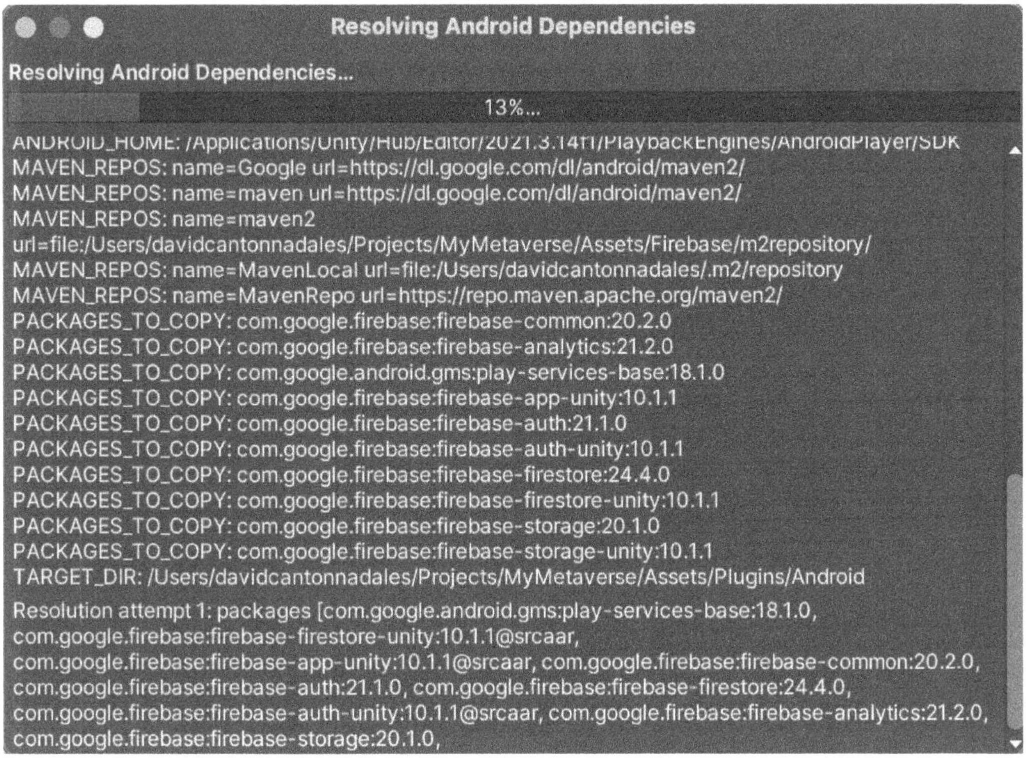

Figure 13.28 – Resolving dependencies in Android

7. If you get an error when automatically resolving dependencies with the plugin, you may have misconfigured the Android SDK. To check this, go to the **Settings** menu in Unity and, in the **External Tools** section, check that you have enabled all the SDK and tools' recommended usage checks. When you activate them, the dependency check will be launched again and this time it will be successful.

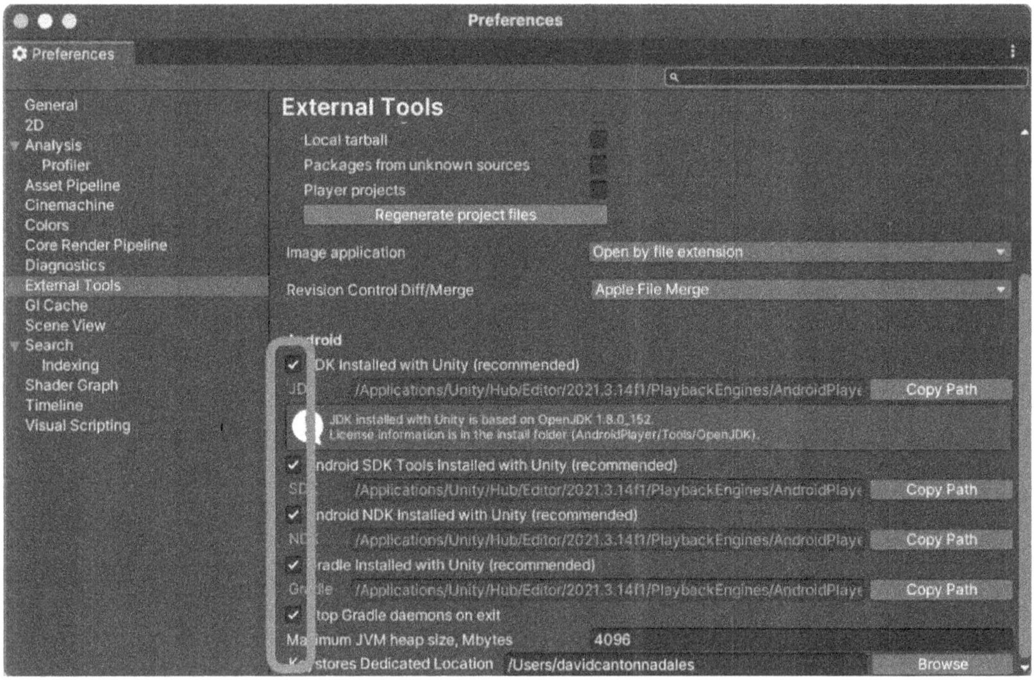

Figure 13.29 – External tools required for proper operation on Android

Finally, we must launch the Oculus test tool to correct possible missing configurations in our project, which may prevent the correct functioning of the Meta Quest 2 glasses.

8. To launch this tool, open **File** | **Build Settings** and click on the **Player Settings...** button.

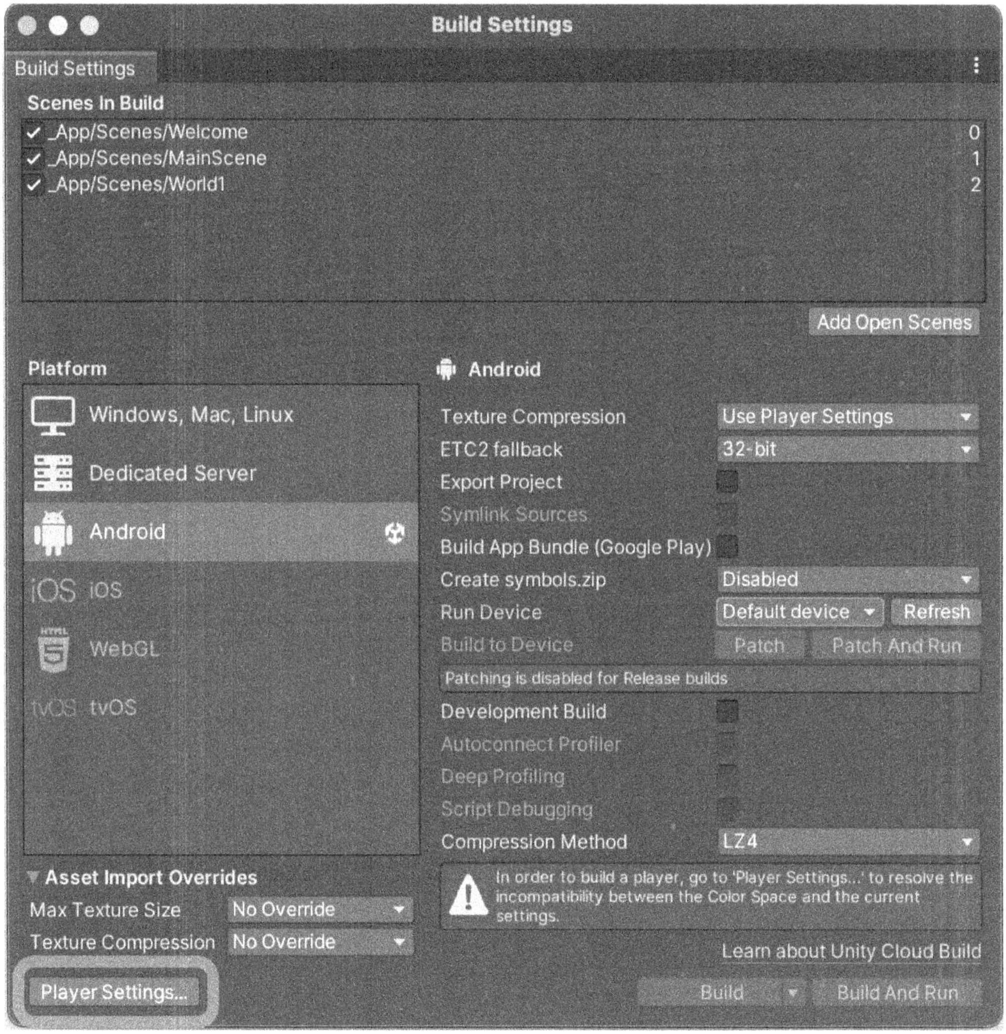

Figure 13.30 – Player Settings button in the Build Settings window

9. Another necessary configuration is to enable the use of the Oculus plugin as an XR service provider in Unity. To activate this option, once the **Player Settings** window is open, look for the **XR Plug-in Management** menu on the right and check the **Initialize XR on Startup** and **Oculus** checkboxes.

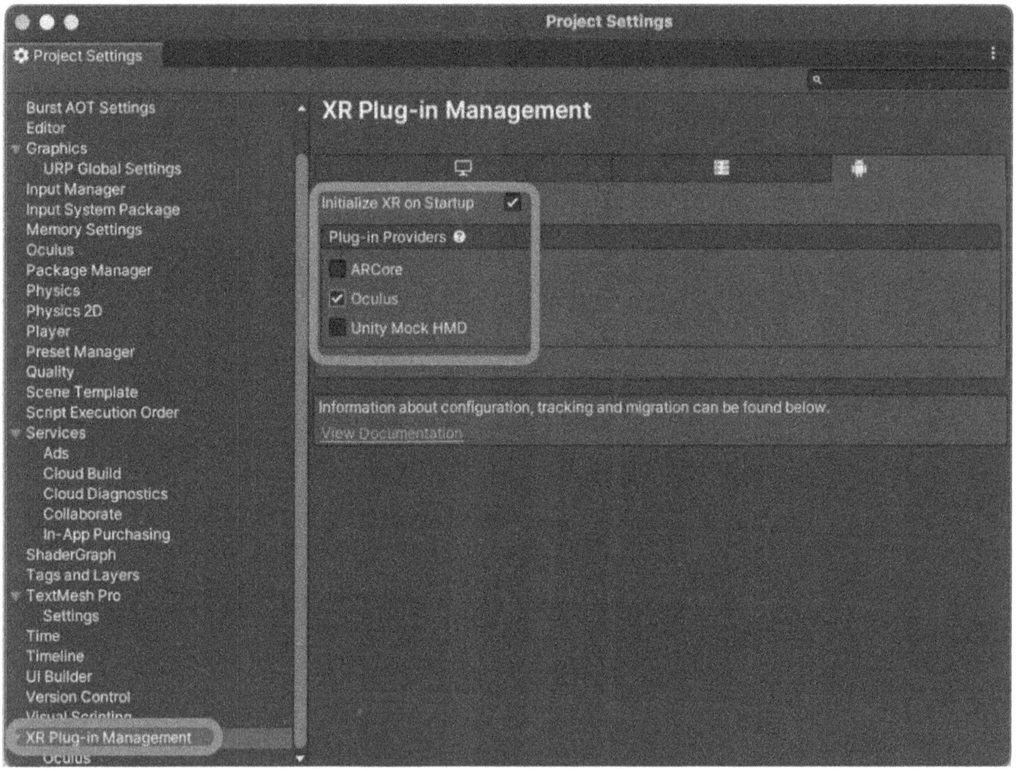

Figure 13.31 – Management of XR plugins for the project

10. Now, with the **Player Settings** window still open, look for the **Oculus** menu and click on the **Fix All** button first, and finally, on the **Apply All** button. This way, Oculus will fix all the wrong settings for us.

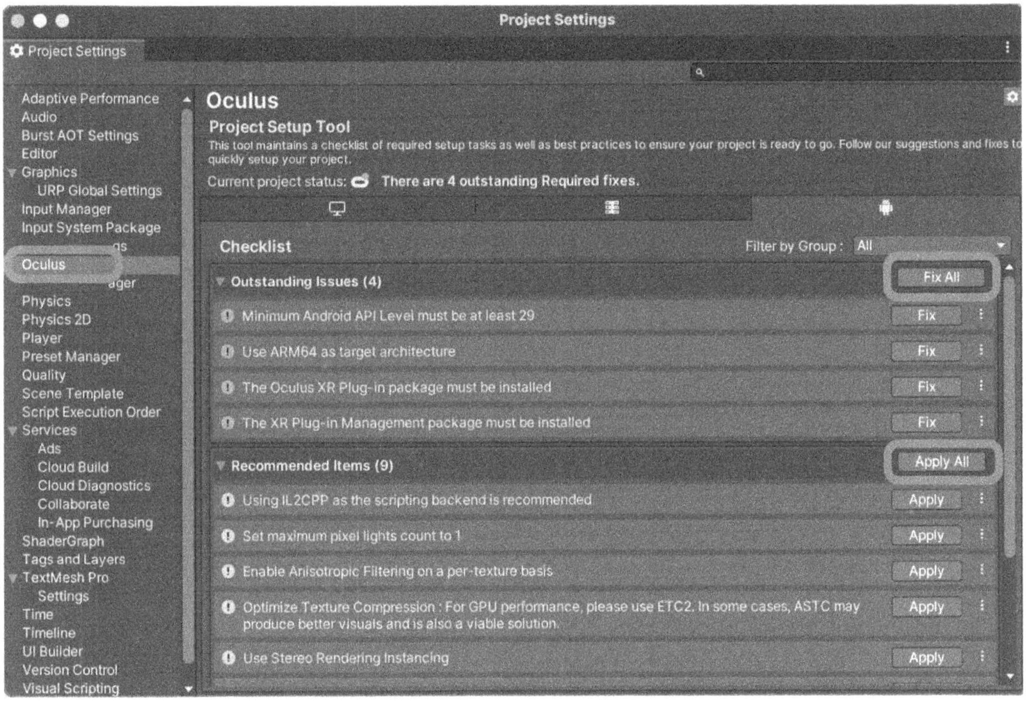

Figure 13.32 – Self-correction window for Oculus SDK-related issues

Although this correction tool has done a great job for us, there is one remaining setting that needs to be done manually.

11. Before closing the **Player Settings** window, find the **Player** menu and, in the **Other Settings** tab, uncheck the **Auto Graphics API** box to manually choose the order in which the graphics APIs are consumed. The **Linear** color space requires OpenGL ES 3. Therefore, we leave only **OpenGLES3** in the list and remove any others that appear.

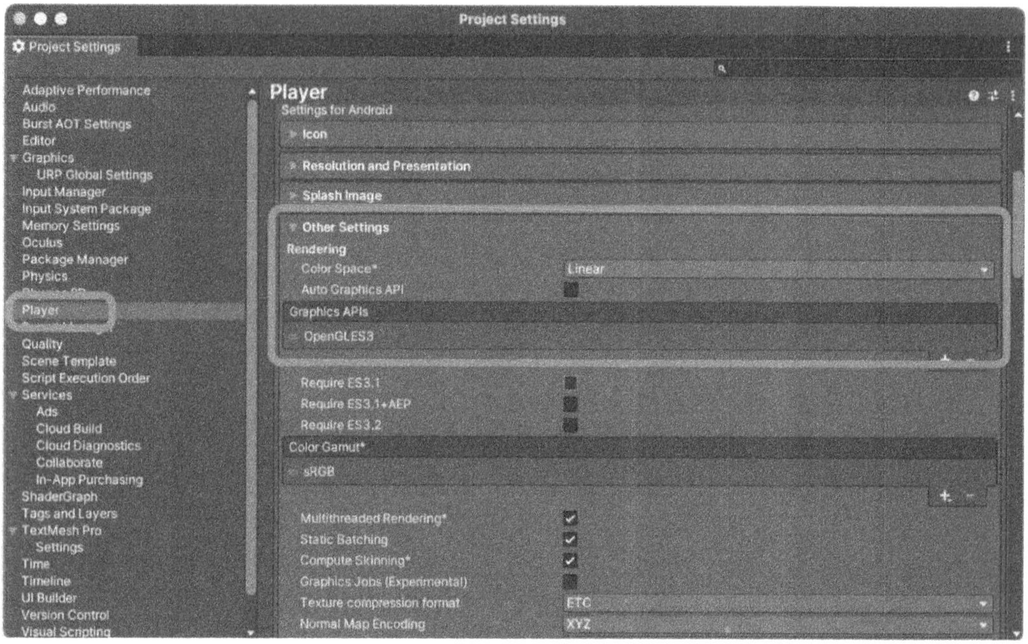

Figure 13.33 – Graphical configuration for the project

> **Tip: important notice about Firebase**
>
> Attention: it is important that you take this action. The Oculus SDK comes with a sample project and, inside it, there is an empty Firebase configuration file that may conflict with our project, so we need to delete it. The file to delete is located in `Assets` | `Oculus` | `SampleFramework` | `Usage` | `Firebase` | `google-services.json`. It is also advisable to re-download our `google-services.json` from the **Firebase Control** panel and drag it to the root of our project, in the `Assets` folder. It is possible that the Oculus file has corrupted the Firebase configuration. If you don't remember how to download and add the `google-firebase.json` file, you can refer to the *Installing the Firebase SDK* section from *Chapter 1, Getting Started with Unity and Firebase*.

Good job – you have completed the remaining configuration to be able to successfully compile our application. We are now ready to get into programming. To keep our project compatible between normal users and VR users, we will create a system that will automatically use either the VR camera or the normal one.

Adding VR compatibility to our project

The Oculus SDK includes a series of ready-made Prefabs to provide all the VR functionality we will need in our project, such as the VR camera, the player controller, or the hand control.

Camera

The great advantage of the Prefabs included in the Oculus SDK is that they have backward compatibility with the non-VR system. For example, the **MainCamera** GameObject will be replaced by **OVRCameraRig** in all scenes, and you will think, *If I replace the normal camera with the VR camera, I will lose compatibility with normal users, right?* The answer is no.

The **OVRCameraRig** Prefab becomes a normal **MainCamera** when running on a non-VR device. Fantastic, isn't it? This allows us to avoid creating complex code to overcome the technical differences between normal devices and VR goggles.

To start, we'll open our main **Welcome** scene, which is located in **Assets** | **_App** | **Scenes**. We'll follow a series of steps to bring VR compatibility to our **Welcome** scene:

1. The Oculus SDK offers a Prefab called **OVRCameraRig**. This Prefab offers a VR camera ready to use just by adding the Prefab to the scene and then removing the normal camera. But first, we need to configure this Prefab to also offer virtual keyboard support. To do this, select the **OVRCameraRig** Prefab found in the **Assets** | **Oculus** | **VR** | **Prefabs** path and modify the following options in the **Inspector** panel:

    I. Check the **Require System Keyboard** option.

    II. Select the **Require** option in **Tracked Keyboard Support**.

    III. Drag the **OVRCameraRig** Prefab into the scene.

    The Prefab we just added gives us a VR camera in our scene, but by itself does not meet the requirements we need to be able to interact with the environment; that is, we also need to add support for controllers to be able to click the buttons in our scene. The Oculus SDK has other Prefabs that solve this problem in an extremely simple way.

2. **OVRInteraction** is the Prefab in charge of instantiating the ability to interact with physical elements of the scene through the two controllers we hold in our hands. This Prefab is located in the **Assets** | **Oculus** | **Interaction** | **OVRInteraction** | **Prefabs** path, and we are going to drag it into the **OVRCameraRig** GameObject of our **Hierarchy** panel.

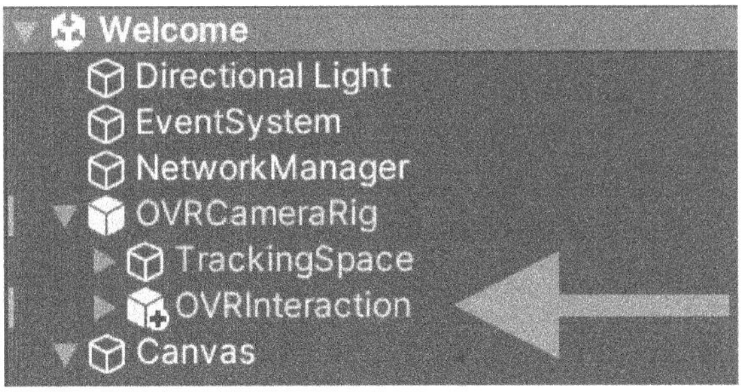

Figure 13.34 – Newly added OVRInteraction Prefab

Fantastic, **OVRInteraction** is the engine that provides the integration for the ability to interact with the environment, as we said before. This interaction can be through the controllers or the hands themselves. We are going to include integration with the controllers.

3. To do this, we will need another Prefab called **OVRControllers**, and that is also found in the **Assets** | **Oculus** | **Interaction** | **OVRIntegration** | **Prefabs** path. We will drag it inside the **OVRInteraction** GameObject.

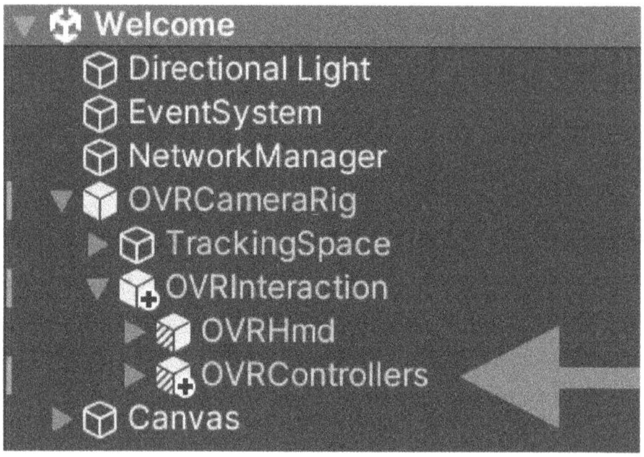

Figure 13.35 – Newly added OVRControllers Prefab

If we were to run the scene now, we would see the virtual controllers reacting to our hand movements. This is great, but we need a further step to be able to *touch* our buttons and inputs.

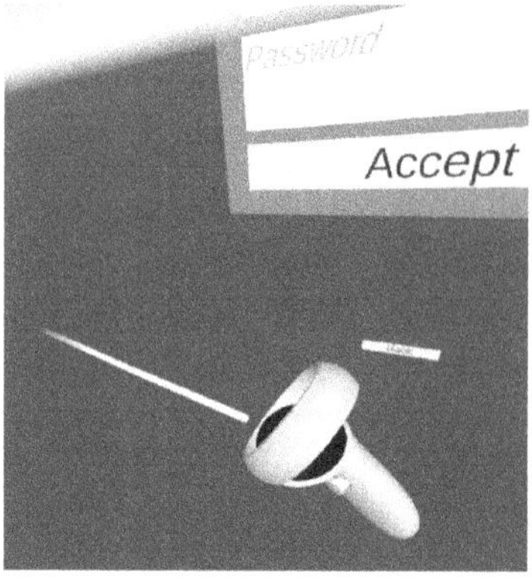

Figure 13.36 – Demonstration of laser raycast

The raycast is a laser that appears from our controllers and allows us to point at elements at different distances to activate or press them.

To add this functionality, again, we will resort to the Prefabs that the Oculus SDK offers us. This time, we will need **ControllerRayInteractor** and we will add it only to the right controller, but if you are left-handed, you can choose the left one – it doesn't matter.

4. This Prefab is located in the **Assets** | **Oculus** | **Interaction** | **Runtime** | **Prefabs** | **Ray** path, and we will drag it to the GameObject node called **ControllerInteractors** inside **RightController** (**LeftController** if you prefer the left hand).

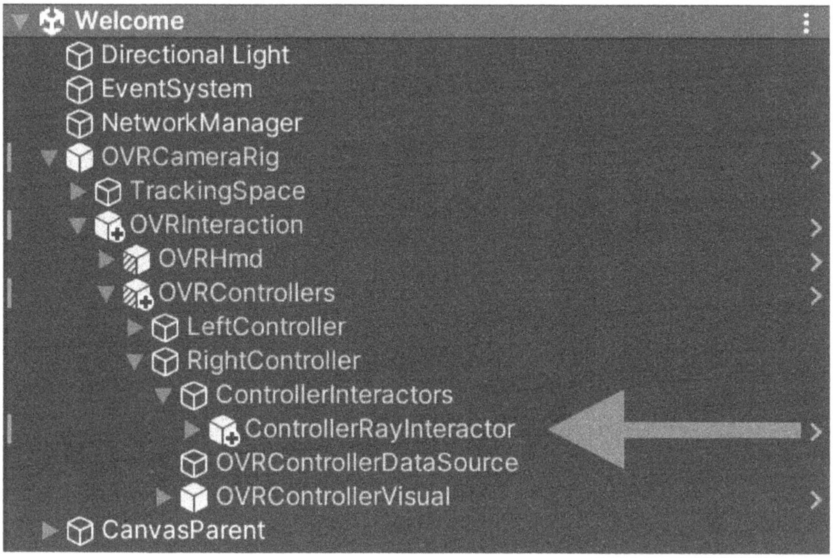

Figure 13.37 – Newly added ControllerRayInteractor Prefab

Fantastic – with this last step, we will have a fully integrated camera in our scene, and in addition, we will see a 3D representation of our two controllers, which we can operate with our hands. On one of the controllers, we will have the laser, with which we will interact with our Canvas later on.

5. In order not to have problems with the distance between the laser and the object that you want to press, I recommend that you use a high value for the **Max Ray Length** property of the **Ray Interactor** component of the **ControllerRayInteractor** Prefab, which you just color in scene; in my case, I have put `100000`.

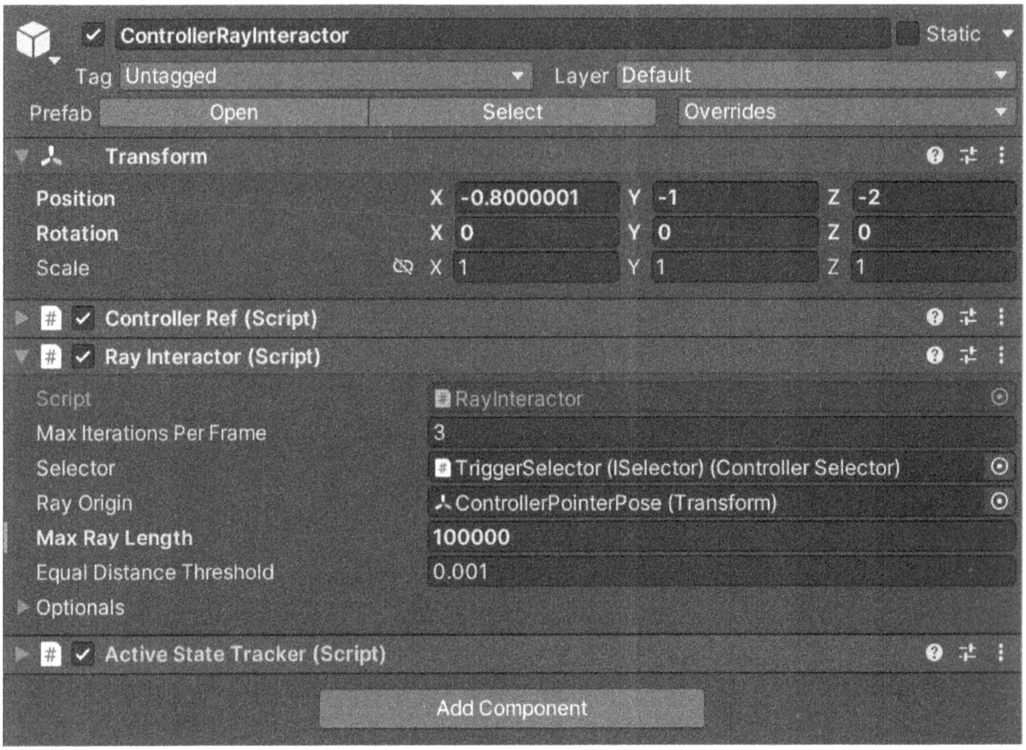

Figure 13.38 – ControllerRayInteractor GameObject Inspector

It is convenient to save the modifications that we have made to **OVRCameraRig** as a new Prefab since it will be useful in the future, and in this way, we will not have to carry out all the steps again.

6. To do this, drag the **OVRCameraRig** GameObject to the same path where the original is (or where you prefer), but give it a new name; for example, I have called it `OVRCameraRigWithControllers`. Unity will ask you whether you want to create an original Prefab or a variant. Select **Original Prefab**.

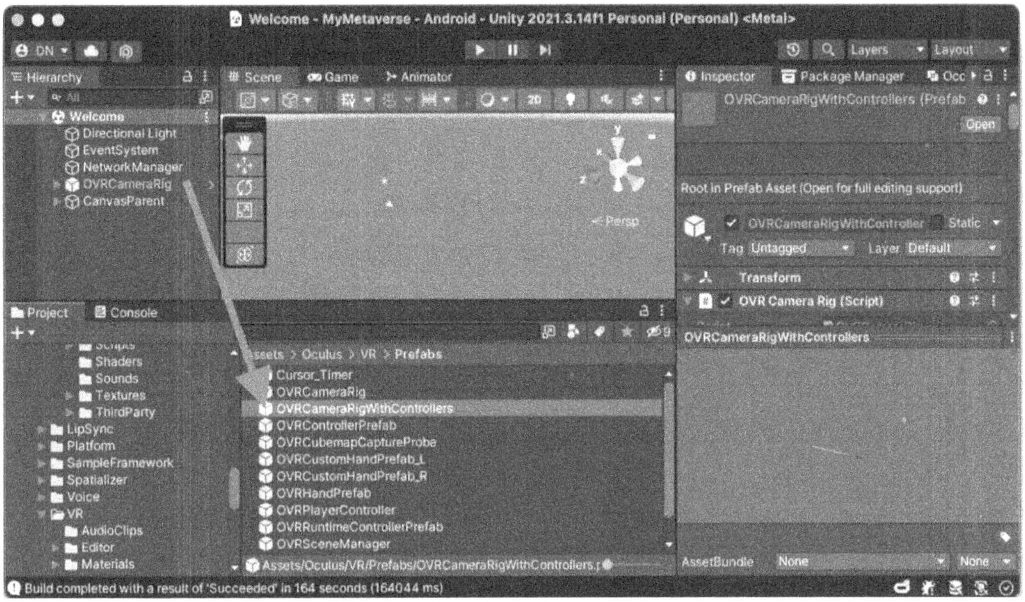

Figure 13.39 – Creating a new Prefab

This Prefab that we have modified and saved will be useful when we create the VR player's Prefab. Now we will continue with the modifications to the Canvas.

Canvas

The Oculus SDK offers integration with Unity's Canvas. We'll just have to make some small adjustments that will allow us to transform the inputs from the controllers into actions intelligible by the Unity UI.

To do this, we will follow these to adapt the Canvas:

1. The first important change to make is to change the rendering type of the Canvas. Now, as the scene becomes a scene where you can turn your head and look around, the Canvas must be an object suspended in space. Previously, it was a fixed plane, and a camera was always focusing on it. There is an easy solution; we just have to select the **Canvas** GameObject in the **Hierarchy** panel and change the **Render Mode** property to **World Space**.

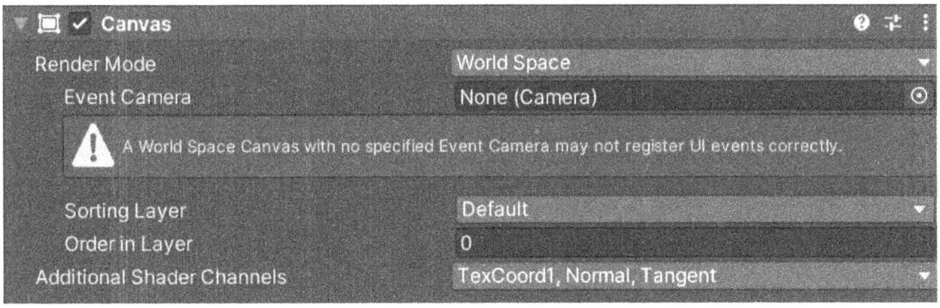

Figure 13.40 – Changing Render Mode of the Canvas

Now you will have an absolute position with X, Y, and Z positions in our scene.

2. When changing the rendering mode of the Canvas, we must now place it in front of the camera so that it can be visualized when starting the scene. To do this, modify its position so that it is right in front of the camera.

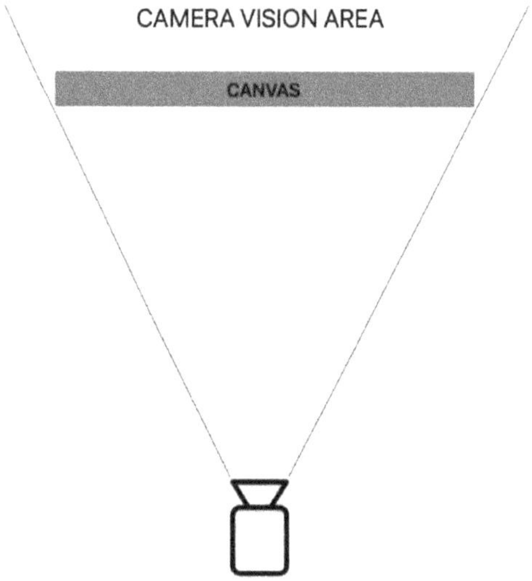

Figure 13.41 – Camera viewing area demonstration

3. Use the **Move** and **Size** tools to place the Canvas within the camera's view area. You can click the **Play** button to see how it looks. For reference, I have placed my Canvas approximately in the position **X**: `0`, **Y**: `0`, **Z**: `865`.

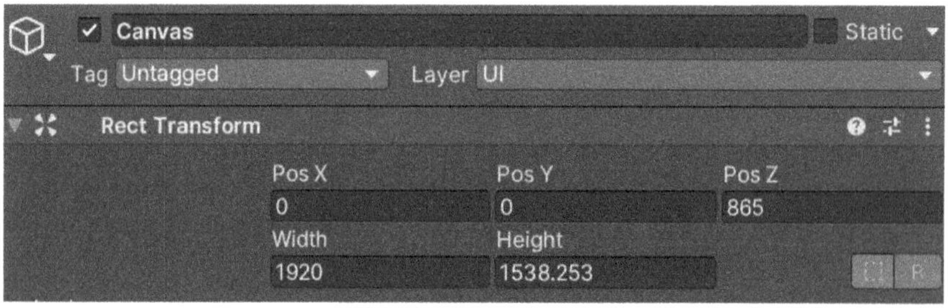

Figure 13.42 – Rect Transform component of the Canvas

Well, as we said before, the Unity UI Canvas must be adapted to transform the actions we take with the controllers into classic events of the Unity UI class. For a Canvas to be interactable with the Oculus Meta Quest, it must meet a series of conditions:

- The Canvas must be inside a GameObject that has the **Pointable Canvas** and **Ray Interactable** components.
- The Canvas must have a Collider and a **Collider Surface** component that will serve as a *shield* where the laser or raycast will go and transmit the point of contact to the interior of the Canvas.
- To add the parent GameObject, simply right-click on the **Canvas** GameObject and select the **Create Empty Parent** option. Rename the new GameObject to your liking. I have called it `CanvasParent`, for example.

4. Then, to add the Collider, we create an empty GameObject, right-click on **CanvasParent,** select **Create Empty**, and rename it `Collider`.

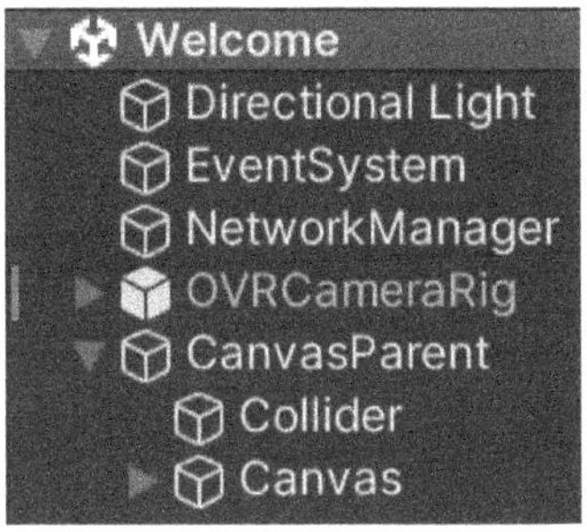

Figure 13.43 –Collider GameObject we just added to CanvasParent

OK, we have the structure, now let's add the necessary components to make the Canvas functional:

I. We select the **Collider** GameObject first and add a **Box Collider** component to it.

II. We change the size of the Collider to cover the whole Canvas. In my case, it's approximately **X**:`1000`, **Y**: `1000`, **Z**: `1`.

III. We add a new component called `Collider Surface`.

IV. Drag the **Collider** GameObject to the empty slot called **Collider** of the **Collider Surface** component.

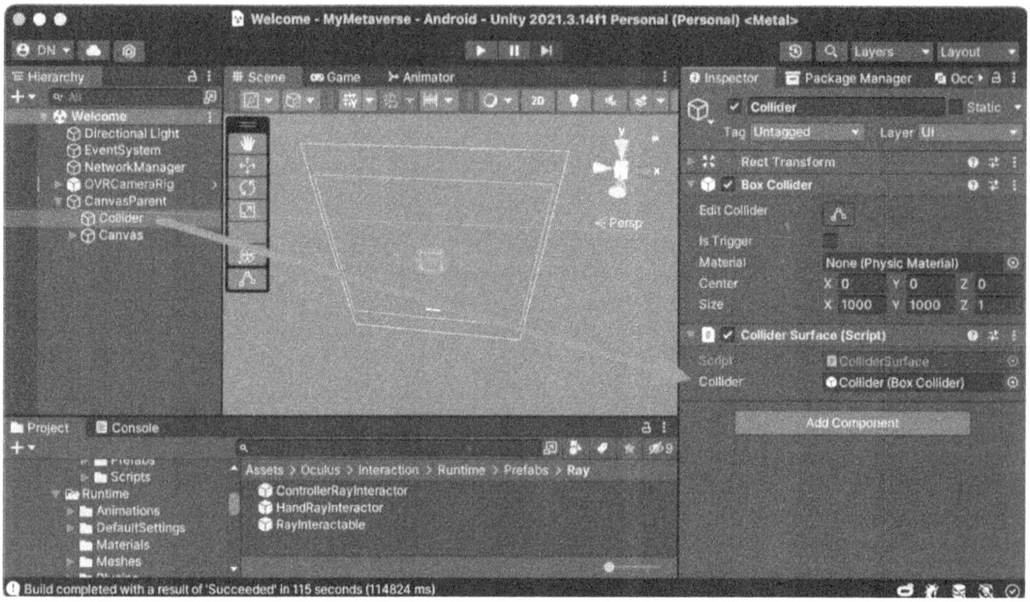

Figure 13.44 – Assigning the Collider GameObject to the Collider Surface Component

5. Now, we select the **CanvasParent** GameObject and add the **Pointable Canvas** component.
6. We will also add the **Ray Interactable** component.
7. Now, drag the **CanvasParent** GameObject to the empty **Pointable Element** slot and the **Collider** GameObject to the empty **Surface** slot.

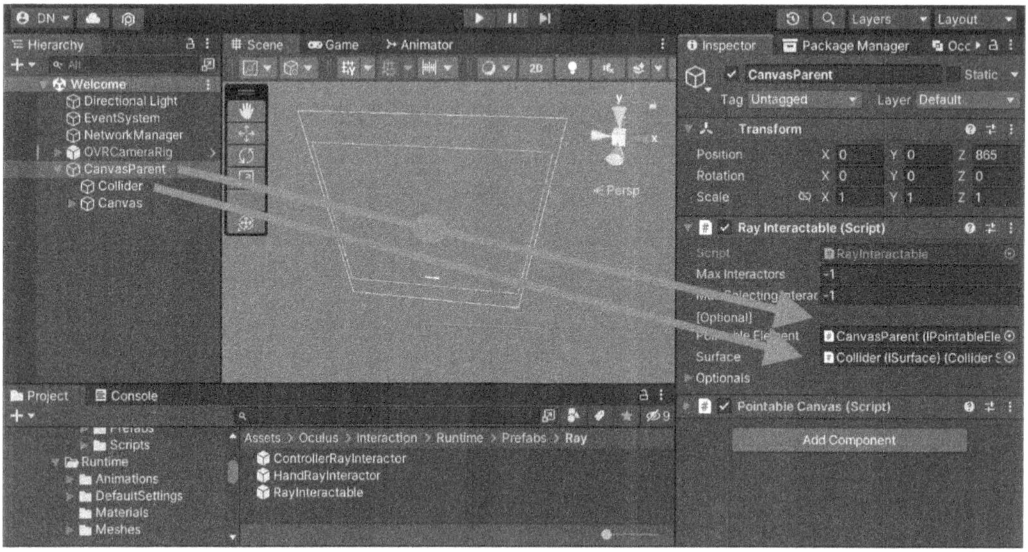

Figure 13.45 – Assigning the CanvasParent and Collider GameObject

8. Almost there. Finally, we have to drag the **Canvas** GameObject to the empty **Canvas** slot of the **Pointable Canvas** component.

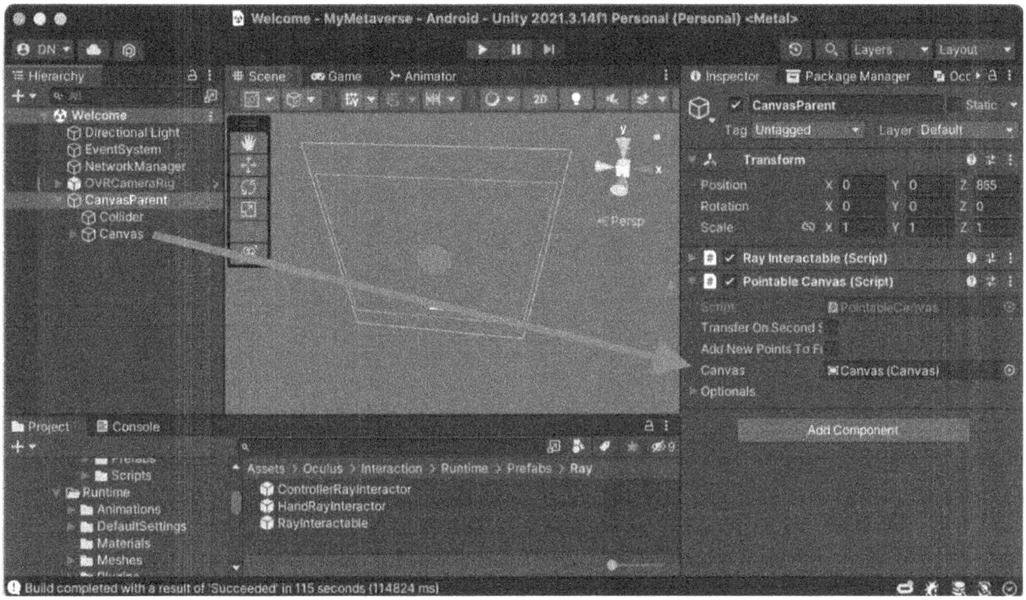

Figure 13.46 – Assigning the Canvas GameObject to the Pointable Canvas component

Fantastic! Excellent work – we have finished the configuration of our Canvas. Remember these steps that we have followed for any future Canvas in your project; otherwise, it may not work as you want.

There is one more step left – the **Input System UI Input Module** component that is in our **EventSystem** GameObject, and we use it in all scenes. This component, which manages the interaction between the user and the UI components, does not work with the Meta Quest 2 headset.

We want our project to be functional for both normal users and VR users. If we leave this component activated, the clicks will not work with our VR controllers; if we remove it, we will be without functionality for normal users. So, what can we do?

The best solution I can think of is to create a small script that enables and disables this component, depending on whether a user is running the project in virtual reality glasses or not.

To do this, right-click on the **Assets** | **_App** | **Scripts** path and select **Create** | **C# Script**, then rename it to `UISystemManager`.

Replace all the auto-generated code with the following:

```
using UnityEngine;
using UnityEngine.InputSystem.UI;
```

```
public class UISystemManager : MonoBehaviour
{
    void Awake()
    {
        try
        {
            var headsetType =
                OVRPlugin.GetSystemHeadsetType();

            // Using the headsetType variable, we can
               detect if we are running the project in
               virtual reality glasses.If it returns None,
               it means that we are not running in virtual
               reality.
            if (headsetType ==
                OVRPlugin.SystemHeadset.None)
            {
                // Access the InputSystemUI component and
                   disable it.
                Getcomponent<InputSystemUIInputModule>()
                    .enabled = true;
            }
            else
            {
                // If we forget to manually activate this
                   component, we activate it
                Getcomponent<InputSystemUIInputModule>()
                    .enabled = false;
            }
        }
        catch (System.Exception ex)
        {
            // It is possible that on some platforms and in
               Unity Editor the OpenVR library is not found
               and causes an error, in that case we will
               also activate the classic UI system.
            Getcomponent<InputSystemUIInputModule>()
                .enabled = true;
        }
    }
}
```

Finally, we add the script we have just created as a component in the **EventSystem** GameObject.

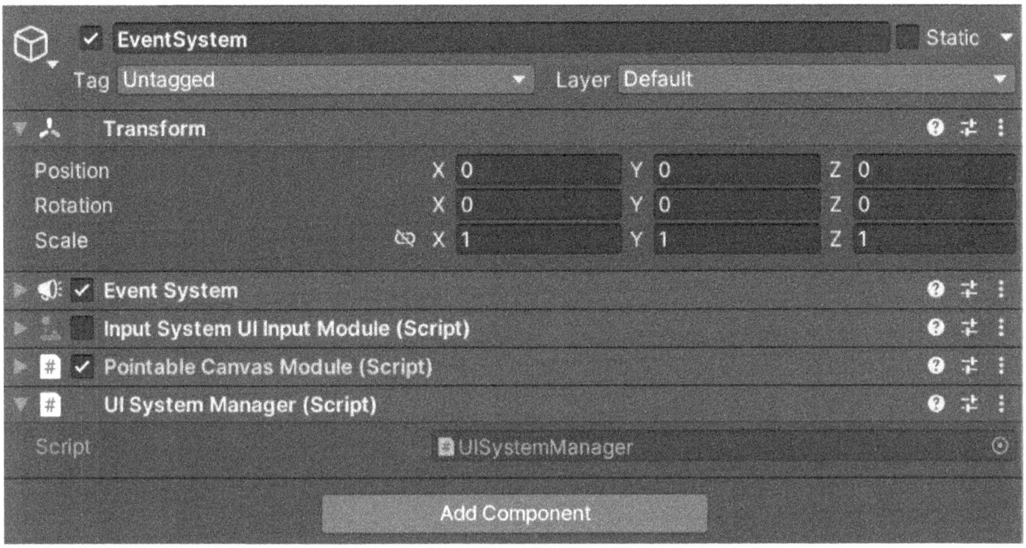

Figure 13.47 – UI system manager component

Good job! With this little script, we keep the compatibility with both types of users. Now, the precious moment has come – let's test it! At the moment, with the changes we have made, we can only test the **Welcome** scene; then we will expand the necessary changes to the other scenes and the character itself.

Testing in Meta Quest 2

Once we have done all the steps to integrate the Meta controllers into our scene, let's proceed to test them. Follow these steps to test it, compiling and deploying the development to run directly on the Meta Quest 2 headset:

1. To test it, first, make sure you have your goggles connected. You can find out whether they are ready to use through the Meta Quest Developer Hub application. In the **Device Manager** section, you should see information about your goggles and their status:

Figure 13.48 – Status of our Meta Quest 2 in Meta Quest Development Hub

2. If the device does not appear active, check that it is correctly turned on and also connected to the computer via cable or Wi-Fi. Now, with the glasses ready to use, open the Unity **Build Settings** window and do the following:

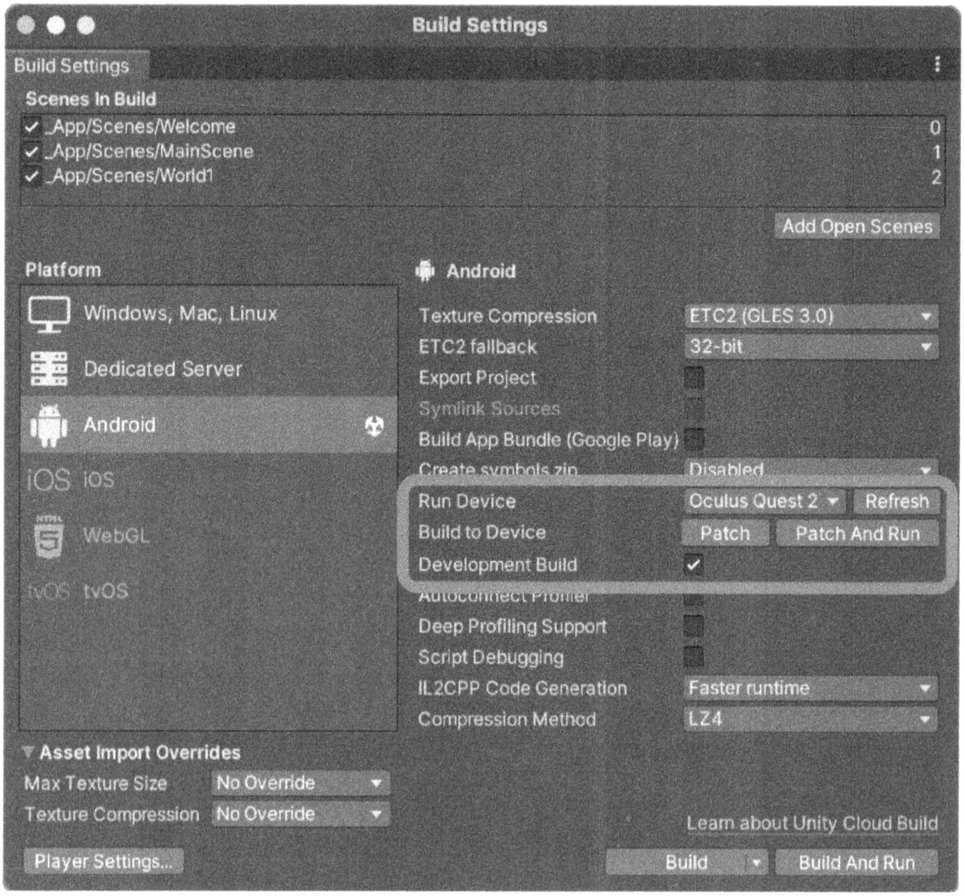

Figure 13.49 – Build Settings window with Oculus Quest 2 selected

3. First, make sure you have the Android platform active, and then select **Oculus Quest 2** in the **Run Device** section. If it doesn't appear, click on the **Refresh** button. It is also important to have the **Development Build** option checked; otherwise, it will not let us send the project to the glasses.

4. Finally, click on the **Patch And Run** button to send a build to the glasses. It may ask for some additional permissions. I recommend you wear the glasses during the process. When the loading bar finishes (it may take several minutes), you will hear a sound in the goggles, which means it has compiled successfully and you can now enjoy the virtual experience on your Oculus Meta Quest 2.

I recommend you play with the sizes of the Canvas buttons until you like the result.

Figure 13.50 – Trying to interact with a virtual keyboard

Congratulations! It turned out great, didn't it? It's very satisfying to see your project working on Meta Quest 2.

The next thing we are going to do is to prepare the other scenes and our NPCs to be compatible with virtual reality as well. Finally, we will convert our player with a VR adaptation.

Preparing the other scenes

Now that we have successfully adapted the **Welcome** scene, we will continue with the other scenes, following the necessary steps so that nothing is left pending. Remember to take these steps for the new scenes you create in your project.

The steps that we are going to take in the following scenes are the same that we have already taken previously, but to bolster our knowledge, we will review – step by step – what it is necessary to do.

MainScene

Let's open our main scene, our lovely little town. The first thing we must do is modify the **EventSystem** GameObject to perform the following actions:

1. Remove (if it exists) the **Standalone Input Module** component.
2. Add the **Input System UI Input Module** component.
3. Add the **UISystemManager** script we created before as a component.
4. Finally, add the **Pointable Canvas Module** component.

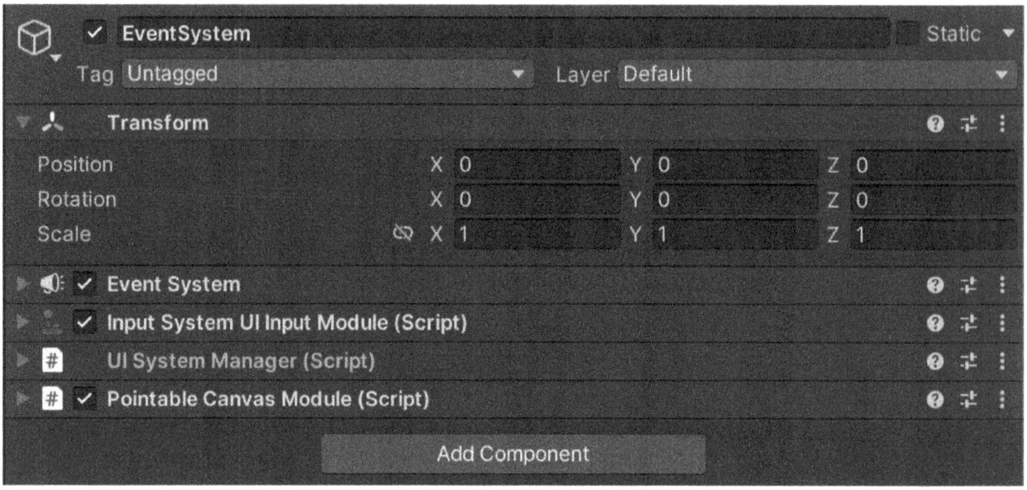

Figure 13.51 – Pointable Canvas Module component

Let's apply the necessary changes to our main Canvas:

1. To do so, right-click on the **Canvas** GameObject and select the **Create Empty Parent** option. Rename the new GameObject `CanvasParent`.

Figure 13.52 – Newly added CanvasParent GameObject

2. Now, add as a child to **CanvasParent** a new GameObject called **Collider**, and in the **Inspector** panel of this **Collider** GameObject, add the **Box Collider** component, with a size that completely covers the Canvas. Also add a **Collider Surface** component.
3. Also remember to drag the **Collider** GameObject to the empty **Collider** slot of the **Collider Surface** component.
4. We'll now select the **CanvasParent** GameObject and add the **Ray Interactable** and **Pointable Canvas** components.
5. We need to drag the **Canvas** GameObject to the empty **Canvas** slot of the **Pointable Canvas** component.
6. Finally, we'll drag the **CanvasParent** GameObject to the empty **Pointable Element** slot and the **Collider** GameObject to the empty **Surface** slot of the **Ray Interactable** component.

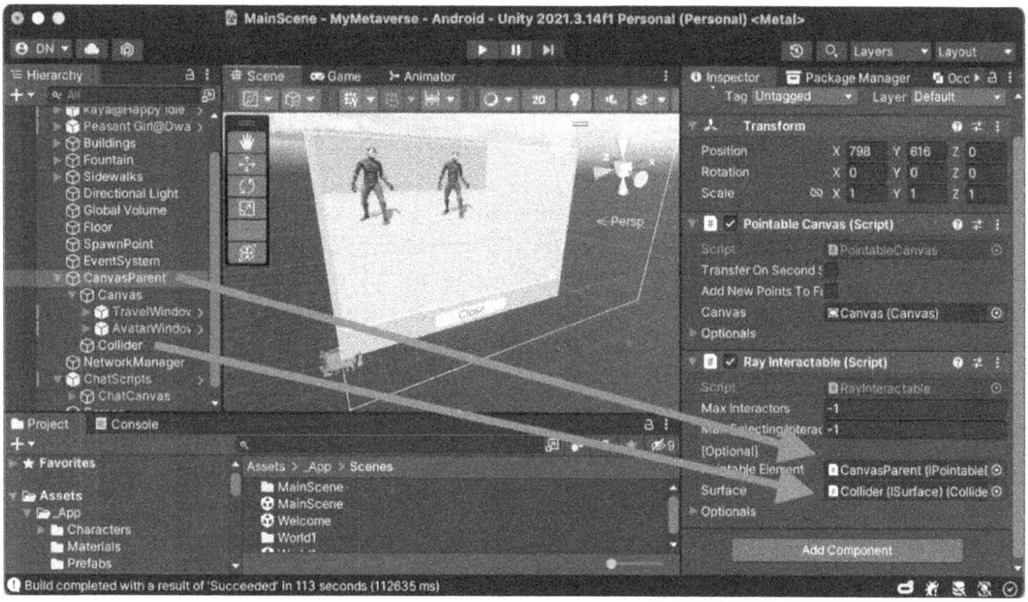

Figure 13.53 – Assigning the CanvasParent and Collider GameObjects to the Ray Interactable component

7. Now, we will change **Rendering Mode** of our **Canvas** to **World Space** and use the **Move**, **Size**, and **Rotate** tools to place it near the NPC. It is important that you apply the size, rotation, and position changes to the **CanvasParent** GameObject so you can apply the changes to the Collider as well.

Figure 13.54 – Transforming the Canvas

8. Currently, NPCs respond to the *E* key to open the Canvas. To add a response compatible with Meta Quest 2 controllers, we must modify the `TravelNPCTrigger` and `AvatarNPCTrigger` scripts and add the highlighted line to the `Update()` function:

```
void Update()
{
    // We check whether the user has pressed the E key
       and is also inside the collider.
    if (Input.GetKeyDown(KeyCode.E) ||
        OVRInput.Get(OVRInput.RawButton.B) &&
            canInteract)
    {
...
```

9. `OVRInput.RawButton.B` is equivalent to the *B* button of the right controller. You can use another one if you prefer. The image found at `https://www.researchgate.net/figure/OVRInput-API-for-unity-uses-to-query-controller-state_fig1_351276981` shows the codes of all the controller buttons.

With this small tweak, we will support our VR players so they can interact with the NPCs.

The next thing we are going to configure is the player's own Prefab. We will create a special variant for VR as there are quite a few changes to make and there would be no (clean) way to keep compatibility for both environments in the same Prefab.

Configuring the player

Unfortunately, the third-person controller system of Unity and its animator, which we have liked so much until now, are not compatible with the Oculus SDK. That is to say, the movements of the player that we make with our controllers will not be reflected in the animations of the Character. It will move on the spot and it will rotate, but the legs will not move, for example.

This is a disadvantage in my opinion, but we will manage to create something decent and functional in this chapter.

Meta has its own Avatar SDK, compatible with animations and VR functionality. You can follow their tutorial when we finish setting up the VR player Prefab as an extra feature not covered in this chapter.

Well, let's get down to work:

1. The first thing we are going to do is create a duplicate of the **PlayerInstance** Prefab found in **Assets** | **Photon** | **PhotonUnityNetworking** | **Resources**, and we will call it `PlayerInstanceVR`.

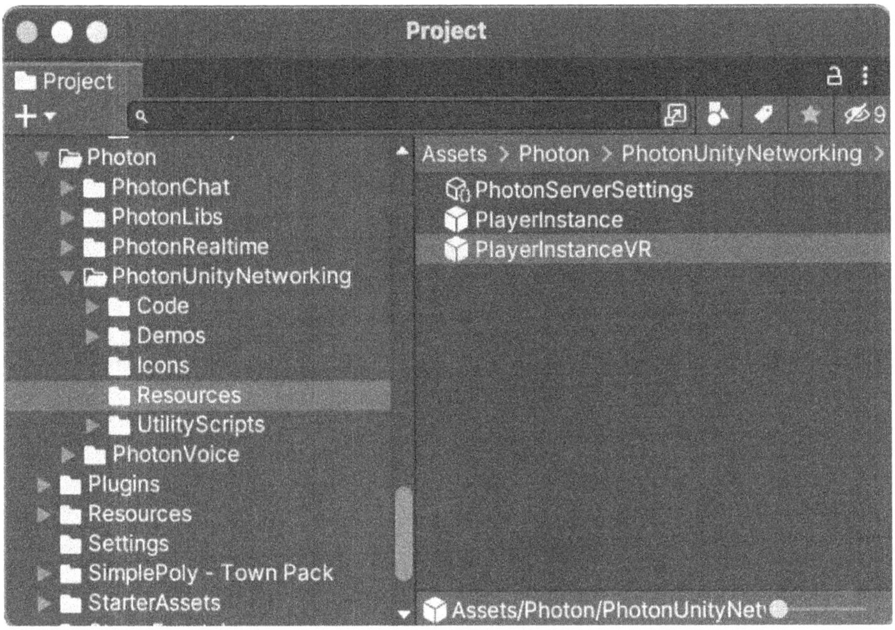

Figure 13.55 – Newly created PlayerInstanceVR Prefab

2. Double-click on **PlayerIntanceVR** to edit it and perform the following steps:

   I. Remove these GameObjects: **Main Camera**, **Player Follow Camera**, **FordwardDirection**, and **PlayerCameraRoot**. We will be left with a view similar to the following screenshot.

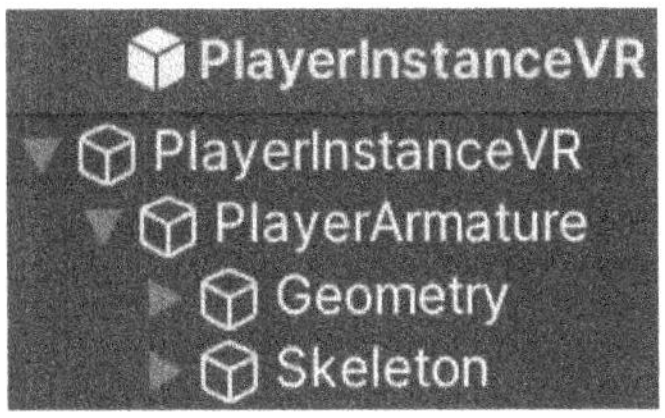

Figure 13.56 – Final view of the PlayerInstanceVR GameObject hierarchy

   II. Select the **PlayerArmature** GameObject and add the **OVRPlayerController** component. Leave the default properties.

   III. Now we will use the **OVRCameraRigWithControllers** Prefab, which includes **Camera** and **Controllers**, that we created before. We will drag it as a child of the **PlayerArmature** GameObject.

   IV. We'll use the **Move** tool to place the **OVRCameraRigWithControllers** GameObject approximately at the place of the character's head.

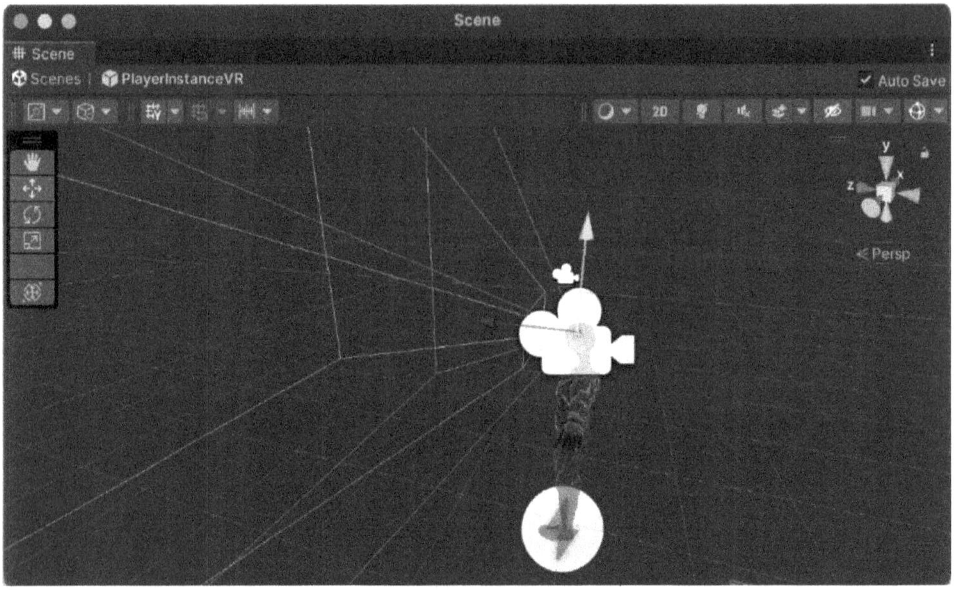

Figure 13.57 – Adjusting camera position – OVRCameraRigControllers

V. Now, we select the **OVRCameraRigWithControllers** GameObject and modify the **Tracking Origin Type** property to **Eye Level**. This way, the camera will stay at the height we want. Otherwise, it will automatically go to the feet, on the ground, when we run the project.

Fantastic – with these changes, we would already have a functional character, but there is a detail that we can't let pass and that we can fix easily. If we now test the project on our glasses (I recommend you do so), the first thing we will see is the following:

Figure 13.58 – Camera preview failed – OVRCameraRigWithControllers

Oh damn, what a scare! What happened here? We simply don't want the character to be shown to us. But we do want it to be shown to other players, so there is an easy solution.

3. We find and edit the **DisableUnneededScriptsForOtherPlayers** script located in the **Assets | _App | Scripts** path and modify the content to adapt it and offer compatibility with Meta Quest 2. We will replace all the content with the following:

```
using System.Collections;
using System.Collections.Generic;
using Photon.Pun;
using StarterAssets;
using UnityEngine;

public class DisableUnneededScriptsForOtherPlayers :
    MonoBehaviourPun
{
    void Start()
    {
...
        OVRPlayerController controllerVR =
            Getcomponent<OVRPlayerController>();
        Transform ovrCameraRig = transform.Find
            ("OVRCameraRigWithControllers");
        Transform geometryParent =
            transform.Find("Geometry");
...
        if (!photonView.IsMine)
        {
...
            if (controllerVR != null)
                Destroy(controllerVR);

            if (ovrCameraRig != null)
                Destroy(ovrCameraRig.gameObject);
...
        }
        else{
            var transparentLayer =
               LayerMask.NameToLayer("TransparentFX");
            SetLayerAllChildren(geometryParent,
               transparentLayer);
        ...
    }
```

```
            void SetLayerAllChildren(Transform root,
                int layer)
            {
                var children =
                    root.GetcomponentsInChildren<Transform>
                        (includeInactive: true);
                foreach (var child in children)
                {
                    child.gameObject.layer = layer;
                }
            }
        }
    }
```

What we achieve by adding the **Geometry** GameObject to the **TransparentFX** layer is to be able to filter these elements in the camera, and thus prevent ourselves from seeing our own character.

4. Once the code is modified, we go back to edit the **PlayerInstanceVR** Prefab, look at the **OVRCameraRigWithControllers** GameObject, and perform the following action:
5. We select the three GameObjects called **LeftEyeAnchor**, **CenterEyeAnchor**, and **RightEyeAnchor**, which represent the three cameras, and modify the **Culling Mask** property. We simply uncheck **TransparentFX** from the list. With this, we avoid rendering in camera objects that belong to this layer.

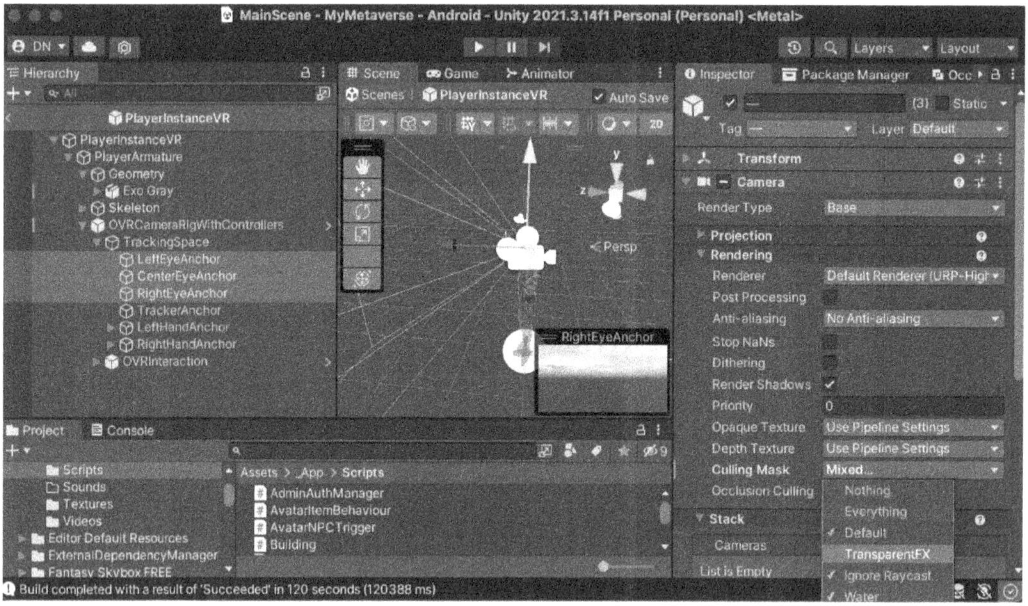

Figure 13.59 – Changing the values of the Culling Mask property

Great – all set. We can again test it on our Oculus Meta Quest 2 to check the result.

Figure 13.60 – Observing the scene from the Meta Quest 2 glasses

Much better! Now we can enjoy an afternoon of cinema in our little town. Now let's apply a modification to the **InstantiatePlayer** script so that it is able to instantiate a normal player or a VR player, depending on where the project is running.

6. To do this, find the **InstantiatePlayer** script located in the **Assets** | **_App** | **Scripts** path and double-click to edit it.

   Replace all the content with the following:

```
using ExitGames.Client.Photon;
using Photon.Pun;
using Photon.Realtime;
using UnityEngine;

public class InstantiatePlayer : MonoBehaviourPun
{
    public GameObject playerPrefab;
    public GameObject playerPrefabVR;


    void Start()
    {

        var headsetType =
            OVRPlugin.GetSystemHeadsetType();
```

```
            // Using the headsetType variable, we can
               detect if we are running the project in
               virtual reality glasses.If it returns None,
               it means that we are not running in virtual
               reality.
            if (headsetType ==
                OVRPlugin.SystemHeadset.None)
            {
                PhotonNetwork.Instantiate(
                    playerPrefab.name, transform.position,
                        Quaternion.identity);
            }
            else
            {
                PhotonNetwork.Instantiate(
                    playerPrefabVR.name,
                        transform.position,
                            Quaternion.identity);
            }
        }
    }
```

7. Now we must assign to each slot the corresponding Prefab. Remember that we have this component in each **SpawnPoint** GameObject, both in **MainScene** and in **World1**.

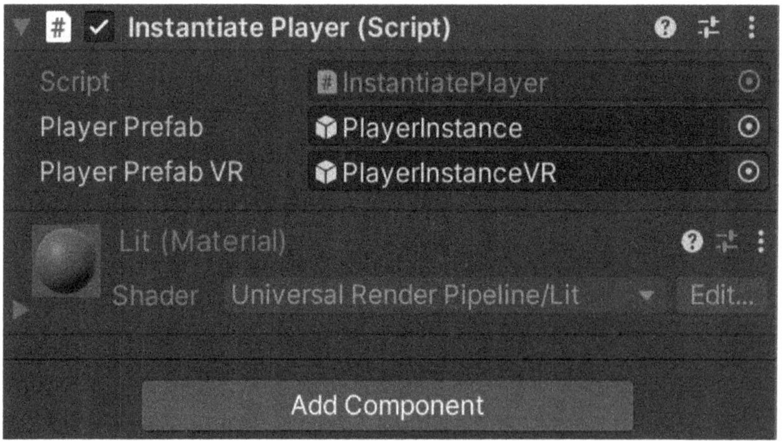

Figure 13.61 – Final result of the Instantiate Player component

Good job! With this last modification, you now have everything you need to enjoy your project in your Meta Quest 2.

Now you have all the knowledge to make the same modifications to the Canvas of the **World1** scene, and to add the recognition of some key of the controller, for interacting with the script for the acquisition of houses of the **World1** scene.

We have reached the end of the chapter, which implemented the last functionality of your metaverse. In the following ones, we will learn what optimizations we must take into account to deploy our project on mobile and desktop devices.

We will also go, step by step, through how to compile for Android, iOS, Windows, and Mac Desktop, as well as how to publish your project to the App Store and Play Store.

Summary

In this extensive chapter, we discovered everything we need to implement the impressive Meta Quest 2 virtual reality goggles in our project. We went through, from start to finish, how to set up a new headset and activate the Developer Mode. We also learned how to implement the Oculus SDK in Unity to enjoy a VR Camera and the possibility to interact with the environment through the controllers.

Finally, we have seen how to add a laser to one of our controllers to interact with our Canvas, allowing us to press buttons, write inputs, and interact with our NPCs.

In the next chapter, we will finally learn how to compile our project for Android, iOS, Windows, Mac Desktop, and Linux. We will go through all the necessary steps to correctly generate executable binary files that can be downloaded to other devices and run normally.

14
Distributing

Welcome to the last chapter of this exciting book about creating your own metaverse with Unity 3D and Firebase. This is undoubtedly one of the most exciting moments for a software developer. Compiling and distributing your project to see it running on another device is without a doubt a magical moment.

Throughout these pages, we will see how to apply another optimization process to make our project lighter. We will also see how to compile for the most popular platforms, and finally, we will see an explanation of what steps to take to publish your game in the Google Play Store, Apple App Store, Windows Store, and Oculus Store.

This chapter is modular, which means that if you are not interested in exporting your project to one of the platforms, you can skip to the part you are interested in. Finally, I will reflect on what we have covered throughout this book and give you my opinion on how you can evolve your project in the future.

We will cover the following topics:

- Optimizing textures
- Compiling for different platforms

Technical requirements

This chapter does not have any special technical requirements, but as we will start programming scripts in C#, it would be advisable to have a basic knowledge of this programming language.

We need an internet connection to browse and download an asset from the Unity Asset Store. We will continue with the project we created in *Chapter 1, Getting Started with Unity and Firebase*. Remember that we have a GitHub repository, `https://github.com/PacktPublishing/Build-Your-Own-Metaverse-with-Unity/tree/main/UnityProject`, which contains the complete project that we will work on here.

Optimizing textures

We could dedicate a whole book to understanding all the possible optimizations we can apply to our Unity project. In *Chapters 3* and *4*, we saw some of them. Now, we will see how to apply one more layer.

By applying some adjustments to the textures that we use in our project, we can achieve the following:

- Improve loading time
- Drastically reduce the final size of the executable binary file
- Save graphics rendering time

The time it takes to apply the settings to the textures is minimal compared to the great optimization result we will obtain.

> **Tip: Advanced guidance on further optimization processes**
>
> Unity offers, in its official documentation, a list of things to take into account for a full advanced optimization. If you are curious about it, you can visit `https://docs.unity3d.com/es/530/Manual/OptimizingGraphicsPerformance.html`.

To understand the optimization process, we will follow these steps:

1. In the **Project** panel, click and select the root **Assets** folder.
2. Click on the **Search by Type** button and select **Texture**. This will filter all the assets of our project to show only the textures, whether they are in folders or subfolders.

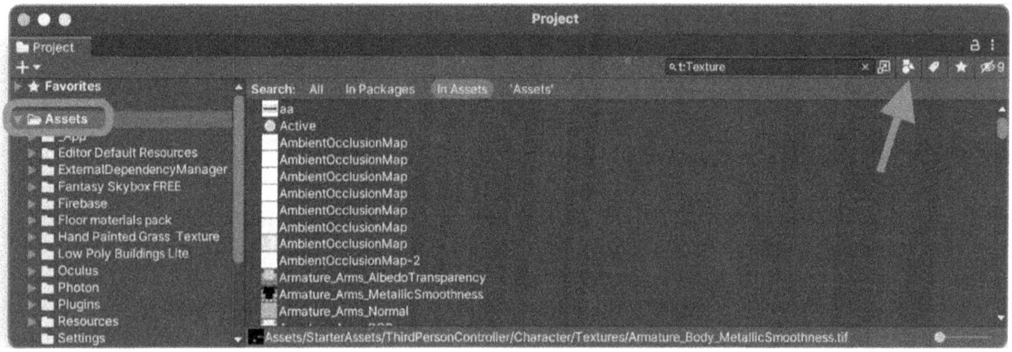

Figure 14.1 – Filter assets by type in the Project panel

3. If we select any of the textures that appear in the result view, we will see the following information in the **Inspector** panel:

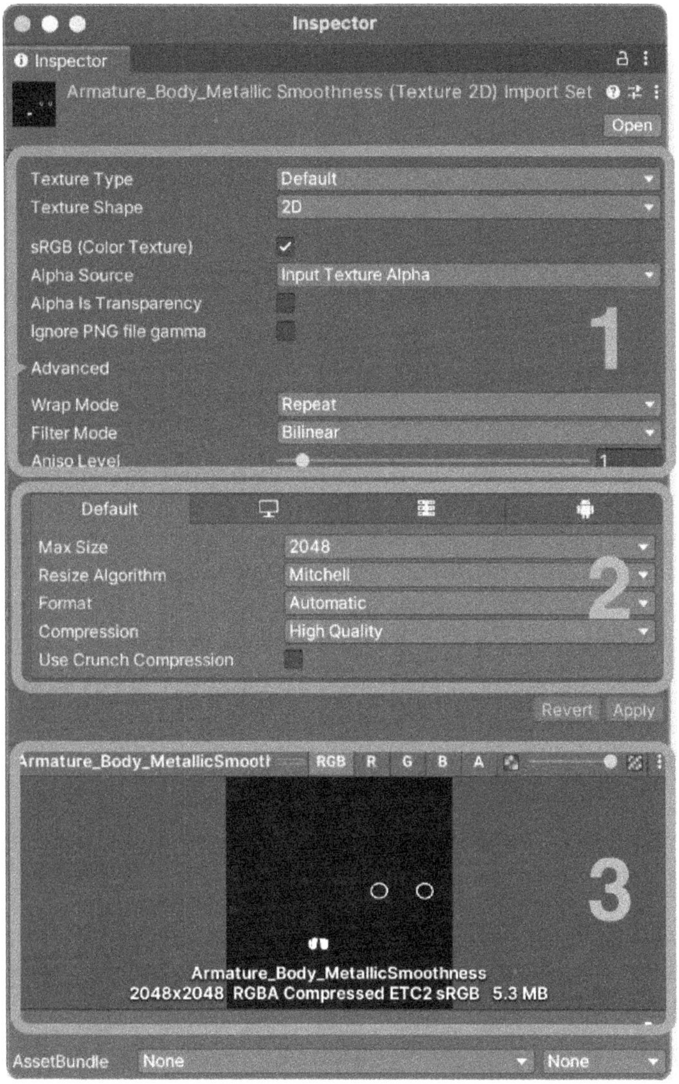

Figure 14.2 – Inspector panel of the selected texture

Considering the preceding screenshot, we have the following:

- In area **1**, we have configuration options for the texture behavior. Generally, the default configuration is valid. However, if, for example, you wanted to use a texture in a Unity UI component, this is where you would have to select the **Sprite (2D and UI)** option for the **Texture Type** property. One of the most important options for texture compression and optimization is in this area. In the **Advanced** section, there is the **Generate Mip Maps** property. By default, this property is deactivated; it is advisable to activate it since it will optimize the performance of our project.

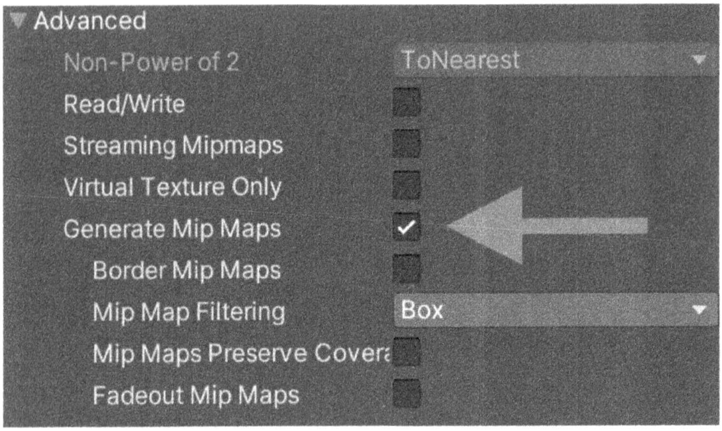

Figure 14.3 – Generate Mip Maps property of the texture's Inspector panel

The **Generate Mip Maps** option allows Unity to create different versions of the texture, in different sizes. This allows the engine to choose the right texture based on the size and distance of the geometry in relation to the camera it is rendering. This is perfect because if an object is small, Unity will load a tiny version of the texture instead of the original.

- In area **2**, we have the options for texture optimization. As you can see, there are different tabs, one for each platform we have in the project. The configuration applied in the **Default** tab is the one that will be used by default in all platforms, unless you have checked the **Override for...** option in the specific platform tabs.
- Finally, in area **3**, we have a preview of the texture with information about its size and weight, as well as the current compression type.

4. To see, with a practical example, how to apply compression to the texture, with the texture you currently have selected, go to area **2** of the **Inspector** panel and select the **Android** tab (this is the platform we currently have active. If you have another one active, select the corresponding tab).
5. To apply compression, we will try to change some of the properties; for example, we will select **512** for **Max Size**, and for the **Format** property, we will select **RGB Compressed ETC2 4 bits**, and then click on the **Apply** button.

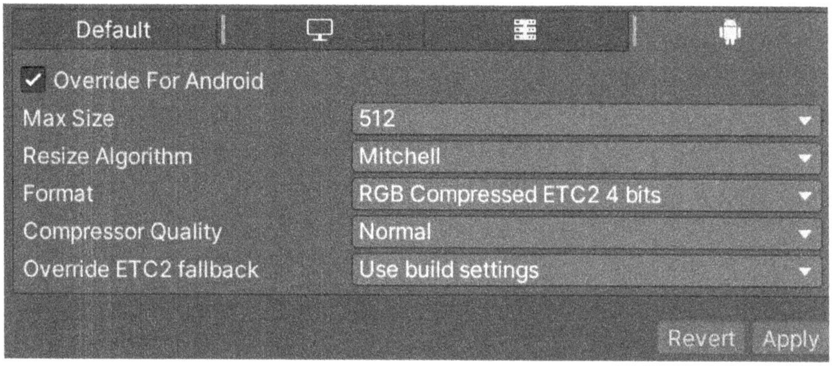

Figure 14.4 – Override configuration tab for the Android platform

6. As you can see, in area **3** of the **Inspector** panel, we have the updated texture information. With the small adjustment we have made, we have drastically optimized the texture weight.

In the following screenshot, we can see the difference between a non-optimized and an optimized texture. As you can see, with a small configuration, we can save approximately 70% of the size per texture. This optimization figure will vary depending on the optimization level we select.

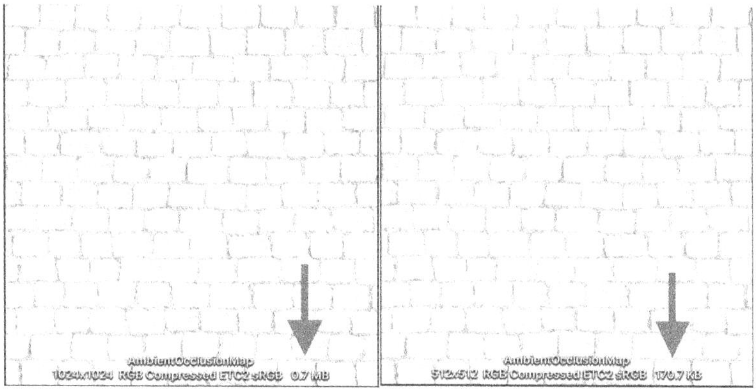

Figure 14.5 – Visual comparison of the result of texture optimization

The pre-optimization size was **0.7 MB** and now it is **170.7 KB**, and we have lost virtually no quality. Impressive, isn't it? By applying this configuration to all the textures in our project, we can reduce the final weight of the executable by more than half.

As you may have noticed, there are hundreds of compression format options available. Some of them may compress more, some less, and many others may cause the texture to lose quality. It can be difficult to choose one option.

As a general rule, I tend to use the **RGB Compressed ETC2 4 bits** option, and **RGBA Compressed ETC2 4 bits** for textures that have transparency.

Tip: Advanced information on textures

If you are curious and want to know more advanced information about texture setting properties and compression types, you can check out the official Unity guide at `https://docs.unity3d.com/es/530/Manual/class-TextureImporter.html`.

As you can see, it has been really easy to optimize the size of the textures and thus optimize the performance of our project. I invite you to put into practice what we have just learned with the textures of your project.

After having learned a new optimization technique, we will finally see how to compile our project for different platforms.

Compiling for different platforms

As we warned at the beginning of the chapter, this part is modular. You can skip directly to the platform of your choice. Before compiling, we need to make some necessary adjustments on each of the platforms.

Before continuing, make sure you have downloaded the platform you want to compile to; for example, in my case, I don't have the iOS platform downloaded.

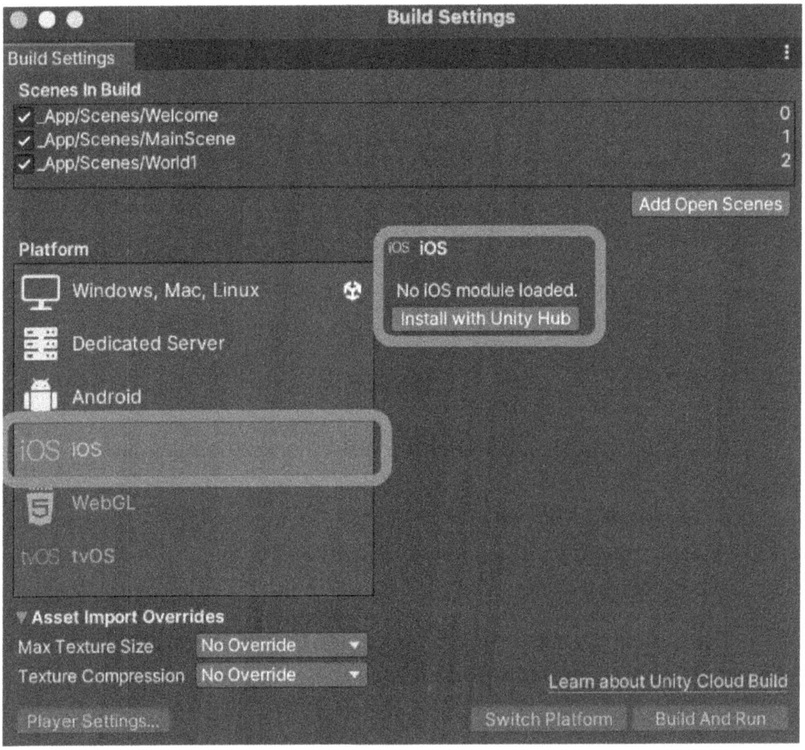

Figure 14.6 – iOS platform in the Build Settings window

If this happens on any of the platforms, click on the **Install with Unity Hub** button. Remember that this step was covered in detail in *Chapter 13, Adding Compatibility for the Meta Quest 2*.

Windows, Mac, and Linux

The process to compile for Windows, Mac, and Linux is identical in all three cases. First, activate the desired platform in the window that can be found by navigating to the **File** | **Build Settings** menu.

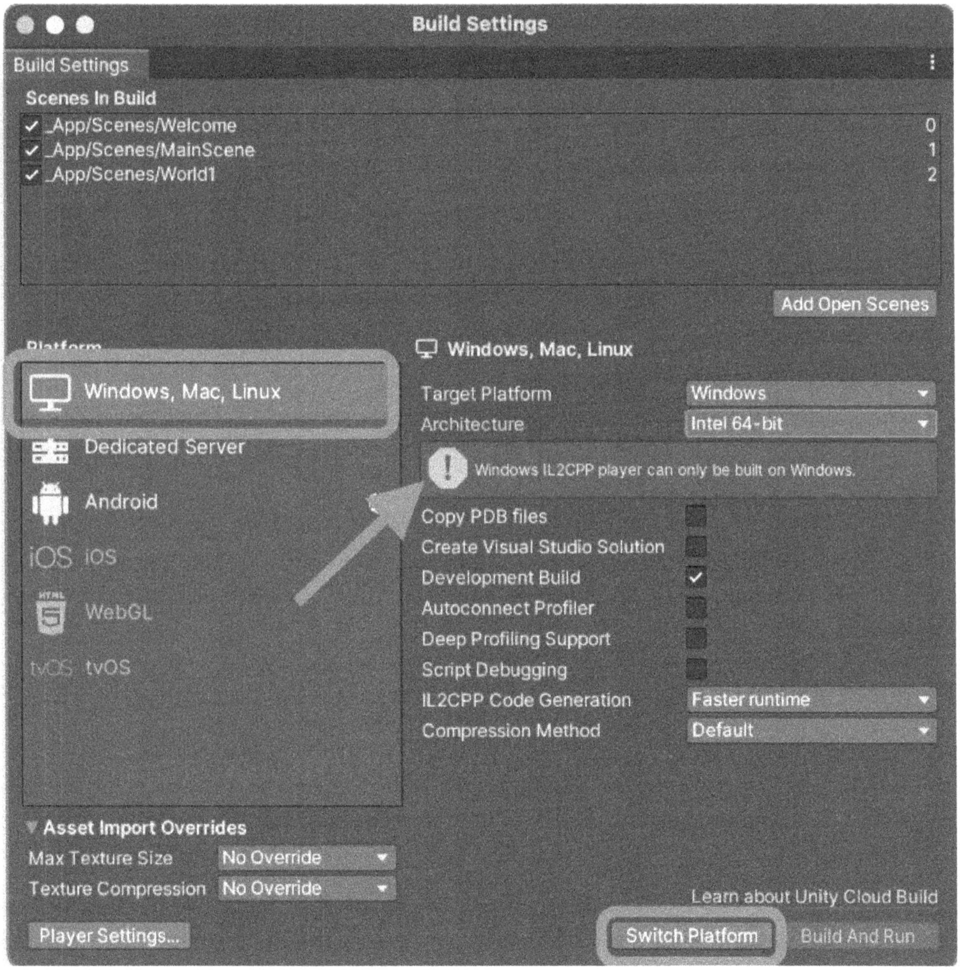

Figure 14.7 – Mac incompatibility warning

Before we continue, let's take a brief look at the most important compilation options we see on this screen:

- **Target Platform**: This is where we choose **Windows**, **Mac**, or **Linux**.
- **Development Build**: For testing on a target device, this offers a small console overlay on the screen, emulating the **Console** panel of the editor, and shows all kinds of logs and errors. It is very useful for test builds.
- **Autoconnect Profiler**: If you have enabled the **Development Build** option, we can connect the **Profile** tool, which generates performance statistics from Unity.
- **Compression Method**: There are three options – **Default**, **LZ4**, and **LZ4HC**:
    - **Default**: On Windows, Linux, and Mac platforms, there is no default compression, so the **Default** option in this case will not perform any compression
    - **LZ4**: Offers fast compression; recommended for development builds.
    - **LZ4HC**: Offers very high compression and increases the compile time considerably, but offers highly optimized executable binaries; ideal for final compilations in production environments.

> **Tip: Advanced information on compression**
> For advanced information on all compilation and compression properties, a page is available in the official Unity guide at `https://docs.unity3d.com/Manual/BuildSettings.html`.

To do this, we select **Window, Mac, Linux** on the left, and on the right, we select one of the three platforms we want to compile in **Target Platform**. Finally, we click on **Switch Platform**. The process may take a few minutes.

Compiling on Windows from Mac

As you can see, if you are trying to compile on Windows from a Mac, you will see the warning message **Windows IL2CPP player can only be built on Windows**. This is because currently compiling Windows with Scripting Backend in IL2CPP from a Mac is currently not supported. You have two options to solve this.

The first is to change **Scripting Backend** to **Mono**. You can do this from the **Player** menu, which can be found by clicking on the **Edit** | **Player Settings** option. Within this menu, in the **Other Settings** section, you can select **Mono** for the **Scripting Backend** property.

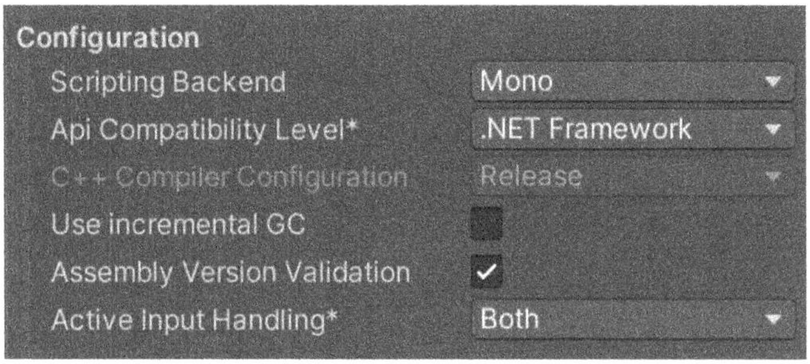

Figure 14.8 – Scripting Backend configuration section

The disadvantage of this solution is that it is possible that some dependencies in your project don't work well with **Mono**. If this is the case and you have failures during the execution, you can go with the second solution. Note, however, that during the development of this project, I compiled Windows from a Mac with the **Mono** option enabled for **Scripting Backend** and it worked correctly on a Windows computer.

The second solution is to open the project on a Windows computer and compile it with **IL2CPP** from there.

Once we have finished the process of switching platforms by clicking the **Switch Platform** button, we are ready to compile. Next, click on the **Build** button, choose a destination folder to save the executable file and its dependencies in, and click on **Accept**.

It is possible that you will get an error when compiling and the process does not finish correctly. This is due to a problem with the Photon Voice SDK libraries. The problem is that there is a library called `libopus_egpv.so` that is available for both x86 and x64 architectures.

This is due to a design flaw in Photon for Linux, as it exists in two folders simultaneously and causes a library collision error. To fix this, we will remove the `libopus_egpv.so` library located in the **Assets** | **Photon** | **PhotonVoice** | **PhotonVoiceLibs** | **Linux** | **x86** path. Once fixed, you can compile again.

This process may take a long time to finish, so be patient. It is a good time to grab a coffee. Once the process is complete, you will see the files in the destination folder you have selected.

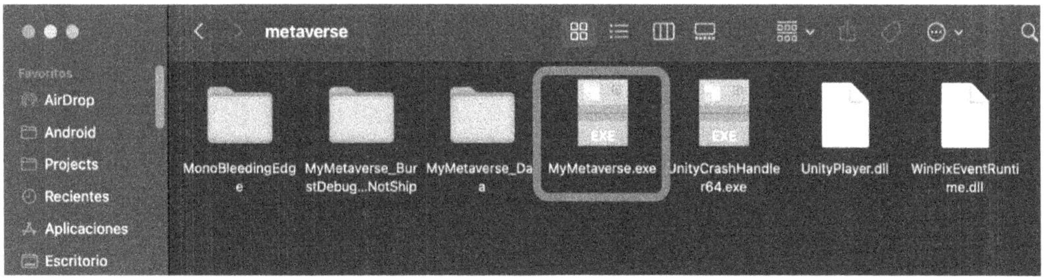

Figure 14.9 – Executable binary file for Windows

As you can see, we have an executable file, with the same name as the project. If we copy all the files to a computer with the target operating system and run it, we will see our artwork running on another device. It's a very comforting feeling. For Windows builds, our executable binary file will be `MyMetaverse.exe`; for Linux, we will find a file similar to `MyMetaverse.x86_64`; and finally, we will find a `MyMetaverse` file in the Mac export.

The following screenshot is of our project running on a Windows computer.

Figure 14.10 – Project running on a Windows computer

As you can see, we have correctly executed our project on a Windows, Linux, or Mac computer. In my case, when I select the **Development Build** option, I can see a magnificent console with logs, which offers very interesting information about what is happening underneath.

You will see the same result on Windows, Linux, and Mac operating systems.

Android

Compiling for mobile systems is a bit more complex than for desktop devices. More configuration steps need to be considered. Between iOS and Android, there is also a huge difference in terms of compilation. Android will always be easier than iOS.

Next, we will see, step by step, how to configure our project to compile on Android. Let's first go through a list of what we need to have ready for a successful Android build:

- Development environment
- Compilation options

We will now take a closer look at the two points mentioned previously.

Development environment

Make sure you have Android Studio installed on your computer. For small projects, Unity allows us to compile directly without going through Android Studio, but when our project start to have a considerable number of dependencies, we may need to have more control over the build, which can only be achieved by using Android Studio. With the project we have built so far, we can compile and export directly from Unity, but it is highly recommended that you install Android Studio.

If you haven't already done so, download and install the **Android Build Support** plugin in Unity Hub. This will allow you to build the Android platform from Unity.

Compilation options

Make sure the Android platform is active. If you haven't already done so, go to the **Build Configuration** window, select the Android platform, and click on **Switch Platform**.

In the **Build Settings** window, click on **Player Settings** to open the Android-specific settings.

The following are the some points that must be taken into account for a correct compilation in Android:

- Make sure that the package name that is set is unique and follows the Android package naming conventions. Remember that the package name was already configured in *Chapter 1*, *Getting Started with Unity and Firebase*. Also, it must be the same as the one we configured in Firebase; otherwise, it will fail to start the application.
- Configure other options, such as **Minimum API Level**, **Target API Level**, and **Graphics APIs**, according to your requirements and desired compatibility.
- It's important to know that if you have the Oculus SDK integrated, it is necessary to disable it if you are going to compile for regular Android devices; otherwise, you will get errors when you start the app, and it will prevent you from running it correctly. To do this, go to the **Edit | Project Settings** menu.

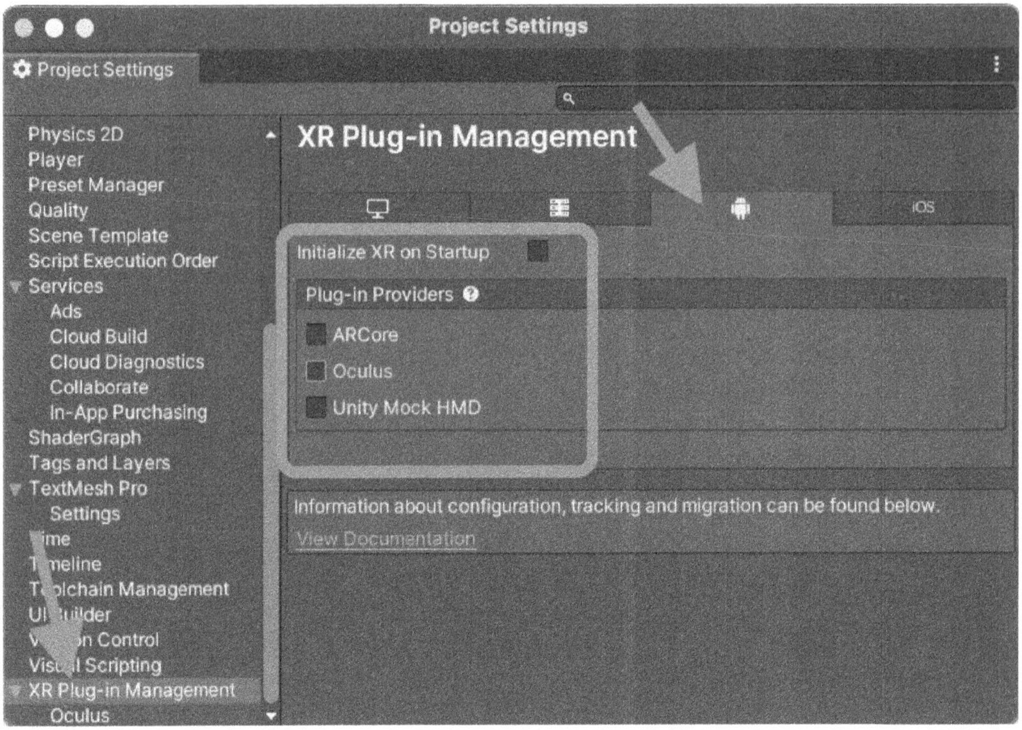

Figure 14.11 – Disabling XR plugins in Project Settings

1. In the menu on the left, scroll down and select the **XR Plug-in Management** option. Finally, select the **Android** tab and uncheck all the options that are checked. Once you have done this, you will be able to compile and run on Android devices without any problems.
2. Once we meet the prerequisites, we can start compiling for Android. It's very easy; just open the **File** | **Build Settings** menu.

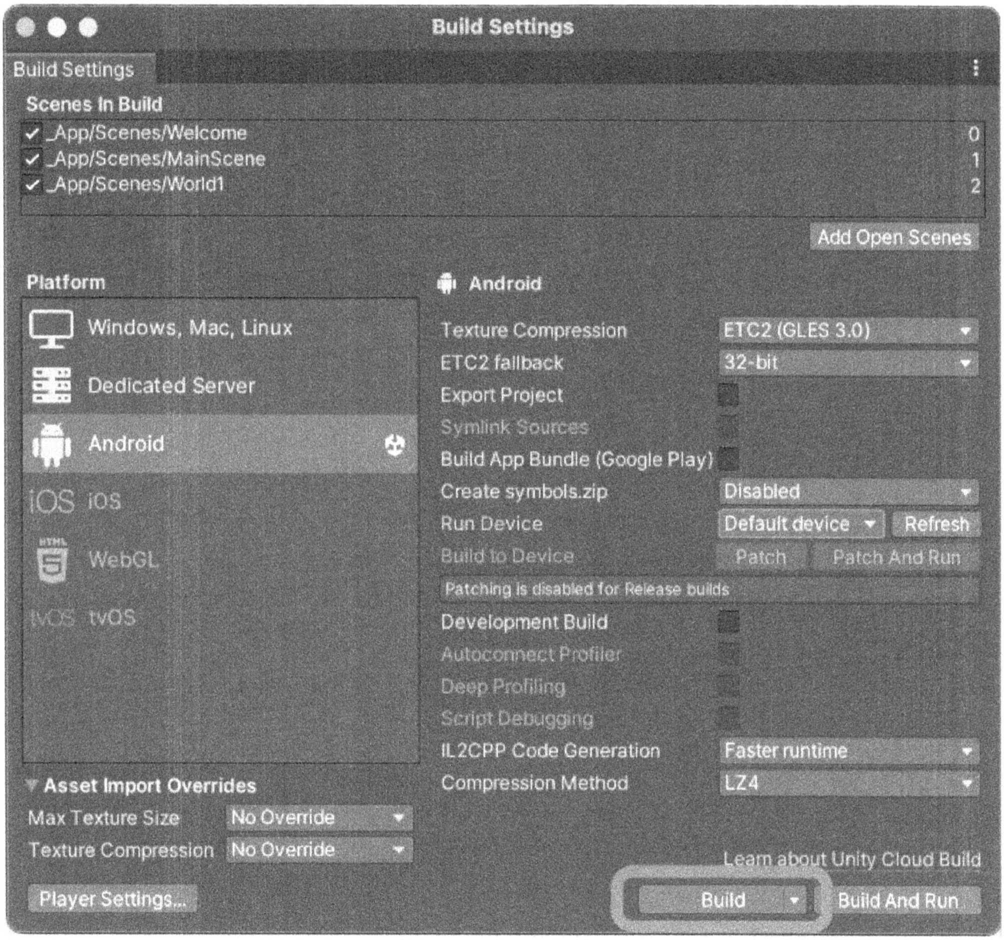

Figure 14.12 – Build button in the Build Settings window

3. Once the **Build Settings** window is open, we can click the **Build** button. Unity will ask us where to save the compiled file; we select the desired destination and click **OK**.

Once the process is finished, we will have a file called `MyMetaverse.apk`. This file can be installed on Android mobiles. You can also distribute it internally with the Firebase **Distribution** tool, which we looked at briefly in *Chapter 1, Getting Started with Unity and Firebase*.

You may have noticed an option in the **Build Settings** window called **Build App Bundle (Google Play)**. It is very important to know what it means. By checking this option, our project will generate an `.aab` file instead of an `.apk` file, but what is the difference?

The main difference between the APK and AAB file formats on Android is in how applications are packaged and distributed:

- **Android Application Package** (**APK**) is the traditional Android application packaging format. It is a single file containing all the resources and code of the application, ready to be installed on Android devices. An APK file can be distributed directly through app stores or installed manually on devices. It can also be distributed with the Firebase **Distribution** tool privately.
- **Android App Bundle** (**AAB**) is a newer, optimized format introduced by Google Play. Instead of being a single file like APK, AAB is a bundle of files that contains the app's modules and resources but is not fully built as an APK ready to be installed. Google Play uses AAB to generate optimized APKs tailored specifically for each device and configuration.

The main advantage of using AAB is that it enables the delivery of optimized and reduced APKs. Google Play can generate device-specific APK variants based on the device characteristics and settings, which can result in faster downloading and the more efficient use of device storage.

This is known as **multiple APK delivery**. The AAB file is a more efficient package that allows Google Play to generate optimized APK files specific to each device at the time of installation. The advantage of AAB lies in the reduced size and more efficient delivery of the application.

Once the compilation process is finished, we can send the APK file to an Android device and run it. The compilation process can take a long time, so please be patient.

In the following figure, we can see our project running on an Android device. It is a very satisfying feeling to see your project running perfectly on another device.

Figure 14.13 – Project running on an Android mobile phone

Android emulators

If you have installed Android Studio, you have the great advantage of being able to create and use advanced Android emulators. This will allow you to have more flexibility and agility to test your project. There is an interesting official guide to creating emulators on the web: `https://developer.android.com/studio/run/managing-avds`.

It is important to know that emulators use the x86-64 architecture and Unity by default compiles for ARM64. This should not be a problem as you can also enable compilation for x86-64 architectures from the **Project Settings | Player** menu.

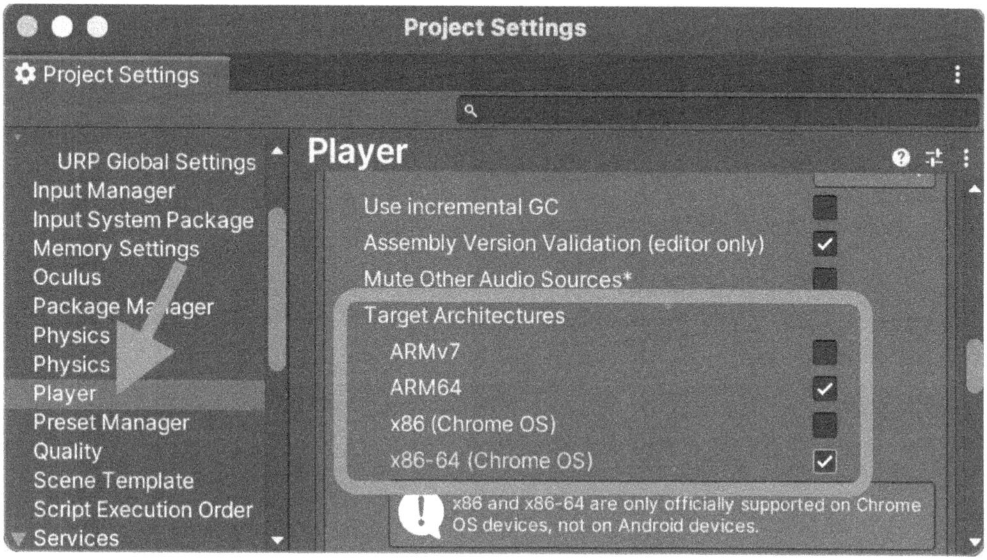

Figure 14.14 – Selecting target architectures in Project Settings

By enabling the x86-64 architecture, you can now compile and launch the project in an emulator created in Android Studio.

iOS

Compiling for iOS is, for me, the most complex (and most expensive) compilation process of all. It is important to know that in order to export your project to publish it on the App Store, you must have a developer account, which costs 99 USD per year.

It is important to know that it is not possible to compile for iOS from a Windows or Linux computer. It is strictly necessary to have a Mac.

Tip: Unity Cloud Build

As we've said, it's strictly necessary to have a Mac to build for iOS, but there are other methods to publish to the App Store if you don't have one. Unity Cloud Build is a paid service from Unity that allows you to build remotely, on your machines for iOS. If you are interested in this point, you can get more information from its official documentation by visiting `https://docs.unity3d.com/2019.4/Documentation/Manual/UnityCloudBuildiOS.html`.

Unlike the other platforms, iOS does not generate an executable file, but rather generates a project in Xcode. That project will be recompiled, then packaged in a process called **Archive**. From there, we can finally send it to the developer control panel, where after a long process of configuration and review by the Apple team, it can be published in the App Store.

In this book, we will learn how to export our project to Xcode, and I will give you some interesting links to follow along the way. As we have seen on Android, there are a number of points to consider before compiling for iOS.

Development environment

Make sure you have Xcode installed on your Mac computer. Download and install the **iOS Build Support** plugin in Unity Hub if you don't already have it installed. This will allow you to build for the iOS platform from Unity.

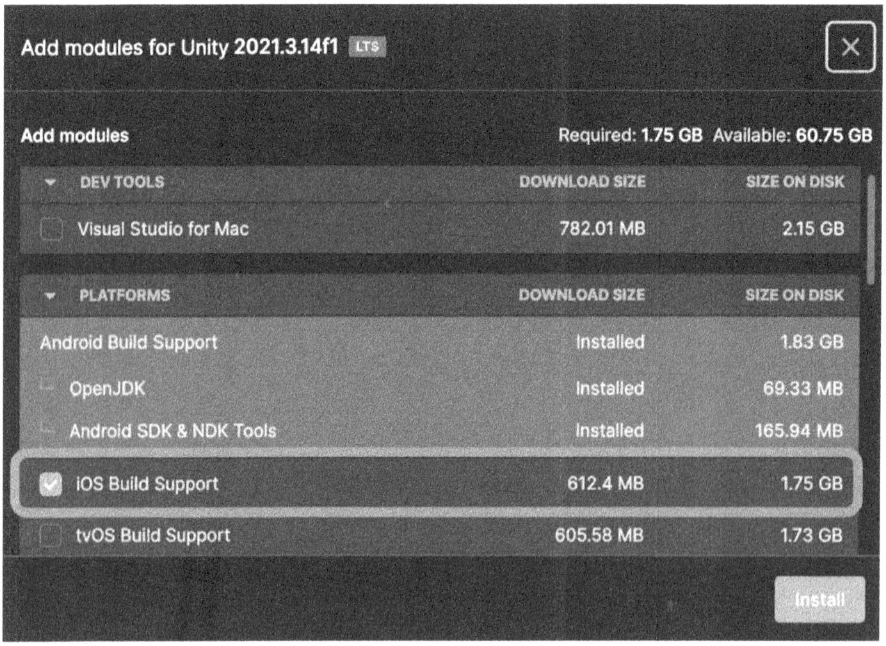

Figure 14.15 – Selecting iOS Build Support in Unity Hub to install it

Compilation options

Make sure you have the iOS platform active. If you haven't already done so, go to the **Build Configuration** window, select the **iOS** platform, and click on **Switch Platform**.

In the **Build Settings** window, click on **Player Settings** to open the iOS-specific settings.

The two points to be taken into account for a correct compilation are detailed here:

- Make sure that the package name that is set is unique and follows the iOS package naming conventions. Remember that the package name was already configured in *Chapter 1, Getting Started with Unity and Firebase*. Also, it is strictly necessary that it is the same as the one we have configured in Firebase; otherwise, it will fail to start the application.
- It is also necessary to write a description to notify **Apple** why we are requesting permission to access the microphone. Remember that by integrating the **Photon Voice SDK**, we allow our users to communicate by voice. To write this description, go to the **Player Settings** window, which can be accessed from the **Player** menu in the **Configuration** section.

In the following info boxes, we refer to two possible issues that you may encounter during compilation. The first one is about a bug that exists in the Oculus SDK version, with which this project has been built.

> **Warning: Possible bug in the Oculus SDK on iOS**
>
> When you switch from the active platform to iOS, it is possible that Unity will show you an error that prevents you from compiling, which looks like this:
>
> **Assets/Oculus/LipSync/Editor/OVRLipSyncBuildPostProcessor.cs(73,51): error CS0619: 'PBXProject.GetUnityTargetName()' is obsolete: 'This function is deprecated. There are two targets now, call GetUnityMainTargetGuid() - for app or GetUnityFrameworkTargetGuid() - for source/plugins to get Guid instead of calling to TargetGuidByName(GetUnityTargetName()).'**
>
> To fix it, we open the `OVRLipSyncBuildPostProcessor` script found by following the **Assets | Oculus | LipSync | Editor** path, or we can simply double-click on the error.
>
> Locate the following line:
>
> ```
> var targetGUID = project.TargetGuidByName(PBXProject.
> GetUnityTargetName());
> ```
>
> Replace it with the following line:
>
> ```
> var targetGUID = project.TargetGuidByName(project.
> GetUnityMainTargetGuid());
> ```
>
> Save the changes and the error should disappear. We can now compile correctly.

The second most frequent issue is a bug with the compilation of shaders in iOS. Fortunately, the solution is very simple. It is detailed as follows.

> **Warning: Error compiling URP shaders**
>
> Another common bug in Unity version 2021 is getting errors related to Shader. The error message states the following:
>
> **maximum ps_5_0 sampler register index (16) exceeded at line 336 (on metal)**
>
> To avoid these errors, we can disable the **Metal API Validation** option in the **Build Settings | Player | Other Settings** window.
>
> Once you have disabled the **Metal API Validation** option, you can relaunch the build process and the error should disappear.

Once we meet the prerequisites, we can start compiling for iOS. It's very easy; just open the **File | Build Settings** menu.

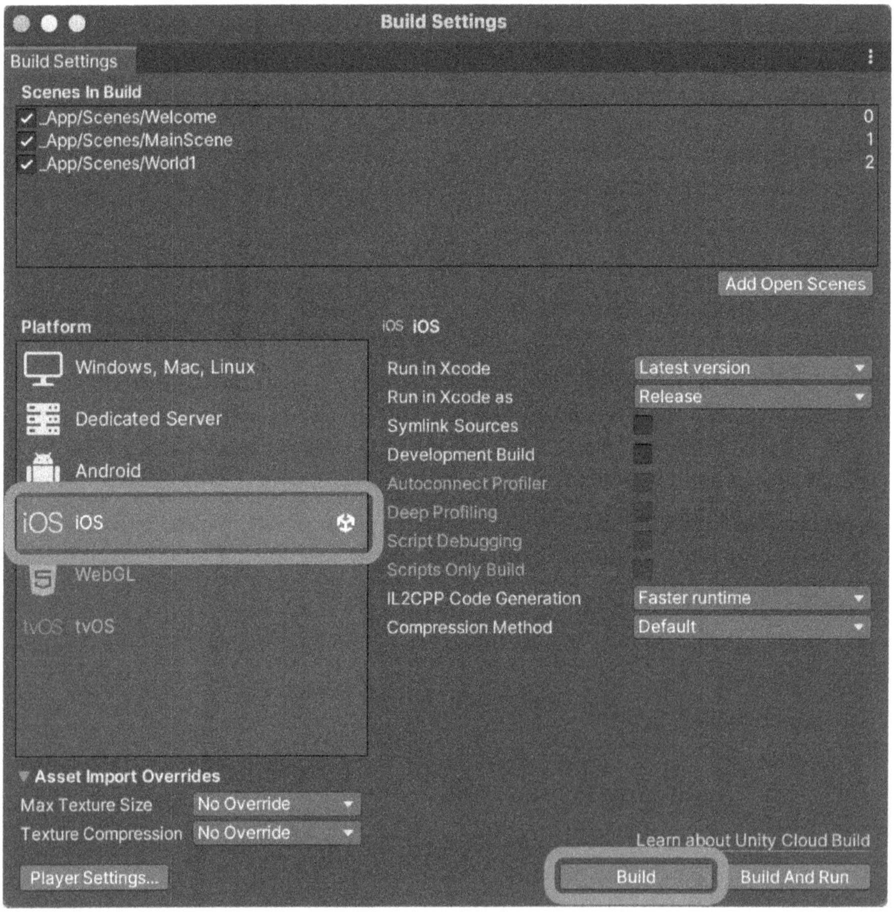

Figure 14.16 – Build button in the Build Settings window

Click on the **Build** button and select the destination folder where you want to save the generated Xcode project. Wait for Unity to compile the project and generate the Xcode project. This may take some time.

Once you have completed these steps, Unity will generate an Xcode project that you can open in Xcode for further configuration. You can then sign the application and build it on an iOS device or generate an IPA file for distribution.

In the following screenshot, we can see the iOS project generated in Unity and opened in Xcode.

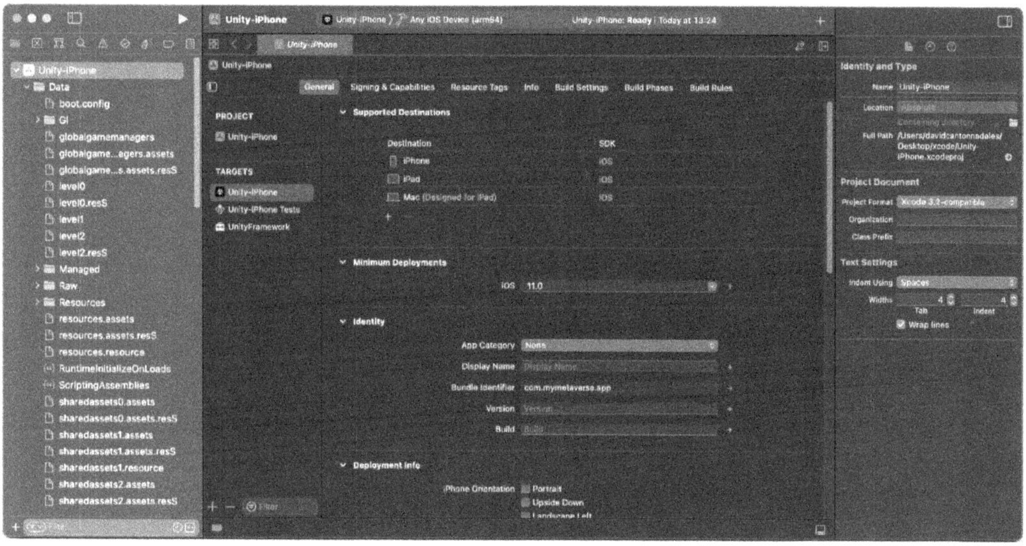

Figure 14.17 – XCode project generated from Unity

> **Tip: More information on compiling for iOS**
>
> Compiling for iOS and publishing to the App Store is so complicated and encompasses so many aspects that a book could be written just for this. I recommend the official Unity documentation if you want to expand your knowledge and find more information to continue your journey toward publishing to the App Store. You can find interesting links on the official website: `https://docs.unity3d.com/Manual/iphone.html`.

Summary

We have reached the end of this wonderful book. In this last chapter, we learned about another optimization technique, texture compression, which offers us the ability to considerably reduce the weight of our project and increase the speed and fluidity of the rendering.

We also learned the necessary steps to compile our project on multiple platforms, such as Windows, Linux, Mac, Android, and iOS. You have successfully exported and tested your metaverse on other devices and are now ready to continue your journey to publication on the app stores.

This chapter closes the circle we started at the beginning of the book; you have learned everything you need to create a virtual universe from conception to export. You have learned how to connect to a Firebase database and how to decorate scenes with buildings and objects. You have created NPCs with logic programmed by you.

We have learned how to make our project compatible with Oculus Quest, and let's not forget fun features such as the cinema screen, buying a home, and the chat system.

Next steps

One more thing...

It is with immense gratitude and satisfaction that I turn to you in these lines to express my heartfelt thanks for choosing to immerse yourself in the pages of this book. It has been an exciting and challenging journey to bring these ideas and insights together in one place, and your decision to join me on this journey means a great deal to me.

My greatest wish in writing this book was to share my passion for the fascinating world of metaverses and to give you the tools and knowledge you need to create your own virtual experiences. I hope that you have found these pages a source of inspiration and knowledge that will push you further into this exciting area.

None of this would have been possible without your support and confidence in this project. Your choice to read this book is a valuable recognition of all the effort and dedication I put into every word written. I hope that the content of these pages has been of great use to you and has provided you with a clear and enriching insight into metaverse development.

Once again, thank you for being part of this literary journey. Your support motivates me to continue exploring new frontiers in the world of technology and creativity. I will always be grateful for the opportunity you have given me by allowing me to accompany you on your journey toward creating unforgettable virtual experiences.

What next? After reading this book, an exciting journey into the world of virtual reality and metaverses awaits you. You are now equipped with valuable skills and knowledge that will enable you to embark on this unique adventure.

You will master fundamental concepts, familiarizing yourself with the terminology and key principles that govern the development of virtual experiences. You will be able to create your own metaverses and immersive multiplayer experiences using the Unity platform and Firebase integration.

Your technical skills will improve, and you will be prepared to tackle challenges and optimize your creations so that they work efficiently and effectively. This book will also inspire you to be creative and explore original new ideas to captivate your users with unique experiences.

By joining the community of readers interested in creating metaverses, you will have the opportunity to share your knowledge, learn from others, and engage in exciting collaborations. The world of metaverses is constantly evolving, but with the solid foundation you have gained here, you will be prepared to keep up with the latest trends and updates in this exciting industry.

You're now ready to delve into the fascinating world of metaverses and create virtual experiences that will amaze and captivate people all over the world – enjoy the journey and the endless possibilities that await you!

Index

A

B

C

D

E

F

T

U

V

W

<packt>

www.packtpub.com

Subscribe to our online digital library for full access to over 7,000 books and videos, as well as industry leading tools to help you plan your personal development and advance your career. For more information, please visit our website.

Why subscribe?

- Spend less time learning and more time coding with practical eBooks and Videos from over 4,000 industry professionals
- Improve your learning with Skill Plans built especially for you
- Get a free eBook or video every month
- Fully searchable for easy access to vital information
- Copy and paste, print, and bookmark content

Did you know that Packt offers eBook versions of every book published, with PDF and ePub files available? You can upgrade to the eBook version at packtpub.com and as a print book customer, you are entitled to a discount on the eBook copy. Get in touch with us at customercare@packtpub.com for more details.

At www.packtpub.com, you can also read a collection of free technical articles, sign up for a range of free newsletters, and receive exclusive discounts and offers on Packt books and eBooks.

Other Books You May Enjoy

If you enjoyed this book, you may be interested in these other books by Packt:

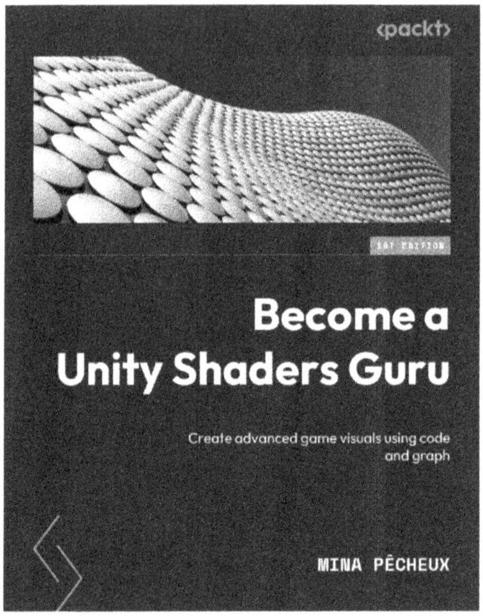

Become a Unity Shaders Guru

Mina Pêcheux

ISBN: 978-1-83763-674-7

- Understand the main differences between the legacy render pipeline and the SRP
- Create shaders in Unity with HLSL code and the Shader Graph 10 tool
- Implement common game shaders for VFX, animation, procedural generation, and more
- Experiment with offloading work from the CPU to the GPU
- Identify different optimization tools and their uses
- Discover useful URP shaders and re-adapt them in your projects

Mind-Melding Unity and Blender for 3D Game Development

Spencer Grey

ISBN: 978-1-80107-155-0

- Transform your imagination into 3D scenery, props, and characters using Blender
- Get to grips with UV unwrapping and texture models in Blender
- Understand how to rig and animate models in Blender
- Animate and script models in Unity for top-down, FPS, and other types of games
- Find out how you can roundtrip custom assets from Blender to Unity and back
- Become familiar with the basics of ProBuilder, Timeline, and Cinemachine in Unity

Packt is searching for authors like you

If you're interested in becoming an author for Packt, please visit `authors.packtpub.com` and apply today. We have worked with thousands of developers and tech professionals, just like you, to help them share their insight with the global tech community. You can make a general application, apply for a specific hot topic that we are recruiting an author for, or submit your own idea.

Share Your Thoughts

Now you've finished *Build Your Own Metaverse with Unity*, we'd love to hear your thoughts! Scan the QR code below to go straight to the Amazon review page for this book and share your feedback or leave a review on the site that you purchased it from.

`https://www.amazon.in/review/create-review/error?asin=1837631735`

Your review is important to us and the tech community and will help us make sure we're delivering excellent quality content.

Download a free PDF copy of this book

Thanks for purchasing this book!

Do you like to read on the go but are unable to carry your print books everywhere?

Is your eBook purchase not compatible with the device of your choice?

Don't worry, now with every Packt book you get a DRM-free PDF version of that book at no cost.

Read anywhere, any place, on any device. Search, copy, and paste code from your favorite technical books directly into your application.

The perks don't stop there, you can get exclusive access to discounts, newsletters, and great free content in your inbox daily

Follow these simple steps to get the benefits:

1. Scan the QR code or visit the link below

`https://packt.link/free-ebook/9781837631735`

2. Submit your proof of purchase
3. That's it! We'll send your free PDF and other benefits to your email directly

www.ingramcontent.com/pod-product-compliance
Lightning Source LLC
Chambersburg PA
CBHW081450050326
40690CB00015B/2741
* 9 7 8 1 8 3 7 6 3 1 7 3 5 *